NON-LINEAR AND DYNAMIC PROGRAMMING

Useful For

Undergraduate and postgraduate students of Mathematics, Statistics, Computer Science, Physical Science, Management, Engineering and other professional courses and competitive examinations

Dr. SUDHIR KUMAR PUNDIR

M.Sc., M.Phil, NET (JRF), Ph.D.

Head
Department of Mathematics
S.D. (P.G.) College
Muzaffarnagar (U.P.)

CBS

CBS Publishers & Distributors Pvt Ltd

New Delhi • Bengaluru • Chennai • Kochi • Kolkata • Mumbai
Bhopal • Bhubaneswar • Hyderabad • Jharkhand • Nagpur • Patna
• Pune • Uttarakhand • Dhaka (Bangladesh) • Kathmandu (Nepal)

ISBN: 978-93-89688-88-7

Copyright © Author and Publisher

First Edition: 2020

Published by Satish-Kumar Jain and produced by Varun Jain for

CBS Publishers & Distributors Pvt Ltd

4819/XI Prahlad Street, 24 Ansari Road, Daryaganj, New Delhi 110 002, India.
Ph: 23289259, 23266861, 23266867   Website: www.cbspd.com
Fax: 011-23243014   e-mail: delhi@cbspd.com; cbspubs@airtelmail.in.

Corporate Office: 204 FIE, Industrial Area, Patparganj, Delhi 110 092

Ph: 4934 4934   Fax: 4934 4935   e-mail: publishing@cbspd.com;
publicity@cbspd.com

Branches

- **Bengaluru:** Seema House 2975, 17th Cross, K.R. Road,
  Banasankari 2nd Stage, Bengaluru 560 070, Karnataka
  Ph: +91-80-26771678/79   Fax: +91-80-26771680   e-mail: bangalore@cbspd.com
- **Chennai:** 7, Subbaraya Street, Shenoy Nagar, Chennai 600 030, Tamil Nadu
  Ph: +91-44-26680620, 26681266   Fax: +91-44-42032115   e-mail: chennai@cbspd.com
- **Kochi:** 68/1534, 35,36, Power House Road, Opp KSEB Power House,
  Ernakulam 682 018, Kochi, Kerala
  Ph: +91-484-4059061-65   Fax: +91-484-4059065   e-mail: kochi@cbspd.com
- **Kolkata:** 6/B, Ground Floor, Rameswar Shaw Road, Kolkata-700 014, West Bengal
  Ph: +91-33-22891126, 22891127, 22891128   e-mail: kolkata@cbspd.com
- **Mumbai:** 83-C, Dr E Moses Road, Worli, Mumbai-400018, Maharashtra
  Ph: +91-22-24902340/41   Fax: +91-22-24902342   e-mail: mumbai@cbspd.com

Representatives

• Bhopal	0-8319310552	• Bhubaneswar	0-9911037372	• Hyderabad	0-9885175004
• Jharkhand	0-9811541605	• Nagpur	0-9421945513	• Patna	0-9334159340
• Pune	0-9623451994	• Uttarakhand	0-9716462459	• Dhaka (Bangladesh)	01912-003485
• Kathmandu (Nepal)	977-9181742655				

Printed at: Glorious Printers, Daryaganj, Delhi

Preface

The book entitled "Non-Linear and Dynamic Programming" meet the needs of Mathematics, Statistics, Computer science, Physical science, Management and Engineering students of UG and PG levels. Besides, it will also be very useful for students preparing for various competitive and professional examinations.

The contents of this book are derived from the curricula offered by various universities across the country. This book consists of thirteen chapters. In each chapter of the book, an ample amount of theory is given which is supported by solved examples followed by exercises along with their answers. The text is organised around mathematical problems, with each chapter devoted to a single type of problem. Within each chapter the presentation begins with the simplest and most basic methods, progressing gradually to more advance topics.

I express my gratitude to the authors and publishers of various books I consulted during the preparation of the book.

I wish to sincerely thank Sh S.K. Jain and Sh Varun Jain, Managing Director, CBS Publishers and Distributors, New Delhi for encouragement and help in bringing out this publication in a present nice form.

My special thanks to Sh. Y.N. Arjuna, Senior director publishing, editorial and publicity and Smt. Ritu Chawla, publishing head, CBS Publishers and Distributors, New Delhi whose encouragement and unstinted support enabled me to complete the book. I also take this opportunity to express my sincere gratitude to Sh. Sunil Dutt, CBS Publishers and Distributors, New Delhi who gave me the inspiration throughout the preparation of the book. Sh. Suresh Sharma and Sh. Nitish Sharma CBS, New Delhi deserve special mention for their kind support and help in this endeavour.

I must also record my appreciation due to my wife Dr. Rimple, daughter Rijuta and son Shrish for their understanding and love during the long period that I have taken to complete this book.

Above all I am thankful to The Almighty God, without whose grace nothing is possible for any one.

Readers are welcomed to point out errors, if any and send their valuable suggestions for improving the quality of the book.

Dr. Sudhir Kumar Pundir
email : skpundir05@yahoo.co.in

Contents

1 Introduction

1.1 INTRODUCTION

Linear Programming problem (LPP) and non-linear programming problem (NLPP) are integral part of Operations Research which is an important branch of mathematics.

So, before discussing the basic concepts of linear programming and non-linear problems, let us recall some mathematical concepts which are very useful in LPP and NLPP.

1.2 MATRIX

A set of mn numbers either real or complex arranged in the form of a rectangular array in which there are m rows and n columns, is called a matrix of order $m \times n$ which is denoted by $[a_{ij}]_{m \times n}$ where $i = 1, 2, 3, ..., m$ represents the number of rows and $j = 1, 2, 3, ..., n$ represents the number of columns and thus a matrix of order $m \times n$ is usually written as

$$[a_{ij}]_{m \times n} = \begin{bmatrix} a_{11} & a_{12} & \cdots & a_{1n} \\ a_{21} & a_{22} & \cdots & a_{2n} \\ \vdots & \vdots & \vdots & \vdots \\ a_{m1} & a_{m2} & \cdots & a_{mn} \end{bmatrix}_{m \times n}$$

☞ REMARK

- Sometimes, a matrix is a rectangular array of numbers enclosed in double straight lines shown as '$\|$ $\|$' or enclosed in parenthesis '( )'.

1.3 TYPE OF MATRICES

1.3.1 NULL MATRIX (OR ZERO MATRIX)

A matrix of order $m \times n$ is called a *null matrix* if it contains all mn elements zero. It is denoted by O and is usually written as

$$O = \begin{bmatrix} 0 & 0 & \cdots & 0 \\ 0 & 0 & \cdots & 0 \\ \vdots & \vdots & \vdots & \vdots \\ 0 & 0 & \cdots & 0 \end{bmatrix} m \times n$$

1.3.2 ROW MATRIX

A matrix having only one row and n columns is called a *row matrix* of order $1 \times n$.

For example : The matrix $A = [a_{11} \quad a_{12} \quad a_{13} \quad \cdots \quad a_{1n}]_{1 \times n}$ is a row matrix.

1.3.3 COLUMN MATRIX

A matrix having m rows and only one column is called a *column matrix* of order $m \times 1$.

For example : The matrix $A = \begin{bmatrix} a_{11} \\ a_{21} \\ a_{31} \\ \vdots \\ a_{m1} \end{bmatrix}_{m \times 1}$ is a column matrix.

1.3.4 HORIZONTAL MATRIX

A matrix having more columns than the number of its rows, is called *Horizontal matrix*.

For example: The matrix $A = \begin{bmatrix} a_{11} & a_{12} & a_{13} \\ a_{21} & a_{22} & a_{23} \end{bmatrix}_{2 \times 3}$ is a horizontal matrix.

1.3.5 VERTICAL MATRIX

A matrix having more number of rows than its columns, is called *vertical matrix*.

For exmaple: The matrix $A = \begin{bmatrix} a_{11} & a_{12} \\ a_{21} & a_{22} \\ a_{31} & a_{32} \end{bmatrix}_{3 \times 2}$ is a vertical matrix.

☛ REMARK
- Row matrix is also a horizontal matrix and column matrix is also a vertical matrix.

1.3.6 SQUARE MATRIX

A matrix having a number of rows equal to number of columns, is called *square matrix*.

For example : The matrix $A = \begin{bmatrix} a_{11} & a_{12} & a_{13} \\ a_{21} & a_{22} & a_{23} \\ a_{31} & a_{32} & a_{33} \end{bmatrix}_{3 \times 3}$ is a square matrix.

Here, the matrix A has 3 rows and 3 columns, so it is a square matrix. Also the elements a_{11}, a_{22}, a_{33} are placed in the diagonal, so these elements are known as *diagonal elements*.

1.3.7 DIAGONAL MATRIX

A matrix of order $n \times n$ is called a *diagonal matrix* if it contains all its off-diagonal elements equal to zero.

Suppose $A = [a_{ij}]_{n \times n}$ and if $a_{ij} = 0$ for all $i \neq j$, then A is a diagonal matrix. Diagonal matrix of order $n \times n$ is usually written as Diag $[a_{11} \quad a_{22} \quad a_{33} \quad \cdots \quad a_{nn}]$

For example: The matrix $A = \begin{bmatrix} 1 & 0 & 0 \\ 0 & 2 & 0 \\ 0 & 0 & 3 \end{bmatrix}_{3 \times 3}$ = Diag [1 2 3] is a diagonal matrix of order 3.

1.3.8 SCALAR MATRIX

A diagonal matrix whose diagonal elements are all equal but not equal to 1 is called a *scalar matrix*.

For example: The matrix $A = \begin{bmatrix} k & 0 & 0 \\ 0 & k & 0 \\ 0 & 0 & k \end{bmatrix}, k \neq 1$ is a sca matrix.

1.3.9 UNIT MATRIX

A square matrix of order $n \times n$ having all off-diagonal elements equal to zero and each of the diagonal elements equal to 1, is called a *unit matrix*. It is usually denoted by I_n and is written as

$$I_n = \begin{bmatrix} 1 & 0 & \dots & 0 \\ 0 & 1 & \dots & 0 \\ 0 & 0 & \dots & 0 \\ \vdots & \vdots & \vdots & \vdots \\ 0 & 0 & \dots & 1 \end{bmatrix}_{n \times n}$$

☞ REMARK
- Unit matrix can also be denoted by I.

1.3.10 TRIANGULAR MATRIX

A matrix in which the elements lying above or below principal diagonal are all zero, is called a *triangular matrix*.

There are two kinds of triangular matrix.

(a) **Upper triangular matrix :** A matrix of order $n \times n$ is called an *upper triangular matrix* if it contains all its elements below the diagonal elements equal to zero.

Suppose $A = [a_{ij}]_{n \times n}$ and if $a_{ij} = 0$ for all $i > j$, then A is an upper triangular matrix.

For example : The matrix $A = \begin{bmatrix} 2 & 3 & 4 \\ 0 & 1 & 5 \\ 0 & 0 & 3 \end{bmatrix}_{3 \times 3}$ is an upper triangular matrix of order 3×3.

(b) **Lower triangular matrix :** A matrix of order $n \times n$ is called a *lower triangular matrix* if it contains all its elements above the diagonal elements equal to zero.

Suppose $A = [a_{ij}]_{n \times n}$ and if $a_{ij} = 0$ for all $i < j$, then A is called lower triangular matrix.

For example : The matrix $A = \begin{bmatrix} 1 & 0 & 0 \\ 3 & 4 & 0 \\ 5 & 6 & 7 \end{bmatrix}_{3 \times 3}$ is a lower triangular matrix of order 3×3.

1.4 OPERATIONS ON MATRICES

1.4.1 ADDITION OF MATRICES

Suppose A and B are two matrices of same order, then the addition of these two matrices is obtained by adding corresponding elements of A and B. It is denoted by $A + B$. If the order of A and B is $m \times n$, then the order of $A+B$ will be $m \times n$.

Suppose $A = [a_{ij}]_{m \times n}$ and $B = [b_{ij}]_{m \times n}$ then $A + B = [a_{ij} + b_{ij}]_{m \times n}$

For example: If $A = \begin{bmatrix} 1 & 2 & 3 \\ 5 & 1 & 4 \\ 7 & 8 & 9 \end{bmatrix}$ and $B = \begin{bmatrix} 1 & 3 & 5 \\ 5 & 0 & 1 \\ 3 & 2 & 12 \end{bmatrix}$

then $A + B = \begin{bmatrix} 1 & 2 & 3 \\ 5 & 1 & 4 \\ 7 & 8 & 9 \end{bmatrix} + \begin{bmatrix} 1 & 3 & 5 \\ 5 & 0 & 1 \\ 3 & 2 & 12 \end{bmatrix} = \begin{bmatrix} 1+1 & 2+3 & 3+5 \\ 5+5 & 1+0 & 4+1 \\ 7+3 & 8+2 & 9+12 \end{bmatrix} = \begin{bmatrix} 2 & 5 & 8 \\ 10 & 1 & 5 \\ 10 & 10 & 21 \end{bmatrix}$

☞ REMARK
- If the orders of the matrices are different, then they are not conformable for addition.

1.4.2 SUBSTRACTION OF MATRICES

Suppose A and B are two matrices of same order, then the substraction of A and B, i.e., $A–B$ is obtained by substracting each element of B from the corresponding element of A. If A and B are of order $m \times n$, then $A - B$ will be of order $m \times n$.

Let $A = [a_{ij}]_{m \times n}$ and $B = [b_{ij}]_{m \times n}$ then $A - B = [a_{ij} - b_{ij}]_{m \times n}$

For example: If $\quad A = \begin{bmatrix} 1 & 2 & 3 \\ 3 & 4 & 5 \\ 5 & 6 & 7 \end{bmatrix}$ and $B = \begin{bmatrix} 0 & 5 & 2 \\ 3 & -2 & 2 \\ 5 & 7 & 8 \end{bmatrix}$

then $\quad A - B = \begin{bmatrix} 1 & 2 & 3 \\ 3 & 4 & 5 \\ 5 & 6 & 7 \end{bmatrix} - \begin{bmatrix} 0 & 5 & 2 \\ 3 & -2 & 2 \\ 5 & 7 & 8 \end{bmatrix} = \begin{bmatrix} 1-0 & 2-5 & 3-2 \\ 3-3 & 4-(-2) & 5-2 \\ 5-5 & 6-7 & 7-8 \end{bmatrix} = \begin{bmatrix} 1 & -3 & 1 \\ 0 & 6 & 3 \\ 0 & -1 & -1 \end{bmatrix}$

☛ REMARK
- If the order of matrices are different, then they are not conformable for substraction.

1.4.3 MULTIPLICATION OF A MATRIX BY A SCALAR

Suppose A is a matrix of order $m \times n$ and k is a scalar, then the multiplication of A by k, i.e. kA is obtained by multiplying each element of A by k.

Let $\quad A = [a_{ij}]_{m \times n} \ \forall \ 1 \le i \le m$ and $1 \le j \le n$, then $kA = [ka_{ij}]_{m \times n}$

For example : If $\quad A = \begin{bmatrix} 1 & 2 & 3 \\ 4 & 5 & 6 \\ 7 & 8 & 9 \end{bmatrix}$ and $k = 3$,

then $\quad 3A = 3\begin{bmatrix} 1 & 2 & 3 \\ 4 & 5 & 6 \\ 7 & 8 & 9 \end{bmatrix} = \begin{bmatrix} 3 \times 1 & 3 \times 2 & 3 \times 3 \\ 3 \times 4 & 3 \times 5 & 3 \times 6 \\ 3 \times 7 & 3 \times 8 & 3 \times 9 \end{bmatrix} = \begin{bmatrix} 3 & 6 & 9 \\ 12 & 15 & 18 \\ 21 & 24 & 27 \end{bmatrix}$

1.4.4 EQUALITY OF MATRICES

Two matrices are said to be equal if both have same order and having same corresponding elements.

For example : The matrices $A = \begin{bmatrix} 1 & 2 \\ -3 & 4 \end{bmatrix}$ and $B = \begin{bmatrix} x & y \\ z & 4 \end{bmatrix}$ are said to be equal if $x = 1$, $y = 2$ and $z = -3$.

1.5 PROPERTIES OF MATRIX ADDITION

1.5.1 COMMUTATIVE LAW

If A and B are two matrices of same order $m \times n$, then $A + B = B + A$

Proof. Let $A = [a_{ij}]_{m \times n}$ and $B = [b_{ij}]_{m \times n}$ where $1 \le i \le m$ and $1 \le j \le n$. Then

$$A + B = [a_{ij}]_{m \times n} + [b_{ij}]_{m \times n}$$

$$= [a_{ij} + b_{ij}]_{m \times n} \quad \text{(By definition of addition)}$$

$$= [b_{ij} + a_{ij}]_{m \times n}$$

$$(\because \text{Addition of real numbers are always commutative})$$

$$= [b_{ij}]_{m \times n} + [a_{ij}]_{m \times n}$$

$$= B + A$$

Hence, $A + B = B + A$

1.5.2 ASSOCIATIVE LAW

If A, B and C are three matrices of same order $m \times n$, then $(A + B) + C = A + (B + C)$

Proof. Let $A = [a_{ij}]_{m \times n}$ and $B = [b_{ij}]_{m \times n}$ where $1 \le i \le m$ and $1 \le j \le n$. Then

$$(A+B)+C = ([a_{ij}]_{m \times n} + [b_{ij}]_{m \times n}) + [c_{ij}]_{m \times n}$$
$$= [a_{ij} + b_{ij}]_{m \times n} + [c_{ij}]_{m \times n}$$
$$= [(a_{ij} + b_{ij}) + (c_{ij})]_{m \times n}$$

($\because$ Addition of numbers are always associative)

$$= [a_{ij}]_{m \times n} + ([b_{ij} + c_{ij}]_{m \times n})$$
$$= [a_{ij}]_{m \times n} + ([b_{ij}]_{m \times n} + [c_{ij}]_{m \times n})$$
$$= A + (B + C)$$

Hence, $(A+B) + C = A + (B+C)$

1.5.3 ADDITIVE IDENTITY

If A is a matrix of order $m \times n$ and O is a null matrix of the same order $m \times n$, then

$$A + O = A = O + A$$

Proof. Let $A = [a_{ij}]_{m \times n}$ and $O = [0]_{m \times n}$, then

$$A + O = [a_{ij}]_{m \times n} + [0]_{m \times n}$$
$$= [a_{ij} + 0]_{m \times n}$$
$$= [a_{ij}]_{m \times n} = A$$

Also $\qquad O + A = [0]_{m \times n} + [a_{ij}]_{m \times n}$
$$= [0 + a_{ij}]_{m \times n}$$
$$= [a_{ij}]_{m \times n} = A$$

Hence $\qquad A + O = A = O + A$

Therefore, the null matrix O is treated as an additive identity.

1.5.4 ADDITIVE INVERSE

If A is a matrix of order $m \times n$ and $-A$ is the negative of A, so its order is also $m \times n$, then

$$-A + A = O \qquad \text{(null matrix)}$$

Here, $-A$ is the additive inverse of A.

1.5.5 CANCELLATION LAW

If A, B and C are three matrices of order $m \times n$ then

(i) $A + B = A + C \Rightarrow B = C$ (Left cancellation law)

(ii) $B + A = C + A \Rightarrow B = C$ (Right cancellation law)

Proof.

(i) It is given that

$$A + B = A + C \qquad \qquad ...(1)$$

Adding $- A$ to the left of both sides, we get

$$-A + (A + B) = -A + (A + C)$$
$\Rightarrow \qquad\qquad (-A + A) + B = (-A + A) + C$ $\qquad$ (By associative law)
$\Rightarrow \qquad\qquad O + B = O + C$ $\qquad$ (By additive inverse)
$\Rightarrow \qquad\qquad B = C$ $\qquad$ (By additive identity)

Similarly, we can prove that if $B + A = C + A$, then $B = C$.

1.6 PROPERTIES OF MULTIPLICATION OF MATRIX BY A SCALAR

(i) **Distributive law of scalar multiplication over matrix addition :** If A and B are two matrices of order $m \times n$ and k is any scalar, then $k(A + B) = kA + kB$

Proof. Let $A = [a_{ij}]_{m \times n}$ and $B = [b_{ij}]_{m \times n}$, then

$$k(A + B) = k([a_{ij}]_{m \times n} + [b_{ij}]_{m \times n})$$

$$= k([a_{ij} + b_{ij}]_{m \times n})$$

$$= [k(a_{ij} + b_{ij})]_{m \times n}$$

$$= [ka_{ij} + kb_{ij}]_{m \times n}$$

$$= [ka_{ij}]_{m \times n} + [kb_{ij}]_{m \times n}$$

$$= k[a_{ij}]_{m \times n} + k[b_{ij}]_{m \times n}$$

$$= kA + kB$$

Hence $\qquad k(A + B) = kA + kB.$

(ii) *If A is a matrix of order m × n and a, b are two scalars, then (a + b)A = aA + bA*

Proof. Let $\qquad A = [a_{ij}]_{m \times n}$, then

$$(a+ b)A = (a + b)[a_{ij}]_{m \times n}$$

$$= [(a + b)a_{ij}]_{m \times n} \qquad \text{(By scalar multiplication)}$$

$$= [aa_{ij} + ba_{ij}]_{m \times n} \qquad (\because \text{ Real numbers are distributive})$$

$$= [aa_{ij}]_{m \times n} + [ba_{ij}]_{m \times n}$$

$$= a[a_{ij}]_{m \times n} + b[a_{ij}]_{m \times n}$$

$$= aA + bA$$

Hence $\qquad (a + b)A = aA + bA$

(iii) *If A is a matrix of order m × n and a, b are two scalars, then a(bA) = (ab)A.*

Proof. Let $A = [a_{ij}]_{m \times n}$, then

$$a(bA) = a(b[a_{ij}]_{m \times n})$$

$$= [a(ba_{ij})]_{m \times n} \qquad \text{(By scalar multiplication)}$$

$$= [(ab)a_{ij}]_{m \times n} \qquad (\because \text{ Numbers are associative})$$

$$= (ab)[a_{ij}]_{m \times n}$$

$$= (ab) A$$

Hence $\qquad a(bA) = (ab)A.$

(iv) *If A is a matrix of order m × n and k is any scalar, then (− k)A = − (kA) = k(− A)*

Proof. Let $A = [a_{ij}]_{m \times n}$, then

$$(-k)A = (-k)[a_{ij}]_{m \times n}$$

$$= [(-k)a_{ij}]_{m \times n} \qquad \text{(By scalar multiplication)}$$

$$= [-ka_{ij}]_{m \times n}$$

$$= -[ka_{ij}]_{m \times n}$$

$$= - (kA)$$

Now $\qquad (-k)A = (-k)[a_{ij}]_{m \times n}$

$$= [(-k)a_{ij}]_{m \times n}$$

$$= [k(-a_{ij})]_{m \times n}$$

$$= k[-a_{ij}]_{m \times n}$$

$$= k(-A)$$

Hence $\qquad (-k)A = -(kA) = k(-A).$

1.7 MULTIPLICATION OF MATRICES

Let A and B be two matrices of order $m \times n$ and $n \times p$ respectively. Then a matrix C of order $m \times p$ is obtained by multiplying each row of A to each column of B.

Suppose $A = [a_{ij}]_{m \times n}$, $B = [b_{jk}]_{n \times p}$, then $C = [c_{ik}]_{m \times p}$ is known as the multiplication of A and B if

$$c_{ik} = \sum_{j=1}^{n} a_{ij} b_{jk}$$

and hence we can write $\qquad C = AB$

 WORKING PROCEDURE

M First we check whether the matrices are conformable for multiplication or not. For this we check that if the number of columns of first matrix is equal to the number of rows of the second matrix, then the matrices can be multiplied. Multiplication is operated by the rule (row × column). In this rule, we first put the first row of the first matrix next to the first column of the second matrix and the corresponding elements are now multiplied and then summed up which gives the first element of the first row of the product matrix. This process runs till the first row of the first matrix is operated to all columns of the second matrix. After that the first process is applied to the second, third etc. rows of the first matrix.

For example : If $\qquad A = \begin{bmatrix} 2 & 1 & 5 \\ 6 & 2 & 3 \end{bmatrix}_{2 \times 3}$ and $B = \begin{bmatrix} 3 & 4 \\ 5 & 6 \\ 7 & 8 \end{bmatrix}_{3 \times 2}$, then

$$AB = \begin{bmatrix} 2 & 1 & 5 \\ 6 & 2 & 3 \end{bmatrix} \begin{bmatrix} 3 & 4 \\ 5 & 6 \\ 7 & 8 \end{bmatrix} = \begin{bmatrix} 2 \times 3 + 1 \times 5 + 5 \times 7 & 2 \times 4 + 1 \times 6 + 5 \times 8 \\ 6 \times 3 + 2 \times 5 + 3 \times 7 & 6 \times 4 + 2 \times 6 + 3 \times 8 \end{bmatrix}$$

$$= \begin{bmatrix} 6 + 5 + 35 & 8 + 6 + 40 \\ 18 + 10 + 21 & 24 + 12 + 24 \end{bmatrix} = \begin{bmatrix} 46 & 54 \\ 49 & 60 \end{bmatrix}$$

☞ **REMARKS**

- If the number of columns of the matrix A is equal to the number of rows of matrix B, then A and B are conformable for the multiplication AB but not for BA.
- Square matrices are always conformable for multiplication in both ways.

1.8 DETERMINANT OF A SQUARE MATRIX

Let A be a square matrix. Then the determinant which is formed by the elements of matrix A is usually denoted by $|A|$.

For example : If $A = \begin{bmatrix} a_{11} & a_{12} & a_{13} \\ a_{21} & a_{22} & a_{23} \\ a_{31} & a_{32} & a_{33} \end{bmatrix}$, then its determinant is

$$A = \begin{vmatrix} a_{11} & a_{12} & a_{13} \\ a_{21} & a_{22} & a_{23} \\ a_{31} & a_{32} & a_{33} \end{vmatrix}$$

☞ REMARK
- The determinant of a matrix is reduced to a number.

1.9 PROPERTIES OF DETERMINANTS

(1) The value of a determinant is zero if all the elements of a row or column are zero.

(2) The value of a determinant remain unchanged when rows are changed into corresponding columns.

(3) If any two rows or columns of a determinant are interchanged, the sign of the determinant is changed.

(4) If any two rows or columns of a determinant are identical, then the value of the determinant is zero.

(5) If every element of same columns or row is the sum of two terms then determinant is equal to the sum of two determinants are containing only the first term and other the second term only in place of each sum.

(6) If each element of a row (or column) is multiplied by a constant k, then the value of the new determinant will be k times the value of original determinant.

(7) If each element of a row (or column) of a determinant multiplied by a constant k and then added to the corresponding elements of some other row (or column) then the value of the determinant remain same.

(8) If the elements of the determinant are the polynomial in a variable x and if by putting $x = a$, the determinant vanishes then $(x - a)$ will be a factor of determinant.

1.10 EVALUATION OF A DETERMINANT BY SARRUS DIAGRAM

$$\begin{vmatrix} a_{11} & a_{12} & a_{13} \\ a_{21} & a_{22} & a_{23} \\ a_{31} & a_{32} & a_{33} \end{vmatrix} = a_{11}(a_{22}a_{33} - a_{32}a_{23}) - a_{12}(a_{21}a_{33} - a_{31}a_{23}) + a_{13}(a_{21}a_{32} - a_{31}a_{22})$$

$$= a_{11}a_{22}a_{33} + a_{12}a_{31}a_{23} + a_{13}a_{21}a_{32} - (a_{11}a_{32}a_{23} + a_{12}a_{21}a_{33} + a_{13}a_{31}a_{22})$$

WORKING PROCEDURE

M Write the columns of the determinant and again write the first and second columns on the right side and draw the lines as shown in the following figure :

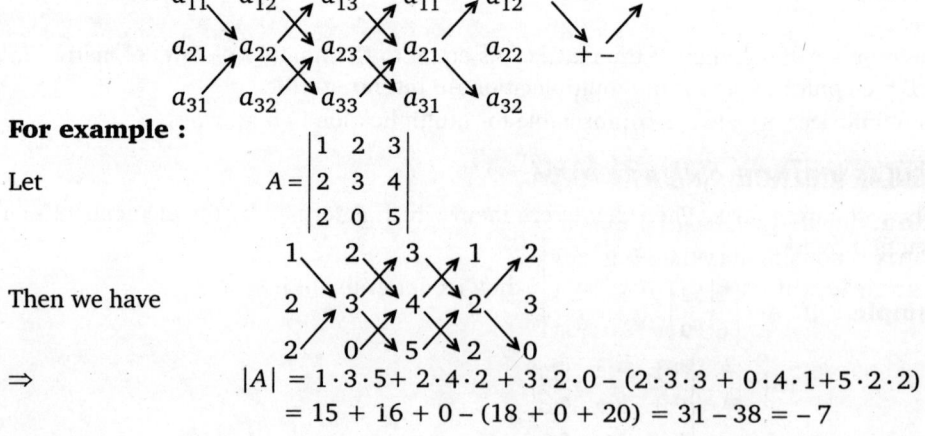

For example :

Let
$$A = \begin{vmatrix} 1 & 2 & 3 \\ 2 & 3 & 4 \\ 2 & 0 & 5 \end{vmatrix}$$

Then we have

$$\Rightarrow \quad |A| = 1 \cdot 3 \cdot 5 + 2 \cdot 4 \cdot 2 + 3 \cdot 2 \cdot 0 - (2 \cdot 3 \cdot 3 + 0 \cdot 4 \cdot 1 + 5 \cdot 2 \cdot 2)$$
$$= 15 + 16 + 0 - (18 + 0 + 20) = 31 - 38 = -7$$

1.11 MINORS AND COFACTORS

1.11.1 MINORS

In determinant, $\quad \Delta = \begin{vmatrix} a_{11} & a_{12} & a_{13} \\ a_{21} & a_{22} & a_{23} \\ a_{31} & a_{32} & a_{33} \end{vmatrix}$ $\qquad \qquad$ ...(1)

if we leave the row and column passing through the element a_{ij} then we obtained the second order determinant, which is called the minor of the element a_{ij}. It is denoted by M_{ij}. Therefore, in a determinant of order 3, we may get 9 minors corresponding to the 9 elements of the determinant.

For example, in determinant (1)

$$\text{Minor of } a_{21} = \begin{vmatrix} a_{12} & a_{13} \\ a_{32} & a_{33} \end{vmatrix} = M_{21}$$

and $\qquad \text{Minor of } a_{32} = \begin{vmatrix} a_{11} & a_{13} \\ a_{21} & a_{23} \end{vmatrix} = M_{32}$

If we expand the determinant along the first row, then

$$\Delta = (-1)^{1+1} a_{11} M_{11} + (-1)^{1+2} a_{12} M_{12} + (-1)^{1+3} a_{13} M_{13}$$
$$= a_{11} M_{11} - a_{12} M_{12} + a_{13} M_{13}$$

Similarly, along second column, we can write

$$\Delta = -a_{12} M_{12} + a_{22} M_{22} - a_{32} M_{32}$$

1.11.2 COFACTOR

If we multiply the minor M_{ij} by $(-1)^{i+j}$. Then resulting value is called cofactor of the element a_{ij}. If A_{ij} is the cofactor of a_{ij}, then we write

$$\text{Cofactor of } a_{ij} = A_{ij} = (-1)^{i+j} M_{ij}$$

$$\text{Cofactor of } a_{21} = A_{21} = (-1)^{2+1} M_{21} = - \begin{vmatrix} a_{12} & a_{13} \\ a_{32} & a_{33} \end{vmatrix}$$

$$\text{Cofactor of } a_{32} = A_{32} = (-1)^{3+2} M_{32} = - \begin{vmatrix} a_{11} & a_{13} \\ a_{21} & a_{23} \end{vmatrix}$$

Hence, cofactor of $a_{ij} = (-1)^{i+j}$ determinant obtained by leaving row and column passing through that element. Therefore, we can write

$$\Delta = a_{11} A_{11} + a_{12} A_{12} + a_{13} A_{13}$$
$$\Delta = a_{21} A_{21} + a_{22} A_{22} + a_{23} A_{23}$$
$$\Delta = a_{31} A_{31} + a_{32} A_{32} + a_{33} A_{33}$$

and $\qquad a_{11} A_{21} + a_{12} A_{22} + a_{13} A_{23} = 0$

$$a_{11} A_{31} + a_{12} A_{32} + a_{13} A_{33} = 0$$

1.12 SINGULAR AND NON-SINGULAR MATRIX

Definition. *A matrix whose determinant value is zero, is said to be singular matrix.*
If the matrix is not singular, then it is said to be non-singular.

For example : If $A = \begin{bmatrix} 2 & 3 \\ 6 & 9 \end{bmatrix}$, then its determinant value is given by

$$|A| = \begin{vmatrix} 2 & 3 \\ 6 & 9 \end{vmatrix} = 2 \times 9 - 3 \times 6 = 18 - 18 = 0$$

Thus the matrix A is singular.

1.13 TRANSPOSE OF A MATRIX

Consider a matrix $A = [a_{ij}]_{m \times n}$. Then a matrix which is obtained by interchanging the rows and columns of A is called the transpose of A. It is denoted by A' or A^T.

That is , if $A = [a_{ij}]_{m \times n}$, then $A' = [a_{ji}]_{n \times m}$.

For example : If $A = \begin{bmatrix} 2 & 3 & 5 \\ 1 & 6 & 7 \end{bmatrix}_{2 \times 3}$, then its transpose is

$$A' = \begin{bmatrix} 2 & 3 & 5 \\ 1 & 6 & 7 \end{bmatrix}' = \begin{bmatrix} 2 & 1 \\ 3 & 6 \\ 5 & 7 \end{bmatrix}_{3 \times 2}$$

☛ REMARKS
- Transpose of row matrix is a column matrix and transpose of a column matrix is a row matrix.
- If a matrix is square then its transpose will be a square matrix of same order.

1.14 PROPERTIES OF TRANSPOSE OF A MATRIX

THEOREM 1. *If A' and B' are the transpose of the matrix A and B respectively, then*
   (i) $(A')' = A$
   (ii) $(A+B)' = A' + B'$, here A and B must be of same order.
   (iii) $(kA')' = kA'$, here k is any scalar.
   (iv) $(AB)' = B'A'$, here AB and $B'A'$ are conformable for multiplication.
Proof.

   (i) Let $A = [a_{ij}]_{m \times n}$, then $A' = [a_{ji}]_{n \times m}$

   Since, (i, j)th element in $(A')' = (j, i)$th element in $A' = (i, j)$th element in A

   Thus by the definition of equality of matrices, we must have $(A')' = A$.

   (ii) Let $A = [a_{ij}]_{m \times n}$, $B = [b_{ij}]_{m \times n}$. So, $A' = [a_{ji}]_{n \times m}$ and $B' = [b_{ji}]_{n \times m}$, then

   (i, j)th element in $(A+B)' = (j, i)$th element in $(A+B)$

   $= (j, i)$th element in $A + (j, i)$th element in B

   $= (i, j)$th element in $A' + (i, j)$th element in B'

   $= (i, j)$th element in $(A'+B')$

   Thus by the definition of equality of matrices, we get

   $(A + B)' = A' + B'$

   (iii) Let $A = [a_{ij}]_{m \times n}$ so that $A' = [a_{ji}]_{n \times m}$ and k be a scalar, then

   (i, j)th element in $(kA)' = (j, i)$th element in (kA)

   $= (i, j)$th element in kA'

   Thus by the definition of equality of matrices, we get

   $(kA)' = kA'$

   (iv) Let $A = [a_{ij}]_{m \times n}$ and $B = [b_{ij}]_{n \times p}$ then AB is conformable for multiplication and having the order $m \times p$. Therefore, the order of $(AB)'$ is $p \times m$. Since the orders of A' and B' are respectively $n \times m$ and $p \times n$ so $B'A'$ is conformable for multiplication and having the order $p \times m$.

Now (k, i)th element in $(AB)' = (i, k)$th element in AB

$$= \sum_{j=1}^{n} a_{ij}b_{jk} \quad \text{[By definition of multiplication of matrices]}$$

But (k, i)th element in $B'A' = \sum_{j=1}^{n} b_{kj}a_{ji} = \sum_{j=1}^{n} a_{ji}b_{kj}$

$$= (i, k)\text{th element in } AB$$

$\therefore$ (k, i)th element in $(AB)' = (k, i)$th element in $B'A'$

Thus by the definition of equality of matrices, we must have

$$(AB)' = B'A'$$

1.15 SYMMETRIC MATRIX

A matrix A is said to be a symmetric matrix if $A' = A$, that is, the transpose of a matrix is equal to the matrix itself.

For exmaple : If $A = \begin{bmatrix} 1 & 2 & 3 \\ 2 & 4 & 5 \\ 3 & 5 & 6 \end{bmatrix}$, then $A' = \begin{bmatrix} 1 & 2 & 3 \\ 2 & 4 & 5 \\ 3 & 5 & 6 \end{bmatrix}$ so that $A' = A$

Hence, A is symmetric.

1.16 SKEW-SYMMETRIC MATRIX

A matrix A is said to be a skew-symmetric matrix if $A' = -A$.

For exmaple : If $A = \begin{bmatrix} 0 & 2 & 3 \\ -2 & 0 & 4 \\ -3 & -4 & 0 \end{bmatrix}$, then

$$A' = \begin{bmatrix} 0 & -2 & -3 \\ 2 & 0 & -4 \\ 3 & 4 & 0 \end{bmatrix} = \begin{bmatrix} 0 & 2 & 3 \\ -2 & 0 & 4 \\ -3 & -4 & 0 \end{bmatrix} = -A$$

Hence A is skew-symmetric matrix.

1.17 RANK OF A MATRIX

Let A be a matrix of order $m \times n$, then a non-negative integer r is said to be the rank of matrix A if it possesses the following two properties :

(i) There exists at least one r-minor of A which is not equal to zero.

(ii) Every s-minor of A for all $s > r$ is zero.

We denote the rank of A by $\rho(A)$.

In other words, the rank of a matrix is the order of any highest order of a non-zero minor of the matrix.

☛ REMARKS

- If the order of a matrix A is $m \times n$, then $\rho(A) \leq$ min. $\{m, n\}$
- A is a null matrix iff $\rho(A) = 0$.
- If A is any non-zero matrix, then $\rho(A) \geq 1$.
- $\rho(A) \geq r$, if there exists a non-zero r-minor of A.
- For any square matrix A of order n, $\rho(A) = n$ iff A is non-singular.
- For any square matrix A of order n, $\rho(A) < n$ iff A is singular.
- $\rho(A) \leq r$ if every s-minor of A is zero, where $s > r$.

Every $(r+1)$- rowed minor of A can be expressed as a linear combination of its r-rowed minors, therefore if every r-minor of A is zero, then its every $(r+1)$-minor is also zero.

1.18 ECHELON FORM OF A MATRIX

A matrix A is said to be in Echelon form if :

(*i*) every row of A has all its entries 0 which occurs below every row having a non-zero entry. and

(*ii*) the number of zeros before the first non-zero entry in a row is less than the number of such zeros in the next row.

☛ REMARK

- The rank of a matrix is equal to the number of non-zero rows in Echelon form of that matrix.

$$\textbf{For example: } \text{Consider a matrix } A = \begin{bmatrix} 0 & 2 & 3 & 5 \\ 0 & 0 & 3 & 2 \\ 0 & 0 & 0 & 0 \end{bmatrix}$$

Clearly, A is in Echelon form which has 2 non-zero rows, hence the rank of A is 2.

THEOREM 1. *The rank of the transpose of a matrix is equal to the rank of that matrix.*

PROOF. Let A be a marix, then A' is its transpose and let $\rho(A) = r$, then there exists an r-rowed minor of A which is not equal to zero and all s-rowed minors of A are zero, where $s > r$. Let $|B|$ be a r-rowed minor of A such that $|B| \neq 0$. Since A' is the transpose of A, then $|B'|$ is the r-rowed minor of A' but $|B'| = |B| \neq 0$, therefore $\rho(A') \geq r$. Suppose there is an s-minor $|C|$ of A' such that $|C| \neq 0$, where $s > r$, then $|C'|$ will be an s-minor of A such that $|C'| = |C| \neq 0$, therefore $\rho(A) > r$ which is a contradiction, hence $\rho(A') = r$.

SOLVED EXAMPLES

EXAMPLE 1. *Find the rank of the following matrices :*

$$\text{(i) } \begin{bmatrix} 3 & 0 & 0 \end{bmatrix} \quad \text{(ii) } \begin{bmatrix} 1 & 2 & 3 \\ 2 & 4 & 5 \end{bmatrix} \quad \text{(iii) } \begin{bmatrix} 1 & 2 & 3 \\ 3 & 4 & 5 \\ 4 & 5 & 6 \end{bmatrix} \quad \text{(iv) } \begin{bmatrix} 1 & 5 & 2 & 4 \\ 0 & 1 & 3 & 1 \\ 0 & 0 & 1 & 3 \end{bmatrix}$$

SOLUTION.

(i) Let $A = \begin{bmatrix} 3 & 0 & 0 \end{bmatrix}$, then A is the non-zero rowed matrix, thereofore $\rho(A) \geq 1$. Also A is a matrix of order 1×3, then $\rho(A) \leq 1$, hence $\rho(A) = 1$.

(ii) Let $A = \begin{bmatrix} 1 & 2 & 3 \\ 2 & 4 & 5 \end{bmatrix}$. The order of A is 2×3, then $\rho(A) \leq 2$.

Also there is a 2-minor $\begin{vmatrix} 2 & 3 \\ 4 & 5 \end{vmatrix}$ of A which is not equal to zero, then $\rho(A) \geq 2$, hence $\rho(A) = 2$.

(iii) Let $A = \begin{bmatrix} 1 & 2 & 3 \\ 3 & 4 & 5 \\ 4 & 5 & 6 \end{bmatrix}$. The order of A is 3×3, then $\rho(A) \leq 3$.

Now $|A| = \begin{vmatrix} 1 & 2 & 3 \\ 3 & 4 & 5 \\ 4 & 5 & 6 \end{vmatrix} = 1(24-25) - 2(18-20) + 3(15-16) = 0$ '

∴ The only 3-minor $|A|$ of A is zero, thus $\rho(A) < 3$. Further, there is a 2-minor $\begin{vmatrix} 1 & 2 \\ 3 & 4 \end{vmatrix}$ of A which is not equal to zero, hence $\rho(A) = 2$.

(iv) Let $A = \begin{bmatrix} 1 & 5 & 2 & 4 \\ 0 & 1 & 3 & 1 \\ 0 & 0 & 1 & 3 \end{bmatrix}$. The order of A is 3×4, then $\rho(A) \le 3$.

Now there is a 3-minor $\begin{vmatrix} 1 & 5 & 2 \\ 0 & 1 & 3 \\ 0 & 0 & 1 \end{vmatrix}$ of A which is not equal to zero, then $\rho(A) \ge 3$.

Hence $\rho(A) = 3$.

1.19 INVERSE OF A MATRIX

Let A be a non-singular matrix of order $n \times n$. Then it is said to be invertible if there exists a non-singular square matrix B of order $n \times n$ such that

$$AB = I_n = BA$$

where I_n is the unit matrix of order $n \times n$.

The matrix B is called the inverse of A and we write $B = A^{-1}$.

SOLVED EXAMPLES

EXAMPLE 1. *By using elementary row-transformations find the inverse of the following matrices:*

(i) $\begin{bmatrix} 1 & 2 \\ 3 & 7 \end{bmatrix}$ (ii) $\begin{bmatrix} 1 & 2 \\ 2 & -1 \end{bmatrix}$

SOLUTION. (i) We write

$$A = I_2 A$$

or

$$\begin{bmatrix} 1 & 2 \\ 3 & 7 \end{bmatrix} = \begin{bmatrix} 1 & 0 \\ 0 & 1 \end{bmatrix} A$$

Applying $R_2 \to R_2 - 3R_1$, we get

$$\begin{bmatrix} 1 & 2 \\ 0 & 1 \end{bmatrix} = \begin{bmatrix} 1 & 0 \\ -3 & 1 \end{bmatrix} A$$

Again applying $R_1 \to R_1 - 2R_2$, we get

$$\begin{bmatrix} 1 & 0 \\ 0 & 1 \end{bmatrix} = \begin{bmatrix} 7 & -2 \\ -3 & 1 \end{bmatrix} A$$

$\Rightarrow \qquad I_2 = BA$

$\Rightarrow \qquad A^{-1} = B = \begin{bmatrix} 7 & -2 \\ -3 & 1 \end{bmatrix}$.

(ii) We write $\qquad A = I_2 A$

or

$$\begin{bmatrix} 1 & 2 \\ 2 & -1 \end{bmatrix} = \begin{bmatrix} 1 & 0 \\ 0 & 1 \end{bmatrix} A$$

Applying $R_2 \to R_2 - 2R_1$, we get

$$\begin{bmatrix} 1 & 2 \\ 0 & -5 \end{bmatrix} = \begin{bmatrix} 1 & 0 \\ -2 & 1 \end{bmatrix} A$$

Applying $R_2 \to -\dfrac{1}{5}R_2$, we get

$$\begin{bmatrix} 1 & 2 \\ 0 & 1 \end{bmatrix} = \begin{bmatrix} 1 & 0 \\ 2/5 & -1/5 \end{bmatrix} A$$

Applying $R_1 \to R_1 - 2R_2$, we get

$$\begin{bmatrix} 1 & 0 \\ 0 & 1 \end{bmatrix} = \begin{bmatrix} 1/5 & 2/5 \\ 2/5 & -1/5 \end{bmatrix} A$$

$$\Rightarrow \qquad I_2 = BA$$

$$\Rightarrow \qquad A^{-1} = B = \begin{bmatrix} 1/5 & 2/5 \\ 2/5 & -1/5 \end{bmatrix}$$

EXAMPLE 2. *Find the inverse of the matrix*

$$A = \begin{bmatrix} 1 & 2 & 1 \\ 3 & 2 & 3 \\ 1 & 1 & 2 \end{bmatrix}$$

by using elementary row-transformation.

SOLUTION. We write $\qquad A = I_3 A$

or

$$\begin{bmatrix} 1 & 2 & 1 \\ 3 & 2 & 3 \\ 1 & 1 & 2 \end{bmatrix} = \begin{bmatrix} 1 & 0 & 0 \\ 0 & 1 & 0 \\ 0 & 0 & 1 \end{bmatrix} A$$

Applying $R_2 \to R_2 - 3R_1$, $R_3 \to R_3 - R_1$, we get

$$\begin{bmatrix} 1 & 2 & 1 \\ 0 & -4 & 0 \\ 0 & -1 & 1 \end{bmatrix} = \begin{bmatrix} 1 & 0 & 0 \\ -3 & 1 & 0 \\ -1 & 0 & 1 \end{bmatrix} A$$

Applying $R_2 \to \dfrac{-1}{4}R_2$, we get

$$\begin{bmatrix} 1 & 2 & 1 \\ 0 & 1 & 0 \\ 0 & -1 & 1 \end{bmatrix} = \begin{bmatrix} 1 & 0 & 0 \\ 3/4 & -1/4 & 0 \\ -1 & 0 & 1 \end{bmatrix} A$$

Applying $R_3 \to R_3 + R_2$, we get

$$\begin{bmatrix} 1 & 2 & 1 \\ 0 & 1 & 0 \\ 0 & 0 & 1 \end{bmatrix} = \begin{bmatrix} 1 & 0 & 0 \\ 3/4 & -1/4 & 0 \\ -1/4 & -1/4 & 1 \end{bmatrix} A$$

Applying $R_1 \to R_1 - 2R_2$, we get

$$\begin{bmatrix} 1 & 0 & 1 \\ 0 & 1 & 0 \\ 0 & 0 & 1 \end{bmatrix} = \begin{bmatrix} -1/2 & 1/2 & 0 \\ 3/4 & -1/4 & 0 \\ -1/4 & -1/4 & 1 \end{bmatrix} A$$

Applying $R_1 \to R_1 - R_3$, we get

$$\begin{bmatrix} 1 & 0 & 0 \\ 0 & 1 & 0 \\ 0 & 0 & 1 \end{bmatrix} = \begin{bmatrix} -1/4 & 3/4 & -1 \\ 3/4 & -1/4 & 0 \\ -1/4 & -1/4 & 1 \end{bmatrix} A$$

$$\Rightarrow \qquad I_3 = BA$$

$$\Rightarrow \qquad A^{-1} = B = \begin{bmatrix} -1/4 & 3/4 & -1 \\ 3/4 & -1/4 & 0 \\ -1/4 & -1/4 & 1 \end{bmatrix}$$

EXERCISE 1.1

1. Are the following pairs of matrices equivalent?

(i) $\begin{bmatrix} 4 & 0 & 2 \\ 3 & 1 & 0 \\ 5 & 2 & 0 \end{bmatrix}, \begin{bmatrix} 3 & 9 & 0 & 2 \\ 7 & -2 & 0 & 1 \\ 8 & 1 & 1 & 5 \end{bmatrix}$

(ii) $\begin{bmatrix} 2 & -1 & 3 & 4 \\ 0 & 3 & 4 & 1 \\ 2 & 3 & 7 & 5 \\ 2 & 5 & 11 & 5 \end{bmatrix}, \begin{bmatrix} 1 & 0 & -5 & 6 \\ 3 & -2 & 1 & 2 \\ 5 & -2 & -9 & 14 \\ 4 & -2 & -4 & 8 \end{bmatrix}$

Determine the rank of the following matrices:

2. $\begin{bmatrix} 1 & 1 & 1 \\ 2 & 2 & 2 \\ 3 & 3 & 3 \end{bmatrix}$

3. $\begin{bmatrix} 2 & 1 & 3 \\ 4 & 7 & 13 \\ 4 & -3 & -1 \end{bmatrix}$

4. $\begin{bmatrix} 4 & 5 & 6 \\ 5 & 6 & 7 \\ 7 & 8 & 9 \end{bmatrix}$

5. $\begin{bmatrix} 1 & 2 & 3 \\ 2 & 3 & 4 \\ 3 & 5 & 7 \end{bmatrix}$

6. $\begin{bmatrix} 2 & 3 & 7 \\ 3 & -2 & 4 \\ 1 & -3 & -1 \end{bmatrix}$

7. $\begin{bmatrix} 3 & -1 & 2 \\ -6 & 2 & -4 \\ -3 & 1 & -2 \end{bmatrix}$

8. $\begin{bmatrix} 1 & 2 & 3 & 1 \\ 2 & 4 & 6 & 2 \\ 1 & 2 & 3 & 2 \end{bmatrix}$

9. $\begin{bmatrix} 1 & 3 & 4 & 3 \\ 3 & 9 & 12 & 9 \\ 1 & 3 & 4 & 1 \end{bmatrix}$

10. $\begin{bmatrix} 1 & 2 & -1 & 4 \\ 2 & 4 & 3 & 5 \\ -1 & -2 & 6 & -7 \end{bmatrix}$

11. $\begin{bmatrix} 1 & 2 & -4 & 5 \\ 2 & -1 & 3 & 6 \\ 8 & 1 & 9 & 7 \end{bmatrix}$

12. $\begin{bmatrix} 1 & -1 & 3 & 6 \\ 1 & 3 & -3 & -4 \\ 5 & 3 & 3 & 11 \end{bmatrix}$

13. $\begin{bmatrix} 1 & 2 & 3 & 0 \\ 2 & 4 & 3 & 2 \\ 3 & 2 & 1 & 3 \\ 6 & 8 & 7 & 5 \end{bmatrix}$

14. $\begin{bmatrix} 2 & 3 & -1 & -1 \\ 1 & -1 & -2 & -4 \\ 3 & 1 & 3 & -2 \\ 6 & 3 & 0 & -7 \end{bmatrix}$

15. $\begin{bmatrix} 1 & 2 & 1 & 2 \\ 1 & 3 & 2 & 2 \\ 2 & 4 & 3 & 4 \\ 3 & 7 & 4 & 6 \end{bmatrix}$

16. $\begin{bmatrix} 3 & -2 & 0 & -1 \\ 0 & 2 & 2 & 1 \\ 1 & -2 & -3 & 2 \\ 0 & 1 & 2 & 1 \end{bmatrix}$

17. $\begin{bmatrix} 0 & 1 & -3 & -1 \\ 1 & 0 & 1 & 1 \\ 3 & 1 & 0 & 2 \\ 1 & 1 & -2 & 0 \end{bmatrix}$

18. $\begin{bmatrix} 1 & 2 & -1 & 3 \\ 4 & 1 & 2 & 1 \\ 3 & -1 & 1 & 2 \\ 1 & 2 & 0 & 1 \end{bmatrix}$

19. $\begin{bmatrix} 1 & 0 & 2 & 1 \\ 0 & 1 & -2 & 1 \\ 1 & -1 & 4 & 0 \\ -2 & 2 & 8 & 0 \end{bmatrix}$

20. $\begin{bmatrix} 8 & 0 & 0 & 1 \\ 1 & 0 & 8 & 1 \\ 0 & 0 & 1 & 8 \\ 0 & 1 & 1 & 8 \end{bmatrix}$

ANSWERS

1. (i) Not equivalent (ii) Not equivalent **2.** 1 **3.** 2 **4.** 2 **5.** 2

6. 1 **7.** 2 **8.** 2 **9.** 2 **10.** 3 **11.** 3 **12.** 3 **13.** 3 **14.** 3

15. 4 **16.** 2 **17.** 3 **18.** 3 **19.** 4 **20.** 2

1.20 CONVEX SET

A set of points is said to be convex if for any two points in the set, the line segment joining these points is also in the set, *i.e.,* the set is said to be convex if convex combinations of any two points in the set is also in the set. [MEERUT–2009, 10, 11, 14, 17; GORAKHPUR–2007, 10, 18; AGRA-2008]

Mathematically, for any two points x_1, x_2 in the set, if every point $x = \lambda x_1 + (1 - \lambda)x_2$, $0 \le \lambda \le 1$ is also in the set, then set is convex.

Following are the examples of convex sets.

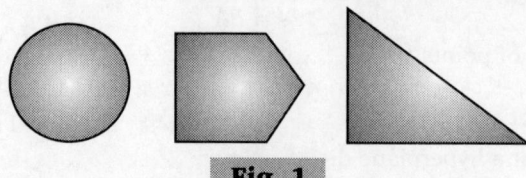

Fig. 1

Similarly some non-convex sets are given below:

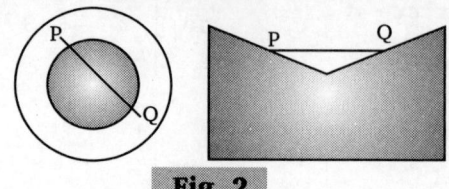

Fig. 2

☞ **REMARKS**

- The convex combination of any number of points in the convex set also belongs to the set.
- A set of one point is always convex.

1.21 SOME RELATED DEFINITIONS

1.21.1 POINT SET

A set whose elements are points or vectors in $\boldsymbol{R}^n$ is said to be point set.

For example:

1. A line $a_1 x_1 + a_2 x_2 = b$ represents a line in two dimensions which may be considered as a set of these points (x_1, x_2). Therefore, set of points can be written as
$$S = \{(x_1, x_2) : a_1 x_1 + a_2 x_2 = b\}$$

2. If we consider the set of points lying inside a circle of unit radius with centre at the origin in two dimensional space. Clearly, the points (x_1, x_2) of this set satisfy the inequality
$$x_1^2 + x_2^2 < 1$$
Therefore, set of points is given by
$$S = \{(x_1, x_2) : x_1^2 + x_2^2 < 1\}$$

1.21.2 HYPERSPHERE

In n-dimensional space, a hypersphere, with centre a and radius $r(> 0)$ is the set of points
$$X = \{x : |x - a| = r\}.$$
The equation of hypersphere in E^n (or R^n) is given by $\Sigma (x_i - a_i)^2 = r^2$

☞ **REMARKS**

- The set of points inside the hypersphere is the set $X = \{x : |x - a| < r\}$
- The set of points lying inside the hypersphere with centre a and radius $\in > 0$ is said to be $\in$-neighbourhood about the point a.

1.21.3 LINES AND LINE SEGMENTS

Let x_1, x_2 be two distinct points in n-dimensional space E^n, then the line through the points x_1 and x_2 is defined to be the set of points given by
$$X = \{x : x = \lambda x_1 + (1 - \lambda)x_2, \text{ for } \lambda \in \boldsymbol{R}\}$$
and the line segment joining two points x_1 and x_2 in E^n is defined to be the set of points given by
$$X = \{x : x = \lambda x_1 + (1 - \lambda)x_2, 0 \le \lambda \le 1\}$$

1.21.4 HYPERPLANE

It is defined as the set of points $(x_1, x_2, \ldots x_n)$ satisfying

$$c_1 x_1 + c_2 x_2 + \ldots + c_n x_n = z, \text{ (not all } c_i = 0)$$

for prescribed values of $c_1, c_2, \ldots, c_n$ and z

We clearly observe that a hyperplane divides the whole space into three mutually disjoint sets as given below:

$$X_1 = \{x : \boldsymbol{cx} > \boldsymbol{z}\} = \{(x_1, x_2, \ldots, x_n) : c_1 x_1 + c_2 x_2 + \ldots + c_n x_n > z\}$$

$$X_2 = \{x : \boldsymbol{cx} = \boldsymbol{z}\} = \{(x_1, x_2, \ldots, x_n) : c_1 x_1 + c_2 x_2 + \ldots + c_n x_n = z\}$$

$$X_3 = \{x : \boldsymbol{cx} < \boldsymbol{z}\} = \{(x_1, x_2, \ldots, x_n) : c_1 x_1 + c_2 x_2 + \ldots + c_n x_n < z\}$$

The set X_1 and X_3 are known as open-half spaces and the sets $\{x : \boldsymbol{cx} \leq \boldsymbol{z}\}$ and $\{x : \boldsymbol{cx} \geq \boldsymbol{z}\}$ are known as closed-half spaces.

☛ REMARKS

- For optimum value of z, the hyperplane $\boldsymbol{cx} = \boldsymbol{z}$ is called optimal hyperplane.
- The vector $\boldsymbol{c}$ is known as vector normal to the hyperplane
- The value $\pm \dfrac{c}{|\boldsymbol{c}|}$ are called unit normals.
- The hyperplanes are always closed sets.

> In an LPP, the objective function represents a hyperplane and each constraints($\leq$ or $\geq$) is a closed-half space produced by the hyperplane given by the constraints by taking ($=$) sign in place of $\leq$ or $\geq$.

1.21.5 PARALLEL HYPERPLANES

Two hyperplanes $\boldsymbol{c_1 x} = z_1$ and $\boldsymbol{c_2 x} = z_2$ are said to be parallel if they have the same unit normals, i.e., if $\boldsymbol{c_1} = \lambda \boldsymbol{c_2}$ for some λ, $(\lambda \neq 0)$

1.21.6 CONVEX COMBINATIONS

A convex combinations of a finite number of points $x_1, x_2, \ldots, x_n$ is defined by the point

$$\boldsymbol{x} = a_1 x_1 + a_2 x_2 + \ldots + a_n x_n, \ a_i \in \boldsymbol{R}, \ a_i \geq 0 \ \forall \ i \text{ and } \Sigma a_i = 1$$

In particular, the convex combination of two points x_1, x_2 be given by $\boldsymbol{x} = a_1 x_1 + a_2 x_2$ such that $a_1, a_2 \geq 0$ and $a_1 + a_2 = 1$ which can also be written as $\boldsymbol{x} = a x_1 + (1-a) x_2, \ 0 \leq a \leq 1$

☛ REMARK

- The line segment of two points x_1 and x_2 is the set of all possible convex combinations of two points x_1 and x_2.

1.21.7 EXTREME POINT OF A CONVEX SET

A point $\boldsymbol{x}$ in a convex set C is an extreme point of C if it does not lie on the line segment of any two points, different from $\boldsymbol{x}$ in the set, i.e., it can not be expressed as a convex combinations of any two distinct points $\boldsymbol{x}_1$ and $\boldsymbol{x}_2$ in C.

Mathematically an extreme point can be defined as follows:

A point $\boldsymbol{x}$ is said to be an extreme point of a convex set if there do not exist other points $\boldsymbol{x}_1, \boldsymbol{x}_2 (\boldsymbol{x}_1 \neq \boldsymbol{x}_2)$ in the set such that $\boldsymbol{x} = \lambda \boldsymbol{x}_1 + (1-\lambda) \boldsymbol{x}_2, \ 0 < \lambda < 1$.

☞ REMARKS
- A convex set may also have infinite number of extreme points.
- The polygons which are convex, have the extreme points as their vertices.
- An extreme point is a boundary point of the set.
- All boundary points of a convex set are not necessarily extreme points.
- A point of a convex set C, which is not an extreme point, is said to be internal point of C.

1.22 CONVEX HULL

The set of all convex combinations of set of points from the set X of points is called convex hull, *i.e.*, the intersection of all convex sets containing X in n-dimensional space is called the convex hull of X. Hence, the convex hull of a set $X \subseteq E^n$ is the smallest convex set containing X.

For example: If X is the boundary of a circle, then the convex hull $C(x)$ is the whole circle.

1.23 CONVEX FUNCTION AND CONVEX POLYHEDRON

Definition 1. *A function $f(x)$ is said to be strictly convex at x if for any two other distinct points x_1 and x_2*

$$f\{\lambda\, x_1 + (1 - \lambda)x_2\} < \lambda f(x_1) + (1 - \lambda) f(x_2),\ 0 < \lambda < 1$$

Definition 2. *The set of all convex combinations of finite number of points is said to be the convex polyhedron generated by these points.*

In other words, we can say that if the set X consist of a finite number of points, the convex hull of X is said to be the convex polyhedron with vertices at those points.

For example: The set of the area of a triangle is a convex polyhedron of its vertices.

☞ REMARK
- A function $f(x)$ is said to be strictly concave if $-f(x)$ is strictly convex.

THEOREM 1. *The hyperplane is a convex set.*

PROOF. Let $X = [x : cx = z]$ be a hyperplane and $x_1, x_2 \in X$ then,

$$cx_1 = z \text{ and } cx_2 = z \qquad \text{(By definition)}$$

Now, if $\quad x_3 = \lambda x_1 + (1 - \lambda)x_2,\ 0 \le \lambda \le 1$

Then, $\quad cx_3 = \lambda c \cdot x_1 + (1 - \lambda)cx_2$

$$= \lambda z + (1 - \lambda)z$$

$$= z$$

$\Rightarrow \quad x_3 = \lambda x_1 + (1 - \lambda)x_2 \in X$

$\Rightarrow x_3$ is also a point in X

Hence, X is a convex set.

THEOREM 2. *The closed half spaces $H_1 = \{x : cx \ge z\}$ and $H_2 = \{x : cx \le z\}$ are convex sets.*

PROOF. Let $x_1 \in H_1$ and $x_2 \in H_2$. Then by definition of H_1, we can write $cx_1 \ge z : cx_2 \ge z$

Now, if $0 \le \lambda \le 1$, then we have

$$c[\lambda x_1 + (1 - \lambda)x_2] = \lambda c \cdot x_1 + (1 - \lambda)cx_2$$

$$\ge \lambda z + (1 - \lambda)z = z$$

Therefore, $x_1, x_2 \in H_1$ and $0 \le \lambda \le 1$ implies $\lambda x_1 + (1 - \lambda)x_2 \in H_1$

Hence, H_1 is a convex set

Similarly, we may prove that H_2 is a convex set.

☞ **REMARK**

- In a similar way (as above) we may prove that the open half spaces $\{x : cx > z\}$ and $\{x : cx < z\}$ are convex sets.

THEOREM 3. *Intersection of two convex sets is also a convex set.* [MEERUT–2007, 08, 12,15, 17, 18]

PROOF. Let X_1 and X_2 be two convex sets. We have to prove that $X_1 \cap X_2$ is also convex.

If $x_1 \in X_1 \cap X_2 \Rightarrow x_1 \in X_1$ and $x_1 \in X_2$

$x_2 \in X_1 \cap X_2 \Rightarrow x_2 \in X_1$ and $x_2 \in X_2$

Now, by definition of convex sets

$x_1, x_2 \in X_1 \Rightarrow \lambda x_1 + (1 - \lambda)x_2 \in X_1$; $0 \le \lambda \le 1$

$x_1, x_2 \in X_2 \Rightarrow \lambda x_1 + (1 - \lambda)x_2 \in X_2$; $0 \le \lambda \le 1$

Therefore, $\lambda x_1 + (1 - \lambda)x_2 \in X_1$ and $\lambda x_1 + (1 - \lambda)x_2 \in X_2 \Rightarrow \lambda x_1 + (1 - \lambda)x_2 \in X_1 \cap X_2$

Hence, $X_1 \cap X_2$ is a convex set.

THEOREM 4. *Finite intersection of convex sets is also a convex set.* [MEERUT–2016]

PROOF. Let $X_1, X_2, ..., X_n$ be n convex sets.

We have to prove that $X_1 \cap X_2 \cap ... \cap X_n$ is also convex.

Let $x_1 \in X_1 \cap X_2 \cap ... \cap X_n \Rightarrow x_1 \in X_i \forall i = 1,2,...,n$

$x_2 \in X_1 \cap X_2 \cap ... \cap X_n \Rightarrow x_2 \in X_i \forall i = 1,2,...,n$

Since each X_i is convex set for $i = 1, 2, ..., n$

Therefore, $x_1, x_2 \in X_i$

$\Rightarrow \lambda x_1 + (1 - \lambda)x_2 \in X_i$ $\forall i = 1,2,...,n, 0 \le \lambda \le 1$ (By definitiion of Convex sets)

$\Rightarrow \lambda x_1 + (1 - \lambda)x_2 \in X_1 \cap X_2 \cap ... \cap X_n$ (By definitiion of Intersection)

So, $x_1 \in X_1 \cap X_2 \cap ... \cap X_n$ and $x_2 \in X_1 \cap X_2 \cap ... \cap X_n$

$\Rightarrow \lambda x_1 + (1 - \lambda)x_2 \in X_1 \cap X_2 \cap ... \cap X_n, 0 \le \lambda \le 1$.

Hence, $X_1 \cap X_2 \cap ... \cap X_n$ is a convex set.

☞ **REMARK**

- In the similar way we may extend the above result as follows:
  "Arbitrary intersection of convex sets is also a convex set"

THEOREM 5. *The set of all convex combinations of a finite number of points* $x_1, x_2, ..., x_n$ *is a convex set.*

PROOF. Let us define the set X of all convex combinations as follows:

$$X = \left\{ x : x = \sum_{i=1}^{n} \lambda_i x_i, \sum_{i=1}^{n} \lambda_i = 1, \lambda_i \ge 0 \right\}$$

We have to prove that X is convex.

Let $\alpha, \beta \in X$ such that

$$\alpha = \sum_{i=1}^{n} a_i x_i ; \sum_{i=1}^{n} a_i = 1, a_i \ge 0 \text{ and } \beta = \sum_{i=1}^{n} b_i x_i ; \sum_{i=1}^{n} b_i = 1, b_i \ge 0$$

Let us consider

$$w = \lambda \alpha + (1 - \lambda)\beta, 0 \le \lambda \le 1$$

$$= \lambda \sum_{i=1}^{n} a_i x_i + (1 - \lambda) \sum_{i=1}^{n} b_i x_i = \sum_{i=1}^{n} \{\lambda a_i + (1 - \lambda)b_i\} x_i$$

$$= \sum_{i=1}^{n} c_i x_i \text{ where } c_i = \lambda a_i + (1 - \lambda) b_i \qquad \ldots(1)$$

Now, we shall prove that

$$\sum_{i=1}^{n} c_i = \sum_{i=1}^{n} \{\lambda a_i + (1 - \lambda) b_i\} = \lambda \sum_{i=1}^{n} a_i + (1 - \lambda) \sum_{i=1}^{n} b_i = \lambda \cdot 1 + (1 - \lambda) \cdot 1 = 1 \qquad \ldots(2)$$

Further, $c_i = \lambda a_i + (1 - \lambda) b_i \geq 0 \ \forall \ i$ $\qquad \ldots(3)$

Hence, from (1), (2) and (3) we conclude that $w = \sum_{i=1}^{n} c_i \cdot x_i$ is a convex combination of $x_1, x_2, \ldots, x_n$. Hence X is convex.

THEOREM 6. *Let S_1 and S_2 be two convex sets in E^n, then for any scalar α, β the set $(\alpha S_1 + \beta S_2)$ is also convex.*

PROOF. Let S_1 and S_2 be two convex sets. We have to prove that for any two scalars α and β, $(\alpha S_1 + \beta S_2)$ is also convex.

Let $x, y \in \alpha S_1 + \beta S_2$. Then these are of the following forms:

$$x = \alpha u_1 + \beta v_1 \text{ and } y = \alpha u_2 + \beta v_2 \quad u_1, u_2 \in S_1 \text{ and } v_1, v_2 \in S_2 \qquad \ldots(1)$$

Now for any scalar λ, $0 \leq \lambda \leq 1$, we have

$$\lambda x + (1 - \lambda) y = \lambda (\alpha u_1 + \beta v_1) + (1 - \lambda)(\alpha u_2 + \beta v_2)$$
$$= \alpha \{\lambda u_1 + (1 - \lambda) u_2\} + \beta \{\lambda v_1 + (1 - \lambda) v_2\} \qquad \ldots(2)$$

Since, S_1 and S_2 both are convex, so by definition, we can write

$$u_1, u_2 \in S_1 \Rightarrow \lambda u_1 + (1 - \lambda) u_2 \in S_1, \ 0 \leq \lambda \leq 1 \qquad \ldots(3)$$
$$v_1, v_2 \in S_2 \Rightarrow \lambda v_1 + (1 - \lambda) v_2 \in S_2, \ 0 \leq \lambda \leq 1 \qquad \ldots(4)$$

Using (2), (3) and (4) we can write

$$\lambda x + (1 - \lambda) y \in \alpha S_1 + \beta S_2, \ 0 \leq \lambda \leq 1$$

Hence, $\alpha \cdot S_1 + \beta \cdot S_2$ is a convex set.

☛ REMARKS

- From the above theorem we may easily prove the following result:
  "The sum $(S_1 + S_2)$ and difference $(S_1 - S_2)$ of two convex sets S_1 and S_2 is again convex."
- The set of all convex combinations of a finite number of points is also convex. [MEERUT–2008, 09]
- Those convex sets which are the intersection of a finite number of closed half spaces are called 'polyhedral convex sets'.

THEOREM 7. *The set of all the internal points of a convex set is convex.* [MEERUT–2011, 12, 15]

PROOF. Let S be the convex set and S_1 be the set of all vertices of S. Then clearly $S - S_1$ is the set of all internal points of S.

If $u, v \in S - S_1$. Then $u, v \in S$

Suppose that z is a point on the line segment joining u and v then we can write

$$z = \lambda u + (1 - \lambda) v, \ 0 \leq \lambda \leq 1$$
$$\in S \qquad (\because S \text{ is convex})$$
$$\Rightarrow \qquad z \in S - S_1$$

Since z is not a vertex of S_1, therefore $S - S_1$ (set of all internal points) is convex.

SOLVED EXAMPLES

EXAMPLE 1. *Show that $S = \{(x_1, x_2) : 2x_1 + 3x_2 = 7\} \subset R^2$ is a convex set.*

[MEERUT–2003, 08; GORAKHPUR–2009; GARHWAL–2014]

SOLUTION.   Let $u, v \in S$

Then we can write $u = (u_1, u_2), v = (v_1, v_2)$

$\therefore$ $2u_1 + 3u_2 = 7$ and $2v_1 + 3v_2 = 7$ ...(1)

Further, suppose that $w = (w_1, w_2)$ is a point on the line segment joining the points u and v. Then we can write

$$w = \lambda u + (1 - \lambda)v, \ 0 \le \lambda \le 1$$

$\therefore$ $(w_1, w_2) = \lambda(u_1, u_2) + (1 - \lambda)(v_1, v_2)$

$= \{\lambda u_1 + (1 - \lambda)v_1, \ \lambda u_2 + (1 - \lambda)v_2\}$

which implies that

$$w_1 = \lambda u_1 + (1 - \lambda)v_1 \text{ and } w_2 = \lambda u_2 + (1 - \lambda)v_2$$

Consider $2w_1 + 3w_2 = 2[\lambda u_1 + (1 - \lambda)v_1] + 3[\lambda u_2 + (1 - \lambda)v_2]$

$= \lambda[2u_1 + 3u_2] + (1 - \lambda)[2v_1 + 3v_2]$

$= \lambda \cdot 7 + (1 - \lambda) \cdot 7$ (By (1))

$= 7$

$\Rightarrow$ $w \in S$

Hence, S is a convex set.

EXAMPLE 2. *Show that the set $S = \{(x_1, x_2, x_3) : 2x_1 - x_2 + x_3 \le 4\} \subset R^3$ is convex.*

[MEERUT–2005, 12, 15]

SOLUTION.   Let $x = (x_1, x_2, x_3)$ and $y = (y_1, y_2, y_3)$ be any two points of the given set S. Then by definition of S, we can write

$2x_1 - x_2 + x_3 \le 4$ and $2y_1 - y_2 + y_3 \le 4$ ...(1)

Let $w = (w_1, w_2, w_3)$ be a point such that

$$w = \lambda x + (1 - \lambda)y, \ 0 \le \lambda \le 1$$

which implies that

$(w_1, w_2, w_3) = \lambda(x_1, x_2, x_3) + (1 - \lambda)(y_1, y_2, y_3)$

$= (\lambda x_1 + (1 - \lambda)y_1, \ \lambda x_2 + (1 - \lambda)y_2, \ \lambda x_3 + (1 - \lambda)y_3)$

$\Rightarrow$ $w_1 = \lambda x_1 + (1 - \lambda)y_1$

$w_2 = \lambda x_2 + (1 - \lambda)y_2$

and $w_3 = \lambda x_3 + (1 - \lambda)y_3$

Consider $2w_1 - w_2 + w_3$

$= \lambda(2x_1 - x_2 + x_3) + (1 - \lambda)(2y_1 - y_2 + y_3)$

$\le 4\lambda + 4(1 - \lambda)$ (By (1))

≤ 4

Thus, $w = (w_1, w_2, w_3) \in S$

Hence, S is convex.

EXAMPLE 3. *Examine the convexity of the set*

$$S = \{(x_1, x_2) \in R^2 : 4x_1 + 3x_2 \le 6, \ x_1 + x_2 \ge 1\}$$

[MEERUT–1997, 2007; 09; DELHI–2009; ASSAM–2011; PATNA–2013]

SOLUTION.   We have $S = \{(x_1, x_2) \in R^2 : 4x_1 + 3x_2 \leq 6, x_1 + x_2 \geq 1\}$

Let $\boldsymbol{u} = (x_1, x_2) \in S$ and $\boldsymbol{v} = (y_1, y_2) \in S$. Then,

$$4x_1 + 3x_2 \leq 6, x_1 + x_2 \geq 1 \text{ and } 4y_1 + 3y_2 \leq 6, y_1 + y_2 \geq 1 \qquad ...(1)$$

Let $\boldsymbol{w} = (w_1, w_2)$ be a point on the line segment joining the points u and v, then

$$\boldsymbol{w} = \lambda\boldsymbol{u} + (1 - \lambda)\boldsymbol{v}$$

$\Rightarrow \qquad (w_1, \dot{w}_2) = (\lambda x_1 + (1 - \lambda)y_1, \lambda x_2 + (1 - \lambda)y_2)$

$\Rightarrow \qquad w_1 = \lambda x_1 + (1 - \lambda)y_1 \text{ and } w_2 = \lambda x_2 + (1 - \lambda)y_2$

Consider, $4w_1 + 3w_2 = \lambda(4x_1 + 3x_2) + (1 - \lambda)(4y_1 + 3y_2)$

$$\leq \lambda \cdot 6 + (1 - \lambda) \cdot 6 \qquad \text{(By (1))}$$

$\Rightarrow \qquad 4w_1 + 3w_2 \leq 6 \qquad\qquad ...(2)$

Also, $\qquad w_1 + w_2 = \lambda(x_1 + x_2) + (1 - \lambda)(y_1 + y_2)$

$$\geq \lambda \cdot 1 + (1 - \lambda) \cdot 1 \qquad \text{(Again by (1))}$$

$\Rightarrow \qquad w_1 + w_2 \geq 1 \qquad\qquad ...(3)$

Hence, from (2) and (3) we conclude that S is a convex set.

EXAMPLE 4.   *Show that the set $S = \{x : x = (x_1, x_2, x_3), x_1^2 + x_2^2 + x_3^2 \leq 1\}$ is a convex set.*

[MEERUT–2006]

SOLUTION.   Let $\boldsymbol{x}, \boldsymbol{y} \in S$ be arbitrary. Then we write

$$\boldsymbol{x} = (x_1, x_2, x_3) \text{ and } \boldsymbol{y} = (y_1, y_2, y_3)$$

Now by definition of S, we have

$$x_1^2 + x_2^2 + x_3^2 \leq \quad \text{and } y_1^2 + y_2^2 + y_3^2 \leq 1 \qquad ...(1)$$

Further, let $\boldsymbol{z} = (z_1, z_2, z_3)$ be a point on the line segment joining the points $\boldsymbol{x}$ and $\boldsymbol{y}$ then we can write

$$\boldsymbol{z} = \lambda\boldsymbol{x} + (1 - \lambda)\boldsymbol{y}, 0 \leq \lambda \leq 1$$

$\Rightarrow \quad (z_1, z_2, z_3) = \lambda(x_1, x_2, x_3) + (1 - \lambda)(y_1, y_2, y_3)$

$$= (\lambda x_1 + (1 - \lambda)y_1, \lambda x_2 + (1 - \lambda)y_2, \lambda x_3 + (1 - \lambda)y_3)$$

which gives $\quad z_1 = \lambda x_1 + (1 - \lambda)y_1, z_2 = \lambda x_2 + (1 - \lambda)y_2, z_3 = \lambda x_3 + (1 - \lambda)y_3$

Now consider $z_1^2 + z_2^2 + z_3^2$

$$= [\lambda x_1 + (1 - \lambda)y_1]^2 + [\lambda x_2 + (1 - \lambda)y_2]^2 + [\lambda x_3 + (1 - \lambda)y_3]^2$$

$$= \lambda^2(x_1^2 + x_2^2 + x_3^2) + (1 - \lambda)^2(y_1^2 + y_2^2 + y_3^2)$$

$$+ 2\lambda(1 - \lambda)(x_1 y_1 + x_2 y_2 + x_3 y_3)$$

$$\leq \lambda^2 \cdot 1 + (1 - \lambda)^2 + 1 + 2\lambda(1 - \lambda)(x_1 y_1 + x_2 y_2 + x_3 y_3) \qquad \text{(By (1))}$$

$$...(2)$$

Using Lagrange's identity

$$(x_1^2 + x_2^2 + x_3^2)(y_1^2 + y_2^2 + y_3^2) - (x_1 y_1 + x_2 y_2 + x_3 y_3)^2 \equiv \Sigma(x_1 y_2 - x_2 y_1)^2 \geq 0$$

we can write

$$x_1 y_1 + x_2 y_2 + x_3 y_3 \leq \sqrt{x_1^2 + x_2^2 + x_3^2} \cdot \sqrt{y_1^2 + y_2^2 + y_3^2} \leq 1 \qquad \text{(By (1))}$$

Using these values in (2) we get

$$z_1^2 + z_2^2 + z_3^2 \leq \lambda^2 + (1 - \lambda)^2 + 2\lambda(1 - \lambda) = 1$$

$\Rightarrow \qquad \boldsymbol{z} = (z_1, z_2, z_3) \in S$

Hence, the given set S is convex.

EXAMPLE 5. *Show that the set $S = \{(x_1, x_2) : 3x_1^2 + 2x_2^2 \leq 6\}$ is convex.*

SOLUTION. Let $u, v \in S$ where $u = (u_1, u_2)$ and $v = (v_1, v_2)$. Then by definition of S, we can write

$$3u_1^2 + 2u_2^2 \leq 6 \text{ and } 3v_1^2 + 2v_2^2 \leq 6 \qquad \qquad …(1)$$

Let $w = (w_1, w_2)$ be a point on the line segment joining the points u and v, then

$$w = \lambda u + (1 - \lambda)v; \ 0 \leq \lambda \leq 1$$

Therefore, $w = (w_1, w_2) = \lambda(u_1, u_2) + (1 - \lambda)(v_1, v_2)$

$$= (\lambda u_1 + (1 - \lambda)v_1, \lambda u_2 + (1 - \lambda)v_2)$$

$\Rightarrow \qquad w_1 = \lambda u_1 + (1 - \lambda)v_1$ and $w_2 = \lambda u_2 + (1 - \lambda)v_2$

Now, consider

$$3w_1^2 + 2w_2^2 = 3\{\lambda u_1 + (1 - \lambda)v_1\}^2 + 2\{\lambda u_2 + (1 - \lambda)v_2\}^2$$

$$= \lambda^2(3u_1^2 + 2u_2^2) + (1 - \lambda)^2(3v_1^2 + 2v_2^2) + 2\lambda(1 - \lambda)(3u_1 v_1 + 2u_2 v_2) \qquad …(2)$$

But we have

$$(3u_1^2 + 2u_2^2)(3v_1^2 + 2v_2^2) - 3(u_1 v_1 + 2u_2 v_2)^2 = 6(u_1 v_2 - u_2 v_1)^2 \geq 0$$

$\Rightarrow \qquad (3u_1 v_1 + 2u_2 v_2)^2 \leq (3u_1^2 + 2u_2^2)(3v_1^2 + 2v_2^2) \leq 6 \times 6$

$\Rightarrow \qquad 3u_1 v_1 + 2u_2 v_2 \leq 6 \qquad \qquad …(3)$

Finally using (1) and (3) in (2) we get

$$3w_1^2 + 2w_2^2 \leq 6\lambda^2 + 6(1 - \lambda)^2 + 2\lambda(1 - \lambda) \times 6$$

$\Rightarrow \qquad 3w_1^2 + 2w_2^2 \leq 6$

$\Rightarrow \qquad w = (w_1, w_2) \in S \ \forall \ 0 \leq \lambda \leq 1$

Hence, S is a convex set.

EXAMPLE 6. *Show that $S = \{(x_1, x_2, x_3) : 2x_1 - x_2 + x_3 \leq 4; \ x_1 + 2x_2 - x_3 \leq 1\}$ is a convex set.*

SOLUTION. Let $x, y \in S$ then by definition of S we can write $x = (x_1, x_2, x_3)$ and $y = (y_1, y_2, y_3)$ such that

$$\left. \begin{array}{l} 2x_1 - x_2 + x_3 \leq 4; x_1 + 2x_2 - x_3 \leq 1 \\ \text{and } 2y_1 - y_2 + y_3 \leq 4; y_1 + 2y_2 - y_3 \leq 1 \end{array} \right\} \qquad …(1)$$

Let $z = (z_1, z_2, z_3)$ be such that

$$z = \lambda x + (1 - \lambda)y, \ 0 \leq \lambda \leq 1$$

$\Rightarrow \qquad (z_1, z_2, z_3) = \lambda(x_1, x_2, x_3) + (1 - \lambda)(y_1, y_2, y_3)$

$$= \{\lambda x_1 + (1 - \lambda)y_1, \lambda x_2 + (1 - \lambda)y_2, \lambda x_3 + (1 - \lambda)y_3\}$$

$\Rightarrow \qquad z_1 = \lambda x_1 + (1 - \lambda)y_1, z_2 = \lambda x_2 + (1 - \lambda)y_2, z_3 = \lambda x_3 + (1 - \lambda)y_3$

Now, consider $2z_1 - z_2 + z_3$

$$= \lambda(2x_1 - x_2 + x_3) + (1 - \lambda)(2y_1 - y_2 + y_3)$$

$$\leq 4\lambda + 4(1 - \lambda) \qquad \qquad \text{(Using (1))}$$

$$= 4$$

$\Rightarrow 2z_1 - z_2 + z_3 \leq 4$

Similarly, $z_1 + 2z_2 - z_3 \leq 1$

Therefore, $z = (z_1, z_2, z_3) \in S$

Hence, S is a convex set.

EXAMPLE 7. *If S_1 and S_2 be two non-empty disjoint convex sets and S be a set such that if $x_1 \in S_1$, $x_2 \in S_2$ then $x_1 - x_2 \in S$. Show that S is also convex and does not contain the origin.*

SOLUTION. Let us write $u = x_1 - x_2$ and $v = y_1 - y_2 \in S$. Then we have $x_1, y_1 \in S_1$ and $x_2, y_2 \in S_2$

If z is a point on the line segment joining u and v then

$$z = \lambda u + (1 - \lambda)v; \ 0 \le \lambda \le 1$$
$$= \lambda(x_1 - x_2) + (1 - \lambda)(y_1 - y_2)$$
$$= \{\lambda x_1 + (1 - \lambda)y_1\} - \{\lambda x_2 + (1 - \lambda)y_2\} \qquad \ldots(1)$$

Further, it is given that S_1 and S_2 both are convex, therefore by definition, we have

$$x_1, y_1 \in S_1 \qquad \Rightarrow \qquad \lambda x_1 + (1 - \lambda)y_1 = z_1 \in S_1; \ (0 \le \lambda \le 1)$$
$$x_2, y_2 \in S_2 \qquad \Rightarrow \qquad \lambda x_2 + (1 - \lambda)y_2 = z_2 \in S_2; \ (0 \le \lambda \le 1)$$

Then from (1)

$$z = z_1 - z_2 \in S \ \forall \ 0 \le \lambda \le 1$$

Hence, S is convex.

Finally, let if possible $0 \in S$ then there exist $x_1 \in S_1$ and $x_2 \in S_2$ such that

$$0 = x_1 - x_2$$
$$\Rightarrow \qquad x_1 = x_2, x_1 \in S_1, x_2 \in S_2$$
$$\Rightarrow \qquad S_1 \text{ and } S_2 \text{ are not disjoint, which is a contradiction.}$$

Hence, $0 \notin S$, *i.e.*, S does not contain the origin.

EXAMPLE 8. *Examine the convexity of the following set*
$$C = \{z \in R^n : z = x + y, x \in A, y \in B\}$$
where A and B are convex sets in R^n.

SOLUTION. Let us suppose $z_1 = x_1 + y_1$ and $z_2 = x_2 + y_2$ be any two points in the set C.

Then $x_1, x_2 \in A$ and $y_1, y_2 \in B$

If u is the point on the line segment joining the points z_1 and z_2

Then we can write

$$u = \lambda z_1 + (1 - \lambda)z_2 : 0 \le \lambda \le 1$$
$$= \lambda(x_1 + y_1) + (1 - \lambda)(x_2 + y_2)$$
$$= (\lambda x_1 + (1 - \lambda)x_2) + (\lambda y_1 + (1 - \lambda)y_2), \ 0 \le \lambda \le 1 \qquad \ldots(1)$$

Further since both sets A and B are convex, therefore, we can write

$$\left. \begin{array}{l} x_1, x_2 \in A \Rightarrow \lambda x_1 + (1 - \lambda)x_2 = u_1 \in A \ \forall 0 \le \lambda \le 1 \\ y_1, y_2 \in B \Rightarrow \lambda y_1 + (1 - \lambda)y_2 = u_2 \in B \ \forall 0 \le \lambda \le 1 \end{array} \right\} \qquad \ldots(2)$$

Using (2) in (1) we get

$$u = u_1 + u_2 \in C$$
$$\Rightarrow \quad \text{every point of the line segment joining } z_1 \text{ and } z_2 \text{ of } C \text{ are also in } C.$$

Hence, C is a convex set

EXAMPLE 9. *Express any point w inside a triangle as a convex combinations of the vertices (extreme points) x_1, x_2, x_3 of the triangle.*

SOLUTION. Consider a triangle ABC with vertices x_1, x_2 and x_3. If P is any point w inside the triangle. Join A and P and extend this line to meet the base BC at $D(u)$, a point on the line BC.

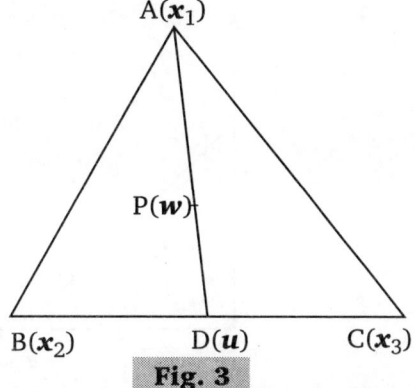

Fig. 3

$\Rightarrow$ $\boldsymbol{u}$ can be expressed as a convex combination of $\boldsymbol{x}_2$ and $\boldsymbol{x}_3$.

Therefore, $\qquad \boldsymbol{u} = \lambda_1 \boldsymbol{x}_2 + (1 - \lambda_1)\boldsymbol{x}_3, \; 0 \le \lambda_1 \le 1$ ...(1)

Further, since P is a point on the line segment AD, therefore,

$$\boldsymbol{w} = \lambda_2 \boldsymbol{x}_1 + (1 - \lambda_2)\boldsymbol{u}, \; 0 \le \lambda_2 \le 1$$
$$= \lambda_2 \boldsymbol{x}_1 + (1 - \lambda_2)[\lambda_1 \boldsymbol{x}_2 + (1 - \lambda)\boldsymbol{x}_3] \qquad \text{(Using (1))}$$
$$= \lambda_2 \boldsymbol{x}_1 + \lambda_1(1 - \lambda_2)\boldsymbol{x}_2 + (1 - \lambda_1)(1 - \lambda_2)\boldsymbol{x}_3$$

$\Rightarrow \qquad \boldsymbol{w} = \mu_1 \boldsymbol{x}_1 + \mu_2 \boldsymbol{x}_2 + \mu_3 \boldsymbol{x}_3$ ...(2)

where $\mu_1 = \lambda_2$, $\mu_2 = \lambda_1(1 - \lambda_2)$, $\mu_3 = (1 - \lambda_1)(1 - \lambda_2)$

Clearly each μ_i lies between 0 and 1, i.e., $0 \le \mu_i \le 1$, $i = 1, 2, 3$

Also, $\mu_1 + \mu_2 + \mu_3 = \lambda_2 + \lambda_1(1 - \lambda_2) + (1 - \lambda_1)(1 - \lambda_2)$
$$= \lambda_2 + \lambda_1 - \lambda_1 \lambda_2 + 1 - \lambda_1 - \lambda_2 + \lambda_1 \lambda_2 = 1$$

Thus, we conclude that $\mu_1 \boldsymbol{x}_1 + \mu_2 \boldsymbol{x}_2 + \mu_3 \boldsymbol{x}_3$ is a convex combination of the points $\boldsymbol{x}_1$, $\boldsymbol{x}_2$, $\boldsymbol{x}_3$.

Hence, the combination given by (2) is the required combination for the point $\boldsymbol{w}$.

EXAMPLE 10. *A hyperplane is given by the equation $3x_1 + 2x_2 + 4x_3 + 7x_4 = 8$. Find in which half spaces do the following points $(-6, 1, 7, 2)$ and $(1, 2, -4, 1)$ lie.*

SOLUTION. We have

The hyperplane: $3x_1 + 2x_2 + 4x_3 + 7x_4 = 8$ ...(1)

Using the values $(-6, 1, 7, 2)$ in (1) we get

$\qquad$ LHS $= 3(-6) + 2(1) + 7(2) = 26 > 8 =$ RHS

$\Rightarrow$ The point $(-6, 1, 7, 2)$ lies in the open half spaces $3x_1 + 2x_2 + 4x_3 + 7x_4 > 8$

Now, substituting $(1, 2, -4, 1)$ in the LHS of (1) we get

$\qquad$ LHS $= 3(1) + 2(2) + 4(-4) + 7(1) = -2 < 8 =$ RHS

Hence, the point $(1, 2, -4, 1)$ lies in the open half space $3x_1 + 2x_2 + 4x_3 + 7x_4 < 8$

EXAMPLE 11. *Sketch the convex polygon spanned by the following points in a two dimensional Euclidean space. Which of these points are vertices? Express the other as the convex linear combination of the vertices*

$(0, 0), (0, 1), (1, 0), \left(\dfrac{1}{2}, \dfrac{1}{4}\right)$

SOLUTION. Clearly, the convex combinations of the points $(0,0)$, $(1,0)$; $(0,0)$, $(0,1)$ and $(1,0)$, $(0,1)$ give the line segments OA, OB and AB respectively.

Therefore, the convex combination of points $(0,0)$, $(1,0)$ and $(0,1)$ is the interior of the $\triangle OAB$.

Thus, the points $O(0,0)$, $A(1, 0)$ and $B(0, 1)$ are the vertices and the point C is the interior point of the convex polygon spanned by the given points.

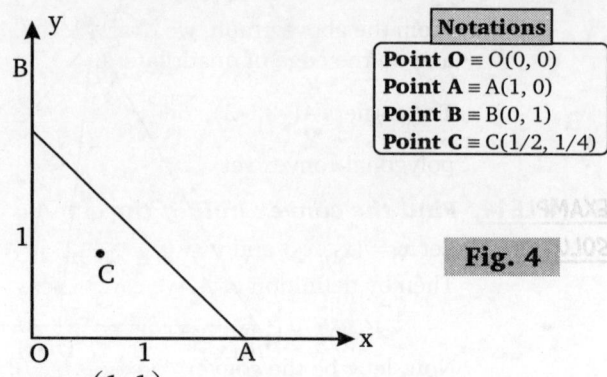

Fig. 4

Now, we have to express the point $\left(\frac{1}{2},\frac{1}{4}\right)$ as the linear combinations of (0,0), (0, 1) and (1, 0)

Write $\left(\frac{1}{2},\frac{1}{4}\right) = \lambda_1(0,0) + \lambda_2(0,1) + \lambda_3(1,0)$ where $\lambda_1 + \lambda_2 + \lambda_3 = 1, \lambda_i \geq 0$

$\Rightarrow \left(\frac{1}{2},\frac{1}{4}\right) = (\lambda_3, \lambda_2) \qquad \Rightarrow \qquad \lambda_2 = 1/4$ and $\lambda_3 = 1/2$

Now, $\lambda_1 = 1 - \lambda_2 - \lambda_3 = = 1 - \frac{1}{4} - \frac{1}{2} = \frac{1}{4}$

Hence, $\left(\frac{1}{2},\frac{1}{4}\right) = \frac{1}{4}(0,0) + \frac{1}{4}(0,1) + \frac{1}{2}(1,0)$

EXAMPLE 12. *Find the extreme points of the polygonal convex set X determined by the system.*

$$2x_1 + x_2 + 9 \geq 0 ; -x_1 + 3x_2 + 6 \geq 0, x_1 + x_2 \leq 0, x_1 + 2x_2 - 3 \leq 0$$

SOLUTION. Draw the graph of the following given equations

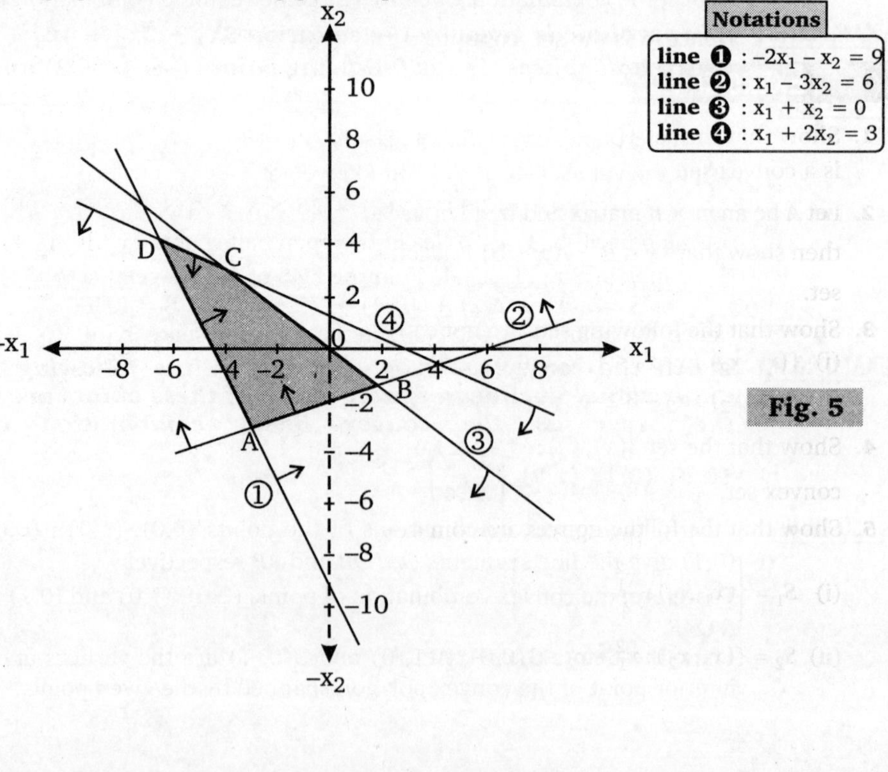

Notations
line ❶ : $-2x_1 - x_2 = 9$
line ❷ : $x_1 - 3x_2 = 6$
line ❸ : $x_1 + x_2 = 0$
line ❹ : $x_1 + 2x_2 = 3$

Fig. 5

From the above graph, we observe that the given inequalities enclose the points within and on the edge of quadrilateral $ABCD$ and it is a convex set.

The corners $A(-3, -3)$, $B\left(\dfrac{3}{2}, \dfrac{-3}{2}\right)$, $C(-3, 3)$ and $D(-7, 5)$ are the extreme points of this polygonal convex set.

EXAMPLE 14. *Find the convex hull of the set* $A = \left\{(x_1, x_2) : x_1^2 + x_2^2 = 1\right\}$

SOLUTION. Let $\boldsymbol{x} = (x_1, x_2)$ and $\boldsymbol{y} = (y_1, y_2)$ be any two points of A

Then by definition of A, we can write

$$x_1^2 + x_2^2 = 1 \text{ and } y_1^2 + y_2^2 = 1 \qquad \qquad ...(1)$$

Now, let $\boldsymbol{z}$ be the convex combination of $\boldsymbol{x}$ and $\boldsymbol{y}$

Then $\boldsymbol{z} = \lambda \boldsymbol{x} + (1 - \lambda)\boldsymbol{y} = \lambda(x_1, x_2) + (1 - \lambda)(y_1, y_2)$; $0 \le \lambda \le 1$

$\Rightarrow \quad \boldsymbol{z} = (z_1, z_2) = (\lambda x_1 + (1 - \lambda)y_1, \lambda x_2 + (1 - \lambda) y_2)$

Therefore, $z_1 = \lambda x_1 + (1 - \lambda)y_1$, $z_2 = \lambda x_2 + (1 - \lambda) y_2$

Now, $z_1^2 + z_2^2 = (\lambda x_1 + (1 - \lambda)y_1)^2 + (\lambda x_2 + (1 - \lambda) y_2)^2$

$$= \lambda^2(x_1^2 + x_2^2) + (1 - \lambda)^2(y_1^2 + y_2^2) + 2\lambda(1 - \lambda)(x_1 y_1 + x_2 y_2) \qquad ...(2)$$

We have, $\quad (x_1^2 + x_2^2)(y_1^2 + y_2^2) - (x_1 y_1 + x_2 y_2)^2 = (x_1 y_2 - x_2 y_1)^2 \ge 0$

$\Rightarrow \quad (x_1 y_1 + x_2 y_2)^2 \le (x_1^2 + x_2^2)(y_1^2 + y_2^2) = 1 \cdot 1$

$\Rightarrow \quad x_1 y_1 + x_2 y_2 \le 1 \qquad \qquad ...(3)$

Using (1) and (3) in (2) we get

$$z_1^2 + z_2^2 \le \lambda^2 \cdot 1 + (1 - \lambda)^2 \cdot 1 + 2\lambda(1 - \lambda) \cdot 1$$

$\Rightarrow \quad z_1^2 + z_2^2 \le 1$

$\Rightarrow \quad \boldsymbol{z} = (z_1, z_2) \in S = \left\{(x_1, x_2) : x_1^2 + x_2^2 \le 1\right\}$

Hence, S is the set, containing the convex combinations of the elements of A.

EXERCISE 1.2

1. Show that the set $\{(x_1, x_2) : x_1 \ge 2, x_1 \le 3)\}$ is a convex set.

2. Let A be an $m \times n$ matrix and $\boldsymbol{b}$, an m-vector, then show that $\{x \in R^n : A\boldsymbol{x} \le \boldsymbol{b}\}$ is a convex set.

3. Show that the following sets are not convex.
(i) $\{(x_1, x_2) : x_1 x_2 \le 1, x_1 \ge 0, x_2 \ge 0\}$
(ii) $\{(x_1, x_2) : x_2 - 3 \ge -x_1^2, x_1 \ge 0, x_2 \ge 0\}$

4. Show that the set $\left\{(x_1, x_2) : x_1^2 + x_2^2 \le 4\right\}$ is a convex set. [MEERUT–2011]

5. Show that the following sets are convex
(i) $S_1 = \left\{(x_1, x_2) : \dfrac{x_1^2}{4} + \dfrac{x_2^2}{9} \le 1\right\}$

(ii) $S_2 = \{(x_1, x_2) : x_1^2 + x_2^2 \le 1, x_1 + x_2 \ge 1\}$

6. If $x_1, x_2 \in S$ implies $1/2(x_1 + x_2) \in S$. Show that S is convex.

7. Show that the set $S = \{x : |x| = 1, x \in E^n\}$ is not convex.

8. Show that union of two convex sets is not necessarily convex.

9. Show that the vector $[7, 0]$ is a convex combination of the vectors $[6, 3]$, $[9, -6]$, $[1, 2]$ and $[1, -1]$.

10. Show that the vector $[2, 1]$ can not be expressed as a convex combinations of $[1, 1]$ and $[-1, 2]$.

11. Find the extreme points of the set $\{(x, y) : |x| \le 1, |y| \le 1\}$.

12. Find the extreme points of the set $\{(x, y) : x^2 + y^2 \le 25\}$.

11. (1, 1), (–1, 1), (–1, –1) and (1, –1)

12. Every point on the circumference is an extreme point.

1.24 FEASIBLE AND BASIC FEASIBLE SOLUTIONS

Definition 1. *For a system AX = B of m equations in n unknown and if n > m and Rank(A) = Rank(A|B) = m then in order to solve the system of given equations, we set (n – m) variables to zero. Thus, a solution to resulting system of equations is said to be basic solution provided, the determinant of the coefficient of the remaining m variables is not zero.*

Definition 2. *If all the basic variables are non-negative then a basic solution is called feasible.*

Definition 3. *If at least one of the basic variables is negative then a basic solution is called infeasible.*

THEOREM 1. **The set of all feasible solutions (if not empty) of a LPP is a convex set.**

[MEERUT-1994, 98, 2008; AGRA-1999, 2002,12; KANPUR-2008; GORAKHPUR-2007]

PROOF. Let X be the set of all feasible solutions of a *LPP given by*

$$Ax = b, x \geq 0 \qquad \ldots(1)$$

Now, we have the following cases:

Case I.

If the set X has only one element, then X is convex. In this case, theorem is true.

Case II.

If the set X has at least two elements

If $x_1, x_2 \in X$ then

$$Ax_1 = b, x_1 \geq 0 \text{ and } Ax_2 = b, x_2 \geq 0$$

Let $$x_3 = \lambda x_1 + (1-\lambda)x_2 : 0 \leq \lambda \leq 1$$

Then $$Ax_3 = A\lambda x_1 + (1-\lambda)Ax_2$$
$$= \lambda b + (1-\lambda)b = b$$

Further, since $x_1 \geq 0, x_2 \geq 0, 1-\lambda \geq 0$ $\qquad (\because 0 \leq \lambda \leq 1)$

So, $$x_3 = \lambda x_1 + (1-\lambda)x_2 \geq 0$$

$\Rightarrow$ x_3 satisfies (1).

Therefore, $x_3 = \lambda x_1 + (1-\lambda)x_2$ is also a feasible solution and hence belongs to X.

$\Rightarrow$ Convex combination of any two points x_1 and x_2 in X belongs to X.

Hence, X is a convex set.

THEOREM 2. **Every basic feasible solution of the system $Ax = b, x \geq 0$ is an extreme point of the convex set of feasible solutions and conversly.**

[MEERUT-1990, 94, 95, 2004]

PROOF. Necessary Part: We have to prove that every B.F.S is an extreme point of the convex set of all feasible solutions.

Let us suppose x be a B.F.S. of $Ax = b$

Also let X_B and B be the vectors of m basic variables and the matrix of vectors associated to basic variables in the B.F.S. respectively, then $x = [X_B, O]$ $\qquad \ldots(1)$

where O is a null vector of $(n\text{-}m)$ components

Also, $Ax = b \Rightarrow B \cdot X_B = b$ $\qquad \ldots(2)$

We want to prove that x is an extreme point. Let if possible x is not an extreme point.

Since $\boldsymbol{x}$ is not an extreme point, then there exist two distinct points $\boldsymbol{x}_1$ and $\boldsymbol{x}_2$ of X (the convex set of all feasible solutions of $A\boldsymbol{x} = \boldsymbol{b}$) such that $\boldsymbol{x} = \lambda\boldsymbol{x}_1 + (1-\lambda)\boldsymbol{x}_2$, $0 < \lambda < 1$ ...(3)

But we can write,

$$\boldsymbol{x}_1 = [\boldsymbol{u}_1, \boldsymbol{v}_1] \text{ and } \boldsymbol{x}_2 = [\boldsymbol{u}_2, \boldsymbol{v}_2] \qquad ...(4)$$

where $\boldsymbol{u}_1$ and $\boldsymbol{u}_2$ are vectors of m components of $\boldsymbol{x}_1$ and $\boldsymbol{x}_2$ respectively and $\boldsymbol{v}_1, \boldsymbol{v}_2$ are $(n-m)$ components vectors.

Now using (1) and (4) in (3) we get

$$[X_B, \boldsymbol{0}] = \lambda[\boldsymbol{u}_1, \boldsymbol{v}_1] + (1-\lambda)[\boldsymbol{u}_2, \boldsymbol{v}_2], 0 < \lambda < 1$$

$$= [\lambda\boldsymbol{u}_1 + (1-\lambda)\boldsymbol{u}_2, \lambda\boldsymbol{v}_1 + (1-\lambda)\boldsymbol{v}_2]$$

$\Rightarrow \qquad \boldsymbol{0} = \lambda\boldsymbol{v}_1 + (1-\lambda)\boldsymbol{v}_2,\ 0 < \lambda < 1 \qquad ...(5)$

Clearly, $1 > \lambda > 0$, $1 - \lambda > 0$ and the components of $\boldsymbol{v}_1$ and $\boldsymbol{v}_2$ are greater than equal to 0. So, (5) is satisfied only when $\boldsymbol{v}_1 = 0$ and $\boldsymbol{v}_2 = 0$

So, $\qquad \boldsymbol{x}_1 = [\boldsymbol{u}_1, 0], \boldsymbol{x}_2 = [\boldsymbol{u}_2, 0]$

Also, $\boldsymbol{x}_1 \in X, \boldsymbol{x}_2 \in X$, then from (2) we have

$$A\boldsymbol{x}_1 = B\boldsymbol{u}_1 = \boldsymbol{b}$$

and $\qquad A\boldsymbol{x}_2 = B\boldsymbol{u}_2 = \boldsymbol{b}$

$\Rightarrow \qquad X_B = \boldsymbol{u}_1 = \boldsymbol{u}_2$

So, $\boldsymbol{x} = x_1 = x_2$, which contradict the fact that $\boldsymbol{x}_1 \neq \boldsymbol{x}_2$

$\Rightarrow \boldsymbol{x}$ cannot be expressed as a convex combinations of any two distinct points in the set of all feasible solutions. Hence, $\boldsymbol{x}$ is an extreme point.

Conversely, let us suppose that $\boldsymbol{x} = (x_1, x_2, ..., x_n)$ be an extreme point. We have to prove that $\boldsymbol{x}$ is a BFS. Here, it is sufficient to prove that the vector associated with the positive elements of $\boldsymbol{x}$ are linearly independent.

Let us suppose p-components in $\boldsymbol{x}$ are non-zero and $(n-p)$ components are zero. We can assume these components as the first p-components of $\boldsymbol{x}$.

$$\text{So, } \sum_{i=1}^{p} \alpha_i x_i = \boldsymbol{b}, x_i > 0, i = 1, 2, ..., p \qquad ...(6)$$

where α_i is the column vector in A associated to the ith variable in $\boldsymbol{x}$

Let if possible the column vectors $\alpha_1, \alpha_2, ..., \alpha_p$ of matrix be linearly dependent. Then by definition, there exist some scalars λ_i ($i = 1, 2, ..., p$) with at least one of them non-zero such that $\sum_{i=1}^{p} \lambda_i \alpha_i = \boldsymbol{0}$ ...(7)

Now, from (6) and (7), for some arbitrary $\delta > 0$ we have

$$\sum_{i=1}^{p} x_i \alpha_i \pm \delta \sum_{i=1}^{p} \lambda_i \alpha_i = \boldsymbol{b}$$

$\Rightarrow \qquad \sum_{i=1}^{p} (x_i \pm \delta\lambda_i)\alpha_i = \boldsymbol{b}$

Then, clearly, two points

$$\boldsymbol{x}_1^* = [x_1 + \delta\lambda_1, x_2 + \delta\lambda_2, ..., x_p + \delta\lambda_p, 0, 0, ..., 0\ (n-p)\text{ in numbers}]$$

and $\qquad \boldsymbol{x}_2^* = [x_1 - \delta\lambda_1, x_2 - \delta\lambda_2, ..., x_p - \delta\lambda_p, 0, 0, ..., 0]$

satisfy $Ax = b$

Further since, $x_i > 0$ therefore taking δ such that

$$0 < \delta < \min_i \left\{ \frac{x_i}{|\lambda_i|} \right\}, \lambda_i \neq 0, i = 1, 2, ..., p$$

$\Rightarrow p$ component of x_1^* and x_2^* are always positive but the remaining components of x_1^* and x_2^* are zero.

$\Rightarrow x_1^*$ and x_2^* are feasible solutions different from x.

Further, $\quad x_1^* + x_2^* = 2[x_1, x_2, ..., x_p, 0, 0, 0, ..., 0]$

$\Rightarrow \qquad \frac{1}{2}x_1^* + \frac{1}{2}x_2^* = [x_1, x_2, ..., x_p, 0, 0, ..., 0] = x$

$\Rightarrow \qquad x = \lambda x_1^* + (1 - \lambda)x_2^* \qquad (\lambda = \frac{1}{2})$

$\Rightarrow x$ can be expressed as a convex combination of two distinct feasible solutions x_1^* and x_2^*, which is a contradiction, because x is an extreme point.

$\Rightarrow \alpha_1, \alpha_2, ..., \alpha_p$ are linearly independent.

$\Rightarrow \alpha_1, \alpha_2, ..., \alpha_p$ can not be more than m

$\Rightarrow$ extreme point x will have atmost m non-zero variables i.e., at least $(n-m)$ variables will be zero.

$\Rightarrow x$ is a BFS.

Hence, every extreme point of the convex set of feasible solution is a BFS.

THEOREM 3. *The extreme point of the convex set of feasible solutions are finite in number.*

PROOF. Using Theorem 2 we have that there is only one extreme point for a given BFS and converlsy.

$\Rightarrow$ there is one-to-one correspondance between the extreme points and the BFS in the absence of degeneracy.

Also, in case of degeneracy, corresponding to an extreme point with the number of non-zero variables less than m, we can form more than one degenerate BFS.

Hence, the number of extreme points of the feasible region is finite and it can not exceed the number of its basic feasible solution.

☞ **REMARKS**
- An extreme point can have atmost m positive x_i's where m is the no. of constraints.
- In an extreme point, vectors associated to the positive x_i's are linearly independent.

THEOREM 4. **(Fundamental Extreme point Theorem)** *If the convex set of the feasible solutions of Ax = b, x ≥ 0 is a convex polyhedron then at least one of the extreme points gives an optimal solution.* [MEERUT 2007, 08, 09, 10, 12]

PROOF. We know that the extreme points of the convex set of feasible solutions of $Ax = b$, $x \geq 0$ are finite. Thus, we suppose that $x_1, x_2, ..., x_k$ are the extreme points of the set X of all feasible solutions of $A x = b$, $x \geq 0$

Further, let z be the objective function such that $z = cx$. We have to be maximized z.

If $x^* \in X$ is the optimal solution, then

$$\text{Max } z = c\, x^*$$

If x^* is the extreme point, then result is obvious.

If x^* is not an extreme point in X, then x^* can be expressed as a convex combinations of the extreme point of X.

i.e., $$x^* = \lambda_1 x_1 + \lambda_2 x_2 + \dots + \lambda_k x_k$$

$$= \sum_{i=1}^{n} \lambda_i x_i, \ \lambda_i \geq 0 \text{ and } \Sigma \lambda_i = 1$$

$\Rightarrow$ $$z^* = cx^* = c(\lambda_1 x_1 + \lambda_2 x_2 + \dots + \lambda_k x_k)$$

$$= (\lambda_1 c x_1 + \lambda_2 c x_2 + \dots + \lambda_k c x_k)$$

Suppose maximum of cx_i is cx_p, then

$$z^* \leq (\lambda_1 + \lambda_2 + \dots + \lambda_k) \cdot cx_p$$

$\Rightarrow$ $$z^* \leq cx_p$$

Since, z^* is the maximum value of z, so

$$z^* = cx_p$$

$\Rightarrow$ $$cx^* = cx_p$$

$\Rightarrow$ $$x^* = x_p, \text{ one of the extreme point}$$

which shows that the optimal solution is attained at the extreme point.

THEOREM 5. *If the objective function of a LPP assumes its optimal value at more than one extreme point, then every convex combinations of these extreme points gives the optimal value of the objective function.*

PROOF. Let us consider an LPP as given below.

$$\text{Max} \cdot z = c \, x$$

such that $$A x = b, x \geq 0$$

Let $x_1, x_2, \dots, x_k$ be the extreme points of the feasible region. If the objective function z assume its optimal value z^* at the extreme points $x_1, x_2, \dots, x_p$ $(p \leq k)$

Then $$z^* = c \, x_1 = c \, x_2 = \dots = c \, x_p$$

If x_0 is the convex combination of the extreme points $x_1, x_2, \dots, x_p$

Then we have

$$x_0 = \lambda_1 x_1 + \lambda_2 x_2 + \dots + \lambda_p x_p \qquad \lambda_i \geq 0, \ \Sigma\lambda_i = 1$$

Thus,

$$cx_0 = c(\lambda_1 x_1 + \lambda_2 x_2 + \dots + \lambda_p x_p)$$

$$= \lambda_1 c \, x_1 + \lambda_2 c \, x_2 + \dots + \lambda_p \, cx_p$$

$$= \lambda_1 z^* + \lambda_2 z^* + \dots + \lambda_p z^*$$

$$= (\lambda_1 + \lambda_2 + \dots + \lambda_p) z^*$$

$$= \Sigma \lambda_i \cdot z^*$$

$$= z^* \qquad\qquad (\because \Sigma\lambda_i = 1)$$

Hence, the optimal value z^* is also attained at x_0 which is the convex combination of the extreme points at which optimal value occur.

SOLVED EXAMPLES

EXAMPLE 1. *Show that the feasible solution $x_1 = 1$, $x_2 = 0$, $x_3 = 1$, $z = 6$ to the system*

$$x_1 + x_2 + x_3 = 2$$

$$x_1 - x_2 + x_3 = 2$$

$$2x_1 + 3x_2 + 4x_3 = z \ (minimized), x_i > 0$$

is not basic.

SOLUTION. Here the objective funtion is given by,

minimized $z = 2x_1 + 3x_2 + 4x_3$

Here, we observe that first two equations in three variables x_1, x_2, x_3, only one variable can be assigned.

The given feasible solution is $x_1 = 1, x_2 = 0, x_3 = 1$ in which the variable x_1 and x_3 are non-zero, therefore, we shall take the vectors α_1 and α_2 associated to these variables, then

$$\alpha_1 = \text{column vector corresponding to } x_1 = \begin{bmatrix} 1 \\ 1 \end{bmatrix}$$

$$\alpha_2 = \text{column vector corresponding to } x_2 = \begin{bmatrix} 1 \\ 1 \end{bmatrix}$$

$\Rightarrow \qquad \alpha_1 = \alpha_2$

$\Rightarrow \qquad 1 \cdot \alpha_1 + (-1)\alpha_2 = 0$

$\Rightarrow \qquad \exists$ two scalars $\lambda_1 = 1, \lambda_2 = -1$ such that $\lambda_1 \alpha_1 + \lambda_2 \alpha_2 = 0$

$\Rightarrow \qquad \alpha_1, \alpha_2$ are linearly dependent

Hence, the given feasible solution is not basic.

EXAMPLE 2. *Find all the basic feasible solution for the system of equations*

$$2x_1 + 6x_2 + 2x_3 + x_4 = 3$$
$$6x_1 + 4x_2 + 4x_3 + 6x_4 = 2$$

and determine the associated general convex combinations of the extreme points solutions. [MEERUT–2007; KANPUR–2010; RAJASTHAN–2012]

SOLUTION. We can write the given system of equations as

$$\begin{bmatrix} 2 & 6 & 2 & 1 \\ 6 & 4 & 4 & 6 \end{bmatrix} \begin{bmatrix} x_1 \\ x_2 \\ x_3 \\ x_4 \end{bmatrix} = \begin{bmatrix} 3 \\ 2 \end{bmatrix}$$

$\Rightarrow \qquad AX = B$

where $A = \begin{bmatrix} 2 & 6 & 2 & 1 \\ 6 & 4 & 4 & 6 \end{bmatrix}$, $X = \begin{bmatrix} x_1 \\ x_2 \\ x_3 \\ x_4 \end{bmatrix}$, $B = \begin{bmatrix} 3 \\ 2 \end{bmatrix}$

If $\alpha_1, \alpha_2, \alpha_3$ and α_4 are the column vectors in A then we have

$$\alpha_1 = \begin{bmatrix} 2 \\ 6 \end{bmatrix} \qquad \alpha_2 = \begin{bmatrix} 6 \\ 4 \end{bmatrix} \qquad \alpha_3 = \begin{bmatrix} 2 \\ 4 \end{bmatrix} \qquad \alpha_4 = \begin{bmatrix} 1 \\ 6 \end{bmatrix}$$

$\Rightarrow \qquad A = [\alpha_1 \quad \alpha_2 \quad \alpha_3 \quad \alpha_4]$

Here, $\qquad n = $ number of unknowns $= 4$

$\qquad m = $ number of equations $= 2$

$\Rightarrow$ There can be atmost $^4C_2 = \dfrac{4!}{2! \, 2!} = 6$ feasible solutions

Now, six set of two vectors out of $\alpha_1, \alpha_2, \alpha_3, \alpha_4$ are given below:

$$|B_1| = \begin{vmatrix} 2 & 6 \\ 6 & 4 \end{vmatrix} = -28 \neq 0 \; ; \; |B_2| = \begin{vmatrix} 2 & 2 \\ 6 & 4 \end{vmatrix} = -4 \neq 0$$

$$|B_3| = \begin{vmatrix} 2 & 1 \\ 6 & 6 \end{vmatrix} = 6 \neq 0; \ |B_4| = \begin{vmatrix} 6 & 2 \\ 4 & 4 \end{vmatrix} = 16 \neq 0$$

$$|B_5| = \begin{vmatrix} 6 & 1 \\ 4 & 6 \end{vmatrix} = 32 \neq 0; \ |B_6| = \begin{vmatrix} 2 & 1 \\ 4 & 6 \end{vmatrix} = 8 \neq 0$$

$\Rightarrow$ all these set of vectors are linearly independent. Hence, all the basic solutions exist. If $x_{B_i} : i = 1, 2, ..., 6$ are the vectors of corresponding basic variables respectively, then the given system of equations reduces in the following forms.

$$B_1 x_1 = b, \qquad B_2 x_2 = b, \qquad B_3 x_3 = b$$
$$B_4 x_4 = b, \qquad B_5 x_5 = b, \qquad B_6 x_6 = b$$

which implies

$$x_{B_1} = B_1^{-1} b = -\frac{1}{28} \begin{bmatrix} 4 & 6 \\ -6 & 2 \end{bmatrix} \begin{bmatrix} 3 \\ 2 \end{bmatrix} = \begin{bmatrix} 0 \\ 1 \\ 2 \end{bmatrix}$$

$$x_{B_2} = B_2^{-1} b = -\frac{1}{4} \begin{bmatrix} 4 & -2 \\ -6 & 2 \end{bmatrix} \begin{bmatrix} 3 \\ 2 \end{bmatrix} = \begin{bmatrix} -2 \\ 7 \\ 2 \end{bmatrix}$$

$$x_{B_3} = B_3^{-1} b = \frac{1}{6} \begin{bmatrix} 6 & -1 \\ -6 & 2 \end{bmatrix} \begin{bmatrix} 3 \\ 2 \end{bmatrix} = \begin{bmatrix} \frac{8}{3} \\ \frac{-7}{3} \end{bmatrix}$$

$$x_{B_4} = B_4^{-1} b = \frac{1}{16} \begin{bmatrix} 4 & -2 \\ -4 & 6 \end{bmatrix} \begin{bmatrix} 3 \\ 2 \end{bmatrix} = \begin{bmatrix} 1 \\ 2 \\ 0 \end{bmatrix}$$

$$x_{B_5} = B_5^{-1} b = \frac{1}{32} \begin{bmatrix} 6 & -1 \\ -4 & 6 \end{bmatrix} \begin{bmatrix} 3 \\ 2 \end{bmatrix} = \begin{bmatrix} 1 \\ 2 \\ 0 \end{bmatrix}$$

and $$x_{B_6} = B_6^{-1} b = \frac{1}{8} \begin{bmatrix} 6 & -1 \\ -4 & 2 \end{bmatrix} \begin{bmatrix} 3 \\ 2 \end{bmatrix} = \begin{bmatrix} 2 \\ -1 \end{bmatrix}$$

Now, we will find the basic solutions

In the basic matrix B_1, basic vectors are α_1 and α_2 and

$$x_{B_1} = \begin{bmatrix} x_1 \\ x_2 \end{bmatrix} = \begin{bmatrix} 0 \\ \frac{1}{2} \end{bmatrix}$$

$\Rightarrow \qquad x_1 = 0, x_2 = 1/2$

These two variables are the basic variables and the remaining x_3, x_4 are non-basic variables. The non-basic variables are zero.

Therefore, the basic solution associated to the basis B_1 is given by (0, 1/2, 0, 0)

In a similar way we can write all other basic solutions as follows:

$$\left(-2, 0, \frac{7}{2}, 0\right), \left(\frac{8}{3}, 0, 0, \frac{-7}{3}\right), \left(0, \frac{1}{2}, 0, 0\right), \left(0, \frac{1}{2}, 0, 0\right) \text{ and } (0, 0, 2, -1)$$

But out of these basic solutions, the BFS are $\left(0, \frac{1}{2}, 0, 0\right), \left(0, \frac{1}{2}, 0, 0\right), \left(0, \frac{1}{2}, 0, 0\right)$

Clearly, the extreme points are $x_1 = \left(0, \frac{1}{2}, 0, 0\right)$, $x_2 = \left(0, \frac{1}{2}, 0, 0\right)$, $x_3 = \left(0, \frac{1}{2}, 0, 0\right)$

$\Rightarrow$ all the extreme points are same.

Hence, there is unique extreme point solution.

EXERCISE 1.3

1. Find all the basic solutions for the following system of linear equations
$$x_1 + 2x_2 + x_3 = 4$$
$$2x_1 + x_2 + 5x_3 = 4$$

2. Find all the basic solutions of the following system of linear equations
$$x_1 + x_2 + x_3 = 4$$
$$2x_1 + 5x_2 - 2x_3 = 0$$

3. Show that the basic solution $x_1 = 1$, $x_2 = 1/2$, $x_3 = x_4 = x_5 = 0$ of the equations
$$x_1 + 2x_2 + x_3 + x_4 = 2$$
$$x_1 + 2x_2 + 1/2\, x_3 + x_5 = 2$$
is not basic.

4. Find a basic feasible solution of the system of the equations
$$x_1 + 2x_3 = 3$$
$$x_2 + x_3 = 4 \text{ and } x_1, x_2, x_3 \geq 0$$

5. Find all basic feasible solutions of the equations:
$$2x_1 + x_2 + 3x_3 = 3$$
$$x_1 + 2x_2 + x_3 = 3 \text{ and } x_1, x_2, x_3 \geq 0$$

6. Show that if $x_1, x_2, \dots x_k$ are k different optimal basic feasible solutions to an LPP then any convex combinations of $x_1, x_2, \dots, x_k$ is also an optimal solution.

ANSWERS

1. $(2,1,0), (5,0,-1), \left(0, \dfrac{5}{3}, \dfrac{2}{3}\right)$

2. $\left(\dfrac{17}{3}, \dfrac{-5}{3}, 0\right), \left(0, \dfrac{11}{7}, \dfrac{17}{7}\right), \left(\dfrac{11}{4}, 0, \dfrac{5}{4}\right)$

4. $x_1 = 1, x_2 = 4, x_3 = 0$

5. $(1,1,0), \left(0, \dfrac{6}{5}, \dfrac{3}{5}\right)$

2 Sequencing

2.1 INTRODUCTION

The selection of an appropriate order in which to service waiting customers or jobs is called sequencing. The total effectiveness which may be time, cost etc. is function of the order or sequence of the jobs in which they are processed. In this chapter we shall consider the problems of determining the optimal sequence of arrivals or jobs that are to be done, which minimizes the total effectiveness which may be cost or time.

2.2 GENERAL SEQUENCING PROBLEM

[MEERUT-2002, 04, 10, 16; AGRA-2001, ROHILKHAND-2000, UPTUMBA-02, 06]

Consider the problem of performing n jobs on each of m machines. We are given the order of the machines for each job, in which it should go to the machines. We also know the actual or expected time required by the jobs on each of the machines. Our problem is to find that sequence out of $(n!)^m$ sequences which minimizes the total elapsed time, *i.e.*, the time from start of the first job upto the completion of the last job.

Mathematically, if we use the notations

A_i = time estimated for the ith job on machine A, $i = 1, 2,..., n$.

(Similarly we can interpret B_i and C_i etc.)

T = the total elapsed time

then we determine a sequence of jobs, *i.e.*, a permutation of numbers 1, 2, ..., n for each machine, which minimizes the time T.

All types of sequencing problems cannot be solved. The satisfactory solutions are available only in few cases. We shall discuss here the following cases:

1. n jobs are to be processed on two machines say A and B in the order AB.
2. n jobs are to be processed on three machines A, B and C in the order ABC.
3. 2 jobs are to be processed on m machines.

2.2.1 BASIC TERMINOLOGY

1. **No. of Machines:** It means that service facilities through which a job must pass before it is completed.
2. **Processing Order:** It means the order in which various machines are required for completing the job.
3. **Processing time:** It refer to the time for each job on each machine.
4. **Total elapsed time:** The time between starting the first job and completing the last job is called total elapsed time.
5. **Idle time on a Machine:** This is the time for which a machine remains idle during the total elapsed time.

6. No passing Rule: It means that passing is not allowed, *i.e.*, the same order of job is maintained over each machine. If each of the *n*-jobs is to be proceed through two machines *A* and *B* in the order *AB*, then this rule state that each job will go to machine *A* first and then go to *B*.

2.2.2 BASIC ASSUMPTIONS

Some principal assumptions in this chapter are as follows:

1. The processing times A_i's etc. are exactly known and are independent of the order of the jobs, in which they are to be processed. Such problems where times are exactly known are called *deterministic problems*.
2. The time taken by the jobs in going from one machine to another is negligible.
3. Each job, once started on a machine, is to be performed upto completion on that machine.
4. A job starts on the machine as soon as the job and the machine both are idle and job is next to the machine and the machine is also next to the job.
5. There is only one machine of each type.
6. No machine may process more than one job at a time.
7. The cost of keeping the jobs in inventory (if needed) during the inprocess is same for all jobs. Also it is too small that it can be neglected.
8. The order of completion of jobs has no significance *i.e.*, no job is to be given priority.
9. Times of jobs are independent of sequence of jobs.

2.3 SEQUENCING DECISION PROBLEM FOR *n* JOBS AND TWO MACHINES

[MEERUT-2002, 04, 07; ROHILKHAND-2002]

Consider the sequencing problem of processing two jobs on two machines. For example, consider the problem of two jobs say 1 and 2 to be processed on each of the two machines *A* and *B* in the order *AB*. Processing times are as follows:

Job\Machine	A	B
1	3	5
2	5	4

There are only 2!, *i.e.*, 2 possible sequences (1, 2) and (2, 1). Corresponding to both these sequences, a total elapsed times are evaluated graphically in the below figures so called Gantt Charts. Obviously optimum sequence is (1, 2) and the total elapsed time is 12 hours.

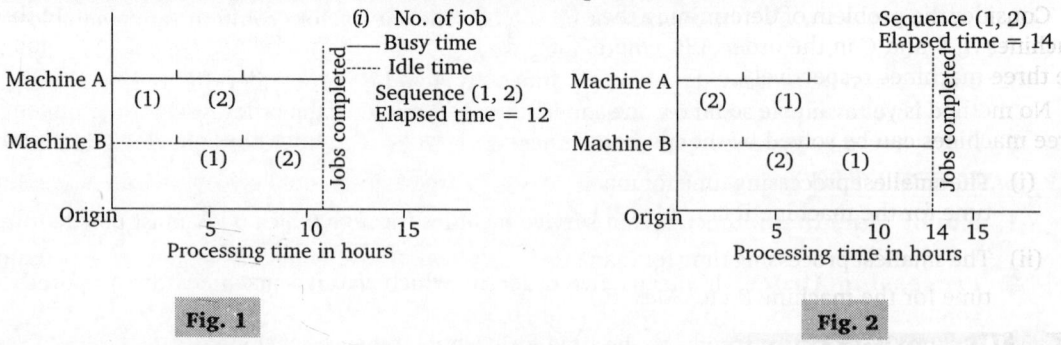

Fig. 1   **Fig. 2**

Above method is not of much practical importance even if n is small, because if there are *n* jobs and even if only two machines are involved and all jobs go over all machines in the same order, there are *n*! possible sequences and it is complicated enough to evaluate each sequence.

2.4 JOHNSON'S METHOD (FOR n-JOBS 2-MACHINES)

Let us suppose n jobs $(1, 2, ..., n)$ are to be processed on two machines say A and B and A_i, B_i, $i = 1, 2, ..., n$ are the respective processing times of ith job on A and B machines respectively.

2.4.1. ASSUMPTIONS

(i) Each job is processed in order AB.

(ii) A_i = Processing time of ith job on machine A $(i = 1, 2, ..., n)$

(iii) B_i = Processing time of ith job on machine B $(i = 1, 2, ..., n)$

We want to find the sequence of jobs to be performed on two machines so that the total time (T) elapsed from the start of the first job to the completion of the last job to be minimized.

WORKING PROCEDURE

STEP 1. Select the smallest processing time in the list $A_1, A_2, ..., A_n$ and $B_1, B_2, ..., B_n$. If there is a tie then either of these smallest processing time may be selected or in this case consider the following cases:

(i) Minimum of all the processing times is A_r which is also equal to B_s. Then $\min\{ A_i, B_i \} = A_r = B_s$. Then do the rth job first and s^{th} job in the end.

(ii) If $\min\{ A_i, B_i \} = A_r$ but also $A_r = A_k$ (say) then do anyone of these jobs for which there is a tie, first.

(iii) If there is a tie for minimum among B_i's, i.e., $\text{Min}\{ A_i, B_i \} = B_s = B_r$ (say) then do any of these jobs in the last.

STEP 2. If the smallest processing time is A_r (i.e., in the list $A_1, ..., A_n$) then do the rth job first. On the other hand if it is B_s (i.e., in the list $B_1, B_2, ..., B_n$). Then do the sth job last.

STEP 3. Delete the times of already assigned job from both the list. If r^{th} job is assigned previously, then delete A_r and B_r both and if s^{th} job is assigned previously then delete A_s and B_s both.

STEP 4. Repeat step 1 to 3 for remaining jobs.

STEP 5. Continuing the same process until all the jobs have been ordered and get optimal sequence of jobs.

2.5 SEQUENCING DECISION PROBLEM OF n-JOB AND THREE MACHINES

[MEERUT-2000, 01, 02, 04, 06, 18 UPTU MBA-2003, 07]

Consider the problem of determining the optimal sequence of n jobs to be performed on the three machines A, B and C in the order ABC where A_i, B_i and C_i are the processing times of the ith job on the three machines respectively.

No method is yet available so far for the solution of the problem as such. However the problem of three machines can be solved by the method developed by Johnson under the following conditions:

(i) The smallest processing time for machine A is greater than or equal to the greatest processing time for the machine B i.e., $\underset{i}{Min}(A_i) \geq \underset{i}{Max}(B_i)$

(ii) The smallest processing time for machine C is greater than or equal to the greatest processing time for the machine B i.e., $\underset{i}{Max}(B_i) \leq \underset{i}{Min}(C_i)$

WORKING PROCEDURE

STEP 1. We replace the three machines by two fictitious machines say G and H with corresponding processing times given by

$$G_i = A_i + B_i, \quad H_i = B_i + C_i, \quad i = 1, 2, ..., n$$

STEP 2. Determine the optimal sequence of jobs for these two machines G and H in the usual manner, *i.e.*, by applying the algorithm meant for problems of n jobs and two machines. The sequences so obtained will be the optimal sequence for the original problem also.

2.6 SEQUENCING DECISION PROBLEMS FOR n-JOBS AND m-MACHINES

[MEERUT 2001; ROHILKHAND-2001; UPTU MBA-2005]

A general sequencing problem of processing n jobs through m machines say $M_1, M_2, ..., M_m$, in the order $M_1, M_2 ... M_m$, can be solved under some conditions explained below.

If $M_{ij}, i = 1, 2, ..., n$, $j = 1, 2, ..., m$ is the processing time of ith job on j^{th} machine, then calculate $\underset{i}{Min} M_{i1}$ and $\underset{i}{Min} M_{im}$ and $\underset{i}{Max} M_{ij}, j = 2, ..., m - 1$, *i.e.*, calculate the minimum processing times for first and the last machines and maximum processing times for all the intermediate machines. The problem can be solved only if either of the following two or both the conditions are satisfied:

(i) $\underset{i}{Min} M_{i1} \geq \underset{i}{Max} M_{ij}$, for all $j = 2, 3, ..., m - 1$ or

(ii) $\underset{i}{Min} M_{im} \geq \underset{i}{Max} M_{ij}$ for all $j = 2, 3, ..., m - 1$.

If at least one of these two conditions is satisfied, then for two fictitious machines say G and H, their processing times, namely G_i and H_i, given by

$$G_i = M_{i1} + M_{i2} + ... + M_{i(m-1)}, \quad i = 1, 2, ..., n$$
$$H_i = M_{i2} + M_{i3} + ... + M_{im}, \quad i = 1, 2, ..., n$$

The sequence which is optimal for the problem for two machines say G and H will give the required optimal sequence for the original problem.

☞ REMARK

- If $M_{i2} + M_{i3} + ... + M_{i(m-1)} = c$ (constant) for all i, where c is a fixed positive quantity, then the required optimal sequence can be obtained by solving a problem involving only two extreme machines *i.e.*, solving the problem of n jobs on two machines M_1, M_m in the order $M_1 M_m$.

SOLVED EXAMPLES

EXAMPLE 1. *We have five jobs, each of which has to go through the machines A and B in the order AB. Processing times are given in the table below:*

Job	Processing time in hours	
	Machines	
	A_i	B_i
1	5	2
2	1	6
3	9	7
4	3	8
5	10	4

Determine a sequence for these jobs that will minimize the total elapsed time T.

[AGRA-2001; UPTU MBA-2006]

SOLUTION. The minimum time in the above table is 1 which is A_2. Hence we shall do the 2nd job first. We list the jobs as shown below

2				

Now we are left with four jobs with the processing times as shown in the given table follow:

Job i	A_i	B_i
1	5	2
3	9	7
4	3	8
5	10	4

Again as the minimum time in this table 2 which is B_1, we shall do the first job in last.

2				1

Now the time for the remaining jobs are as shown in the following table:

Job i	A_i	B_i
3	9	7
4	3	8
5	10	4

Similarly using the prescribed criterion, we conclude that the optimal sequence of jobs is

2	4	3	5	1

Further the minimum elapsed time can be calculated as follows :

Job	Machine A		Idle time of A	Machine B		Idle time of B
	Time in	Time out		Time in	Time out	
2	0	1	—	1	7	1
4	1	4	—	7	15	—
3	4	13	—	15	22	—
5	13	23	—	23	27	1
1	23	28	2	28	30	1

From the above table it is clear that the total time elapsed is 30 hours and the idle time for the machine B is 3 hours. Note that the total elapsed time is equal to the sum of the idle time of B and the total processing time on machine B.

The total elapsed time can also be calculated by using Gantt Chart as follows

From the Fig. 3, it can be seen that the total elapsed time is 30 hours and the idle time of the machine B is 3 hours.

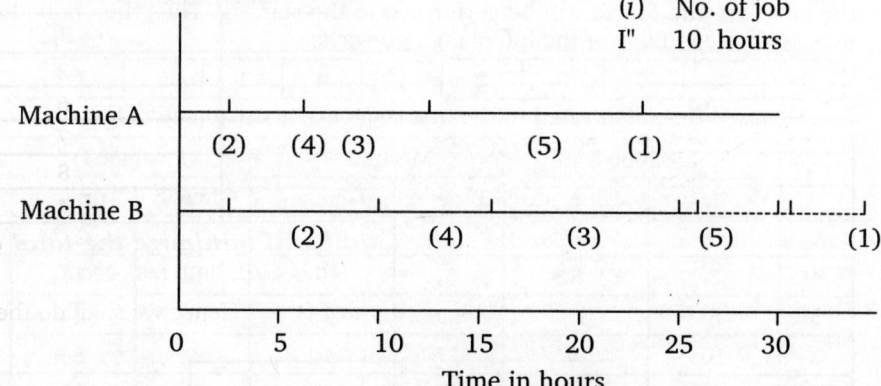

Fig. 3

☞ **REMARK**

- In the above problem it is to be noted that a job may be held in inventory before going to the machine. For instance 4th job will be free on machine A after 4th hour and will start on machine B after 7th hour. Therefore, it will be kept in inventory for 3 hours. So it is assumed that the storage space is available and the cost of holding the inventory for each job is either same or negligible. For short duration process problems generally it is negligible. Second general assumption is that the order of completion of jobs has no significance *i.e.*, no job claims the priority.

EXAMPLE 2. *Determine the optimal sequence of jobs which minimizes the total elapsed time based on the following information.*

Job	Processing times on the machines A, B, C		
	A_i	B_i	C_i
1	3	3	5
2	8	4	8
3	7	2	10
4	5	1	7
5	2	5	6

[MEERUT-1999; DELHI-1996]

SOLUTION. We have Min. $A_i = 2$, Max. $B_i = 5$, Min. $C_i = 5$.

Since Max. $B_i \leq$ Min. C_i, the problem can be solved by the above procedure. The times for the fictitious machines G and H are given by the following table:

Job	Processing times	
	$G_i = A_i + B_i$	$H_i = B_i + C_i$
1	6	8
2	12	12
3	9	12
4	6	8
5	7	11

Note that here minimum time is 6, which is both G_1 and G_4. As there is a tie, any of the jobs first and fourth can be performed in the starting. Thus the optimal sequence may be formed in any of the following two ways:

1	4	5	3	2

4	1	5	3	2

Total elapsed time associated to the first sequence is calculated below :

Job	Machine A		Machine B		Machine C		Idle time of C
	Time in	Time out	Time in	Time out	Time in	Time out	
1	0	3	3	6	6	11	6
4	3	8	8	9	11	18	0
5	8	10	10	15	18	24	0
3	10	17	17	19	24	34	0
2	17	25	25	29	34	42	0

Hence the total elapsed time is 42 hours. Similarly we can show that other sequence also takes total elapsed time 42 hours.

EXAMPLE 3. *Find the optimal sequence for processing 4 jobs A, B, C, D on four Machines A_1, A_2, A_3, A_4 in the order A_1 A_2 A_3 A_4. Processing times are as given below:*

	Processing times (M_{ij}) in hours			
Job\Machine	$A_1(M_{i1})$	$A_2(M_{i2})$	$A_3(M_{i3})$	$A_4(M_{i4})$
A	15	5	4	14
B	12	2	10	12
C	13	3	6	15
D	16	0	3	19

SOLUTION. From the above table we get for extreme machines

Min. (M_{i1}) = Min. processing time on first machine = 12
Min. (M_{i4}) = Min. processing time on last machine = 12

and for intermediate machines

Max. (M_{i2}) = Max. processing time on 2nd machine = 5
Max. (M_{i3}) = Max. processing time on 3rd machine = 10

Since Min. (M_{i1}) > Max. a_{i2} and Max. a_{i3} (both), the problem can be reduced to a problem involving only two machines G and H with processing times as

Jobs	Processing times	
	G_i	H_i
A	15 + 5 + 4 = 24	5 + 4 + 14 = 23
B	12 + 2 + 10 = 24	2 + 10 + 12 = 24
C	13 + 3 + 6 = 22	3 + 6 + 15 = 24
D	16 + 0 + 3 = 19	0 + 3 + 19 = 22

Using the algorithm for solving a sequencing problem of n jobs and 2 machines, we get the optimal sequence as given below.

D	C	B	A

Total elapsed time can be calculated as follows

Job\Machine	A_1		A_2		A_3		A_4	
	in	out	in	out	in	out	in	out
D	0	16	16	16	16	19	19	38
C	16	29	29	32	32	38	38	53
B	29	41	41	43	43	53	53	65
A	41	56	56	61	61	65	65	79

∴ Total elapsed time = 79 hours.

EXAMPLE 4. *Consider the problem of five jobs, each of which must go through the machines A, B, C in the order ABC. Processing time are:*

Job	A	B	C
1	4	5	8
2	9	6	10
3	8	2	6
4	6	3	7
5	5	4	11

Find a sequence for the five jobs that will minimize the elapsed time T.

[MEERUT 2003]

SOLUTION.   We have

Min. $A_i = 4$, Max. $B_i = 6$, Min. $C_i = 6$

Since Max $B_i \leq$ Min. C_i. Then we proceed as follows

Jobs	1	2	3	4	5
$G_i = A_i + B_i$	9	15	10	9	9
$H_i = B_i + C_i$	13	16	8	10	15

Then proceeding same as previous examples, we get the following optimal sequencing

5	1	4	2	3

4	1	5	2	3

1	4	5	2	3

1	5	4	2	3

4	5	1	2	3

5	4	1	2	3

EXAMPLE 5.   *We have five jobs, each of which must go through the machines A, B and C in order ABC*

Job	1	2	3	4	5
Machine A	5	7	6	9	5
Machine B	2	1	4	5	3
Machine C	3	7	5	6	7

Determine a sequence for the jobs that will minimize the total elapsed time.   [MEERUT-1993; 98, 2001, 03, 2012, 2016; ROHILKHAND-2001]

SOLUTION.   We have

Min. $A_i = 5$, Max. $B_i = 5$, Min. $C_i = 3$.

Clearly, Min $A_i \geq$ Max. B_i.

Jobs	1	2	3	4	5
$G_i = A_i + B_i$	7	8	10	14	8
$H_i = B_i + C_i$	5	8	9	11	10

The minimum time in table is 5 which is H_1. Therefore the job 1 will be done last. Thus, we get the following optimal sequence.

2	5	4	3	1

5	4	3	2	1

5	2	4	3	1

To find the elapsed time, we prepare the following table.

Job	Machine A		Machine B		Machine C		Idle time of B	Idle time of C
	Time in	Time out	Time in	Time out	Time in	Time out		
2	0	7	7	8	8	15	7	8
5	7	12	12	15	15	22	4	–
4	12	21	21	26	26	32	6	4
3	21	27	27	31	32	37	1	
1	27	32	32	34	37	40	1 + 6	

Hence, the minimum elapsed time is 40 hours. We may easily verify the time for other alternative sequencing also. Idle time for machines A, B and C are respectively given by 8, 25 and 12 hours.

EXAMPLE 6. **(a)** *A book binder has one printing press, one binding machine and the manuscripts of a number of different books. The times required to perform the printing and binding operations for each book are known. Determine the order in which the books should be processed in order to minimize the total time required. Find also the total time required to process all the books.*

Processing time in minutes					
Book	1	2	3	4	5
Printing time	40	90	80	60	50
Binding time	50	60	20	30	40

[ROHILKHAND-1995; UPTU MBA-2005, 08]

(b) *Suppose that an additional operation is added to the process described in (a) : finishing. The times required for operation are given below*

Book	1	2	3	4	5
Finishing time	80	100	60	70	110

What is the order in which the books should be processed? Find also the minimum total elapsed time.

SOLUTION. (a) The optimal sequence obtained by the sequencing algorithm, meant for n jobs and 2 machines is

1	2	5	4	3

and the total minimum elapsed time is 340 minutes.

(b) Now let the three machines be P (for printing), B (for printing) and F (for finishing) then clearly min $P_i = 40$, max $B_i = 60$, min $F_i = 60$.

Since Min. $F_i \geq$ Max. B_i, the given problem can be converted into n jobs and 2 machines problem. If the new machines are G and H with processing times G_i and $G_i = P_i + B_i$, $H_i = B_i + F_i$.

Thus the new problem will become as follows:

Book	1	2	3	4	5
G_i	90	150	100	90	90
H_i	130	160	80	100	150

The optimal sequence for this problem is given below :

4	1	5	2	3

The total elapsed time can be calculated as follows:

Book	4		1		5		2		3	
	in	out	in	out	in	out	in	out	in	out
Printing	0	60	60	100	100	150	150	240	240	320
Binding	60	90	100	150	150	190	240	320	320	340
Finishing	90	160	160	240	240	350	350	450	450	510

$\Rightarrow$ Total elapsed time is 510 minutes.

EXAMPLE 7. *A readymade garments manufacturer has to process 7 items through two stages of production, i.e., cutting and sewing. The time taken for each of these items at the different stages are given below in appropriate units.*

Item		1	2	3	4	5	6	7
Processing	Cutting	5	7	3	4	6	7	12
Time	Sewing	2	6	7	5	9	5	8

(a) **Find an order in which these seven items are to be processed so as to minimise the total processing time.** [ROHILKHAND-2003]

(b) **Suppose a third stage of production is added, say pressing and packing, with processing times as follows :**

Processing time	10	12	11	13	12	10	11

Find an order in which these seven items are to processed so as to minimize the time taken to process all the items through all the three stages.

SOLUTION. (a) The optimal sequence by the sequencing algorithm is :

3	4	5	7	2	6	1

and the total elapsed time = 46 hours.

(b) Proceed as in the previous example. The optimal sequence is

1	4	3	6	2	5	7

and the total elapsed time is 86 hours.

EXAMPLE 8. *Find an optimal sequence for the following sequencing problem of four jobs and five machines when passing is not allowed, of each processing time (in hours) is given below*

	Job	1	2	3	4
Machine	M_1	6	5	4	7
Machine	M_2	4	5	3	2
Machine	M_3	1	3	4	2
Machine	M_4	2	4	5	1
Machine	M_5	8	9	7	5

Also find the total elapsed time. [MEERUT-2002 BP]

SOLUTION. Here, Min $M_{1i} = 4$, Max. $M_{2i} = 5$, Max. $M_{3i} = 4$, Max. $M_{4i} = 5$ and Min. $M_{5i} = 5$.

Since Min. $M_{5i} \geq$ Max. M_{2i}, Max. M_{3i}, Max. M_{4i} i.e., minimum of one extreme machine is greater than or equal to the maximums of all the intermediate machines, the problem can be converted into 4 jobs and two machines problem. If the two fictitious machines are G and H, then there processing times can be calculated by

$$G_i = M_{1i} + M_{2i} + M_{3i} + M_{4i}, \quad H_i = M_{2i} + M_{3i} + M_{4i} + M_{5i}.$$

The new problem will be as follows :

Job i	1	2	3	4
Times of machine G (G_i)	13	17	16	12
Times of machine $H(H_i)$	15	21	19	10

The optimal sequence is given below

1	3	2	4

Evaluation of total elapsed time si given in the following table :

Job	1		3		2		4	
	in	out	in	out	in	out	in	out
M_1	0	6	6	10	10	15	15	22
M_2	6	10	10	13	15	20	22	24
M_3	10	11	13	17	20	23	24	26
M_4	11	13	17	22	23	27	27	28
M_5	13	21	22	29	29	38	38	43

Total minimum elapsed time = 43 hours.

2.7 SEQUENCING PROBLEM INVOLVING TWO JOBS AND m MACHINES

Here, we consider the problem in which
  (i) there are only two jobs say 1 and 2 to be performed,
  (ii) there are m machines say A, B, C, ..., M.
  (iii) exact or expected processing times of jobs on the machines are known.
  (iv) the technological ordering of each job through machines is known in advance.

Then we are to determine the sequence of jobs for each machines so that the total elapsed time is minimum.

2.8 GRAPHICAL METHOD   [MEERUT-2003; UPTU (MBA)-2003]

This method is applicable to the problems involving two jobs and m machines. Job 1 and Job 2 are represented by two horizontal and vertical axes respectively. On these two axes we mark the processing times of jobs on different machines in given order. Thus a horizontal line in the graph will represent the work on job 1 only while job 2 remains idle. Similarly a vertical line in the graph will represent only the work on job 2 while job 1 remains idle. A line inclined at an angle 45° with the horizontal, will represent the work on both the jobs simultaneously. Note that a job will be processed on a machine if machine is idle and is next for this job. A horizontal or vertical line will occur whenever some job is idle but the machine which is the next to this job is not idle. Also note that both the jobs cannot be processed on the same machine. Now we start with zero time (origin 0) and go on doing the jobs avoiding these shaded rectangular blocks. A best path is that which minimizes the idle time for job 1 and job 2, i.e., minimizes the horizontal and vertical lines in the path. Thus we try to move along the line inclined at 45° as much as we can.

SOLVED EXAMPLES

EXAMPLE 1. *Use graphical method to minimize the time needed to process the following jobs on the machines shown, i.e., for each machine find the job which should be done first. Also calculate the total time needed to complete both the jobs.*

		Machines				
Job 1	Sequence	A	B	C	D	E
	Time	3	4	2	6	2
Job 2	Sequence	B	C	A	D	E
	Time	5	4	3	2	6

[AGRA-1999, 2000 MEERUT-2004]

SOLUTION. We use the following steps:

STEP 1. Mark the processing times for first and second jobs on the horizontal and vertical axis according to the technological ordering of the machines.

STEP 2. Construct the rectangular blocks by pairing the same machines as shown in Fig. 4.

STEP 3. Now mark a path from origin 0 to the point of finish by moving along the 45° line as much as possible avoiding rectangular blocks.

STEP 4. Find the total elapsed time by adding the idle time of job 1 to its processing time or adding idle time of job 2 to its processing time. These two times will be equal.

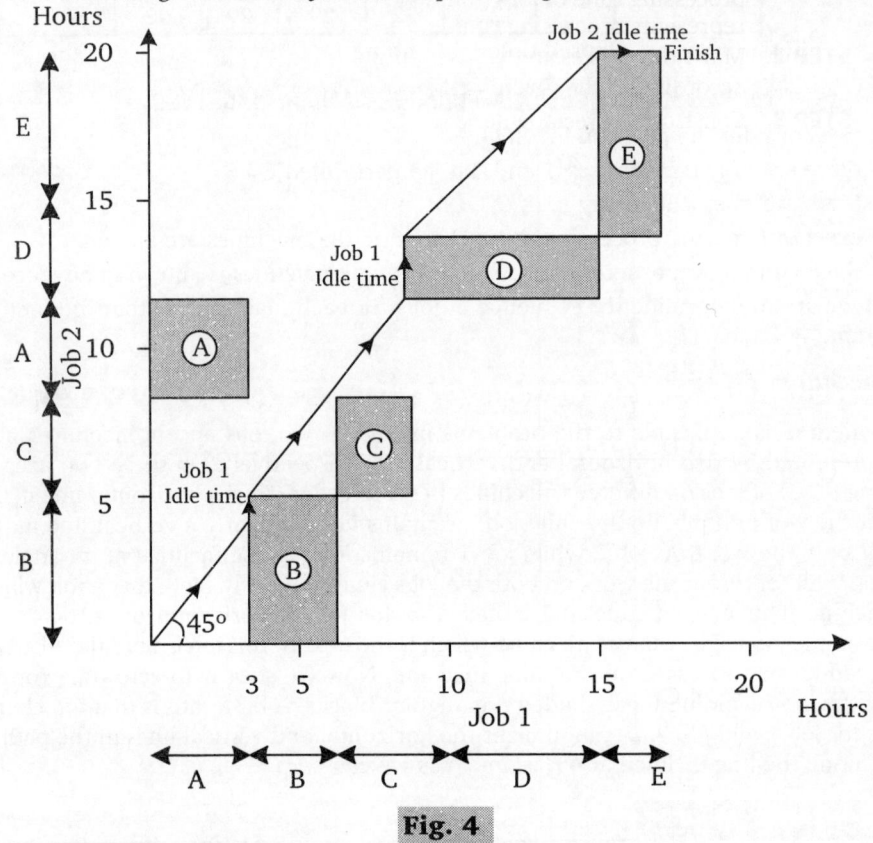

Fig. 4

The 'best' path is shown in Fig. 4 by arrows. The elapsed time is

"Processing time of Job 1 + idle time for Job 1 = 17 + 5 = 22 hours"

"Processing time of Job 2 + idle time of Job 2 = 20 + 2 = 22 hours"

Obviously in this route we have processed,

"Job 1 before 2 on machine *A*"

"Job 2 before 1 on machine *B*"

"Job 2 before 1 on machine *C*"

"Job 2 before 1 on machine *D*"

and "Job 2 before 1 on machine *E*"

EXAMPLE 2. *Use the graphical method to minimize the time needed to process the following jobs on the machine shown, i.e., for each machine, find the job which should be done first. Also calculate the total elapsed time to complete both jobs.*

							Total
Job-1	Machine sequence	A	B	C	D	E	17
	Time	2	3	4	6	2	
Job-2	Machine sequence	C	A	D	E	B	20
	Time	4	5	3	2	6	

[MEERUT-2002, 06, 09]

SOLUTION. To solve the above problem by graphical method we proceed as follows:

STEP 1. Draw the set of axes at right angles to each other where x-axis represents the processing time of job-1 on different machine, while job-2 remains idle and y-axis represents processing time of job-2 while job-1 remains idle.

STEP 2. Mark the processing time for first and second job on x and y-axis respectively according to the given order of machine.

STEP 3. Construct the rectangular block by pairing the same machine as shown in the figure.

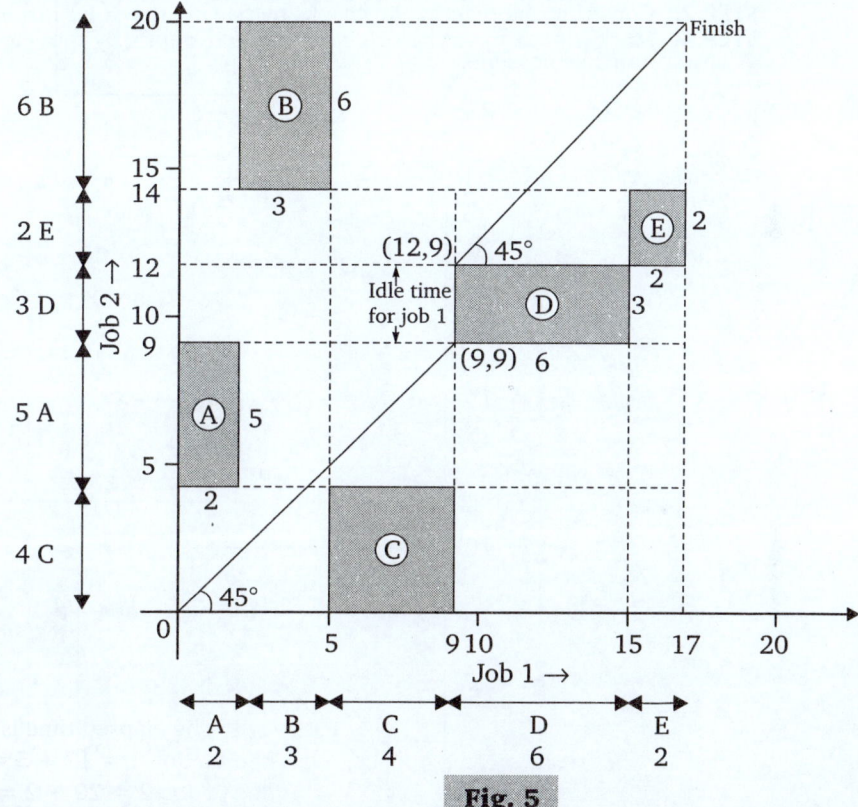

Fig. 5

STEP 4. Make a program by starting from the origin O and moving through various states of completion until we reach the point 'finish' by moving along the 45° line as much as possible avoiding rectangular blocks. Moving to the right means that job-1 is proceeding while job-2 is idle and moving upward indicate that job-2 is proceeding while job-1 remains idle and moving diagonally means simultaneous work on both jobs.

STEP 5. The optimum path is one which coincides with 45° are to the maximum extent. Further both jobs can not be processed simultaneously on one machine. Graphically, it means that diagonal movement through the blocked out areas is not allowed.

STEP 6. Total elapsed time is obtained by adding the idle time for either job to the processing time for that job. The idle time for the chosen path is found to be 3 for job-1 and 0 for job-2.

Hence, total elapsed time = 20 hours.

EXAMPLE 3. *A company has five machines A, B, C, D and E. Two jobs 1 and 2 must be processed through each of these machine. The processing time (in hours) for each job on different machine are as follows:*

Job-1 Machine						Job-2 Machine				
Sequence	A	B	C	D	E	Sequence	D	E	A	C	B
Time (h)	2	4	5	1	2	Time (h)	6	4	2	3	6

Use the graphical method to determine the total elapsed time.

SOLUTION. Proceeding as in previous examples, we have the following steps:

STEP 1. Mark the processing times for job 1 and job 2 on the x and y-axis respectively according to the given sequential order of five machines.

STEP 2. Construct the rectangular blocks by pairing the same machines.

STEP 3. Mark a path from the origin to the end points by moving along the 45° line as much as possible.

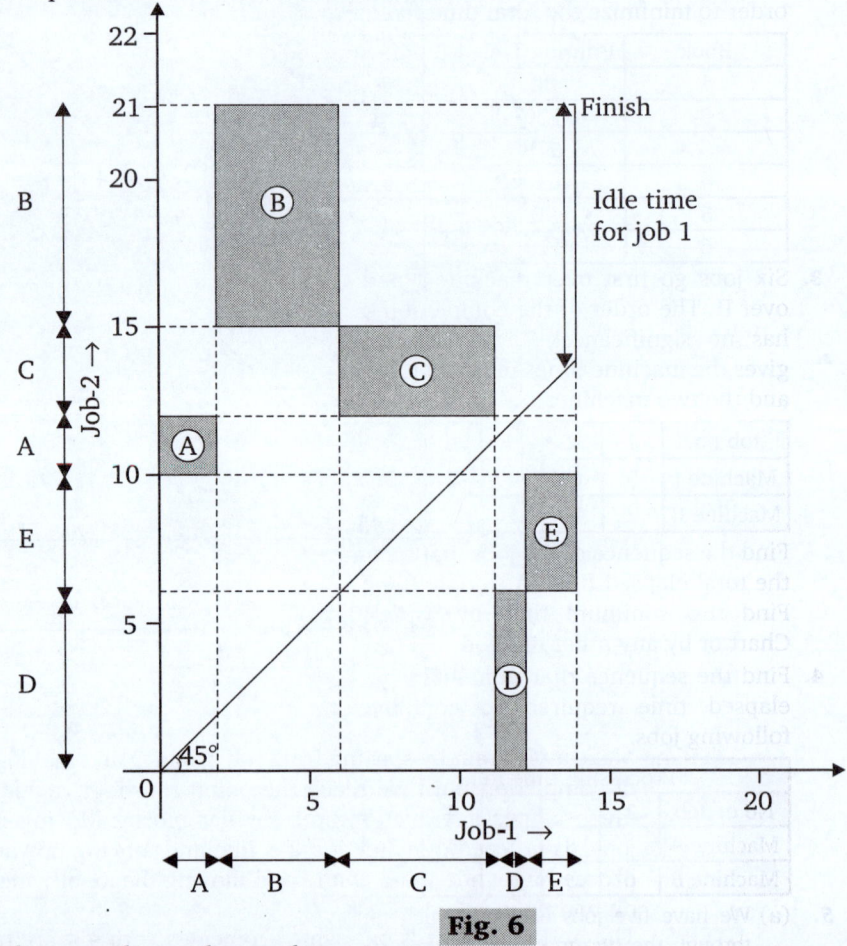

Fig. 6

From the above graph, we observe that

Idle time for job-1 = 7 hrs., Processing time for job-1 = 14 hrs. and the total elapsed time = 14 + 7 = 21 hrs.

EXERCISE 2.1

1. Find the sequence that minimizes the total elapsed time required to complete the following jobs.

Job	1	2	3	4	5	6
Processing A_i	2	5	4	3	2	1
Time B_i	6	8	1	2	3	5

2. A book binder has one printing press, one binding machine and the manuscripts of a number of different books. The time required to perform the printing and binding operations for each book are known. We wish to determine the order in which books should be processed on the machines, in order to minimize the total time required

Book	Printing Time	Binding time
1	30	80
2	120	100
3	50	90
4	20	60
5	90	30
6	110	10

3. Six jobs go first over machine I and then over II. The order of the completion of jobs has no significance. The following table gives the machine times in hours for six jobs and the two machines:

Job no.	1	2	3	4	5	6
Machine I	5	9	4	7	8	6
Machine II	7	4	8	3	9	5

Find the sequence of the jobs that minimizes the total elapsed time to complete the jobs. Find the minimum time by using Gantt Chart or by any other method.

4. Find the sequence that minimizes the total elapsed time required to complete the following jobs.

Processing time in hours						
No of Job	1	2	3	4	5	6
Machine A	4	8	3	6	7	5
Machine B	6	3	7	2	8	4

5. (a) We have five jobs each of which must go through the two machines in the order AB. Processing times are given in the table below :

Job No.	1	2	3	4	5
Machine A	10	2	18	6	20
Machine B	4	12	14	16	8

Determine a sequence for the five jobs that will minimize the total elapsed time.

(b) Find the optimal sequence of the job.

Job No.	1	2	3	4	5
Machine A	3	7	4	5	7
Machine B	6	2	7	3	4

Passing not allowed.   (ROHILKHAND-1997)

6. (a) Seven jobs each of which has to go through two machines M_1 and M_2 in order M_1M_2, take time on the machines as follows. Find in which order the jobs should be performed to minimize total time.

Job No.	1	2	3	4	5	6	7
Machine M_1	3	12	15	6	10	11	9
Machine M_2	8	10	10	6	12	1	3

[MEERUT-2002,06; AGRA-1998, 2003]

(b) Find the sequence that minimizes the total elapsed time required to complete the following jobs

Jobs	A	B	C	D	E	F	G	H	I
Machine I	2	5	4	9	6	8	7	5	4
Machine II	6	8	7	4	3	9	3	8	11

7. Find the sequence that minimizes the total elapsed time (in hours) required to complete all the following jobs on machines M_1, M_2, M_3 in the order $M_1M_2M_3$.

Machine	Jobs				
	A	B	C	D	E
M_1	4	9	8	6	5
M_2	5	6	2	3	4
M_3	8	10	6	7	11

8. State the conditions under which the problem of processing of n jobs through three machines has been solved. Describe the corresponding algorithm.
Find the sequence that minimizes the total time required to complete the following tasks:

Task	A	B	C	D	E	F	G
Machine I	3	8	7	4	9	8	7
Machine II	4	3	2	5	1	4	3
Machine III	6	7	5	11	5	6	12

9. Find the sequence that minimizes the total time required for performing the following jobs on three machines in the order *ABC*.

Processing times in hours

Jobs	1	2	3	4	5	6
Machine A	8	3	7	2	5	1
Machine B	3	4	5	2	1	6
Machine C	8	7	6	9	10	9

10. Solve the following sequencing problems:

(i) Processing times in hours

Jobs/Machine	M_1	M_2	M_3	M_4
A	10	3	5	14
B	12	2	6	7
C	8	4	4	12
D	15	1	7	8
E	16	5	3	10

[MEERUT 1993, 2005; ROHILKHAND-1998, 99]

(ii) Processing times in hours

Jobs/Machine	M_1	M_2	M_3	M_4
A	13	8	7	14
B	12	6	8	19
C	9	7	5	15
D	8	5	6	15

11. Find the sequence that minimizes the total elapsed time required to complete the following tasks. Each task is processed in the order *ACB*.

Processing times

Jobs	1	2	3	4	5	6	7
Machine A	12	6	5	11	5	7	6
Machine B	7	8	9	4	7	8	3
Machine C	3	4	1	5	2	3	4

[MEERUT 2001 BP, ROHILKHAND-2001]

ANSWERS

1. Alternative sequences exist.

| 6 | 1 | 5 | 2 | 4 | 3 | or | 6 | 5 | 1 | 2 | 4 | 3 |

Time = 26 hours.

2. Sequence

| 4 | 1 | 3 | 2 | 5 | 6 |

Time = 430 hours

3. Sequence

| 3 | 1 | 5 | 6 | 2 | 4 |

Total time = 42 hours

4. Sequence is

| 3 | 1 | 5 | 6 | 2 | 4 |

Total time = 35 hours

5. (a) Sequence

| 2 | 4 | 3 | 5 | 1 | Total time = 60 hours

(b) $1 \to 3 \to 5 \to 4 \to 2$, Time = 28 hours

6. (a) Sequence $1 \to 4 \to 5 \to 3 \to 2 \to 7 \to 6$, Time = 67 hours or $1 \to 4 \to 5 \to 2 \to 3 \to 7 \to 6$.
(b) Optimal sequences :
(i) $A \to C \to I \to B \to H \to F \to D \to E \to G$ (ii) $A \to I \to C \to H \to B \to F \to D \to G \to E$
(iii) $A \to C \to I \to H \to B \to F \to D \to G \to E$ (iv) $A \to I \to C \to B \to H \to F \to D \to E \to G$
(v) $A \to C \to I \to B \to H \to F \to D \to G \to E$ (vi) $A \to I \to C \to H \to B \to F \to D \to E \to G$
(vii) $A \to I \to C \to B \to H \to F \to D \to G \to E$ (viii) $A \to C \to I \to H \to B \to F \to D \to E \to G$
Total min. elapsed time = 61 hours.

7. Optimal sequences :
(i) $A \to D \to E \to B \to C$ (ii) $A \to E \to D \to B \to C$ (iii) $D \to A \to E \to B \to C$
(iv) $D \to E \to A \to B \to C$ (v) $E \to D \to A \to B \to C$ (vi) $E \to A \to D \to B \to C$
Time = 51 hours

8. Optimal sequences are

| A | D | G | F | B | C | E | or | A | D | G | B | F | C | E |

Time = 59 hours.

9. Sequence is as follows

| 4 | 5 | 2 | 6 | 1 | 3 | Min. time = 53 hours

10. (i) Sequence is as follows   (ii) Sequence is as follows

| C | A | E | D | B | | D | C | B | A | Total time = 76 hours

11. Sequence is as follows $3 \to 5 \to 2 \to 6 \to 1 \to 4 \to 7$ or $3 \to 5 \to 6 \to 2 \to 1 \to 4 \to 7$,
Time = 59 hours.

3 Classical Optimization Techniques

3.1 INTRODUCTION

If $y = f(x)$ be a continuous function. At a point $x = x_1$, if $f(x)$ does not increase and begins to decrease, then $f(x)$ has its maximum value at $x = x_1$ and if at a point $x = x_2$, $f(x)$ does not decrease and begins to increase, then $f(x)$ has its minimum value at $x = x_2$.

If $f(x)$ is maximum at a point $x = x_1$ then $f(x)$ is an increasing function for the preceding values of x_1 and decreasing for those value of x just below x_1 or we can say derivative of the function $\left(i.e., \dfrac{dy}{dx}\right)$ will be positive before $x = x_1$ and will be negative after $x = x_1$. But $\dfrac{dy}{dx}$ is a continuous function and $\dfrac{dy}{dx}$ changes the sign from positive to negative. So, $\dfrac{dy}{dx}$ will be zero at any point.

Therefore, for a maximum value of $y = f(x)$ at a point, we have $\dfrac{dy}{dx} = 0$ and $\dfrac{dy}{dx}$ changes the sign from positive to negative. On the other hand, for a minimum value of $y = f(x)$ we have $\dfrac{dy}{dx} = 0$ and $\dfrac{dy}{dx}$ changes the sign negative to positive.

☞ REMARKS

- If $\dfrac{dy}{dx}$ changes the sign positive to negative; it means that $f(x)$ is a decreasing function of x, i.e., $\dfrac{d^2y}{dx^2} < 0$.

- If $\dfrac{dy}{dx}$ changes the sign from negative to positive, it means that $f(x)$ is an increasing function of x, i.e., $\dfrac{d^2y}{dx^2} > 0$.

- A function may have more than one maximum and minimum value.
- Any minimum value of the function $f(x)$ can be greater than any maximum value.
- Maximum and minimum values of the function occur alternately.
- Maximum and minimum values of the function are sometimes known as extreme values.
- From the definition of maxima and minima, it is clear that $\dfrac{dy}{dx} = 0$ is the necessary condition for maximum or minimum.

- $\dfrac{d^2y}{dx^2} < 0$ is sufficient condition for maximum and $\dfrac{d^2y}{dx^2} > 0$ is sufficient condition for minimum.

WORKING PROCEDURE

STEP 1. Find the derivative i.e., $\dfrac{dy}{dx}$ of y of the given function.

STEP 2. Put $\dfrac{dy}{dx} = 0$ and find all the real values of x_i. (say x_1, x_2, x_3 ...).

STEP 3. Find $\dfrac{d^2y}{dx^2}$.

STEP 4. Put $x = x_i$ in $\dfrac{d^2y}{dx^2}$ and find the result. If result is negative then the function $f(x)$ is maximum at $x = x_i$ and max. $f(x)=f(x_i)$. On the other hand, if result is positive then the function $f(x)$ is minimum at $x = x_i$ and minimum $f(x)= f(x_i)$.

☛ REMARKS

- In a continuous function, maxima and minima values occur alternately, i.e., between two successive maxima there is one minimum and between two successive minima, there is one maximum.
- If $\dfrac{d^2y}{dx^2}$ is equal to 0 at any point $x = x_i$ then find $\dfrac{d^3y}{dx^3}, \dfrac{d^4y}{dx^4}$, and find the values of these derivatives at $x = x_i$ successively and check the sign.

SOLVED EXAMPLES

EXAMPLE 1. *Find the value of x for which $f(x)=y= x^4+2x^3- 3x^2- 4x+4$ is maximum or minimum and also find those value of f(x).*

SOLUTION. Here, the given function is

$$y = f(x) = x^4 + 2x^3 -3x^2 - 4x + 4 \qquad \qquad ...(1)$$

So $\dfrac{dy}{dx} = 4x^3 +6x^2 -6x -4 = 2(x+2)(2x+1)(x-1)$

Now, put $\dfrac{dy}{dx} = 0$, we have

$$2\,(x+2)(2x+1)(x-1) = 0 \Rightarrow x = -2, -\frac{1}{2}, 1$$

Again differentiating (2) w.r.t. to x, we get

$$\dfrac{d^2y}{dx^2} = 12x^2 +12x -6$$

At $x = -2$, we have

$$\dfrac{d^2y}{dx^2} = 12(-2)^2 +12(-2) -6 = 48 - 24 - 6 = 18 > 0$$

Since, $\dfrac{d^2y}{dx^2} > 0$ (i.e., positive). So $f(x)$ is minimum at $x = -2$. The minimum value of $f(x)$ at $x=-2$ is given by

$$f(-2) = (-2)^4 + 2(-2)^3 - 3(-2)^2 - 4(-2) + 4 = 0$$

Now, at $x = -\dfrac{1}{2}$, we have

$$\frac{d^2y}{dx^2} = 12\left(-\frac{1}{2}\right)^2 + 12\left(-\frac{1}{2}\right) - 6 = 3 - 6 - 6 = -9 < 0$$

Since, $\frac{d^2y}{dx^2} < 0$ (*i.e.*, negative). So, $f(x)$ is maximum at $x = -\frac{1}{2}$ and maximum value of $f(x)$ at $x = -\frac{1}{2}$ is

$$f\left(-\frac{1}{2}\right) = \left(-\frac{1}{2}\right)^4 + 2\left(-\frac{1}{2}\right)^3 - 3\left(-\frac{1}{2}\right)^2 - 4\left(-\frac{1}{2}\right) + 4 = \frac{81}{16}$$

Similarly, at $x = 1$, we have

$$\frac{d^2y}{dx^2} = 12(1)^2 + 12(1) - 6 = 12 + 12 - 6 = 18 > 0$$

Since, $\frac{d^2y}{dx^2} > 0$ (*i.e.*, positive). So $f(x)$ is minimum at $x = 1$ and minimum value of $f(x)$ at $x = 1$ is

$$f(1) = (1)^4 + 2(1)^3 - 3(1)^2 - 4(1) + 4 = 0$$

EXAMPLE 2. *Find the maximum and minimum value of the function*
$$y = f(x) = x^3 - 12x^2 + 36x + 21$$

SOLUTION. Here, the given function is
$$y = x^3 - 12x^2 + 36x + 21$$

Now, differentiating *w.r.t.* x, we get $\frac{dy}{dx} = 3x^2 - 24x + 36$

Putting $\frac{dy}{dx} = 0$, we get $3x^2 - 24x + 36 = 0$

$\Rightarrow$ $\qquad\qquad x^2 - 8x + 12 = 0$

$\Rightarrow$ $\qquad\qquad (x - 2)(x - 6) = 0$　or　$x = 2, 6$

Again, differentiating *w.r.t.* x, we get

$$\frac{d^2y}{dx^2} = 6x - 24$$

At $x = 2$, we have

$$\frac{d^2y}{dx^2} = 6(2) - 24 = -12 < 0$$

Since, $\frac{d^2y}{dx^2} < 0$ so $f(x)$ is maximum at $x = 2$. The maximum value of $f(x)$ at $x = 2$ is given by

$$f(2) = (2)^3 - 12(2)^2 + 36(2) + 21 = 53.$$

Similarly, at $x = 6$, we have

$$\frac{d^2y}{dx^2} = 6 \times 6 - 24 = 36 - 24 = 12 > 0$$

Since, $\frac{d^2y}{dx^2} > 0$ so, $f(x)$ is minimum at $x = 6$ and minimum value of $f(x)$ at $x = 6$ is

$$f(6) = (6)^3 - 12(6)^2 + 36(6) + 21 = 21$$

EXAMPLE 3. *Investigate for maximum and minimum values, the function* (sin x + cos 2x).

[MEERUT–2005, 12]

SOLUTION.   Let $y = \sin x + \cos 2x,$

$$\Rightarrow \qquad \frac{dy}{dx} = \cos x - 2\sin 2x = \cos x - 4\sin x \cos x$$

For stationary point

$$\frac{dy}{dx} = 0 \quad \Rightarrow \quad \cos x (1 - 4\sin x) = 0$$

or $\cos x = 0$ or $1 - 4\sin x = 0 \Rightarrow x = \dfrac{\pi}{2}$ or $\sin x = \dfrac{1}{4}$

For maxima or minima

$$\frac{d^2y}{dx^2} = -\sin x - 4\cos 2x = -\sin x - 4(1 - 2\sin^2 x)$$

$$= -\sin x - 4 + 8\sin^2 x$$

(i)   At $x = \dfrac{\pi}{2}$,

$$\left(\frac{d^2y}{dx^2}\right) = -1 - 4 + 8 = 3 \qquad\qquad \text{(which is positive)}$$

So, given function is minimum at $x = \dfrac{\pi}{2}$ and min. value of y at $x = \dfrac{\pi}{2}$ is given by

$$\sin\frac{\pi}{2} + \cos 2 \times \frac{\pi}{2} = 1 - 1 = 0.$$

(ii)   At $\sin x = \dfrac{1}{4}$,

$$\left(\frac{d^2y}{dx^2}\right) = -\frac{1}{4} - 4 + 8 \cdot \frac{1}{16} = \frac{-15}{4} \qquad\qquad \text{(which is negative)}$$

Hence, given function is maximum at $x = \sin^{-1}\dfrac{1}{4}$ and max. value at $\sin x = \dfrac{1}{4}$ is,

$$\frac{1}{4} + \left[1 - 2 \times \left(\frac{1}{4}\right)^2\right] = \frac{7}{8}.$$

EXAMPLE 4. *Find the maximum value of (x – 1)(x – 2)(x – 3).*   [ROHILKHAND–2005]

SOLUTION.   Let $f(x) = (x-1)(x-2)(x-3) = x^3 - 6x^2 + 11x - 6$

then $f'(x) = 3x^2 - 12x + 11$

For a maximum or minimum value of $f(x)$, we must have $f'(x) = 0$

$\Rightarrow \qquad 3x^2 - 12x + 11 = 0$

i.e., $x = \dfrac{12 \pm \sqrt{144 - 4 \times 3 \times 11}}{6} = 2 \pm \dfrac{1}{\sqrt{3}}$

Also $f''(x) = 6x - 12$

Now $f''[2 + (1/\sqrt{3})] = +$ve, therefore $f(x)$ has minimum value at $x = 2 + (1/\sqrt{3})$.

Again $f''[2 - (1/\sqrt{3})] = -$ve, therefore $f(x)$ has a maximum value at $x = 2 - (1/\sqrt{3})$

$= f[2 - (1/\sqrt{3})] = 2/3\sqrt{3}$

EXAMPLE 5. *Show that sin x(1 + cos x) is a maximum at x = π/3.*

[MEERUT–2009, 13; KANPUR–2004, 08, 10, 15]

SOLUTION.   Let $f(x) = \sin x(1 + \cos x) = \sin x + \dfrac{1}{2}\sin 2x$

Then $f'(x) = \cos x + \cos 2x$

For a maximum or a minimum value of $f(x), f'(x) \quad 0$

i.e., $\cos x + \cos 2x = 0 \Rightarrow 2\cos^2 x + \cos x - 1 = 0$

$\Rightarrow \quad (2\cos x - 1)(\cos x + 1) = 0$

$\therefore \qquad\qquad \cos x = 1/2, -1 \Rightarrow x = \pi/3, \pi$

Now $f''(x) = -\sin x - 2\sin 2x$

$\therefore \qquad\qquad f''\left(\dfrac{\pi}{3}\right) = -\sin\left(\dfrac{\pi}{3}\right) - 2\sin\left(\dfrac{2\pi}{3}\right) = -ve$

Hence $f(x)$ is maximum at $x = \pi/3$.

EXAMPLE 6. *Find the maximum value of* $(1/x)^x$.   [ROHILKHAND–2005, 06]

SOLUTION.   Let $\qquad\qquad\qquad y = (1/x)^x$

$\Rightarrow \qquad\qquad\qquad \log y = x(\log 1 - \log x) = -x \log x$

$\Rightarrow \qquad\qquad\qquad \dfrac{1}{y}\dfrac{dy}{dx} = -1\log x - x(1/x) = -(1 + \log x)$

$\Rightarrow \qquad\qquad\qquad \dfrac{dy}{dx} = -y(1 + \log x) = -(1/x)^x(1 + \log x)$

For a maximum or a minimum of y, we must have $\dfrac{dy}{dx} = 0$

$\Rightarrow \qquad -(1/x)^x(1 + \log x) = 0 \Rightarrow 1 + \log x = 0 \Rightarrow x = 1/e$

$\Rightarrow \qquad\qquad\qquad \dfrac{d^2 y}{dx^2} = -\dfrac{dy}{dx}(1 + \log x) - y(1/x)$

$\qquad\qquad\qquad\qquad = -\dfrac{dy}{dx}(1 + \log x) - (1/x)^x.(1/x)$

Therefore,   when $x = 1/e$,

$\qquad\qquad\qquad \dfrac{d^2 y}{dx^2} = 0 - (e)^{1/e}.e = -ve$

$\Rightarrow \qquad y$ is maximum at $x = 1/e$.

Thus the maximum value of y is given by $e^{1/e}$.

EXAMPLE 7. *Show that the semi-vertical angle of the right circular cone of given total surface (including area of the base) and maximum value is* $\sin^{-1}(1/3)$.

[KANPUR-2006; MEERUT-2002, 03, 11; ROHILKHAND-2004, 09]

SOLUTION.   Let x be the radius of the base, h be the height and y, the slant height of the cone. Then the total surface of the cone = constant

$\Rightarrow \qquad \pi x^2 + \pi x y = $ constant   ...(1)

Now $\qquad\qquad\qquad V = $ volume of the cone

$\qquad\qquad\qquad\qquad = \dfrac{1}{3}\pi x^2 h = \dfrac{1}{3}\pi x^2 (y^2 - x^2)^{1/2}$

Since, $h = \sqrt{y^2 - x^2}$, therefore

$\qquad\qquad\qquad V^2 = \dfrac{1}{9}\pi^2 x^4 (y^2 - x^2)$.

Now, V is maximum or minimum according as V^2 or $\dfrac{9V^2}{\pi^2}$ is maximum or minimum.

Let $\qquad\qquad S = \dfrac{9V^2}{\pi^2} = x^4(y^2 - x^2)$.

Then S can be regarded as a function of x because y is connected with x by (1).

We have $\qquad \dfrac{dS}{dx} = 4x^3(y^2 - x^2) + x^4 \left\{ 2y\left(\dfrac{dy}{dx}\right) - 2x \right\}$ ...(2)

Differentiating (1) w.r. to x, we get

$$\pi\left(2x + y + x\dfrac{dy}{dx} \right) = 0 \quad \text{or} \quad \dfrac{dy}{dx} = -\dfrac{2x + y}{x}$$

Substituting this value of $\dfrac{dy}{dx}$ in (2), we get

$$\dfrac{dS}{dx} = 4x^3 y^2 - 4x^5 + x^4 \left[-2y\dfrac{(2x + y)}{x} - 2x \right]$$

$$= 2x^3 y^2 - 6x^5 - 4x^4 y$$

For a maximum or a minimum of S, we must have $\dfrac{dS}{dx} = 0$

Now $\qquad \dfrac{dS}{dx} = 0$

$\Rightarrow \ 2x^3(y^2 - 2xy - 3x^2) = 0 \ \Rightarrow \ 2x^3(y - 3x)(y + x) = 0$

i.e., $\ y = 3x$ since $x \ne 0$ and $y \ne -x$.

Again $\qquad \dfrac{d^2 S}{dx^2} = 6x^2 y^2 + 4x^3 y \dfrac{dy}{dx} - 30x^4 + 16x^3 y - 4x^4 \dfrac{dy}{dx}$

When $\ y = 3x, \dfrac{dy}{dx} = -5,\ $ so when $y = 3x$, we have

$$\dfrac{d^2 S}{dx^2} < 0$$

Therefore S is maximum when $y = 3x$.

<u>EXAMPLE 8.</u> ***In a submarine telegraph cable the speed of signalling varies as*** $\log x^2 \log(1/x)$***, where x is the ratio of the radius of the core to that of the covering. Show that the greatest speed is attained when this ratio is*** $1 : \sqrt{e}$. [MEERUT–2002, 04, 14]

<u>SOLUTION.</u> Let S be the speed of signalling. Then

$$S = \mu x^2 \log(1/x) = -\mu x^2 \log x \text{ , where } \mu \text{ is a constant.}$$

For a maximum or a minimum of S, we have $\dfrac{dS}{dx} = 0$

i.e., $\qquad x(2 \log x + 1) = 0$

$\Rightarrow \qquad\qquad\qquad x = 0 \ \text{ or } \ \log x = -1/2$

But $x = 0$ is inadmissible. Therefore

$$\log x = -1/2 \text{ or } x = e^{-1/2} = 1/\sqrt{e}$$

Now $\qquad \dfrac{d^2 S}{dx^2} = -\mu(2\log x + 1) - \mu x(2/x)$

$$= -\mu(2\log x + 1) = -2\mu$$

When $x = 1/\sqrt{e}$, we have $2 \log x + 1 = 0$, when $x = 1/\sqrt{e}$, we have $\dfrac{d^2 S}{dx^2} = -2\mu$ which is negative.

Hence, S is maximum, when $x = 1/\sqrt{e}$.

<u>EXAMPLE 9.</u> ***Show that maximum rectangle that can be inscribed in a circle is square.***

[MEERUT–2005, 09, 16]

SOLUTION. Let $PQRS$ be the rectangle inscribed in circle with centre O and radius a. Also, let $PQ = 2x$ and $QR = 2y$. Then

$$a^2 = x^2 + y^2 \qquad \qquad ...(1)$$

Area of rectangle $PQRS$

$$A = (2x)(2y) = 4xy = 4x\sqrt{a^2 - x^2} \qquad \text{[From (1)]}$$

For maximum or minimum area, $\dfrac{dA}{dx} = 0$

$$\Rightarrow \quad 4\left\{\sqrt{a^2 - x^2} - \frac{x^2}{\sqrt{a^2 - x^2}}\right\} = 0 \Rightarrow 4\left\{\frac{a^2 - 2x^2}{\sqrt{a^2 - x^2}}\right\} = 0$$

$$\Rightarrow \qquad \qquad a^2 - 2x^2 = 0 \Rightarrow x = \frac{a}{\sqrt{2}}$$

Now $\dfrac{d^2A}{dx^2} = 4\left\{(-4x)(a^2 - x^2)^{1/2} + (a^2 - 2x^2)\left(-\frac{1}{2}\right)(a^2 - x^2)^{-3/2}(-2x)\right\}$

$$= 4\left[\frac{-4x}{\sqrt{a^2 - x^2}} + \frac{x(a^2 - 2x^2)}{(a^2 - x^2)^{3/2}}\right]$$

$$\Rightarrow \qquad \left(\frac{d^2A}{dx^2}\right)_{x = a/\sqrt{2}} = -16 \qquad \qquad \text{(which is negative.)}$$

Thus, A is max. when $x = \dfrac{a}{\sqrt{2}}$.

Then, from (1), $\qquad y = \dfrac{a}{\sqrt{2}}$. Therefore, $x = y = \dfrac{a}{\sqrt{2}}$

Hence, area is maximum when $x = y = \dfrac{a}{\sqrt{2}}$ *i.e.*, rectangle is square.

EXAMPLE 10. *Show that the height of the closed cylinder of given surface and greatest volume is equal to its diameter.* [MEERUT–2005]

SOLUTION. Let r be radius of base and h the height of a closed cylinder of given surface S, then

$$S = 2\pi r^2 + 2\pi rh \Rightarrow h = \frac{S - 2\pi r^2}{2\pi r} \qquad \qquad ...(1)$$

If V be volume of cylinder then

$$V = \pi r^2 h = \pi r^2\left(\frac{S - 2\pi r^2}{2\pi r}\right) = \frac{rS - 2\pi r^3}{2}$$

$$\Rightarrow \qquad \frac{dV}{dr} = \frac{S}{2} - 3\pi r^2 \qquad \qquad ...(2)$$

For max or min we have $\dfrac{dV}{dr} = 0$

$$\frac{S}{2} - 3\pi r^2 = 0 \Rightarrow S = 6\pi r^2$$

$$\Rightarrow \qquad 2\pi r^2 + 2\pi rh = 6\pi r^2$$

$$\Rightarrow \qquad h = 2r$$

From (2) $\dfrac{d^2V}{dr^2} = -6\pi r$, (–ve) for any positive value of r.

Hence V is maximum when $h = 2r$, *i.e.*, when the height of cylinder is equal is diameter of base.

EXAMPLE 11. *Prove that a conical tent of a given capacity will required the least amount of canvas when the height is $\sqrt{2}$ times the radius of the base.* [MEERUT–2006]

SOLUTION. Let us suppose h be the height, r be the radius of the base and l is the slant height of the conical tent. Let V be the given capacity (*i.e.* volume) and S denote the area of the curved surface of the tent.

We know that
$$V = \frac{1}{3}\pi r^2 h \qquad \qquad ...(1)$$

and
$$S = \pi l r = \pi(\sqrt{h^2 + r^2})r$$
$$\Rightarrow \qquad S^2 = \pi^2 r^2(h^2 + r^2) = u \,(\text{say}) \qquad ...(2)$$

From (1) and (2), we get
$$u = \pi^2 r^2 \left[\frac{9V^2}{\pi^2 r^4} + r^2\right] = \frac{9V^2}{r^2} + \pi^2 r^4$$

$$\therefore \qquad \frac{du}{dr} = -\frac{18V^2}{r^3} + 4\pi^2 r^3$$

and
$$\frac{d^2u}{dr^2} = \frac{54V^2}{r^4} + 12\pi^2 r^2$$

Now
$$\frac{du}{dr} = 0 \Rightarrow V = \left(\frac{2}{3\sqrt{2}}\right)\pi r^3$$

for
$$V = \left(\frac{2}{3\sqrt{2}}\right)\pi r^3, \frac{d^2u}{dr^2} > 0$$

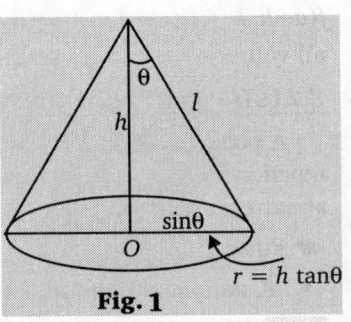

Fig. 1

i.e., u is minimum when $V = \frac{2}{3\sqrt{2}}\pi r^3$

i.e., when $\frac{2}{3\sqrt{2}}\pi r^3 = \frac{1}{3}\pi r^2 h$ *i.e.*, $h = r\sqrt{2}$

Hence, u is minimum, when $h = r\sqrt{2}$.

EXAMPLE 12. *Show that the radius of the right circular cylinder of greatest curved surface which can be inscribed in a given cone is half that of the cone.*

SOLUTION. Let r be the radius and H, the height of the given cone
i.e., $OB = r$, $OA = H$
where O is the centre of the base circle.
Suppose x is the radius and h the height of the cylinder inscribed in the given cone.
Now triangles AOB and ADE are similar, therefore
$$\frac{AD}{AO} = \frac{DE}{OB} \text{ or } \frac{H-h}{H} = \frac{x}{r}$$

$$\Rightarrow \qquad 1 - \frac{h}{H} = \frac{x}{r} \Rightarrow \frac{h}{H} = 1 - \frac{x}{r}$$

$$\Rightarrow \qquad h = H\left(1 - \frac{x}{r}\right)$$

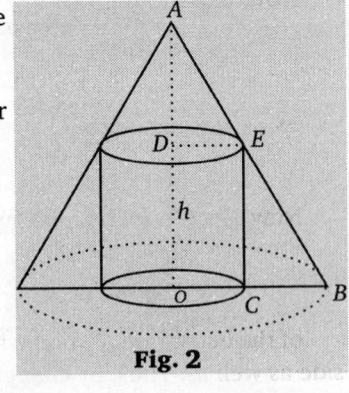

Fig. 2

Now the curved surface of the cylinder

$$\Rightarrow \qquad S = 2\pi.x.h = 2\pi x H\left(1 - \frac{x}{r}\right) = \frac{2\pi H}{r}(rx - x^2)$$

$$\Rightarrow \qquad \frac{dS}{dx} = \frac{2\pi H}{r}(r - 2x)$$

So $\dfrac{dS}{dx} = 0$, we get $r - 2x = 0$ or $x = \dfrac{r}{2}$

Also $\dfrac{d^2S}{dx^2} = \dfrac{2\pi H}{r}(-2) < 0 \Rightarrow S$ is greatest, when $x = \dfrac{r}{2}$

i.e., when radius of the cylinder is half of that can be inscribed in a sphere of radius x.

3.2 MAXIMA AND MINIMA OF A FUNCTION OF SEVERAL INDEPENDENT VARIABLES

Let $f(x, y, z, ...)$ be a function of several independent variables $x, y, z....$ If f is continuous and finite for all values of $x, y, z, ...$ in the neighbourhood of $x = a, y = b, z = c, ...$ respectively, then the value of $(a, b, c, ...)$ is said to be a maximum or minimum if $f(a+h, b+k, c+l, ...)$ is less than or greater than $f(a, b, c, ...)$ for all values of $h, k, l, ...$ (where $h, k, l, ...$ are sufficiently small, may be positive or negative provided they are not all zero.)

In other words we can say, the value of $f(a, b, c, ...)$ is said to be a maximum or minimum if $f(a+h, b + k, c + l, ...) - f(a, b, c, ...)$ maintain an invariant sign (may be positive or negative) for all values of $h, k, l, ...$ positive or negative provided they are taken sufficiently small and finite.

3.2.1 STATIONARY AND EXTREME POINTS

A point $(a_1, a_2, ..., a_n)$ is called a stationary point, if all the first order partial derivatives of the function $f(x_1, x_2, ..., x_n)$ vanish at the point. A stationary point, if it is maximum or minimum is known as extreme point and the value of the function at an extreme point is known as an extreme value.

☛ REMARK

- A stationary point may be a maximum or minimum or neither of these two.

3.3 NECESSARY CONDITION FOR THE EXISTENCE OF MAXIMA OR MINIMA

Let $f(x, y, z, ...)$ be a function of several independent variables $x, y, z,...$ It is clear from the definition of maxima and minima that maximum or minimum of $f(x, y, z, ..)$ will occur for those values of $x, y, z, ...$, for which the expression $f(x+h, y +k, z+l, ...) - f(x, y, z, ...)$ maintain an invariant sign for all sufficiently small and finite values of $h, k, l, ...$ positive or negative.

Now, expanding $f(x+h, y+k, z+l, ...)$ by Taylor's theorem, we have

$$f(x+h, y+k, z+l...) = f(x, y, z) + \left(h\dfrac{\partial f}{\partial x} + k\dfrac{\partial f}{\partial y} + l\dfrac{\partial f}{\partial z} + ... \right) + \text{terms of second and higher order.}$$

$$\Rightarrow \quad f(x+h, y+k, z+l...) - f(x, y, z,...) = \left(h\dfrac{\partial f}{\partial x} + k\dfrac{\partial f}{\partial y} + l\dfrac{\partial f}{\partial z} + ... \right)$$

$$+ \text{terms of second and higher orders.} \qquad ...(1)$$

Now, since $h, k, l, ...$ are sufficiently small, the first degree expression

$$\left(h\dfrac{\partial f}{\partial x} + k\dfrac{\partial f}{\partial y} + l\dfrac{\partial f}{\partial z} + ... \right)$$

of the equation (1) can be made to govern the sign of right hand side and hence, of the left hand side as well as. Thus, by changing the sign of the left hand side of the equation (1) will also change.

Since, left hand side is to preserve an invariable sign for maxima or minima, therefore, as a necessary condition for maximum and minimum values, we must have

$$h\dfrac{\partial f}{\partial x} + k\dfrac{\partial f}{\partial y} + l\dfrac{\partial f}{\partial z} + ... = 0 \qquad ...(2)$$

Now, since $h, k, l, ...$ are arbitrary and independent of each other, we must have

$$\frac{\partial f}{\partial x} = 0, \frac{\partial f}{\partial y} = 0, \frac{\partial f}{\partial z} = 0, \text{ etc.} \qquad ...(3)$$

If the number of independent variables be n, we shall get n simultaneous equations in these n variables, which will give the values $a, b, c, ...$ of the n variables $x, y, z, ...$ respectively for which $f(x, y, z, ...)$ will have a maximum or a minimum values.

☞ **REMARKS**

- The necessary condition for a function $f(x, y, z, ...)$ of the independent variables $x, y, z, ...$ to be maximum or minimum is given by

$$\frac{\partial f}{\partial x} = 0, \frac{\partial f}{\partial y} = 0, \frac{\partial f}{\partial z} = 0,$$

- The conditions given above is only a necessary condition for the maxima and minima of the function $f(x, y, z, ...)$. These conditions are not sufficient.

3.3.1 MAXIMA AND MINIMA FOR A FUNCTION OF TWO INDEPENDENT VARIABLES

(1) *To find the condition which governs the sign of a quadratic expression.*
   Consider, a binary expression
   $$I = ax^2 + 2hxy + by^2$$
   of two variables x and y. Then I can be written as
   $$I = ax^2 + 2hxy + by^2 = \frac{1}{a}[(ax + hy)^2 + (ab - h^2)y^2].$$

   If $(ab - h^2)$ is positive, the sign of I will be the same as that of a.

   But if $(ab - h^2)$ is negative, then, the expression within the brackets may be positive or negative and therefore we cannot say anything about the sign of expression I.

(2) *Stationary and extreme points (For the function of two independent variables):*
   Let $f(x, y)$ be a function of two independent variables x and y. A point (a, b) is called a stationary point, if both the first order partial derivatives $\left(\frac{\partial f}{\partial a} \text{ and } \frac{\partial f}{\partial b}\right)$ of the function $f(x, y)$ at (a, b) vanish.
   A stationary point which is either a maximum or minimum is called an extreme point.

☞ **REMARKS**

- A stationary point is not necessarily an extreme point, hence a stationary point may be a maximum or a minimum or neither of these two.
- The value of the function at extreme point is called extreme value.
- A point at which function is neither maximum nor minimum, is known as saddle point.

3.4 NECESSARY CONDITION FOR MAXIMA AND MINIMA

Let $f(x, y)$ be a function of two independent variables x and y.Then, we have the maximum or minimum of $f(x, y)$ at $x = a$ and $x = b$ if the expression $f(a + h, b + k) - f(a, b)$ is of invariable sign for all sufficiently small independent variables h and k provided both of them are not equal to zero.

We observe that,

(i) If the sign of $f(a+h, b+k) - f(a, b)$ is negative, then we have a maximum of $f(x, y)$ at $x = a, y = b$.

(ii) If the sign of $f(a+h, b+k) - f(a, b)$ is positive, then we have a minimum of $f(x, y)$ at $x = a$, $y = b$.

Expand $f(a+h, b+k)$ by Taylor's theorem, we have

$$f(a+h,b+k) = f(a,b) + \left(h\frac{\partial f}{\partial x} + k\frac{\partial f}{\partial y} \right)_{\substack{x=a\\y=b}}$$

$$+ \frac{1}{2!}\left(h^2 \frac{\partial^2 f}{\partial x^2} + 2hk\frac{\partial^2 f}{\partial x\,\partial y} + k^2 \frac{\partial^2 f}{\partial y^2} \right)_{\substack{x=a\\y=b}} + \dots \qquad \dots(1)$$

$$\Rightarrow f(a+h,b+k) - f(a,b) = h\left(\frac{\partial f}{\partial x} \right)_{\substack{x=a\\y=b}} + k\left(\frac{\partial f}{\partial y} \right)_{\substack{x=a\\y=b}} + \text{ term of the second and higher orders in}$$

h and k.

Now, since h and k are sufficiently small, the expression $h\left(\frac{\partial f}{\partial x} \right)_{\substack{x=a\\y=b}} + k\left(\frac{\partial f}{\partial y} \right)_{\substack{x=a\\y=b}}$ of the equation (1) can be made to govern the sign of right hand side and hence of the left hand side as well. Thus by changing the sign of h and k, the sign of the left hand side of the equation (1) will also change.

Since L.H.S. is to preserve an invariable sign for maximum or minimum, therefore as a necessary condition for maximum and minimum values, we must have

$$h\left(\frac{\partial f}{\partial x} \right)_{\substack{x=a\\y=b}} + k\left(\frac{\partial f}{\partial y} \right)_{\substack{x=a\\y=b}} = 0. \qquad \dots(2)$$

If $k = 0$, we find that if $\left(\frac{\partial f}{\partial x} \right)_{\substack{x=a\\y=b}} \neq 0$, the R.H.S. of (2) changes sign when h changes sign.

Therefore $f(x, y)$ cannot have a maximum or minimum at $x = a$, $y = b$ if $\left(\frac{\partial f}{\partial x} \right)_{\substack{x=a\\y=b}} \neq 0$.

Similarly, taking $h = 0$, we see that $f(x, y)$ cannot have a maximum or a minimum at $x = a$, $y = b$ if $\left(\frac{\partial f}{\partial y} \right)_{\substack{x=a\\y=b}} \neq 0$.

Thus, a set of necessary conditions that $f(x, y)$ should have a maximum or minimum at $x = a$, $y = b$ is that

$$\left(\frac{\partial f}{\partial x} \right)_{\substack{x=a\\y=b}} = 0 \text{ and } \left(\frac{\partial f}{\partial y} \right)_{\substack{x=a\\y=b}} = 0.$$

3.5 SUFFICIENT CONDITION FOR MAXIMA AND MINIMA: THE LAGRANGE'S CONDITION

Let $f(x, y)$ be a function of two variables x and y.

Let $$r = \frac{\partial^2 f}{\partial x^2}, s = \frac{\partial^2 f}{\partial x\,\partial y}, t = \frac{\partial^2 f}{\partial y^2} \text{ at } x = a \text{ and } y = b.$$

As a set of necessary conditions for a maximum or minimum at (a, b) we have

$$\frac{\partial f}{\partial x} = 0 \text{ and } \frac{\partial f}{\partial y} = 0 \text{ at } (a, b)$$

then $$f(a + h, b + k) - f(a, b) = \frac{1}{2!}[rh^2 + 2shk + tk^2] + R \qquad \dots(1)$$

Where R consists of terms of third and higher order of small quantities h and k.

Now, by taking h and k sufficiently small, the second degree terms in R.H.S. of (1) may be made to govern the sign of R.H.S. and therefore of the L.H.S. also *i.e.*, for sufficiently small values of h and k, the sign of $\frac{1}{2}(rh^2 + 2shk + tk^2) + R$ is same as that of $rh^2 + 2shk + tk^2$.

If the sign is negative, then the function is maximum at (a, b) and if the sign is positive, then the function is minimum at (a, b).

Case (i) If $(rt - s^2) > 0$.

Then, neither r nor t can be zero. Hence, we can write

$$rh^2 + 2shk + tk^2 = \frac{1}{2}[r^2h^2 + 2rshk + rtk^2] = \frac{1}{2}[(rh + sk)^2 + (rt - s^2)k^2]$$

since $rt - s^2 > 0$, therefore $(rh + sk)^2 + (rt - s^2)k^2 > 0$ for all values of h and k except when $rh + sk = 0$, $k = 0$ *i.e.*, at $h = 0$, $k = 0$, which is not possible.

Hence, in this case the expression $rh^2 + 2shk + tk^2$ will have the same sign for all values of h and k, and the sign is determined by the sign of r.

Thus, the function $f(x, y)$ will have a maximum or minimum at $x = a$ and $y = b$. If $rt - s^2 > 0$. The function $f(x, y)$ is maximum or minimum according as r is negative or positive.

Case (ii) If $(rt - s^2) < 0$.

If $rt - s^2$ is negative, we are not sure about the sign of second degree term of R.H.S. of (1) and hence there is neither a maximum nor a minimum value.

Case (iii) If $(rt - s^2) = 0$.

If $rt = s^2$, then quadratic expression $rh^2 + 2shk + tk^2$ becomes $\frac{1}{r}(hr + ks)^2$.

So that, the quadratic expression will be of the same sign as that of r or t unless

$$\frac{h}{k} = -\frac{s}{r} = \alpha \text{ (say)} \text{ i.e., } rh + sk = 0.$$

If this condition is satisfied, then the second degree expression in R.H.S. of (1) vanishes and hence, the sign of the R.H.S. of (1) depends upon third degree expression in h and k, which change sign with the change of sign of h and k and hence, the sign of L.H.S. of (1) will also change and hence, there will be neither maximum nor minimum.

Thus, the necessary condition for the existence of maxima and minima now is that the cubic terms must vanish collectively in R.H.S. of (1) when $\frac{h}{k} = -\frac{s}{r} = \alpha$; and then the biquadratic terms of R.H.S. of (1) must collectively be of the same sign as r and t, when

$$\frac{h}{k} = -\frac{s}{r} = \alpha \text{ i.e., } hr + ks = 0$$

Hence, the case is doubtful.

Thus, if $rt - s^2 = 0$, the case is doubtful and further, investigation is needed to determine the maxima and minima of $f(x, y)$ at (a, b).

WORKING PROCEDURE

M To discuss the maxima and minima at $x = a$, $y = b$, we must find

$$r = \left(\frac{\partial^2 u}{\partial x^2}\right)_{\substack{x=a \\ y=b}}, \quad s = \left(\frac{\partial^2 u}{\partial x \partial y}\right)_{\substack{x=a \\ y=b}}, \quad t = \left(\frac{\partial^2 u}{\partial y^2}\right)_{\substack{x=a \\ y=b}}$$

Then, calculate $rt - s^2$.

Now following cases arise :

(i)   If $rt - s^2 > 0$, then

    (A) If r is negative then, $f(x, y)$ is maximum at $x = a$, $y = b$.

    (B) If r is positive then, $f(x, y)$ is minimum at $x = a$, $y = b$.

(ii) If $rt - s^2 < 0$, $f(x, y)$ is neither maximum nor minimum at $x = a$, $y = b$.

(iii) If $rt - s^2 = 0$, the case is doubtful, and further investigation will be required.

☛ REMARK

- While solving problems, we frequently used the identity, given by Lagrange.

$$\{(a^2 + b^2 + c^2)(p^2 + q^2 + r^2) - (ap + bq + cr)^2\} = \{(br - cq)^2 + (cp + ar)^2 + (aq - bp)^2\}.$$

🏛 SOLVED EXAMPLES

EXAMPLE 1. *Find all maximum or minimum values of the function :*

$$f(x,y) = y^2 + x^2 y + x^4.$$

SOLUTION. Since, we have

$$f(x,y) = y^2 + x^2 y + x^4.$$

$$\therefore \quad \frac{\partial f}{\partial x} = 2xy + 4x^3 \text{ and } \frac{\partial f}{\partial y} = 2y + x^2.$$

For a maximum or minimum of $f(x, y)$, we must have $\frac{\partial f}{\partial x} = 0$ and $\frac{\partial f}{\partial y} = 0$

$$\therefore \quad \frac{\partial f}{\partial x} = 0 \Rightarrow 2xy + 4x^3 = 0 \quad \Rightarrow \quad 2x\,(y + 2x^2) = 0 \qquad ...(1)$$

$$\frac{\partial f}{\partial y} = 0 \Rightarrow 2y + x^2 = 0$$

Solving (1) and (2), we get $x = 0$, $y = 0$.

Thus (0, 0) is the only point of maximum or minimum.

Now $r = \left(\frac{\partial^2 f}{\partial x^2}\right)_{(0,0)} = [2y + 12x^2]_{(0,0)} = 0$; $s = \left(\frac{\partial^2 f}{\partial x \partial y}\right)_{(0,0)} = [2x]_{(0,0)} = 0$

and $t = \left(\frac{\partial^2 f}{\partial y^2}\right)_{(0,0)} = [2]_{(0,0)} = 2$

$\therefore rt - s^2 = 0\,(2) - 0^2 = 0.$

Thus, the case is doubtful and further investigation will be required.

EXAMPLE 2. *Find the maximum or minimum values of the function $x^3 y^2 (1 - x - y)$.*

(ANNA–2009, JNTU–2006, 08, BHOPAL–2012)

SOLUTION. Let $u = x^3 y^2 (1 - x - y)$

$$\Rightarrow \frac{\partial u}{\partial x} = 3x^2 y^2 (1 - x - y) - x^3 y^2 \text{ and } \frac{\partial u}{\partial y} = 2x^3 y(1 - x - y) - x^3 y^2.$$

For a maximum or minimum of u, we must have $\frac{\partial u}{\partial x} = 0$ and $\frac{\partial u}{\partial y} = 0$

$$\Rightarrow \quad 3x^2 y^2 (1 - x - y) - x^3 y^2 = 0 \qquad\qquad ...(1)$$

and $\quad 2x^3 y(1 - x - y) - x^3 y^2 = 0.$ $\qquad\qquad ...(2)$

Now, subtracting (2) from (1), we have

$$x^2 y(1 - x - y)(3y - 2x) = 0$$

which gives $y = \frac{2}{3}x.$

Putting the value of y in (1), we get $x = \frac{1}{2}$

So $\left(\dfrac{1}{2},\dfrac{1}{3}\right)$ be the point of maxima or minima.

Now $\qquad r = \dfrac{\partial^2 u}{\partial x^2} = 6xy^2 - 12x^2 y^2 - 6xy^3 \ = -\dfrac{1}{9},\ \text{at}\ \left(\dfrac{1}{2},\dfrac{1}{3}\right)$

$\qquad\qquad t = \dfrac{\partial^2 u}{\partial y^2} = 2x^3 - 2x^4 - 6x^3 y \ = -\dfrac{1}{8},\ \text{at}\ \left(\dfrac{1}{2},\dfrac{1}{3}\right)$

$\qquad\qquad s = \dfrac{\partial^2 u}{\partial x \partial y} = 6x^2 y - 8x^3 y - 9x^2 y^2 \ = -\dfrac{1}{12}\ \text{at}\ \left(\dfrac{1}{2},\dfrac{1}{3}\right).$

Now, $rt - s^2 =$ positive.

Also, r is negative, hence the function u has a maximum at $x = \dfrac{1}{2},\ y = \dfrac{1}{3}$.
The maximum value is

$$= \left(\frac{1}{2}\right)^3 \left(\frac{1}{3}\right)^2 \left(1 - \frac{1}{2} - \frac{1}{3}\right) = \frac{1}{432}.$$

EXAMPLE 3. *Discuss the maximum or minimum values of u, where*
$$u = 2a^2 xy - 3ax^2 y - ay^3 + x^3 y + xy^3.$$

SOLUTION. We have
$$u = 2a^2 xy - 3ax^2 y - ay^3 + x^3 y + xy^3$$

which gives
$$\frac{\partial u}{\partial x} = 2a^2 y - 6axy + 3x^2 y + y^3$$

and
$$\frac{\partial u}{\partial y} = 2a^2 x - 3ax^2 - 3ay^2 + x^3 + 3xy^2$$

For a maximum and minima of u, we have
$$\frac{\partial u}{\partial x} = 0,\ \frac{\partial u}{\partial y} = 0$$

which gives,
$$y(2a^2 - 6ax + 3x^2 + y^2) = 0 \qquad\qquad\qquad \text{...(1)}$$
and $\quad 2a^2 x - 3ax^2 - 3ay^2 + x^3 + 3xy^2 = 0 \qquad\qquad \text{...(2)}$

Equation (1) and (2) gives the following values of x and y :
$$x = 0,\ y = 0;\ x = a,\ y = 0;$$
$$x = 2a,\ y = 0;\ x = \frac{3}{2}a,\ y = \pm\frac{1}{2}a;$$
$$x = a,\ y = a,\ x = \frac{1}{2}a,\ y = \frac{1}{2}a;$$
$$x = a,\ y = -a;\ x = \frac{1}{2}\,a,\ y = -\frac{1}{2}a.$$

Then, we get the following pairs of values of x and y which make the function u stationary.

$$(0,0),\ (a,0),\ (2a,0),\ \left(\frac{3}{2}a,\frac{1}{2}a\right),\ \left(\frac{3}{2}a,-\frac{1}{2}a\right),\ (a,a),\ \left(\frac{1}{2}a,\frac{1}{2}a\right),\ (a,-a),,\ \left(\frac{1}{2}a,-\frac{1}{2}a\right).$$

Also $\qquad r = \dfrac{\partial^2 u}{\partial x^2} = -6ay + 6xy,$

$$s = \frac{\partial^2 u}{\partial x \, \partial y} = 2a^2 - 6ax + 3x^2 + 3y^2,$$

and
$$t = \frac{\partial^2 u}{\partial y^2} = -6ay + 6xy.$$

For (0, 0).
$$r = 0, s = 2a^2, t = 0$$
$\Rightarrow \quad rt - s^2$, is negative.

Therefore, we have neither maximum nor a minimum of u at (0, 0).

Similarly, we can easily shown that u has neither a maximum nor a minimum at $(a, 0)$, $(2a, 0)$, (a, a), $(a, -a)$.

For $\left(\dfrac{3a}{2}, \dfrac{a}{2}\right)$. $r = \dfrac{3}{2}a^2, s = \dfrac{1}{2}a^2, t = \dfrac{3}{2}a^2,$

$\Rightarrow \quad rt - s^2$ is positive.

Here, since r is positive, therefore u has minimum at $\left(\dfrac{3a}{2}, \dfrac{a}{2}\right)$.

Similarly, we can check the maxima and minima at all other points.

☛ **REMARK**

• The point $\left(\dfrac{x_1 + x_2 + x_3}{3}, \dfrac{y_1 + y_2 + y_3}{3}\right)$ is the centroid of the given triangle.

EXAMPLE 4. *Show that the minimum value of* $u = xy + \left(\dfrac{a^3}{x}\right) + \left(\dfrac{a^3}{y}\right)$ *is* $3a^2$.

SOLUTION. We have $u = xy + \left(\dfrac{a^3}{x}\right) + \left(\dfrac{a^3}{y}\right)$

$\Rightarrow \quad \dfrac{\partial u}{\partial x} = y - \dfrac{a^3}{x^2}$ and $\dfrac{\partial u}{\partial y} = x - \dfrac{a^3}{y^2}.$

For a maximum or minimum of u, we have $\dfrac{\partial u}{\partial x} = 0$ and $\dfrac{\partial u}{\partial y} = 0$

Now, $\dfrac{\partial u}{\partial x} = 0 \Rightarrow y - \dfrac{a^3}{x^2} = 0$ ...(1)

and $\dfrac{\partial u}{\partial y} = 0 \Rightarrow x - \dfrac{a^3}{y^2} = 0.$ ...(2)

Solving (1) and (2), we get, $x = a, y = a$

Now $r = \dfrac{\partial^2 u}{\partial x^2} = \dfrac{2a^3}{x^3}, s = \dfrac{\partial^2 u}{\partial x \, \partial y} = 1$ and $t = \dfrac{\partial^2 u}{\partial y^2} = \dfrac{2a^3}{y^3}.$

At $x = y = a$, we have $r = 2, s = 1, t = 2$

$\Rightarrow rt - s^2 = 3 > 0.$

Thus, at (a, a), $rt - s^2 > 0$ and $r > 0$. Therefore u is minimum at $x = a$, $y = a$.

The minimum value of $u = a.a + \left(\dfrac{a^3}{a}\right) + \left(\dfrac{a^3}{a}\right) = 3a^2.$

EXAMPLE 5. *Determine the points where a function $x^3 + y^3 - 3axy$ has maximum or minimum.*

SOLUTION. Here, we have $u = x^3 + y^3 - 3axy$

$$\Rightarrow \qquad \frac{\partial}{\partial} = 3x - 3ay \text{ and } \frac{\partial u}{\partial y} = 3y^2 - 3ax.$$

For a maximum or minimum of u, we must have $\dfrac{\partial u}{\partial x} = 0$ and $\dfrac{\partial u}{\partial y} = 0$

which gives, $\qquad x^2 - ay = 0$ ...(1)

and $\qquad\qquad y^2 - ax = 0$ ...(2)

Solving (1) and (2), we get

$$x = 0, y = 0; x = a, y = a.$$

Thus $(0, 0)$ and (a, a) are the stationary points of u.

Now $r = \dfrac{\partial^2 u}{\partial x^2} = 6x, s = \dfrac{\partial^2 u}{\partial x \, \partial y} = -3a, t = \dfrac{\partial^2 u}{\partial y^2} = 6y.$

For x = 0, y = 0

$\qquad\qquad r = 0, s = -3a \text{ and } t = 0$

$\therefore \qquad rt - s^2 = -9a^2 < 0$, for all values of a.

$\Rightarrow u$ is neither maximum nor minimum at $x = 0, y = 0$.

For x = a, y = a

$\qquad\qquad r = 6a, s = -3a \text{ and } t = 6a$

$\Rightarrow \qquad rt - s^2 = 27a^2 > 0$, for all values of a.

Also $r = 6a$, which is positive if $a > 0$.

Thus (i) u is maximum at $x = a, y = a$ if $a < 0$

and (ii) u is minimum at $x = a, y = a$ if $a > 0$.

EXAMPLE 6. *Discuss the maxima and minima of the function $u = \sin x \sin y \sin (x+y)$.*

[UPTU–2009]

SOLUTION. Here, we have $u = \sin x \sin y \sin (x + y)$

$$\Rightarrow \qquad \frac{\partial u}{\partial x} = \sin y [\sin x \cos(x + y) + \cos x \sin(x + y)]$$

and $\qquad \dfrac{\partial u}{\partial y} = \sin x [\sin y \cos(x + y) + \cos y \sin(x + y)].$

For a maxima and minima of u, we must have

$$\frac{\partial u}{\partial x} = 0 \text{ and } \frac{\partial u}{\partial y} = 0.$$

$\Rightarrow \sin y [\sin x \cos (x + y) + \cos x \sin (x+y)] = 0$

and $\sin x [\sin y \cos (x + y) + \cos y \sin (x+y)] = 0$.

Equation (1) and (2) gives

$\qquad \tan (x + y) = -\tan x$ ...(1)

$\Rightarrow \qquad \tan x = \tan y$

and $\tan (x + y) = -\tan y$ ...(2)

$\Rightarrow \qquad x = y$

From (1) and (2), we have

$\qquad \tan 2x = -\tan x = \tan (\pi - x) \Rightarrow 2x = \pi - x$

$\Rightarrow \qquad 3x = \pi \quad \Rightarrow x = \dfrac{\pi}{3} = y.$

Moreover, $\qquad \dfrac{\partial u}{\partial x} = 0$, gives $\sin y = 0 \Rightarrow y = 0$

and $\quad\quad \dfrac{\partial u}{\partial y} = 0$, gives $\sin x = 0 \Rightarrow x = 0$.

Thus, we get the following pair of values, which makes the function u stationary

$(0,0), \left(\dfrac{\pi}{3}, \dfrac{\pi}{3}\right)$.

Now $\quad r = \dfrac{\partial^2 u}{\partial x^2} = 2\sin y \cos(2x + y), \; s = \dfrac{\partial^2 u}{\partial x\,\partial y} = \sin 2(x + y)$,

and $\quad\quad t = \dfrac{\partial^2 u}{\partial y^2} = 2\sin x \cos(2y + x)$.

For (0, 0).

$$r = 0, s = 0, t = 0 \;\Rightarrow\quad rt - s^2 = 0.$$

$\therefore$ this case is doubtful and need further investigation.

For $\left(\dfrac{\pi}{3}, \dfrac{\pi}{3}\right)$.

$$r = 2\sin\dfrac{1}{3}\pi.\cos\pi = -\sqrt{3}, \; s = \sin\left(\dfrac{4\pi}{3}\right) = -\sin\dfrac{\pi}{3} = -\dfrac{\sqrt{3}}{2},$$

and $\quad\quad t = 2\sin\dfrac{1}{3}\pi\cos\pi = -\sqrt{3}$.

$\therefore \quad\quad rt - s^2 = \dfrac{9}{4} = \text{positive}.$

Also $\quad\quad r = -\sqrt{3}.$

Hence, u has a maximum value at $\left(\dfrac{\pi}{3}, \dfrac{\pi}{3}\right)$.

EXAMPLE 7. *Find a point within a triangle such that the sum of the squares of its distances from the vertices is a minimum.*

SOLUTION. Let us suppose $[(x_r, y_r) : r = 1, 2, 3]$ be the vertices of the triangle and (x, y) be any point inside the triangle.

Now, let us define a function

$$u = \sum_{r=1}^{3} [(x - x_r)^2 + (y - y_r)^2].$$

Then, we have

$$\dfrac{\partial u}{\partial x} = \Sigma 2(x - x_r) = 2[(x - x_1) + (x - x_2) + (x - x_3)]$$

and $\quad\quad \dfrac{\partial u}{\partial y} = \Sigma 2(y - y_r) = 2[(y - y_1) + (y - y_2) + (y - y_3)].$

For a maximum or minimum of u, we must have

$$\dfrac{\partial u}{\partial x} = 0 \;\Rightarrow\; (x - x_1) + (x - x_2) + (x - x_3) = 0$$

$\Rightarrow \quad\quad x = \dfrac{x_1 + x_2 + x_3}{3}$

and $\quad\quad \dfrac{\partial u}{\partial y} = 0 \;\Rightarrow\; (y - y_1) + (y - y_2) + (y - y_3) = 0$

$\Rightarrow \quad\quad y = \dfrac{y_1 + y_2 + y_3}{3}.$

Thus, we have

$$\left(\frac{x_1 + x_2 + x_3}{3}, \frac{y_1 + y_2 + y_3}{3} \right)$$

is the only point at which u have a maximum or minimum.

Now $r = \dfrac{\partial^2 u}{\partial x^2} = 6, s = \dfrac{\partial^2 u}{\partial x \partial y} = 0, t = \dfrac{\partial^2 u}{\partial y^2} = 6.$

At $\left[\dfrac{x_1 + x_2 + x_3}{3}, \dfrac{y_1 + y_2 + y_3}{3} \right] \Rightarrow r = 6, s = 0, t = 6$

$\Rightarrow \qquad rt - s^2 = 36 > 0.$

Also, since $\qquad r > 0.$

Therefore u have a minimum value at $\left[\dfrac{x_1 + x_2 + x_3}{3}, \dfrac{y_1 + y_2 + y_3}{3} \right].$

Hence, the point $\left(\dfrac{x_1 + x_2 + x_3}{3}, \dfrac{y_1 + y_2 + y_3}{3} \right)$ is the required point at which u is minimum.

EXAMPLE 8. *Show that distance l of any point (x, y, z) on the plane $2x + 3y - z = 12$ from the origin is given by*

$$l = \sqrt{[x^2 + y^2 + (2x + 3y - 12)^2]}.$$

Hence, find the point on the plane that is nearest to the origin.

SOLUTION. If l is the distance from $(0, 0, 0)$ of any point (x, y, z) then $l = \sqrt{(x^2 + y^2 + z^2)}.$ If the point (x, y, z) lies on the plane $2x + 3y - z = 12$, then

$$l = \sqrt{[x^2 + y^2 + (2x + 3y - 12)^2]}$$

$$[\because z = 2x + 3y - 12, \text{ from the equation of the plane}]$$

$\therefore \qquad l^2 = x^2 + y^2 + (2x + 3y - 12)^2$

$$= 5x^2 + 10y^2 + 12xy - 48x + 72y + 144 = u \,(\text{say}).$$

Now l is maximum or minimum according as l^2 i.e., u is maximum or minimum.

For a maximum or minimum of u, we get

$$\frac{\partial u}{\partial x} = 10x + 12y - 48 = 0 \text{ and } \frac{\partial u}{\partial y} = 20y + 12x - 72 = 0$$

Solving these equations, we get $x = \dfrac{12}{7}$ and $y = \dfrac{18}{7}.$

Also $\qquad r = \dfrac{\partial^2 u}{\partial x^2} = 10, s = \dfrac{\partial^2 u}{\partial x \partial y} = 12 \text{ and } t = \dfrac{\partial^2 u}{\partial y^2} = 20.$

Therefore $rt - s^2 = 10 \times 20 - (12)^2 = + \text{ ve, since } rt - s^2 > 0$

and $r > 0$, then u is minimum and hence l is minimum.

When $x = \dfrac{12}{7}$ and $y = \dfrac{18}{7}.$ Putting these values of x and y in the equation of the plane, we get

$$z = 2 \cdot \left(\frac{12}{7}\right) + 3 \cdot \left(\frac{18}{7}\right) - 12 = -\frac{6}{7}.$$

Hence, the required point is $\left(\frac{12}{7}, \frac{18}{7}, -\frac{6}{7}\right)$.

EXAMPLE 9. *Find the points on $z^2 = xy + 1$ nearest to the origin.*

SOLUTION. Let l be the distance from the origin $(0, 0, 0)$ of any point (x, y, z) on the surface $z^2 = xy + 1$
...(1)

Then $\qquad l = \sqrt{x^2 + y^2 + z^2} = \sqrt{(x^2 + y^2 + xy + 1)}$ [Using equation (1)]

Since l is always greater than zero, therefore l is maximum or minimum according as l^2, *i.e.*, u is maximum or minimum, where $u = l^2$.

For a maximum or minimum of u, we must have

$$\frac{\partial u}{\partial x} = 2x + y = 0 \qquad\qquad ...(2)$$

and $\qquad\qquad \dfrac{\partial u}{\partial y} = 2y + x = 0. \qquad\qquad ...(3)$

Solving the equation (2) and (3), we get $x = 0, y = 0$

Also $\qquad\qquad r = \dfrac{\partial^2 u}{\partial x^2} = 2, \; s = \dfrac{\partial^2 u}{\partial x \, \partial y} = 1, \; t = \dfrac{\partial^2 u}{\partial y^2} = 2.$

$\therefore \qquad\qquad rt - s^2 = 2 . 2 - 1 = 3 > 0.$

Since at $x = 0, y = 0$, then $rt - s^2 > 0$ and $r > 0$.

Therefore u is minimum at $x = 0, y = 0$. Hence l is minimum, when $x = 0, y = 0$.

Putting $x = 0, y = 0$ in the equation (1), we get $z^2 = 1$ *i.e.*, $z = \pm 1$.

Hence, the required points are $(0, 0, 1)$ and $(0, 0, -1)$.

EXERCISE 3.1

1. Discuss the maxima and minima of the function $f(x, y) = x^2 + y^2 + \dfrac{2}{x} + \dfrac{2}{y}$.

2. Find the values of x and y for which the expression
$$(a_1 x + b_1 y + c_1)^2 + (a_2 x + b_2 y + c_2)^2$$
$$+ ... + (a_n x + b_n y + c_n)^2$$
is minimum.

3. Examine for maximum and minimum values of the function $f(x, y) = x^2 - 3xy + y^2 + 2x$.

4. Examine the function $f(x, y) = x^2 y - y^2 x - x + y$ for maxima and minima.

5. Discuss the maxima and minima of the function
$$f(x, y) = 2\sin\frac{1}{2}(x + y)\cos\frac{1}{2}(x - y) + \cos(x + y).$$

6. Find the maximum and minimum values of $u = 6xy + (47 - x - y)(4x + 3y)$.

7. Examine for extreme values
   (i) $x^2 + y^2 + 6x + 12$ [GBTU–2012]
   (ii) $x^3 + y^3 - 63(x + y) + 12xy$ [UKTU–2011]

ANSWERS

1. $f(x, y)$ is minimum at $(1, 1)$. 2. $f(x, y)$ is minimum for the value of x and y which are obtained by $\Sigma(a_1^2)x + (a_1 b_1)y + a_1 c_1 = 0$ and $\Sigma(a_1 b_1)x + (b_1^2)y + b_1 c_1 = 0$.

3. Stationary point is $x = \dfrac{4}{5}, y = \dfrac{6}{5}$. The function $f(x, y)$ is neither maximum nor minimum at $\left(\dfrac{4}{5}, \dfrac{6}{5}\right)$. 4. At $(1, 1)$ and $(-1, -1)$ function is neither maximum nor minimum.

5. $x = y = 2n\pi \pm \pi/2$; neither maximum nor minimum ; $x = y = n\pi + (-1)^n \pi/6$; f is maximum.

6. Maximum value is 3384.

7. (i) At $x = -3, y = 0$, minimum (ii) max at $(-7, -7)$ min. at $(3, 3)$ neither max nor min. at $(5, -1)$ and $(-1, 5)$.

3.6 MAXIMA AND MINIMA OF THE FUNCTION OF THREE INDEPENDENT VARIABLES

(1) *To find the condition, which governs the sign of the quadratic equation of three independent variables.*

Let I be the expression of three independent variables x, y and z given by

$$I = ax^2 + by^2 + cz^2 + 2fyz + 2gzx + 2hxy$$

I can be written as

$$I = \frac{1}{a}\left[a^2x^2 + aby^2 + acz^2 + 2afyz + 2agzx + 2ahxy \right] (a \neq 0)$$

$$= \frac{1}{a}\left[a^2x^2 + 2ax(gz + hy) + aby^2 + acz^2 + 2afyz \right]$$

$$= \frac{1}{a}\left[(ax + hy + gz)^2 + aby^2 + acz^2 + 2afyz - (gz + hy)^2 \right]$$

$$= \frac{1}{a}\left[(ax + hy + gz)^2 + \left(ab - h^2\right)y^2 + 2yz(af - gh) + \left(ac - g^2\right)z^2 \right]$$

Here, we observe that I be of the same sign as provided the expression within the square brackets is positive which will of course be so if $ab-h^2$ and $\{(ab-h^2)(ac-g^2)-(af-gh)^2\}$ are positive *i.e.*, if $ab-h^2$ and $a[abc+2fgh-af^2-bg^2-ch^2]$ are both positive.

Hence, I will be positive if a, $\begin{vmatrix} a & h \\ h & b \end{vmatrix}$, $\begin{vmatrix} a & h & g \\ h & b & f \\ g & f & c \end{vmatrix}$ be all positive and will be negative if these three expression are alternately negative and positive.

3.7 MAXIMA AND MINIMA FOR A FUNCTION OF THREE INDEPENDENT VARIABLES : THE LAGRANGE'S CONDITION

Let $f(x, y, z)$ be a given function of three independent variables x, y and z.

Let A, B, C, F, G, H stand for $\dfrac{\partial^2 f}{\partial x^2}, \dfrac{\partial^2 f}{\partial y^2}, \dfrac{\partial^2 f}{\partial z^2}, \dfrac{\partial^2 f}{\partial y \partial z}, \dfrac{\partial^2 f}{\partial z \partial x}, \dfrac{\partial^2 f}{\partial x \partial y}$ respectively.

Let a set of the values of x, y, z obtained by solving the equations

$$\frac{\partial f}{\partial x} = \frac{\partial f}{\partial y} = \frac{\partial f}{\partial z} = 0 \quad \text{be } a, b, c.$$

By Taylor's theorem, we have

$$f(a+h, b+k, c+l), -f(a,b,c) = \frac{1}{2!}\left[Ah^2 + Bk^2 + Cl^2 + 2Fkl + 2Glh + 2Hhk \right] + R \quad \text{...(1)}$$

where, remainder term R consist of third and higher order of same quantity (*i.e.*, h, k, l).

Now, by taking h, k, l sufficiently small, the second term of R.H.S. of (1) can be made to govern the sign of R.H.S. and therefore of L.H.S. also.

If for all such values of h, k and l, these terms be of permanent sign, then we shall have a maximum or minimum of $f(x, y, z)$ according as that sign is negative or positive.

Hence, the function will be minimum if the expression A, $\begin{vmatrix} A & H \\ H & B \end{vmatrix}$, $\begin{vmatrix} A & H & G \\ H & B & F \\ G & F & C \end{vmatrix}$ be all positive.

The function will have a maximum value, if the above three quantities are alternately negative and positive. If these conditions are not satisfied, we have neither a maximum nor a minimum.

WORKING PROCEDURE

Let $f(x, y, z)$ be a function of three independent variables x, y and z. Find the values of triads (a,b,c) of the value x, y and z by putting $\frac{\partial f}{\partial x} = 0, \frac{\partial f}{\partial y} = 0, \frac{\partial f}{\partial z} = 0$. The values of triads (a,b,c) will give the stationary values of $f(x, y, z)$.

Now, to discuss maximum and minimum values, at (a, b, c) we find the following six partial derivatives of second order

$$A = \frac{\partial^2 f}{\partial x^2}, B = \frac{\partial^2 f}{\partial y^2}, C = \frac{\partial^2 f}{\partial z^2}, F = \frac{\partial^2 f}{\partial y \partial z}, G = \frac{\partial^2 f}{\partial z \partial x}, \text{ and } H = \frac{\partial^2 f}{\partial x \partial y}$$

Now, we have the following cases :

Case (i) The function $f(x,y,z)$ will be minimum at (a,b,c) if the expressions

$$A, \begin{vmatrix} A & H \\ H & B \end{vmatrix}, \begin{vmatrix} A & H & G \\ H & B & F \\ G & F & C \end{vmatrix} \text{ be all positive at } (a, b, c).$$

Case (ii) The function $f(x, y, z)$ will be maximum at (a, b, c) if the expressions

$$A, \begin{vmatrix} A & H \\ H & B \end{vmatrix}, \begin{vmatrix} A & H & G \\ H & B & F \\ G & F & C \end{vmatrix}$$

be alternately negative and positive.

Case (iii) If the expression, using in case (i) and (ii) neither be all positive nor having alternately negative and positive sign at (a,b,c). Then $f(x, y, z)$ is neither maximum nor minimum at (a, b, c).

☛ REMARK

- To find the maximum and minimum of the function at stationary point, it is sufficient to find the value of a second order partial derivative of function with respect to any of the independent variables. Then, the value of the function is maximum or minimum according as the value of this second order partial derivative at the stationary point under consideration is negative or positive.

SOLVED EXAMPLES

EXAMPLE 1. *Find the maximum value of u, where* $u = \dfrac{xyz}{(a+x)(x+y)(y+z)(z+b)}$.

SOLUTION. We have $\qquad u = \dfrac{xyz}{(a+x)(x+y)(y+z)(z+b)}$

Taking, log of both the sides, we have

$$\log u = \log x + \log y + \log z - \log(a+x) - \log(x+y) - \log(y+z) - \log(z+b)$$

Differentiating w.r.t. x, we have

$$\frac{1}{u}\frac{\partial u}{\partial x} = \frac{1}{x} - \frac{1}{a+x} - \frac{1}{x+y} = \frac{ay - x^2}{x(a+x)(x+y)}$$

$$\Rightarrow \qquad \frac{\partial u}{\partial x} = \frac{\left(ay - x^2\right)u}{x(a+x)(x+y)}$$

Similarly $\qquad \dfrac{\partial u}{\partial y} = \dfrac{\left(xz - y^2\right)u}{y(x+y)(y+z)}$ and $\dfrac{\partial u}{\partial z} = \dfrac{\left(by - z^2\right)u}{z(y+z)(z+b)}$

For, a maxima and minima of u, we must have

$$\frac{\partial u}{\partial x} = 0 \implies ay - x^2 = 0 \; ; \; \frac{\partial u}{\partial y} = 0 \implies xz - y^2 = 0$$

and

$$\frac{\partial u}{\partial z} = 0 \implies by - z^2 = 0$$

Here, we observe that $x^2 = ay$, $y^2 = xz$, $z^2 = by$ which implies that a, x, y, z and b are in G.P. Let r be the common ratio of this G.P.

Then
$$ar^4 = b \quad \text{or} \quad r = \left(\frac{b}{a}\right)^{1/4}$$

Also
$$x = ar, y = ar^2, z = ar^3.$$

Hence, we have

$$u = \frac{ar.ar^2.ar^3}{a(1+r)ar(1+r)ar^2(1+r)ar^3(1+r)}$$

$$= \frac{1}{a(1+r)^4} = \frac{1}{a\left[1 + \left(\dfrac{b}{a}\right)^{1/4}\right]^4}$$

$$= \frac{1}{\left(a^{1/4} + b^{1/4}\right)^4}$$

which gives a stationary value of u. Now, to decide whether this value of u is a maximum or a minimum, we proceed to find the second order partial derivative of u.

Here $\dfrac{\partial^2 u}{\partial x^2} = \dfrac{-2ux}{x(a+x)(x+y)} + (ay - x^2)\dfrac{\partial}{\partial x}\left[\dfrac{u}{x(a+x)(x+y)}\right]$

When $x = ar$, $y = ar^2$, $z = ar^3$, we have

$$A = \frac{\partial^2 u}{\partial x^2} = -\frac{2u}{a^2 r(1+r)^2} < 0$$

Hence, the above stationary value of u is maximum.

EXAMPLE 2. ***Find the maxima and minima value of the function***
         $u = \sin x \sin y \sin z$
where x, y and z are the vertex angles of a triangle.

SOLUTION. Here, we have
$$u = \sin x \sin y \sin z \; ; \; \text{where } x + y + z = \pi \qquad \qquad \text{...(1)}$$

$\therefore$ $u = \sin x \sin y \sin[\pi - (x+y)]$
$$= \sin x \sin y \sin(x+y)$$

$\therefore$ $\dfrac{\partial u}{\partial x} = \cos x \sin y \sin(x+y) + \sin x \sin y \cos(x+y)$
$$= \sin y \sin(2x+y). \qquad \qquad \text{...(2)}$$

Similarly $\dfrac{\partial u}{\partial y} = \sin x \sin(2y+x)$ ...(3)

For a maxima and minima, we must have

$$\frac{\partial u}{\partial x} = 0, \frac{\partial u}{\partial y} = 0$$

So, $\dfrac{\partial u}{\partial x} = 0 \implies \sin y \sin(2x+y) = 0$

$\Rightarrow$ $\qquad$ $\sin y=0$ or $\sin(2x+y)=0$ $\Rightarrow$ $y=0$ or $\sin(x+x+y)=0$

$\Rightarrow$ $\qquad$ $y=0$

or $\sin x \cos(x+y)+\cos x \sin(x+y)=0$

$\Rightarrow$ $\tan(x+y) = -\tan x$

$\Rightarrow$ $\tan(x+y) = \tan(-x) = \tan(\pi - x)$ $\qquad$ ...(4)

$\Rightarrow$ $\qquad$ $x+y=\pi -x$

$\Rightarrow$ $\qquad$ $2x+y=\pi$ $\qquad$ ...(5)

Similarly, from (3), $\qquad$ $x=0$

or $\tan(x+y) = -\tan y$ $\qquad$ ...(6)

Now, by (4) and (6), we have

$\qquad$ $\tan x = \tan y$ $\Rightarrow$ $\qquad$ $x=y$.

Hence, by (5), we have

$\qquad$ $3y=\pi$ $\Rightarrow$ $y = \pi/3$ and $x = \pi/3$

Therefore, the stationary points are $\left(\dfrac{\pi}{3},\dfrac{\pi}{3}\right)$ and $(0, 0)$.

For (0,0): $u=0$.

For $\left(\dfrac{\pi}{3},\dfrac{\pi}{3}\right)$

$$r = \frac{\partial^2 u}{\partial x^2} = 2\sin y \cos(2x + y) = 2\sin\frac{\pi}{3}\cos\left(\frac{2\pi}{3}+\frac{\pi}{3}\right)$$

$$= -\sqrt{3} < 0$$

and $\qquad$ $s = \dfrac{\partial^2 u}{\partial x \partial y} = \sin(2x + 2y)$

$$= \sin\left(\frac{2\pi}{3}+\frac{2\pi}{3}\right) = \sin\left(\frac{4\pi}{3}\right) = -\frac{\sqrt{3}}{2} < 0$$

$$t = \frac{\partial^2 u}{\partial y^2} = 2\sin x \cos(x + 2y) = 2\sin\frac{\pi}{3}\cos\pi = -\sqrt{3} < 0$$

Now $\qquad$ $rt-s^2 = \left(-\sqrt{3}\right)\left(-\sqrt{3}\right)-\left(\dfrac{\sqrt{3}}{2}\right)^2 = \dfrac{9}{4} > 0$. Thus, $rt-s^2 > 0$ and $r<0$.

Hence, the function u will be maximum at $\left(\dfrac{\pi}{3},\dfrac{\pi}{3}\right)$.

EXERCISE 3.2

1. Prove that the function $u = x^2 + y^2 +x - 2z - xy$ is minimum at $\left(-\dfrac{2}{3},-\dfrac{1}{3},1\right)$.

2. Find the maximum and minimum values of $u = y^2+2z^2 - 5x^4 + 4x^5$.

3. Find the maximum or minimum values of the function u, where $u = axy^2z^3 - x^2y^2z^3 - xy^3z^3 - xy^2z^4$

4. Find the maximum value of
$$(ax+by+cz)\, e^{-\left(\alpha^2 x^2+\beta^2 y^2+\gamma^2 z^2\right)}.$$

5. A rectangle box is placed on x-y plane. The one end of the box is at the origin. If the vertex opposite to the origin be on the plane $6x + 4y + 3z = 24$, then find the maximum value of this box.

6. In a plane triangle xyz, find the maximum value of $\sin x \sin y \sin z$.

7. A rectangular box, open at the top is to have a given capacity. Show that the domain of the box requiring least material for its construction $x = y = (2v)^{1/3}$, where $v = xyz$.

ANSWERS

2. Minimum at $(1,0,0)$, neither maximum nor minimum at $(0,0,0)$.

3. Maximum at $\left(\dfrac{a}{7}, \dfrac{2a}{7}, \dfrac{3a}{7}\right)$, max. value $= \dfrac{108a^7}{7^7}$

4. Maximum at $\left(\dfrac{a}{2\alpha^2 k}, \dfrac{b}{2\beta^2 k}, \dfrac{c}{2\gamma^2 k}\right)$ where $k = \sqrt{\left\{\dfrac{1}{2}\left(\dfrac{a^2}{\alpha^2} + \dfrac{b^2}{\beta^2} + \dfrac{c^2}{\gamma^2}\right)\right\}}$,

   Maximum value $= \sqrt{\left\{\dfrac{1}{2e}\left(\dfrac{a^2}{\alpha^2} + \dfrac{b^2}{\beta^2} + \dfrac{c^2}{\gamma^2}\right)\right\}}$

5. Maximum at $\left(\dfrac{4}{3}, 2\right)$. maximum value $= \dfrac{64}{9}$ cube units. Neither maximum nor minimum at $(0,0)$.

6. Maximum at $\left(\dfrac{\pi}{3}, \dfrac{\pi}{3}, \dfrac{\pi}{3}\right)$, maximum value $= \dfrac{3\sqrt{3}}{8}$

3.8 LAGRANGE'S METHOD OF UNDETERMINED MULTIPLIERS

Let $u = f(x_1, x_2, ..., x_n)$ be a function of n variables $x_1, x_2, ..., x_n$.

Let us suppose these variables $x_1, x_2, ..., x_n$ are connected by k equations

$$g_1(x_1, x_2, ..., x_n) = 0$$
$$g_2(x_1, x_2, ..., x_n) = 0$$
$$\vdots \quad ... \quad ... \quad ... \quad \vdots$$
$$g_k(x_1, x_2, ..., x_n) = 0$$

so, that there are $n-k$ independent variables out of these n variables. For the maxima and minima of u, we find

$$du = \frac{\partial u}{\partial x_1} dx_1 + \frac{\partial u}{\partial x_2} dx_2 + ... + \frac{\partial u}{\partial x_n} dx_n = 0 \qquad ...(1)$$

Also

$$dg_1 = \frac{\partial g_1}{\partial x_1} dx_1 + \frac{\partial g_1}{\partial x_2} dx_2 + ... + \frac{\partial g_1}{\partial x_n} dx_n = 0 \qquad ...(2)$$

$$dg_2 = \frac{\partial g_2}{\partial x_1} dx_1 + \frac{\partial g_2}{\partial x_2} dx_2 + ... + \frac{\partial g_2}{\partial x_n} dx_n = 0 \qquad ...(3)$$

$$\vdots \qquad \vdots \qquad \qquad \vdots \qquad \qquad \vdots$$

$$dg_k = \frac{\partial g_k}{\partial x_1} dx_1 + \frac{\partial g_k}{\partial x_2} dx_2 + ... + \frac{\partial g_k}{\partial x_n} dx_n = 0 \qquad ...(k+1)$$

Multiplying equation (1), (2), (3)...(k+l) by $1, l_1, l_2, ..., k$ respectively and adding, we get the result, which can be written as

$$P_1 dx_1 + P_2 dx_2 + P_3 dx_3 + ... + P_n dx_n = 0 \qquad ...(4)$$

where

$$P_k = \frac{\partial u}{\partial x_k} + l_1 \frac{\partial g_1}{\partial x_k} + l_2 \frac{\partial g_2}{\partial x_k} + ... + l_k \frac{\partial g_k}{\partial x_k}$$

Now we have at our choice k multiple viz $l_1, l_2, ..., l_k$ and can be chosen such that

$$P_1 = 0, P_2 = 0, ..., P_k = 0$$

Then, the equation (4) reduces to

$$P_{k+1} dx_{k+1} + P_{k+2} dx_{k+2} + P_{k+3} dx_{k+3} + ... + P_n dx_n = 0 \qquad ...(5)$$

Now, let us suppose that out of n variables, the $(n-k)$ variables $x_{k+1}, x_{k+2}, ..., x_n$ are independent.

Then, since $n-k$ quantities $dx_{k+1}, dx_{k+2}, ..., dx_n$ are independent so their coefficients must be separately zero. Hence, we have

$$P_{k+1}=0, P_{k+2}=0, ..., P_n=0$$

Thus, we have $k+n$ equations

$$P_1=0, P_2=0, ..., P_n=0$$

and $$g_1=0, g_2=0, ..., g_k=0.$$

Hence, we get $(n+k)$ equations which determine the k multipliers $l_1, l_2, ..., l_k$ and get the possible value of u.

☛ **REMARKS**

- The Lagrange's method of undetermined multipliers is very convenient to apply. It gives the maximum and minimum values of the function without actually determining the values of the multipliers $l_1, l_2, ..., l_k$.
- It does not determine the nature of stationary point, which is the only drawback of this method.

3.8.1 APPLICATIONS OF THE METHOD OF UNDETERMINED MULTIPLIERS

The Lagrange's method of undetermined multipliers can be applied to determine the extreme values of the given functions, it does not determine the nature of stationary point. Now, it is more convenient to find out the extreme values of a function F with the help of new function, given by

$$V=g+l_1f_1+l_2f_2+...+l_mf_m$$

and use the following method. Here, we give the method for four variables x, y, u, v connected by the following two relations.

Let $F=g(x, y, u, v)$ be subjected to the conditions

$$f_1(x,y,u,v)=0 \qquad \qquad ...(1)$$

and $$f_2(x,y,u,v)=0. \qquad \qquad ...(2)$$

For the maxima and minima of F, we have

$$dF = \frac{\partial g}{\partial x}dx + \frac{\partial g}{\partial y}dy + \frac{\partial g}{\partial u}du + \frac{\partial g}{\partial v}dv = 0 \qquad \qquad ...(3)$$

Now, from (1) and (2), we have

$$df_1 = \frac{\partial f_1}{\partial x}dx + \frac{\partial f_1}{\partial y}dy + \frac{\partial f_1}{\partial u}du + \frac{\partial f_1}{\partial v}dv = 0 \qquad \qquad ...(4)$$

and $$df_2 = \frac{\partial f_2}{\partial x}dx + \frac{\partial f_2}{\partial y}dy + \frac{\partial f_2}{\partial u}du + \frac{\partial f_2}{\partial v}dv = 0 \qquad \qquad ...(5)$$

Multiplying (4) by l_1, (5) by l_2 and adding their sum to (3), we get

$$\left(\frac{\partial g}{\partial x}+l_1\frac{\partial f_1}{\partial x}+l_2\frac{\partial f_2}{\partial x}\right)dx + \left(\frac{\partial g}{\partial y}+l_1\frac{\partial f_1}{\partial y}+l_2\frac{\partial f_2}{\partial y}\right)dy$$

$$+\left(\frac{\partial g}{\partial u}+l_1\frac{\partial f_1}{\partial u}+l_2\frac{\partial f_2}{\partial u}\right)du + \left(\frac{\partial g}{\partial v}+l_1\frac{\partial f_1}{\partial v}+l_2\frac{\partial f_2}{\partial v}\right)dv = 0 \qquad ...(6)$$

Here, we have l_1 and l_2 are arbitrary, therefore we can choose them to satisfy the two linear equations

$$\frac{\partial g}{\partial x}+l_1\frac{\partial f_1}{\partial x}+l_2\frac{\partial f_2}{\partial x}=0 \qquad \qquad ...(7)$$

and $$\frac{\partial g}{\partial y}+l_1\frac{\partial f_1}{\partial y}+l_2\frac{\partial f_2}{\partial y}=0 \qquad \qquad ...(8)$$

Using (7) and (8), equation (6) reduces to

$$\left(\frac{\partial g}{\partial u}+l_1\frac{\partial f_1}{\partial u}+l_2\frac{\partial f_2}{\partial u}\right)du + \left(\frac{\partial g}{\partial v}+l_1\frac{\partial f_1}{\partial v}+l_2\frac{\partial f_2}{\partial v}\right)dv = 0$$

Since, the given function contains four variables (namely x, y, u and v) and we are given two equations of conditions, therefore, only two of the variables are independent and it is immaterial which two of the four variables are regarded as independent. Let them be u and v then du and dv are also independent, therefore, their coefficients must be zero separately. Thus

$$\frac{\partial g}{\partial u} + l_1 \frac{\partial f_1}{\partial u} + l_2 \frac{\partial f_2}{\partial u} = 0 \qquad \qquad ...(9)$$

$$\frac{\partial g}{\partial v} + l_1 \frac{\partial f_1}{\partial v} + l_2 \frac{\partial f_2}{\partial v} = 0 \qquad \qquad ...(10)$$

Now, we have six equations namely (1),(2),(7),(8),(9) and (10) to determine the two multipliers l_1, l_2 and values of the four variables x, y, u and v for which maximum and minimum values of F are possible.

Now, defined a new function $V(x, y, u, v)$ such that

$$V(x,y,u,v) = g(x, y, u, v) + l_1 f_1(x, y, u, v) + l_2 f_2(x, y, u, v).$$

Assuming that x, y, u, v are now all independent variables. Hence, for the maxima and minima of V, we must have

$$\frac{\partial V}{\partial x} = \frac{\partial g}{\partial x} + l_1 \frac{\partial f_1}{\partial x} + l_2 \frac{\partial f_2}{\partial x} = 0 \qquad \qquad ...(11)$$

$$\frac{\partial V}{\partial y} = \frac{\partial g}{\partial y} + l_1 \frac{\partial f_1}{\partial y} + l_2 \frac{\partial f_2}{\partial y} = 0 \qquad \qquad ...(12)$$

$$\frac{\partial V}{\partial u} = \frac{\partial g}{\partial u} + l_1 \frac{\partial f_1}{\partial u} + l_2 \frac{\partial f_2}{\partial u} = 0 \qquad \qquad ...(13)$$

and $$\frac{\partial V}{\partial v} = \frac{\partial g}{\partial v} + l_1 \frac{\partial f_1}{\partial v} + l_2 \frac{\partial f_2}{\partial v} = 0 \qquad \qquad ...(14)$$

Equations (11), (12), (13) and (14) are exactly the same as the equations (7), (8), (9) and (10). Hence, the maxima and minima of $V(x, y, u, v)$ are same as those of $F(x, y, u, v)$ assuming that $V(x, y, u, v)$ the variables x, y, u, v are now all independent.

Now, we proceed to find whether the values of F obtained with the help of above equations are maximum or minimum. For this, adopt the procedure, which is discussed ahead.

From (3), we get

$$d^2F = \left(\frac{\partial}{\partial x}dx + \frac{\partial}{\partial y}dy + \frac{\partial}{\partial u}du + \frac{\partial}{\partial y}dy\right)^2 g + \left(\frac{\partial g}{\partial x}d^2x + \frac{\partial g}{\partial y}d^2y + \frac{\partial g}{\partial u}d^2u + \frac{\partial g}{\partial y}d^2v\right) ... \qquad ...(15)$$

Also

$$d^2f_1 = \left(\frac{\partial}{\partial x}dx + \frac{\partial}{\partial y}dy + \frac{\partial}{\partial u}du + \frac{\partial}{\partial v}dv\right)^2 f_1 + \frac{\partial f_1}{\partial x}d^2x + \frac{\partial f_1}{\partial y}d^2y + \frac{\partial f_1}{\partial u}d^2u + \frac{\partial f_1}{\partial v}d^2v = 0 \qquad ...(16)$$

and $$d^2f_2 = \left(\frac{\partial}{\partial x}dx + \frac{\partial}{\partial y}dy + \frac{\partial}{\partial u}du + \frac{\partial}{\partial v}dv\right)^2 f_2 + \frac{\partial f_2}{\partial x}d^2x + \frac{\partial f_2}{\partial y}d^2y + \frac{\partial f_2}{\partial u}d^2u + \frac{\partial f_2}{\partial v}d^2v = 0 \qquad ...(17)$$

Multiplying (16) by l_1 and (17) by l_2 and adding their sum to (15) and using the result (11), (12),(13) and (14), we have

$$d^2F = \left(\frac{\partial}{\partial x}dx + \frac{\partial}{\partial y}dy + \frac{\partial}{\partial u}du + \frac{\partial}{\partial v}dv\right)^2 (g + l_1 f_1 + l_2 f_2)$$

$$= \left(\frac{\partial}{\partial x}dx + \frac{\partial}{\partial y}dy + \frac{\partial}{\partial u}du + \frac{\partial}{\partial v}dv\right)^2 V = d^2V.$$

Hence d^2F is equal to d^2V, where d^2V is obtained by assuming all the variables x, y, u and v as independent. Therefore, it is clear that d^2F and d^2V have the same sign. Hence, F will be minimum or maximum according as V is minimum or maximum.

☛ REMARK

- This method has the advantage over the Lagrange's methods that it enables us to decide whether the values are maximum or minimum.

SOLVED EXAMPLES

EXAMPLE 1. *Find the maxima and minima of* $x^2+y^2+z^2$ *subject to the conditions :*

$$ax^2+by^2+cz^2 = 1 \text{ and } lx+my+nz = 0$$ [UKTU–2011]

SOLUTION. Here, we have $\qquad u = x^2 + y^2 + z^2$ ...(1)

$$ax^2+by^2+cz^2 = 1$$ ...(2)

and $\qquad lx + my + nz = 0$ ...(3)

For the maxima and minima of u, we must have

$$du = 0 \Rightarrow 2xdx + 2ydy + 2zdz = 0$$

$$\Rightarrow \qquad xdx + ydy + zdz = 0$$ ...(4)

From (2) and (3), we get

$$ax\,dx + by\,dy + cz\,dz = 0$$ ...(5)

$$ldx + mdy + ndz = 0$$ ...(6)

Now, multiplying (4) by 1, (5) by l_1 and (6) by l_2 and adding, we get

$$(x\,dx + y\,dy + z\,dz) + l_1(ax\,dx + by\,dy + cz\,dz) + l_2(l\,dx + m\,dy + n\,dz) = 0$$

$$\Rightarrow (x + al_1x + ll_2)dx + (y + bl_1y + ml_2)\,dy + (z + cl_1z + nl_2)dz = 0$$

Now equating the coefficient of dx, dy, dz to zero, we get

$$x + l_1ax + l_2l = 0$$ ...(7)

$$y + bl_1y + ml_2 = 0$$ ...(8)

and $\qquad z + cl_1z + nl_2 = 0$ ...(9)

Multiplying the equations (7), (8) and (9) by x, y and z respectively, and adding we get

$$x^2+y^2+z^2 + l_1(ax^2+by^2+cz^2) + l_2(lx+my+nz) = 0$$

or $\qquad u + l_1.1 + l_2.0 = 0$ [By using (1), (2) and (3)]

$$\Rightarrow \qquad l_1 = -u$$

Substituting for l_1 in the equations (7), (8) and (9), we get

$$x = \frac{l_2l}{au-1}, y = \frac{l_2m}{bu-1}, z = \frac{l_2n}{cu-1}$$ ...(10)

Now from (10) and (3), we get

$$\frac{l_2l^2}{au-1} + \frac{l_2m^2}{bu-1} + \frac{l_2n^2}{cu-1} = 0$$

or $\qquad \dfrac{l^2}{au-1} + \dfrac{m^2}{bu-1} + \dfrac{n^2}{cu-1} = 0$ ...(11)

which gives the maximum and minimum of $u = x^2 + y^2 + z^2$.

☛ REMARKS

- Equation (11) is a quadratic in u. So it gives two stationary values of u.
- Geometrically, the surface $ax^2+by^2+cz^2 = 1$ represents an ellipsoid whose centre is origin, and $lx+my+nz=0$ represents a plane passing through the origin. The points (x, y, z) satisfying both the conditions (2) and (3) lies on the conic in which (2) and (3) intersect. $x^2+y^2+z^2$ gives the square of the distance (x, y, z) from the origin, which is also the centre of the conic of intersection. The maximum value of this distance is the major axis of this conic, and the minimum value of this distance is the minor axis of this conic. Hence, equation (11) gives the squares of the lengths of the semi-axis of the conic of intersection.

EXAMPLE 2. *Find the maxima and minima of* $x^2+y^2+z^2$, *where* $ax^2 + by^2 + cz^2 + 2fyz + 2gzx + 2hxy = 1.$

SOLUTION. Let
$$u = x^2+y^2+z^2 \qquad \text{...(1)}$$
where the relation between the variables x,y and z is
$$ax^2+by^2+cz^2+2fyz+2gzx+2hxy=1. \qquad \text{...(2)}$$
For a maximum or minima of u, we must have $du=0$
$$\Rightarrow \qquad x\,dx+y\,dy+z\,dz=0. \qquad \text{...(3)}$$
From (2), we have
$$2ax\,dx+2by\,dy+2cz\,dz+2fy\,dz+2fz\,dy+2gz\,dx+2gx\,dz+2hx\,dy+2hy\,dx=0$$
$$\Rightarrow (ax+hy+gz)dx+(hx+by+fz)dy+(gx+fy+cz)dz=0. \qquad \text{...(4)}$$
Now, multiplying (3) by 1 and (4) by l_1, adding, and then equating the coefficient of dx, dy, dz to zero, we have
$$x+l_1(ax+hy+gz)=0. \qquad \text{...(5)}$$
$$y+l_1(hx+by+fz)=0. \qquad \text{...(6)}$$
$$z+l_1(gx+fy+cz)=0. \qquad \text{...(7)}$$
Multiplying (5) by x, (6) by y, (7) by z and adding, we get
$$x^2+y^2+z^2+l_1(ax^2+by^2+cz^2+2fyz+2gzx+2hxy)=0$$
$$\Rightarrow \qquad u+l_1.1=0 \qquad \qquad \text{[From (1) and (2)]}$$
$$\therefore \qquad l_1 =-u.$$
Hence, from (5), we have
$$x-u(ax+hy+gz)=0$$
$$\Rightarrow \qquad \left(a-\frac{1}{u}\right)x+hy+gz = 0 \qquad \text{...(8)}$$
Similarly from (6) and (7), we get
$$hx +\left(b-\frac{1}{u}\right)y+fz = 0 \qquad \text{...(9)}$$
and
$$gx + fy +\left(c-\frac{1}{u}\right)z = 0 \qquad \text{...(10)}$$
Eliminating x, y, z from (8), (9) and (10), we get
$$\begin{vmatrix} \left(a-\dfrac{1}{u}\right) & h & g \\[2mm] h & \left(b-\dfrac{1}{u}\right) & f \\[2mm] g & f & \left(c-\dfrac{1}{u}\right) \end{vmatrix} = 0 \qquad \text{...(11)}$$
Hence, the maximum or minimum values of u are the roots of the equation (11).

EXAMPLE 3. *Find the maxima and minima of* $u=x^2+y^2$ *subject to the condition*
$$ax^2+2hxy+by^2=1.$$

SOLUTION. Here, we have
$$u = x^2+y^2 \qquad \text{...(1)}$$
where the relation between the variables x and y is
$$ax^2+2hxy+by^2 =1. \qquad \text{...(2)}$$
For the maxima and minima of u, we must have
$$du = 0 \Rightarrow 2x\,dx+2y\,dy = 0$$

$\Rightarrow$ $\qquad x\,dx + y\,dy = 0.$ $\qquad$ ...(3)

Now, from (2), we get

$\qquad 2ax\,dx + 2hx\,dy + 2hy\,dx + 2by\,dy = 0$

$\Rightarrow$ $\qquad (ax+hy)dx + (hx+by)dy = 0$ $\qquad$ ...(4)

Now, multiplying (3) by 1, (4) by l_1, adding and then equating the coefficients of dx, dy to zero, we have

$\qquad x + l_1(ax+hy) = 0$ $\qquad$ ...(5)

and $\qquad y + l_1(hx+by) = 0$ $\qquad$ ..(6)

Multiplying (5) by x, (6) by y and adding, we get

$\qquad x^2 + y^2 + l_1(ax^2 + 2hxy + by^2) = 0$

$\Rightarrow$ $\qquad u + l_1.1 = 0$ $\qquad\qquad$ [Using (1) and (2)]

$\Rightarrow$ $\qquad\qquad u = -l_1$

Therefore, from (5), we have

$\qquad x - u(ax+hy) = 0$

$\Rightarrow$ $\qquad \left(a - \dfrac{1}{u}\right)x + hy = 0$ $\qquad$ ...(7)

Similarly from (6), we have

$\qquad hx + \left(b - \dfrac{1}{u}\right)y = 0$ $\qquad$ ...(8)

Eliminating x and y from (7) and (8), we get

$$\begin{vmatrix} a - \dfrac{1}{u} & h \\ h & b - \dfrac{1}{u} \end{vmatrix} = 0 \qquad ...(9)$$

Hence, the maximum or minimum values of u are the roots of the equation (9).

EXAMPLE 4. ***Find the maximum value of $u = x^m y^n z^p$ subject to the condition $x+y+z=a$.***

[ANNA–2009]

SOLUTION. Here, we have $\qquad u = x^m y^n z^p$ $\qquad$ ...(1)

and x, y, z connected by the relation given by $x+y+z = a$ $\qquad$ ...(2)

Taking log of both the sides of (1), we get

$\qquad \log u = m\log x + n\log y + p\log z.$

On differentiating, we get

$$\frac{1}{u}du = \frac{m}{x}dx + \frac{n}{y}dy + \frac{p}{z}dz$$

For the maxima and minima of u, we must have $du = 0$

$\Rightarrow$ $\qquad \dfrac{m}{x}dx + \dfrac{n}{y}dy + \dfrac{p}{z}dz = 0$ $\qquad$ ...(3)

Now, differentiating (2), we get

$\qquad dx + dy + dz = 0.$ $\qquad$ ...(4)

Now, multiplying (3) by 1 and (4) by l, and equating the coefficient of dx, dy, dz to zero (after adding), we get

$\qquad \dfrac{m}{x} + l = 0,\ \ \dfrac{n}{y} + l = 0\ \text{ and }\ \dfrac{p}{z} + l = 0$

which implies

$$x = -\frac{m}{l},\, y = -\frac{n}{l},\, z = -\frac{p}{l}$$

Putting the values of x, y and z in (2), we get $l = -\left(\dfrac{m+n+p}{a}\right)$ therefore, we can say that, u is stationary when

$$x = \frac{am}{m+n+p}, y = \frac{an}{m+n+p}, z = \frac{ap}{m+n+p}$$

Now, we find the nature of this stationary value of u.

Let us regard x and y as independent variable and z is a function of x and y given by (2) [It is justify, because the variables x, y and z are connected by the relation (2), any two of them may be regarded as independent].

Now from (1), we get

$$\log u = m \log x + n \log y + p \log z$$

$$\therefore \qquad \frac{1}{u}\frac{\partial u}{\partial x} = \frac{m}{x} + \frac{p}{z}\frac{\partial z}{\partial x} \qquad \qquad \dots(5)$$

Now, differentiating (2) partially w.r.t x (treating y as constant), we get

$$1 + \frac{\partial z}{\partial x} = 0 \quad \Rightarrow \quad \frac{\partial z}{\partial x} = -1$$

Put this value in (5), we get

$$\frac{1}{u}\frac{\partial u}{\partial x} = \frac{m}{x} - \frac{p}{z}$$

$$\Rightarrow \quad \frac{1}{u}\frac{\partial^2 u}{\partial x^2} - \frac{1}{u^2}\left(\frac{\partial u}{\partial x}\right)^2 = -\frac{m}{x^2} + \frac{p}{z^2}\frac{\partial z}{\partial x} = -\frac{m}{x^2} - \frac{p}{z^2}$$

At stationary point, $\dfrac{\partial u}{\partial x} = 0$

Therefore, $\qquad \dfrac{1}{u}\dfrac{\partial^2 u}{\partial x^2} = \dfrac{-m}{x^2} - \dfrac{p}{z^2}$

$$\Rightarrow \quad \frac{\partial^2 u}{\partial x^2} = u\left[-\frac{m}{x^2} - \frac{p}{z^2}\right] = -x^m y^n z^p\left[-\frac{m}{x^2} - \frac{p}{z^2}\right]$$

which is negative for the obtained values of x, y and z.

Hence, at the stationary point, u is maximum and maximum value is

$$= \left(\frac{am}{m+n+p}\right)^m \left(\frac{an}{m+n+p}\right)^n \left(\frac{ap}{m+n+p}\right)^p$$

EXAMPLE 5. *In a plane triangle ABC, find the maximum value of u = cos A cosB cosC.*

[VTU–2010; ANNA–2006]

SOLUTION. Here, we have $\qquad u = \cos A \cos B \cos C \qquad \qquad \dots(1)$

Since, we know that the sum of the angles of a triangle is always 180°.

$\therefore$ The variables A, B and C are connected by the relation

$$A + B + C = \pi \qquad \qquad \dots(2)$$

From (1), we get

$$\log u = \log \cos A + \log \cos B + \log \cos C$$

$$\Rightarrow \qquad \frac{1}{u}du = -\tan A \, dA - \tan B \, dB - \tan C \, dC.$$

For the maxima and minima of u, we must have $du = 0$

$$\Rightarrow \tan A \, dA + \tan B \, dB + \tan C \, dC = 0 \qquad \qquad \dots(3)$$

Also from (2),

$$dA + dB + dC = 0 \qquad \qquad \dots(4)$$

Now, multiply (3) by 1, (4) by l, adding, and equating the coefficients of dA, dB and dC to zero, we get

$$\tan A + l = 0; \quad \tan B + l = 0; \quad \tan C + l = 0$$

$$\Rightarrow \qquad l = -\tan A = -\tan B = -\tan C$$

$$\Rightarrow \qquad A = B = C.$$

Now from (2), $A = B = C = \dfrac{\pi}{3}$ *i.e.,* the triangle is equilateral.

Now to show that the stationary value of u given by $A = B = C = \pi/3$ is maximum. Let C be a function of A and B, regarding A and B as independent variables. From (1),

$$\log u = \log \cos A + \log \cos B + \log \cos C$$

$$\Rightarrow \qquad \frac{1}{u}\frac{\partial u}{\partial A} = -\tan A - \tan C \frac{\partial C}{\partial A}$$

Now, differentiating (2), partially w.r.t. A, we get

$$1 + \frac{dC}{dA} = 0 \quad \Rightarrow \quad \frac{\partial C}{\partial A} = -1$$

$$\therefore \qquad \frac{1}{u}\frac{\partial u}{\partial A} = -\tan A + \tan C$$

$$\Rightarrow \qquad \frac{1}{u}\frac{\partial^2 u}{\partial^2 A} - \frac{1}{u^2}\left(\frac{\partial u}{\partial A}\right)^2 = -\sec^2 A + \sec^2 C.\frac{\partial C}{\partial A} = -(\sec^2 A + \sec^2 C)$$

At stationary point $\dfrac{\partial u}{\partial A} = 0$

$$\because \qquad \frac{\partial^2 u}{\partial^2 A} = -u\left(\sec^2 A + \sec^2 C\right) = -\text{ve} \ \text{ for } A = B = C = \pi/3.$$

Hence, u is maximum at $A = B = C = \dfrac{\pi}{3}$ and the maximum value is given by

$$u = \left(\cos\frac{\pi}{3}\right)^3 = \left(\frac{1}{2}\right)^3 = \frac{1}{8}.$$

EXERCISE 3.3

1. Find the maximum and minimum values of
$$\frac{x^2}{a^4} + \frac{y^2}{b^4} + \frac{z^2}{c^4}$$
where $lx + my + nz = 0$ and $\dfrac{x^2}{a^2} + \dfrac{y^2}{b^2} + \dfrac{z^2}{c^2} = 1$.

2. Find the maximum and minimum values of
$$f = a^2 x^2 + b^2 y^2 + c^2 z^2$$
where $x^2 + y^2 + z^2 = 1$ and $lx + my + nz = 0$.

3. Show that the maximum and minimum values of $u = x^2 + y^2 + z^2$ subject to the conditions
$$px + qy + rz = 0 \text{ and } \frac{x^2}{a^2} + \frac{y^2}{b^2} + \frac{z^2}{c^2} = 1$$
are given by $\dfrac{a^2 p^2}{u - a^2} + \dfrac{b^2 q^2}{u - b^2}$.

4. Find the minimum value of $u = x + y + z$

subject to the condition $\dfrac{a}{x} + \dfrac{b}{y} + \dfrac{c}{z} = 1$.

5. Find the minimum value of $u = x^2 + y^2 + z^2$, subject to the condition $ax + by + cz = p$.
   (UKTU–2012, UPTU–2009)

6. Find the minimum value of $x + y + z$ where $xyz = c^3$.

7. Find the extreme values of $x^p y^q z^r$ subject to the condition $\dfrac{a}{x} + \dfrac{b}{y} + \dfrac{c}{z} = 1$.

8. Show that the maximum and minimum values of the radii vectors of the sections of the surface $(x^2 + y^2 + z^2)^2 = \dfrac{x^2}{a^2} + \dfrac{y^2}{b^2} + \dfrac{z^2}{c^2}$ by the plane $\lambda x + \mu y + \nu z = 0$ are given by
$$\frac{a^2\lambda^2}{1 - a^2 r^2} + \frac{b^2\mu^2}{1 - b^2 r^2} + \frac{c^2\nu^2}{1 - c^2 r^2} = 0.$$

9. Find the stationary points of the function $u = ax^p + by^q + cz^r$ subject to the condition

$$x^l + y^m + z^n = k.$$

10. If two variables x and y are connected by the relation $ax^2 + by^2 = ab$, show that the maximum and minimum values of the function $u = x^2 + y^2 + xy$ will be the roots of the equation $4(u-a)(u-b) = ab$.

11. Prove that of all rectangular parallelopipeds of the same volume, the cube has the least surface.   [KURUKSHETRA–2006, UPTU–2004]

12. Prove that if $x + y + z = 1$, $ayz + bzx + cxy$ has an extreme value equal to

$$\frac{abc}{2bc + 2ca + 2ab - a^2 - b^2 - c^2}$$

Also, prove if a, b, c are all positive and c lies between $a + b - 2\sqrt{ab}$ and $a + b + 2\sqrt{ab}$ this value is true maximum and that if a, b, c are all negative and c lies between $a + b \pm 2\sqrt{ab}$. It is true minimum.

13. Find the maximum value of u, when
$$u = \sin x \sin y \sin z$$
and x,y,z are the angles of a triangle.

14. Find the triangle of maximum area inscribed in a circle.

ANSWERS

1. The maximum and minimum values of the given function is given by the equation

$$\frac{l^2 a^4}{a^2 u - 1} + \frac{m^2 b^4}{b^2 u - 1} + \frac{n^2 c^4}{c^2 u - 1} = 0$$

2. The maximum and minimum values of the given function is given by the equation

$$\frac{l^2}{u - a^2} + \frac{m^2}{u - b^2} + \frac{m^2}{u - c^2} = 0$$

4. Stationary points are $x = \sqrt{a}\left(\sqrt{a} + \sqrt{b} + \sqrt{c}\right), y = \sqrt{b}\left(\sqrt{a} + \sqrt{b} + \sqrt{c}\right), z = \sqrt{c}\left(\sqrt{a} + \sqrt{b} + \sqrt{c}\right),$ minimum value is $\left(\sqrt{a} + \sqrt{b} + \sqrt{c}\right)^2$.

5. Minimum value is $\dfrac{p^2}{\left(a^2 + b^2 + c^2\right)}$ 6. u is minimum at the point $x = y = z = c$. Value is $= 3c^4$.

7. u is stationary when $\dfrac{px}{a} = \dfrac{qy}{b} = \dfrac{rc}{c} = p + q + r$, Minimum value is $\dfrac{a^p b^q c^r}{p^p q^q r^r}(p+q+r)^{p+q+r}$.

9. Stationary points are given by $\dfrac{x^{p-1}}{l/pa} = \dfrac{y^{q-m}}{m/qb} = \dfrac{z^{r-n}}{n/rc}$

13. u is maximum, when $x = y = z = \pi/3$. Maximum value is $\dfrac{3\sqrt{3}}{8}$.

14. Equilateral.

3.9 LAGRANGIAN MULTIPLIERS METHOD IN NON-LINEAR PROGRAMMING

In non-linear programming problem, if the objective function is continuous and differentiable and having all equally constraints, then we have the following necessary and sufficient conditions for the optimality of objective function by Lagrange's method.

Necessary Condition:

Case-I : Two decision variables and one equally constraint

Let us consider the following non-linear programming problem.

Optimize (max. or min.) $Z = f(x_1, x_2)$

subject to the constraints

$$g(x_1, x_2) = b, \ b \text{ is a constant}$$

and $x_1, x_2 \geq 0$

We can write the above problem as follows :
$$\text{Optimize } Z = f(x_1, x_2)$$
subject to the constraints
$$h(x_1, x_2) = g(x_1, x_2) - b = 0$$
and
$$x_1, x_2 \geq 0$$
For an unknown function λ, let us define a function
$$L(x_1, x_2, \lambda) = f(x_1, x_2) - \lambda h(x_1, x_2) \qquad \ldots(1)$$

The function $L(x_1, x_2, \lambda)$ defined above in (1) is called the Lagrangian function and λ is called the Lagrangian multiplier.

Further, suppose that L, f and h are differentiable functions w.r.t. x_1, x_2 and λ.

Therefore, the necessary conditions for maxima and minima of f, subject to the condition $h(x_1, x_2) = 0$ are given as below :
$$\frac{\partial L}{\partial x_1} = \frac{\partial f}{\partial x_1} - \lambda \frac{\partial h}{\partial x_1} = 0$$
$$\frac{\partial L}{\partial x_2} = \frac{\partial f}{\partial x_2} - \lambda \frac{\partial h}{\partial x_2} = 0$$
and
$$\frac{\partial L}{\partial \lambda} = -h = 0$$

Hence, we conclude that the necessary and sufficient conditions for maxima and minima of $f = f(x_1, x_2)$ subject to $h(x_1, x_2) = g(x_1, x_2) - b = 0$ are that
$$\frac{\partial f}{\partial x_1} = \lambda \frac{\partial h}{\partial x_1}, \frac{\partial f}{\partial x_2} = \lambda \frac{\partial h}{\partial x_2} \text{ and } h(x_1, x_2) = 0$$

Case II. n-decision variables and one equality constraint

Consider the non-linear programming of n variables.
$$\text{Optimize } Z = f(x_1, x_2, \ldots, x_n) = f(X) \text{ where } X = (x_1, x_2, .., x_n)$$
subject to the constraints $\qquad \ldots(1)$
$$g(x_1, x_2, \ldots, x_n) = g(X) = b, \, b \text{ is a constant}$$
and
$$x_1, x_2, \ldots, x_n \geq 0$$
The NLPP (1) can also be written as
$$\text{Optimize } Z = f(x_1, x_2, \ldots, x_n)$$
subject to the constraints
$$h(x_1, x_2, \ldots, x_n) = g(x_1, x_2, \ldots, x_n) - b$$
i.e.,
$$h(X) = g(X) - b$$
and
$$x_1, x_2, \ldots, x_n \geq 0$$
Let λ be an unknown quantity (called Lagrangian multiplier). Then consider the Lagrangian function.
$$L(x_1, x_2, \ldots, x_n; \lambda) = f(x_1, x_2, \ldots, x_n) - \lambda h(x_1, x_2, \ldots, x_n)$$
i.e.,
$$L(X, \lambda) = f(X) - \lambda h(X)$$
Since, L, f and h are all differentiable, the necessary conditions for maxima and minima of $f(X)$ subject to $h(X) = 0$ are given by
$$\frac{\partial L}{\partial x_j} = \frac{\partial f}{\partial x_j} - \lambda \frac{\partial h}{\partial x_j} = 0 \; \forall \; j = 1, 2, \ldots, n$$
and
$$\frac{\partial L}{\partial \lambda} = -h = 0$$

Hence, we conclude that 'the necessary condition for maximum or minimum of $f(x_1, x_2, \ldots, x_n)$ subject of the constraints

$$h(x_1, x_2, .., x_n) = g(x_1, x_2, ..., x_n) - b = 0$$

are that

$$\frac{\partial f}{\partial x_j} = \lambda \frac{\partial h}{\partial x_j} \ \forall \ j = 1,2,....,n \ \text{ and } h(x_1, x_2, ..., x_n) = 0$$

Case-III : n-decision variables and two equality constraints

Consider the non-linear programming of n variables given by

$$\text{Optimize } Z = f(x_1, x_2, ..., x_n)$$

subject to the constraints

$$g_1(x_1, x_2, ..., x_n) = b_1,$$
$$g_2(x_1, x_2, ..., x_n) = b_2,$$

and $\qquad x_1, x_2, ..., x_n \geq 0, \quad b_1 \text{ and } b_2 \text{ are constants}$

The above problem can also be written as

$$\text{Optimize } Z = f(x_1, x_2, ..., x_n)$$

subject to the constraints

$$h_1(x_1, x_2, ..., x_n) = g_1(x_1, x_2, ..., x_n) - b_1 = 0$$
$$h_2(x_1, x_2, ..., x_n) = g_2(x_1, x_2, ..., x_n) - b_2 = 0$$

and $\qquad x_1, x_2, ..., x_n \geq 0$

Now, taking the Lagrangian multipliers λ_1 and λ_2, the Lagrangian function is given by

$$L(x_1, x_2, ..., x_n; \lambda_1, \lambda_2) = f(x_1, x_2,..., x_n) - \lambda_1 h_1(x_1, x_2, ..., x_n)$$
$$- \lambda_2 h_2(x_1, x_2, ..., x_n)$$

which can also be written as

$$L(X, \lambda_1, \lambda_2) = f(X) - \lambda_1 h_1(X) - \lambda_2 h_2(X)$$
$$X = (x_1, x_2,..., x_n)$$

Further since L, f and h_1, h_2 are all differentiable partially, $w.r.t.$ $x_1, x_2, ..., x_n$ and λ_1, λ_2, therefore, the necessary conditions for maximum or minimum of $f(X)$ subject to the constraints $h_i(X) = 0$, $i = 1, 2$ are given by

$$\frac{\partial L}{\partial x_j} = \frac{\partial f}{\partial x_j} - \lambda_1 \frac{\partial h_1}{\partial x_j} - \lambda_2 \frac{\partial h_2}{\partial x_j} \ \forall \ j = 1,2,....,n$$

$$\frac{\partial L}{\partial \lambda_1} = -h_1(X) = 0 \ \text{ and } \ \frac{\partial L}{\partial \lambda_2} = -h_2(X) = 0$$

Hence, we conclude that the necessary and sufficient conditions for maximum or minimum of $f = f(x_1, x_2, ..., x_n)$ subject to the constraints

$$h_1 = h_1(x_1, x_2, ..., x_n) = g_1(x_1, x_2,..., x_n) - b_1 = 0$$

and $\qquad h_2 = h_2(x_1, x_2, ..., x_n) = g_2(x_1, x_2,..., x_n) - b_2 \text{ are that}$

$$\frac{\partial f}{\partial x_j} = \lambda_1 \frac{\partial h_1}{\partial x_j} + \lambda_2 \frac{\partial h_2}{\partial x_j} \quad \forall \ j = 1,2,...,n$$

$$h_1(x_1, x_2, ..., x_n) = 0 \text{ and } h_2(x_1, x_2, ..., x_n) = 0$$

Case-IV : n-decision variables and m equality constraints ($m < n$)

Consider the non-linear programming problem given by

$$\text{Optimize } Z = f(x_1, x_2, ..., x_n)$$

subject to the constraints

$$g_i(x_1, x_2, ..., x_n) = b_i$$

and $\qquad x_i \geq 0 ; \quad i = 1, 2, ..., m \ (< n)$

The above problem can also be written in the following form

$$\text{Optimize } Z = f(x_1, x_2, ..., x_n)$$

subject to the constraint

$$h_i(x_1, x_2, ..., x_n) = g_i(x_1, x_2, ..., x_n) - b_i = 0$$

and $\qquad x_i \geq 0, \ i = 1, 2, ..., m \ (< n)$

Now, take the Lagrangian multiplier $\lambda = (\lambda_1, \lambda_2, ..., \lambda_m)$, the Lagrangian function $L(x_1, x_2,..., x_n \ ; \ \lambda)$ is given by

$$L(X,\lambda) = f(X) - \sum_{i=1}^{m} \lambda_i h_i(X) \qquad X = (x_1, x_2, ..., x_n)$$

Assuming that L, f and h_i are all differentiable partially w.r.t., $x_1, x_2, ..., x_n$ and $\lambda_1, \lambda_2, ..., \lambda_m$, the necessary conditions for maximum or minimum of $f(X)$ subject to $h_i(X) = 0$ ($i = 1, 2,..., n$) are given by

$$\frac{\partial L}{\partial x_j} = \frac{\partial f}{\partial x_j} - \sum_{i=1}^{m} \lambda_i \frac{\partial h_i}{\partial x_j} = 0 \ \forall \ j = 1, 2, ..., n$$

and

$$\frac{\partial L}{\partial \lambda_i} = -h_i(X) = 0; \ \ i = 1, 2, .., m(< n)$$

Hence, we conclude that the necessary condition for maximum or minimum of $f(X)$ subject to $h_i(X) = g_i(X) - b_i = 0; \ i = 1, 2, ..., n$ are that

$$\frac{\partial f}{\partial x_j} = \sum_{i=1}^{m} \lambda_i \frac{\partial h_i}{\partial x_j} \forall \ j = 1, 2, ..., n$$

and

$$h_i(X) = 0, \ i = 1, 2, ..., m \ (< n)$$

3.10 SUFFICIENT CONDITIONS FOR MAXIMUM OR MINIMUM OF THE OBJECTIVE FUNCTION

Consider the following non-linear programming problem of n decision variables and m equality constraints

$$\text{Optimize } Z = f(x_1, x_2, ..., x_n)$$

subject to the constraints

$$h_i \ (x_1, x_2, ..., x_n) = 0 \ ; \ \ i = 1, 2,..., m \ (<n)$$

Now introduce m Lagrangian multipliers $\lambda = (\lambda_1, \lambda_2, ..., \lambda_m)$ the Lagrangian function is given by

$$L(x_1, x_2, ..., x_n; \ \lambda) = f(x_1, x_2, ..., x_n) - \sum_{i=1}^{m} \lambda_i h_i(x_1, x_2, ..x_n) \quad m < n$$

or

$$L(X, \ \lambda) = f(X) - \sum_{i=1}^{m} \lambda_i h_i(X) \quad m < n$$

where

$$X = (x_1, x_2, ..., x_n) \in R^n$$

We know that the necessary conditions for stationary points of $f(X)$ may be maximum or minimum are given by

$$\frac{\partial L(X, \lambda)}{\partial x_j} = \frac{\partial f(X)}{\partial x_j} - \sum_{i=1}^{m} \lambda_i \frac{\partial h_i(X)}{\partial x_j} = 0 \ \text{ for } j = 1, 2, .., n$$

and

$$\frac{\partial L(X, \lambda)}{\partial \lambda_i} = -\frac{\partial h_i(X)}{\partial \lambda_i} = 0 \qquad \text{for } i = 1, 2, ..., m \ (m < n)$$

The above equations will give the stationary point of $f(X)$ at which the function $f(X)$ may be maximum or minimum.

Now, let us define $\qquad U = \left[\dfrac{\partial h_i(X)}{\partial x_j} \right]_{m \times n} = \begin{bmatrix} \dfrac{\partial h_1}{\partial x_1} & \dfrac{\partial h_1}{\partial x_2} ,....., & \dfrac{\partial h_1}{\partial x_n} \\ \dfrac{\partial h_2}{\partial x_1} & \dfrac{\partial h_2}{\partial x_2} ,...., & \dfrac{\partial h_2}{\partial x_n} \\ \vdots & & \\ \dfrac{\partial h_m}{\partial x_1} & \dfrac{\partial h_m}{\partial x_2} ,...., & \dfrac{\partial h_m}{\partial x_n} \end{bmatrix}_{m \times n}$

and
$$V = \left[\frac{\partial^2 L(X,\lambda)}{\partial x_i x_j}\right]_{n\times n} = \begin{bmatrix} \dfrac{\partial^2 L}{\partial x_1^2} & \dfrac{\partial^2 L}{\partial x_1 \partial x_2} & ,...., & \dfrac{\partial^2 L}{\partial x_1 \partial x_n} \\[2mm] \dfrac{\partial^2 L}{\partial x_2 \partial x_1} & \dfrac{\partial^2 L}{\partial x_2^2} & ,...., & \dfrac{\partial^2 L}{\partial x_2 \partial x_n} \\[1mm] \vdots & \vdots & & \vdots \\[1mm] \dfrac{\partial^2 L}{\partial x_n \partial x_1} & \dfrac{\partial^2 L}{\partial x_n \partial x_2} & ,...., & \dfrac{\partial^2 L}{\partial x_n^2} \end{bmatrix}_{n\times n}$$

and
$$O = \begin{bmatrix} 0 & 0 & .... & 0 \\ 0 & 0 & .... & 0 \\ \vdots & & & \\ 0 & 0 & .... & 0 \end{bmatrix}_{m\times m}$$ is a null matrix.

Then the square matrix H_B of order $(m+n) \times (m+n)$ (called bordered Hessian matrix) is given as below

$$H_B = \left[\begin{array}{c|c} O & U \\ \hline U^T & V \end{array}\right]_{(m+n)\times(m+n)}$$

Then the required sufficient conditions for maximum and minimum stationary points are as follows :

"If (x_0, λ_0) be a stationary point for the function $L(x_0, \lambda)$ and H_{OB} is the value of the corresponding bordered Hessian matrix H_B at this stationary point, then

(i) The point x_0 gives the maximum value of the objective function if starting with principal minor of order $(2m+1)$, the last $(n–m)$ principal minor of H_{OB} are of alternating signs starting with $(-1)^{m+n}$ sign.

(ii) The point x_0 gives the maximum value of the objective function, starting with the principal minor of order $(2m+1)$, the last $(n–m)$ principal minors of H_{OB} are of the sign of $(-1)^m$.

3.10.1 SOME PARTICULAR CASES

Case I : If $n = 2$, $m = 1$: In this case, the order of H_B is 3×3 ($\because n + m = 2 + 1 = 3$)

Since $2m + 1 = 3$, $(-1)^{m+n} = (-1)^3 = -1, n - m = 1$

$$(-1)^m = (-1)^1 = -1$$

Hence, the extreme point x_0 gives maximum value of the objective function if $\Delta_3 = |H_B| < 0$ and minimum value of the obejctive function if $\Delta_3 = |H_B| > 0$

Case II : If $m = 1$, $n = 3$, then order of H_B is 4×4.

Therefore $n + m = 4$

Since $2m + 1 = 3$, $(-1)^{n+m} = 1, (-1)^m = -1$ and $n - m = 2$

Therefore, the extreme point x_0 gives the maximum value of the objective function of $\Delta_3 > 0$ and $\Delta_4 < 0$ and minimum if $\Delta_3 < 0$ and $\Delta_4 < 0$.

Case III : If $n = 3$ and $m = 2$, then order of H_B is 5×5.

Since $n + m = 5$, therefore

$$2m + 1 = 5, (-1)^{n+m} = -1$$
$$(-1)^m = 1, n - m = 1$$

Thus, the extreme point x_0 gives maximum value of the objective function if $\Delta_5 = |H_B| < 0$ and minimum value of the objective function if $\Delta_5 = |H_B| > 0$.

☞ REMARKS

- A stationary point may be an extreme point without satisfying the above conditions.
- If $f(X)$ is a real valued continuous and differentiable function of $X = (x_1, x_2, ..., x_n)$, then Hessian matrix of $f(X)$ is given by

$$H_B(X) = \begin{vmatrix} \dfrac{\partial^2 f}{\partial x_1^2} & \dfrac{\partial^2 f}{\partial x_1 \partial x_2} & ..... & \dfrac{\partial^2 f}{\partial x_1 \partial x_n} \\ \dfrac{\partial^2 f}{\partial x_2 . \partial x_1} & \dfrac{\partial^2 f}{\partial x_2^2} & ..... & \dfrac{\partial^2 f}{\partial x_2 \partial x_n} \\ \vdots & & & \\ \dfrac{\partial^2 f}{\partial x_n \partial x_1} & \dfrac{\partial^2 f}{\partial x_n \partial x_2} & .... & \dfrac{\partial^2 f}{\partial x_n^2} \end{vmatrix}$$

- If all the leading principal minors of Hessian matrix $H(X)$ of $f(X)$ are positive in sign, then the function $f(X)$ is convex, while if the signs of leading principal minors of Hessian matrix $H(X)$ of $f(X)$ are alternatively negative and positive, then $f(X)$ is concave.

> The necessary conditions for maximum or minimum of the objective function in non-linear programming problem with equality constraints also become the sufficient conditions for a maximum of the objective function if it is concave and for a minimum of the objective function if it is convex.

SOLVED EXAMPLES

EXAMPLE 1. *Determine the set of necessary conditions for the non-linear programming problem given below*

$$\text{Maximize } Z = x_1^2 + 3x_2^2 + 5x_3^2$$

subject to the constraints

$$x_1 + x_2 + 3x_3 = 2$$
$$5x_1 + 2x_2 + x_3 = 5$$

and $\quad x_1, x_2, x_3 \geq 0$

SOLUTION. We have $f(X) = Z = x_1^2 + 3x_2^2 + 5x_3^2$, $X = (x_1, x_2, x_3)$

$$g_1(X) = x_1 + x_2 + 3x_3 = 2$$

and $\quad g_2(X) = 5x_1 + 2x_2 + x_3 = 5$

Define $h_1(X)$ and $h_2(X)$ such that

$$h_1(X) = g_1(X) - 2 = 0 \text{ and } h_2(X) = g_2(X) - 5 = 0$$

Let λ_1, λ_2 be the multipliers then Lagrangian function $L(X, \lambda)$ is given by

$$L(X, \lambda) = f(X) - \lambda_1 h_1(X) - \lambda_2 h_2(X)$$

$$= (x_1^2 + 3x_2^2 + 5x_3^2) - \lambda_1(x_1 + x_2 + 3x_3 - 2) - \lambda_2(5x_1 + 2x_2 + x_3 - 5)$$

Therefore, the required necessary conditions for optimality can be obtained as follows :

$$\frac{\partial L}{\partial x_1} = 2x_1 - \lambda_1 - 5\lambda_2 = 0; \qquad \frac{\partial L}{\partial x_2} = 6x_2 - \lambda_1 - 2\lambda_2 = 0$$

$$\frac{\partial L}{\partial x_3} = 10x_3 - 3\lambda_1 - \lambda_3 = 0; \qquad \frac{\partial L}{\partial \lambda_1} = -(x_1 + x_2 + 3x_3 - 2) = 0$$

and $\quad \dfrac{\partial L}{\partial \lambda_2} = -(5x_1 + 2x_2 + x_3 - 5) = 0$

EXAMPLE 2. *Solve the following non-linear programming problem by the method of Lagrangian multipliers.*

$$\text{Maximize } Z = 6x_1 + 8x_2 - x_1^2 - x_2^2$$

subject to the constraints

$$4x_1 + 3x_2 = 16; \qquad 3x_1 + 5x_2 = 15$$

and $\qquad x_1, x_2 \geq 0$

SOLUTION. We have

$$f(X) = Z = 6x_1 + 8x_2 - x_1^2 - x_2^2$$
$$h_1(X) = 4x_1 + 3x_2 - 16 = 0$$
$$h_2(X) = 3x_1 + 5x_2 - 15 = 0$$

Taking λ_1, λ_2 as the multipliers, the Lagrangian function is given by

$$L(X, \lambda) = 6x_1 + 8x_2 - x_1^2 - x_2^2 - \lambda_1(4x_1 + 3x_2 - 16) - \lambda_2(3x_1 + 5x_2 - 15)$$

For the optimality, the necessary conditions are given by

$$\frac{\partial L}{\partial x_1} = 6 - 2x_1 - 4\lambda_1 - 3\lambda_2 = 0 \qquad \qquad ...(1)$$

$$\frac{\partial L}{\partial x_2} = 8 - 2x_2 - 3\lambda_1 - 5\lambda_2 = 0 \qquad \qquad ...(2)$$

$$\frac{\partial L}{\partial \lambda_1} = -(4x_1 + 3x_2 - 16) = 0 \qquad \qquad ...(3)$$

and $\qquad \dfrac{\partial L}{\partial \lambda_2} = -(3x_1 + 5x_2 - 15) = 0 \qquad \qquad ...(4)$

From the above equations, we can easily find

$$x_1 = \frac{35}{11}, x_2 = \frac{12}{11}, \lambda_1 = -\frac{212}{121}, \lambda_2 = \frac{268}{121}$$

Therefore, the stationary point is given by

$$x_0 = (x_1, x_2) = \left(\frac{35}{11}, \frac{12}{11} \right)$$

$$\lambda_1 = -\frac{212}{121}, \lambda_2 = \frac{268}{121}$$

Here, $n = 2, m = 2$, so Hessian matrix of the objective function $f(X)$ at x_0 is given by

$$H(X) = \begin{bmatrix} \dfrac{\partial^2 f}{\partial x_1^2} & \dfrac{\partial^2 f}{\partial x_1 \partial x_2} \\ \dfrac{\partial^2 f}{\partial x_2 \partial x_1} & \dfrac{\partial^2 f}{\partial x_2^2} \end{bmatrix} = \begin{bmatrix} -2 & 0 \\ 0 & -2 \end{bmatrix}$$

and principal minors of $H(x_0)$ are

$$D_1 = -2, D_2 = \begin{vmatrix} -2 & 0 \\ 0 & -2 \end{vmatrix} = 4$$

$\Rightarrow$ signs are alternatively negative and positive
$\Rightarrow$ $f(X)$ is negative definite
$\Rightarrow$ $f(X)$ is concave
Hence $f(X)$ is minimum at the point $\left(\dfrac{35}{11}, \dfrac{12}{11} \right)$

and maximum value of $f(X)$ is given by

$$\frac{1997}{121} = 16.5$$

EXAMPLE 3. *Solve the following non-linear programming problem by using the Lagrangian multiplier method*

$$\text{Min}.\, Z = 2x_1^2 + x_2^2 + 3x_3^2 + 10x_1 + 8x_2 + 6x_3 - 100$$

subject to the constraints

$$x_1 + x_2 + x_3 = 20$$

and $\quad x_1, x_2, x_3 \geq 0$

SOLUTION. Let λ be the Lagrangian multiplier, then define the Lagrangian function such that

$$L(X, \lambda) = f(X) - \lambda h(X)$$

$$= (2x_1^2 + x_2^2 + 3x_3^2 + 10x_1 + 8x_2 + 6x_3 - 100) - \lambda(x_1 + x_2 + x_3 - 20)$$

Clearly, the necessary conditions for $f(X) = Z$ to be maximum or minimum are given by

$$\frac{\partial L}{\partial x_1} = 0 \Rightarrow 4x_1 + 10 - \lambda = 0 \qquad \qquad \dots(1)$$

$$\frac{\partial L}{\partial x_2} = 0 \Rightarrow 2x_2 + 8 - \lambda = 0 \qquad \qquad \dots(2)$$

$$\frac{\partial L}{\partial x_3} = 0 \Rightarrow \frac{\partial L}{\partial x_3} = 0 \Rightarrow 6x_3 + 6 - \lambda = 0 \qquad \qquad \dots(3)$$

and $\quad \dfrac{\partial L}{\partial \lambda} = 0 \Rightarrow -(x_1 + x_2 + x_3 - 20) = 0 \qquad \qquad \dots(4)$

On solving the above equations (1) to (3), we can easily find that

$$x_1 = \frac{-10 + \lambda}{4}, \; x_2 = \frac{-8 + \lambda}{2} \text{ and } x_3 = \frac{-6 + \lambda}{6}$$

Using all these values in (4), we get $\lambda = 30$

Hence, $\quad x_1 = 5, x_2 = 11, x_3 = 4$

$\therefore$ The stationary point is $(x_1, x_2, x_3) = (5, 11, 4)$

Now, we have to check the maximum or minimum of $f(X)$ at (5, 11, 4)

We proceed as usual, we have

$$H_B = \begin{bmatrix} 0 & \frac{\partial h}{\partial x_1} & \frac{\partial h}{\partial x_2} & \frac{\partial h}{\partial x_3} \\ \frac{\partial h}{\partial x_1} & \frac{\partial^2 L}{\partial x_1^2} & \frac{\partial^2 L}{\partial x_1 \partial x_2} & \frac{\partial^2 L}{\partial x_1 \partial x_3} \\ \frac{\partial h}{\partial x_2} & \frac{\partial^2 L}{\partial x_2 \partial x_1} & \frac{\partial^2 L}{\partial x_2^2} & \frac{\partial^2 L}{\partial x_2 \partial x_3} \\ \frac{\partial h}{\partial x_3} & \frac{\partial^2 L}{\partial x_3 \partial x_1} & \frac{\partial^2 L}{\partial x_3 \partial x_2} & \frac{\partial^2 L}{\partial x_3^2} \end{bmatrix} = \begin{bmatrix} 0 & 1 & 1 & 1 \\ 1 & 4 & 0 & 0 \\ 1 & 0 & 2 & 0 \\ 1 & 0 & 0 & 6 \end{bmatrix}$$

Here, we have

$$n = 3, m = 1 \Rightarrow 2m + 1 = 3 \text{ and } n - m = 3 - 1 = 2$$

So, we have to consider last two principal minors given by

$$D_3 = \begin{vmatrix} 0 & 1 & 1 \\ 1 & 4 & 0 \\ 1 & 0 & 2 \end{vmatrix} = -6 \text{ and } D_4 = \begin{vmatrix} 0 & 1 & 1 & 1 \\ 1 & 4 & 0 & 0 \\ 1 & 0 & 2 & 0 \\ 1 & 0 & 0 & 6 \end{vmatrix} = -44$$

Since signs of D_3 and D_4 are of sign $(-1)^m = (-1)^1 = -1$

$\Rightarrow$ $f(X)$ is minimum at the point (5, 11, 4) and the minimum value of $f(X)$ is given by $Z = 281$.

EXAMPLE 4. *Using Lagrangian multiplier method solve the following non-linear programming problem.*

$$\text{Min}. \, Z = x_1^2 + x_2^2 + x_3^2$$

subject to the constraints

$$4x_1 + x_2^2 + 2x_3 = 14$$

and $\qquad x_1, x_2, x_3 \geq 0$

SOLUTION. We have

$$f(X) = Z = x_1^2 + x_2^2 + x_3^2$$

$$h(X) = 4x_1 + x_2^2 + 2x_3 - 14 = 0$$

If λ is the multiplier, then Lagrangian function is given by

$$L(X, \lambda) = f(X) - \lambda h(X)$$
$$= x_1^2 + x_2^2 + x_3^2 - \lambda(4x_1 + x_2^2 + 2x_3 - 14)$$

For the minimum of $f(X)$, the necessary conditions are

$$\frac{\partial L}{\partial x_1} = 0 \Rightarrow 2x_1 - 4\lambda = 0 \qquad\qquad\qquad ...(1)$$

$$\frac{\partial}{\partial} = 0 \Rightarrow 2x_2 - 2\lambda x_2 = 0 \qquad\qquad\qquad ...(2)$$

$$\frac{\partial L}{\partial x_3} = 0 \Rightarrow 2x_3 - 2\lambda = 0 \qquad\qquad\qquad ...(3)$$

and $\qquad \frac{\partial L}{\partial \lambda} = 0 \Rightarrow -(4x_1 + x_2^2 + 2x_3 - 14) = 0 \qquad\qquad ...(4)$

From (2), we have

$$x_2(1 - \lambda) = 0 \Rightarrow x_2 = 0 \text{ or } \lambda = 1$$

From (1) and (3), we have

$$x_1 = 2\lambda \text{ and } x_3 = \lambda$$

So, when $x_2 = 0$, then from (4), $\lambda = 1.4$. Thus $x_1 = 2.8$ and $x_3 = 1.4$ and when $\lambda = 1$ then from (1) and (3), we get $x_2 = 2, x_3 = 1$

Then from (4)

$$x_2 = 2$$

Therefore, we have the following stationary points:

$$x_0 = (x_1, x_2, x_3) = (2.8, 0, 1.4), \lambda = 1.4$$

and $\qquad x_0 = (x_1, x_2, x_3) = (2, 2, 1), \lambda = 1$

Now, for the sufficient condition of maxima or minima of $f(X)$ at these stationary points, we proceed as follows.

Let

$$H_B = \begin{bmatrix} 0 & \frac{\partial h}{\partial x_1} & \frac{\partial h}{\partial x_2} & \frac{\partial h}{\partial x_3} \\ \frac{\partial h}{\partial x_1} & \frac{\partial^2 L}{\partial x_1^2} & \frac{\partial^2 L}{\partial x_1 \partial x_2} & \frac{\partial^2 L}{\partial x_1 \partial x_3} \\ \frac{\partial h}{\partial x_2} & \frac{\partial^2 L}{\partial x_2 \partial x_1} & \frac{\partial^2 L}{\partial x_2^2} & \frac{\partial^2 L}{\partial x_2 \partial x_3} \\ \frac{\partial h}{\partial x_3} & \frac{\partial^2 L}{\partial x_3 \partial x_1} & \frac{\partial^2 L}{\partial x_3 \partial x_2} & \frac{\partial^2 L}{\partial x_3^2} \end{bmatrix} = \begin{bmatrix} 0 & 4 & 2x_2 & 2 \\ 4 & 2 & 0 & 0 \\ 2x_2 & 0 & 2 & 0 \\ 2 & 0 & 0 & 2 \end{bmatrix}.$$

Now,

$$H_B \text{ at } x_0 = (2.8, 0, 1.4) = \begin{bmatrix} 0 & 4 & 0 & 2 \\ 4 & 2 & 0 & 0 \\ 0 & 0 & 2 & 0 \\ 2 & 0 & 0 & 2 \end{bmatrix}$$

Here, $n = 3, m = 1 \Rightarrow n - m = 3 - 1 = 2$ and $2m + 1 = 3$

Now, we have to check the sign of two principal minors D_3 and D_4

$$\therefore \quad D_3 = \begin{vmatrix} 0 & 4 & 0 \\ 4 & 2 & 0 \\ 0 & 0 & 2 \end{vmatrix} = -32 \text{ and } D_4 = \begin{vmatrix} 0 & 4 & 0 & 2 \\ 4 & 2 & 0 & 0 \\ 0 & 0 & 2 & 0 \\ 2 & 0 & 0 & 2 \end{vmatrix} = -80$$

$\Rightarrow$ The sign of D_3 and D_4 are same as the sign of $(-1)^m = (-1)^1 = -1$

$\Rightarrow f(X)$ has local minimum at the point $(2.8, 0, 1.4)$ and minimum value of $Z = 9.8$

Similarly

$$H_B \text{ at } x_0 = (2, 2, 1) = \begin{bmatrix} 0 & 4 & 4 & 2 \\ 4 & 2 & 0 & 0 \\ 4 & 0 & 2 & 0 \\ 2 & 0 & 0 & 2 \end{bmatrix}$$

and $D_3 = -64 < 0$ and $D_4 = -144 < 0$

$\Rightarrow \quad f(X)$ has local minimum at the point $(2, 2, 1)$ and $\min.Z = 9$

Finally since, $\quad 9 < 9.8$. Therefore, minimum value of Z is 9 at $(2, 2, 1)$

3.11 SOLUTION OF NON-LINEAR PROGRAMMING PROBLEMS WHEN CONSTRAINTS ARE NOT EQUALITY CONSTRAINTS: KUHN-TUCKER CONDITIONS

3.11.1 KUHN-TUCKER NECESSARY CONDITIONS

Consider the non-linear programming problem

$$\text{Max.} Z = f(X), \quad X = (x_1, x_2, ..., x_n)$$

subject to the constraints

$$g_i(X) \le b_i \quad i = 1, 2, ..., m(< n) \, ; \, b_i\text{'s are constants}$$

Define $\qquad h_i(X) = g(X) - b_i \le 0$

Now, introducing the slack variables $s_1, s_2, ..., s_m$, then above inequality constraints reduces to

$$h_i(X) + s_i^2 = 0 \qquad i = 1, 2, ..., m$$

Here, addition of s_i^2 ensure the quantity added to $h_i(x)$ to be non-negative

Then the given problem becomes

$$\text{Max.} Z = f(X), \, X = (x_1, x_2, ..., x_n)$$

subject to the constraints

$$h_i(X) + s_i^2 = 0 \quad \forall i = 1, 2, ..., m$$

and $\qquad\qquad X \ge 0$

which is clearly a non-linear programming problem in $(n + m)$ variables $x_j, s_i \, i = 1, 2, .., m;$ $j = 1, 2, .., n$ with m equality constraints.

Now introduce the Lagrangian multipliers $\lambda_1, \lambda_2, ..., \lambda_m$, then Lagrangian function is defined as follows:

$$L(X,S,\lambda) = f(S) - \sum_{i=1}^{m} \lambda_i [h_i(X) + s_i^2]$$

$$X = (x_1, x_2,...,x_n) \quad S = (s_1, s_2,...s_m) \text{ and } \lambda = (\lambda_1, \lambda_2,...,\lambda_m)$$

If L, f and h_i are all differentiable partially w.r.t. $x_1, x_2, ..., x_n$; $\lambda_1, \lambda_2, ..., \lambda_m$; $s_1, s_2,..., s_m$ then the necessary conditions for the stationary points are given by

$$\frac{\partial L}{\partial x_j} = \frac{\partial f}{\partial x_j} - \sum_{i=1}^{m} \lambda_i \frac{\partial h_i}{\partial x_j} = 0 \ \forall \ j = 1,2,...,n \qquad ...(1)$$

$$\frac{\partial L}{\partial \lambda_i} = -[h_i(X) + s_i^2] = 0 \ ; \quad i = 1, 2, ..., m \qquad ...(2)$$

and $$\frac{\partial L}{\partial s_i} = -2s_i \lambda_i = 0 \ ; \qquad i = 1, 2, ..., m \qquad ...(3)$$

From (3), we have either $\quad s_i = 0$ or $\lambda_i = 0$

Now, if $s_i = 0$, then (2) implies

$$h_i(X) = 0 \Rightarrow \lambda_i = 0 \text{ or } s_i = 0$$

$$\Rightarrow \qquad\qquad \lambda_i = 0 \text{ or } h_i(X) = 0$$

$$\therefore \qquad\qquad \lambda_i h_i(X) = 0 \qquad\qquad\qquad ...(4)$$

Since, $\qquad\qquad s_i^2 \geq 0$; therefore (2) implies $h_i(x) \leq 0$

When $h_i(X) < 0$ then (4) implies $\lambda_i = 0$ and when $\lambda_i > 0$, $h_i(X) = 0$

But from (4), $h_i(X) = 0$ therefore λ_i is unrestricted in sign

If $\lambda_i \neq 0$, then clearly $s_i = 0$, then from (2)

$$h_i(X) = g_i(X) - b_i = 0 \Rightarrow g_i(X) = b_i$$

Here λ_i represent the rate of change of f w.r.t. b_i

Therefore, $$\frac{\partial f}{\partial b_i} = \lambda_i$$

Now, as the RHS of $h_i(X) \leq 0$ increases about zero, the solution space becomes less constraints $\Rightarrow f$ can not decreases

$$\Rightarrow \qquad\qquad \frac{\partial f}{\partial b_i} = \lambda_i \nleq 0$$

$$\Rightarrow \qquad\qquad \lambda_i \geq 0$$

Therefore, when $h(X) < 0$, $\lambda = 0$ and when $\lambda > 0$, $h(X) = 0$

Hence, we conclude that the Kuhn-Tucker necessary conditions for X to be a point of maximum for $f(X)$ subject to $h_i(X) = g_i(X) - b_i \leq 0$ are given by

$$\frac{\partial f}{\partial x_j} - \sum_{i=1}^{m} \lambda_i \frac{\partial h_i}{\partial x_j} = 0 \qquad \forall \ j = 1,2,...,n$$

and $\lambda_i h_i(X) = 0$, $h_i(X) \leq 0$, $\lambda_i \geq 0$, $i = 1,2,...,m$

and the Kuhn-Tucker conditions for the point X to be a point of minimum for $f(X)$ subject to $h_i(X) = g_i(X) - b_i \geq 0$ are given by

$$\frac{\partial f}{\partial x_j} - \sum_{i=1}^{m} \lambda_i \frac{\partial h_i}{\partial x_j} = 0 \qquad \forall \ j = 1,2,...,n$$

and $\lambda_i h_i(X) = 0$, $h_i(X) \geq 0$ and $\lambda_i \geq 0$, $i = 1,2,...,m$.

☞ **REMARKS**

- In case of maximization, NLPP convert all the constraints in $\leq$ type and in case of minimization, convert all the constraints in $\geq$ type.
- The Lagrangian multiplier, λ_i corresponding to $h_i(X) = 0$ must be unrestricted in sign.
- If the given problem is of minimization (having the constraints of the form $g_i(X) \geq 0$) then $\lambda_i \leq 0$ and if the problem is of maximization (having the constraints of the form $g_i(X) \leq 0$) then $\lambda_i \geq 0$.

3.11.2 KUHN-TUCKER SUFFICIENT CONDITIONS

"The Kuhn-Tucker necessary conditions for a non-linear programming problem given by

Max. $f(X)$ subject to the constraints

$$h_i(X) = g_i(X) - b_i \leq 0 \qquad i = 1, 2, ..., m \text{ and } X \geq 0$$

will also be the sufficient condition for maximum of $f(X)$ if,

(i) $f(X)$ is concave and

(ii) $h_i(X)$ are convex *i.e.*, $-h_i(X)$ is also concave.

The Lagrangian function of the problem can be written as

$$L(X, s, \lambda) = f(X) - \sum_{i=1}^{m} \lambda_i \left[h_i(X) + s_i^2 \right]$$

where $X = (x_1, x_2, ..., x_n)$, $S = (s_1, s_2, ..., s_m)$ and $\lambda = (\lambda_1, \lambda_2, ... \lambda_m)$ and s_i are slack variables used to convert inequality constraints to equality and λ_i are the Lagrangian multipliers.

Using the slack variables s_i, we get

$$h_i(X) + s_i^2 = 0, \, i = 1, 2, ... m$$

and from the necessary conditions, we have $\lambda_i h_i(X) = 0, \, i = 1, 2, ..., m$

Therefore,

$$\lambda_i s_i^2 = -\lambda_i h_i(X) = 0, \, i = 1, 2, ..., m$$

Now, since $h_i(X)$ are convex functions of X and $\lambda_i \geq 0$

therefore, $\lambda_i h_i(X)$ is convex, *i.e.*, $-\lambda_i h_i(X)$ is concave.

$\Rightarrow \quad -\sum_{i=1}^{m} \lambda_i h_i(X)$ is concave function of X.

$\Rightarrow \quad f(X) - \sum_{i=1}^{m} \lambda_i h_i(X)$ is concave function of X.

$\Rightarrow \quad f(X) - \sum_{i=1}^{m} \lambda_i \left[h_i(X) + s_i^2 \right]$ is concave function of X. $\qquad [\because \lambda_i s_i^2 = 0]$

$\Rightarrow \quad L(X, S, \lambda)$ is concave.

Thus, we conclude that the necessary conditions for maximum of $f(X)$ at an extreme point implies that $L(X, S, \lambda)$ also have the same extreme point. Since $L(X, S, \lambda)$ is concave, its derivative must be zero at one point, which give the absolute maximum value of $f(X)$.

Also, the Kuhn-Tucker necessary conditions for the minimization of non-linear programming problem given by

Min. $f(X)$.

subject to the constraints

$$h_i(X) = g_i(X) - b_i \geq 0, \, i = 1, 2, ..., m$$

and $X \geq 0$

are also the sufficient conditions for minimum of $f(X)$ if

(i) $f(X)$ is convex, and

(ii) $h_i(X)$ are also convex *i.e.*, $-h_i(X)$ is concave.

If in the minimizing non-linear programming problem, the constraints are taken of the type $h_i(X) \leq 0$ then the Kuhn-Tucker necessary conditions for the optimality of the objective function will be given by

$$\frac{\partial f}{\partial x_j} - \sum_{i=1}^{m} \lambda_i \frac{\partial h_i(X)}{\partial x_j} = 0 \qquad \forall\, j = 1, 2, \dots, n$$

$$\lambda_i h_i(X) = 0,\ h_i(X) \leq 0 \ \text{and} \ \lambda_i \leq 0 \,\forall\, i = 1, 2, \dots, m$$

and these conditions will be sufficient if $f(X)$ and all $h_i(X)$ are convex function of X.

 ## SOLVED EXAMPLES

EXAMPLE 1. *Find the value of x_1 and x_2 which*

   Maximize $Z = 12x_1 + 21x_2 + 2x_1x_2 - 2x_1^2 - 2x_2^2$

subject to the constraints

$$x_2 \leq 8$$
$$x_1 + x_2 \leq 10$$

and $x_1, x_2 \geq 0$

SOLUTION. We have $f(X) = f(x_1, x_2) = 12x_1 + 21x_2 + 2x_1x_2 - 2x_1^2 - 2x_2^2$

$$h_1(X) = h_1(x_1, x_2) = x_2 - 8 \leq 0$$
$$h_2(X) = h_2(x_1, x_2) = x_1 + x_2 - 10 \leq 0$$

Then the Lagrangian function $L(X, S, \lambda)$ is given by

$$L(X, S, \lambda) = f(x) - \lambda_1 \left[h_1(x) + s_1^2 \right] - \lambda_2 \left[h_2(x) + s_2^2 \right]$$

Then we have the following Kuhn-Tucker necessary conditions

(i) $\dfrac{\partial f}{\partial x_j} - \sum\limits_{i=1}^{2} \lambda_i \dfrac{h_i(X)}{\partial x_j} = 0$

   $\Rightarrow$ $12 + 2x_2 - 4x_1 - \lambda_2 = 0$ and $21 + 2x_1 - 4x_2 - \lambda_1 - \lambda_2 = 0$

(ii) $\lambda_i h_i(X) = 0,\ i = 1, 2$

   $\Rightarrow$ $\lambda_1(x_2 - 8) = 0$ and $\lambda_2(x_1 + x_2 - 10) = 0$

(iii) $h_i(X) \leq 0$

   $\Rightarrow$ $x_2 - 8 \leq 0$ and $x_1 + x_2 - 10 \leq 0$

(iv) $\lambda_i \geq 0 \ \ i = 1, 2$

Then we have the following cases:

Case 1: If $\lambda_1 = 0,\ \lambda_2 = 0$ then from (i) we have

$$12 + 2x_2 - 4x_1 = 0 \text{ and } 21 + 2x_1 - 4x_2 = 0$$

On solving we get $x_1 = \dfrac{15}{2}, x_2 = 9$

But x_1 and x_2 do not satisfy (iii) and therefore it may be discarded.

Case 2: If $\lambda_1 \neq 0,\ \lambda_2 = 0$ then from (i) and (ii), we have

$$x_1 + x_2 = 10$$
$$2x_2 - 4x_1 = -12$$

$$2x_1 - 4x_2 = -12 + \lambda_1$$

On solving the above equations, we get $x_1 = 2$, $x_2 = 8$, $\lambda_1 = -16$

which violates the condition (iv) and therefore it may also be discarded.

Case 3: **If** $\lambda_1 \neq 0, \lambda_2 \neq 0$ **then from (ii)**

$$x_2 - 8 = 0 \qquad \Rightarrow \quad x_2 = 8$$
$$x_1 + x_2 - 10 = 0 \Rightarrow \quad x_1 = 2$$

Putting these values in (i) we get $\lambda_1 = -27$ and $\lambda_2 = 20$

which violates the condition (iv) and therefore may be discarded.

Case 4: **If** $\lambda_1 = 0, \lambda_2 \neq 0$ **then from (i) and (ii) we have**

$$2x_2 - 4x_1 = -12 + \lambda_2$$
$$2x_1 - 4x_2 = -21 + \lambda_2$$
$$x_1 + x_2 = 10$$

On solving the above equations, we get

$$x_1 = \frac{17}{4}, x_2 = \frac{23}{4}, \lambda_2 = \frac{13}{4}$$

which does not violate any of the Kuhn-Tucker conditions and therefore accepted. Hence, the optimum solution of the given NLPP is

$$x_1 = \frac{17}{4}, x_2 = \frac{23}{4}, \lambda_1 = 0 \text{ and } \lambda_2 = \frac{13}{4}$$

and $\qquad \text{Max.} Z = \dfrac{1734}{16}$

EXAMPLE 2. *Solve the following non-linear programming problem.*

$$\text{Min } Z = (x_1 - 2)^2 + (x_2 - 1)^2$$

subject to the constraints

$$x_1^2 - x_2 \leq 0, \ x_1 + x_2 \leq 2$$

and $\qquad x_1, x_2 \geq 0$

SOLUTION. We have $\qquad Z = f(X) = f(x_1, x_2) = (x_1 - 2)^2 + (x_2 - 1)^2$

$$h_1(X) = -x_1^2 + x_2 \geq 0, \ h_2(X) = 2 - x_1 - x_2 \geq 0$$

and $\qquad X = (x_1, x_2) \geq 0$

Since it is a problem of minimization, therefore the signs in inequality constraints are taken as $\geq$.

Now, the Hessian matrix

$$H_B = \begin{bmatrix} \dfrac{\partial^2 f}{\partial x_1^2} & \dfrac{\partial^2 f}{\partial x_1 \partial x_2} \\ \dfrac{\partial^2 f}{\partial x_2 \partial x_1} & \dfrac{\partial^2 f}{\partial x_2^2} \end{bmatrix} = \begin{bmatrix} 2 & 0 \\ 0 & 2 \end{bmatrix}$$

whose principal minors are $D_1 = 2$, $D_2 = \begin{vmatrix} 2 & 0 \\ 0 & 2 \end{vmatrix} = 4$, clearly both are positive.

$\Rightarrow \quad f(X)$ is a convex function.

Also, $h_1(X)$ and $h_2(X)$ are convex functions of X.

Therefore the Kuhn-Tucker necessary conditions for minimum of $f(X)$ will also be the sufficient conditions.

Here, the Kuhn-Tucker necessary conditions for the minimum of $f(X)$ are given as below:

$$\frac{\partial f}{\partial x_j} = \sum_{i=1}^{2} \lambda_i \frac{\partial h_i}{\partial x_j}, \ j = 1,2$$

$$\lambda_1 h_1(X) = 0, \quad \lambda_2 h_2(X) = 0$$

$$h_1(X) \geq 0, \quad h_2(X) \geq 0 \text{ and } \lambda_1, \lambda_2 \geq 0.$$

which implies

$$2(x_1 - 2) = -2x_1\lambda_1 - \lambda_2 \qquad \qquad \text{...(1)}$$

$$2(x_2 - 1) = \lambda_1 - \lambda_2 \qquad \qquad \text{...(2)}$$

$$\lambda_1(-x_1^2 + x_2) = 0 \qquad \qquad \text{...(3)}$$

$$\lambda_2(2 - x_1 - x_2) = 0 \qquad \qquad \text{...(4)}$$

$$-x_1^2 + x_2 \geq 0 \qquad \qquad \text{...(5)}$$

$$2 - x_1 - x_2 \geq 0 \qquad \qquad \text{...(6)}$$

and

$$\lambda_1, \lambda_2 \geq 0 \qquad \qquad \text{...(7)}$$

Now, we have the following cases:

Case 1: If $\lambda_1 = \lambda_2 = \mathbf{0}$: Here, from (1) and (2), we get $x_1 = 2, x_2 = 1$

These values do not satisfy (5) and (6) and so are discarded.

Case 2: If $\lambda_1 = \mathbf{0}, \lambda_2 \neq \mathbf{0}$: In this case from (1) and (2)

$$2(x_1 - 2) = -\lambda_2 \text{ and } 2(x_2 - 1) = -\lambda_2 \Rightarrow x_1 - x_2 - 1 = 0 \qquad \text{...(8)}$$

and from (4), $\qquad -x_1 - x_2 + 2 = 0 \qquad \qquad \text{...(9)}$

On solving (8) and (9), we get $x_1 = \dfrac{3}{2}, x_2 = \dfrac{1}{2}$

which does not satisfy (5) and hence discarded.

Case 3: If $\lambda_1 \neq \mathbf{0}, \lambda_2 = \mathbf{0}$: Here, from (1) and (2) we have

$$x_1 - 2 = -x_1\lambda_1 \text{ and } 2(x_2 - 1) = \lambda_1$$

$$\Rightarrow \quad -x_1 + 2x_1x_2 - 2 = 0 \qquad \qquad \text{...(10)}$$

Again from (3) $\qquad -x_1^2 + x_2 = 0 \qquad \qquad \text{...(11)}$

From (10) and (11) we get

$$2x_1^3 - x_1 - 2 = 0$$

$$\Rightarrow \qquad \qquad x_1 = 1.52 \text{ and } x_2 = 2.31$$

These values does not satisfy (6) and hence discarded.

Case 4: If $\lambda_1 \neq \mathbf{0}$ and $\lambda_2 \neq \mathbf{0}$: Here, from (3) and (4) we have

$$-x_1^2 + x_2 = 0 \text{ and } 2 - x_1 - x_2 = 0$$

$$\Rightarrow \qquad \qquad x_2 = x_1^2 \text{ and } x_1^2 + x_1 - 2 = 0$$

$$\Rightarrow \qquad \qquad x_2 = x_1^2 \text{ and } (x_1 - 1)(x_1 + 2) = 0$$

$$\Rightarrow \qquad \qquad x_1 = 1, x_2 = 1$$

Putting the values of x_1 and x_2 in (1) and (2), we get,

$$2\lambda_1 + \lambda_2 = 2 \text{ and } \lambda_1 - \lambda_2 = 0$$

On solving, we get,

$$\lambda_1 = \frac{2}{3} \text{ and } \lambda_2 = \frac{2}{3} \geq 0$$

Hence, the optimal solution is given by
$$x_1 = 1, \ x_2 = 1, \ Min.Z = 1$$

EXAMPLE 3. *Solve the following non-linear programming problem:*
$$Max.Z = f(x_1, x_2) = 3.6x_1 - 0.4x_1^2 + 1.6x_2 - 0.2x_2^2$$
subject to the constraints
$$2x_1 + x_2 \le 10 \text{ and } x_1, x_2 \ge 0$$

SOLUTION. We have,
$$f(X) = f(x_1, x_2) = 3.6x_1 - 0.4x_1^2 + 1.6x_2 - 0.2x_2^2$$
$$h(X) = 2x_1 + x_2 - 10 \le 0$$
$$X = (x_1, x_2) \ge 0$$
The Hessian matrix is given by
$$H_B = \begin{bmatrix} \dfrac{\partial^2 f}{\partial x_1^2} & \dfrac{\partial^2 f}{\partial x_1 \partial x_2} \\ \dfrac{\partial^2 f}{\partial x_2 \partial x_1} & \dfrac{\partial^2 f}{\partial x_2^2} \end{bmatrix} = \begin{bmatrix} -0.8 & 0 \\ 0 & -0.4 \end{bmatrix}$$
whose principal minors are given by
$$D_1 = -0.8 < 0, \ D_2 = \begin{vmatrix} -0.8 & 0 \\ 0 & -0.4 \end{vmatrix} = 0.32 > 0$$
which are clearly have alternate signs.
$$\Rightarrow \quad f(X) \text{ is a concave function.}$$
and $h(X) = 2x_1 + x_2 - 10$ is a convex function
Therefore, the Kuhn-Tucker necessary conditions for maximum of $f(X)$ are given by
$$\frac{\partial f}{\partial x_1} = \lambda \frac{\partial h}{\partial x_1}, \frac{\partial f}{\partial x_2} = \lambda \frac{\partial h}{\partial x_2}$$
$$\lambda h(X) = 0, \ h(X) \le 0 \text{ and } \lambda \ge 0$$
which implies

$3.6 - 0.8 \, x_1 = 2\lambda$	...(1)
$1.6 - 0.4 \, x_2 = \lambda$	...(2)
$\lambda(2x_1 + x_2 - 10) = 0$	...(3)
$2x_1 + x_2 - 10 \le 0$	...(4)
and $\qquad \lambda \ge 0$	...(5)

Now, from (3) we have either $\lambda = 0$ or $2x_1 + x_2 - 10 = 0$
If $\lambda = 0$ then from (1) and (2), we have,
$$3.6 - 0.8x_1 = 0 \quad \Rightarrow \ x_1 = 4.5$$
and $\qquad 1.6 - 0.4 \, x_2 = 0 \quad \Rightarrow \ x_2 = 4$
These values does not satisfy (4) therefore $\lambda \ne 0$ which implies $2x_1 + x_2 - 10 = 0$...(5)
Now, from (1) and (2), we have
$$x_1 = \frac{1.8 - \lambda}{0.4}, \ x_2 = \frac{1.6 - \lambda}{0.4}$$
Then, from (5) we get $\lambda = 0.4$
$$\Rightarrow \qquad x_1 = 3.5 \text{ and } x_2 = 3$$
Clearly, these values of x_1 and x_2 also satisfy (4)
Also, $\qquad \lambda = 0.4 > 0$

$\Rightarrow$ (5) is also satisfied

Hence, the optimum solution of the given problem is

$$x_1 = 3.5, \ x_2 = 3 \text{ and Max. } Z = 10.7$$

EXAMPLE 4. *Solve the following non-linear programming problem*

$$\textbf{Max .Z} = -\boldsymbol{x_1^2 - x_2^2 - x_3^2 + 4x_1 + 6x_2}$$

subject to the constraints

$$\boldsymbol{x_1 + x_2 \leq 2; \ 2x_1 + 3x_2 \leq 12 \text{ and } x_1, x_2 \geq 0}$$

SOLUTION. We have, $Z = f(X) = f(x_1, x_2, x_3) = -x_1^2 - x_2^2 - x_3^2 + 4x_1 + 6x_2$

$$h_1(X) = x_1 + x_2 - 2 \leq 0$$

$$h_2(X) = 2x_1 + 3x_2 - 12 \leq 0$$

Now, the Hessian matrix is given by

$$H_B = \begin{bmatrix} \dfrac{\partial^2 f}{\partial x_1^2} & \dfrac{\partial^2 f}{\partial x_1 \partial x_2} & \dfrac{\partial^2 f}{\partial x_1 \partial x_3} \\[2ex] \dfrac{\partial^2 f}{\partial x_2 \partial x_1} & \dfrac{\partial^2 f}{\partial x_2^2} & \dfrac{\partial^2 f}{\partial x_2 \partial x_3} \\[2ex] \dfrac{\partial^2 f}{\partial x_3 \partial x_1} & \dfrac{\partial^2 f}{\partial x_3 \partial x_2} & \dfrac{\partial^2 f}{\partial x_3^2} \end{bmatrix} = \begin{bmatrix} -2 & 0 & 0 \\ 0 & -2 & 0 \\ 0 & 0 & -2 \end{bmatrix}$$

whose principle minors are $D_1 = -2 < 0$, $D_2 = \begin{vmatrix} -2 & 0 \\ 0 & -2 \end{vmatrix} = 4 > 0$

and $D_3 = \begin{vmatrix} -2 & 0 & 0 \\ 0 & -2 & 0 \\ 0 & 0 & -2 \end{vmatrix} = -8 < 0$ which are clearly of alternate sign.

Therefore, $f(X)$ is a concave function.

Also, $h_1(X), h_2(X)$ are convex function of $X = (x_1, x_2)$.

$\Rightarrow$ Kuhn-Tucker necessary conditions for max. $f(X)$ will also be the sufficient conditions and are given by

$$\frac{\partial f}{\partial x_j} = \sum_{i=1}^{2} \lambda_i \frac{\partial h_i(X)}{\partial x_j}, \quad j = 1, 2, 3$$

$\lambda_1 h_1(X) = 0, \ \lambda_2 h_2(X) = 0, \ h_1(X) \leq 0, \ h_2(X) \leq 0 \text{ and } \lambda_1, \lambda_2 \geq 0$

which implies

$$-2x_1 + 4 = \lambda_1 + 2\lambda_2 \qquad \qquad \dots(1)$$

$$-2x_2 + 6 = \lambda_1 + 3\lambda_2 \qquad \qquad \dots(2)$$

$$-2x_3 = 0 \qquad \qquad \dots(3)$$

$$\lambda_1(x_1 + x_2 - 2) = 0 \qquad \qquad \dots(4)$$

$$\lambda_2(2x_1 + 3x_2 - 12) = 0 \qquad \qquad \dots(5)$$

$$x_1 + x_2 - 2 \leq 0 \qquad \qquad \dots(6)$$

$$2x_1 + 3x_2 - 12 \leq 0 \qquad \qquad \dots(7)$$

and $\qquad \qquad \lambda_1, \lambda_2 \geq 0 \qquad \qquad \dots(8)$

Now, we have the following cases:

Case 1: If $\lambda_1 = \lambda_2 = 0$ Here, from (1) and (2) we have

$$-2x_1 + 4 = 0 \text{ and } -2x_2 + 6 = 0$$
$$\Rightarrow \qquad x_1 = 2 \text{ and } x_2 = 3$$

These values do not satisfy (6) and (7) and hence discarded.

Case 2: **If $\lambda_1 \neq 0, \lambda_2 = 0$** In this case, from (1) and (2), we have

$$-2x_1 + 4 = \lambda_1 \text{ and } -2x_2 + 6 = \lambda_1$$
$$\Rightarrow \qquad x_1 - x_2 + 1 = 0 \qquad\qquad\qquad \dots(9)$$

If $\lambda_1 \neq 0$ then from (4),

$$x_1 + x_2 - 2 = 0 \qquad\qquad\qquad \dots(10)$$

On solving (9) and (10) we get

$$x_1 = \frac{1}{2}, x_2 = \frac{3}{2}$$

Clearly, these values satisfy (6) and (7)

Also, for these values of x_1 and x_2, from (1) and (2), we get

$$\lambda_1 = 3 > 0, \ \lambda_2 = 0$$

For this solution, $x_1 = \dfrac{1}{2}, x_2 = \dfrac{3}{2}$ and max. $Z = \dfrac{17}{2}$

Case 3: **If $\lambda_1 = 0, \lambda_2 \neq 0$** Here, from (1) and (2), we get

$$-2x_1 + 4 = 2\lambda_2 \text{ and } -2x_2 + 6 = 3\lambda_2$$
$$\Rightarrow \qquad 3x_1 - 2x_2 = 0 \qquad\qquad\qquad \dots(11)$$

If $\lambda_2 \neq 0$, then from (5), we get $2x_1 + 3x_2 - 12 = 0$ $\qquad\qquad \dots(12)$

On solving (11) and (12), we get

$$x_1 = \frac{24}{13}, x_2 = \frac{36}{13}$$

Also, from (3), $\qquad\qquad x_3 = 0$

These values does not satisfy (6) and hence discarded.

Case 4: **If $\lambda_1 \neq 0, \lambda_2 \neq 0$** From (4) and (5) we get

$$x_1 + x_2 - 2 = 0 \text{ and } 2x_1 + 3x_2 - 12 = 0$$
$$\Rightarrow \qquad x_1 = -6 \text{ and } x_2 = 8$$

Since, $x_1 = -6 < 0 \qquad \Rightarrow \quad$ solution is discarded.

Hence, the optimal solution of the given problem is

$$x_1 = \frac{1}{2}, x_2 = \frac{3}{2}, \ x_3 = 0 \text{ and Max.} Z = \frac{17}{2}$$

EXAMPLE 5. *Solve the following non-linear programming problem*

$$\textbf{Max.} Z = f(X) = f(x_1, x_2)$$

$$= (200x_1 - 2x_1^2) + (500x_2 - 3x_2^2)$$

subject to the constraints

$$2x_1 + x_2 \le 140; \ 2x_1 + 3x_2 \le 180 \text{ and } x_1, x_2 \ge 0$$

SOLUTION. We have,

$$f(X) = (200x_1 - 2x_1^2) + (500x_2 - 3x_2^2)$$

$$h_1(X) = 2x_1 + x_2 - 140 \le 0$$

$$h_2(X) = 2x_1 + 3x_2 - 180 \le 0$$

$$X = (x_1, x_2) \ge 0$$

Here, the Hessian matrix is given by

$$H_B = \begin{bmatrix} \dfrac{\partial^2 f}{\partial x_1^2} & \dfrac{\partial^2 f}{\partial x_1 \partial x_2} \\[2mm] \dfrac{\partial^2 f}{\partial x_2 \partial x_1} & \dfrac{\partial^2 f}{\partial x_2^2} \end{bmatrix} = \begin{bmatrix} -4 & 0 \\ 0 & -6 \end{bmatrix}$$

whose principal minors are $D_1 = -4$, $D_2 = \begin{vmatrix} -4 & 0 \\ 0 & -6 \end{vmatrix} = 24 > 0$ are of alternate sign.

$\Rightarrow$ $f(X)$ is a concave function.

Further, $h_1(x)$ and $h_2(x)$ are convex functions.

$\Rightarrow$ The kuhn-Tucker necessary conditions for max.$f(X)$ will also be the sufficient conditions.

The Kuhn-Tucker necessary conditions for maximum of $f(X)$ are given by

$$\frac{\partial f}{\partial x_1} = \sum_{i=1}^{2} \lambda_i \frac{\partial h_i(X)}{\partial x_i}; \frac{\partial f}{\partial x_2} = \sum_{i=1}^{2} \lambda_i \frac{\partial h_i(X)}{\partial x_2}$$

$$\lambda_1 h_1(X) = 0, \lambda_2 h_2(X) = 0, h_1(X) \le 0, h_2(X) \le 0 \text{ and } \lambda_1, \lambda_2 \ge 0$$

which implies

$$200 - 4x_1 = 2\lambda_1 + 2\lambda_2 \qquad \text{...(1)}$$

$$500 - 6x_2 = \lambda_1 + 3\lambda_2 \qquad \text{...(2)}$$

$$\lambda_1(2x_1 + x_2 - 140) = 0 \qquad \text{...(3)}$$

$$\lambda_2(2x_1 + 3x_2 - 180) = 0 \qquad \text{...(4)}$$

$$2x_1 + x_2 - 140 \le 0 \qquad \text{...(5)}$$

$$2x_1 + 3x_2 - 180 \le 0 \qquad \text{...(6)}$$

and $\qquad \lambda_1, \lambda_2 \ge 0 \qquad \text{...(7)}$

Now, we have the following cases:

Case 1: If $\lambda_1 = 0, \lambda_2 = 0$ Here, from (1) and (2), we get

$$200 - 4x_1 = 0 \text{ and } 500 - 6x_2 = 0$$

$\Rightarrow$ $\qquad x_1 = 50$ and $x_2 = \dfrac{250}{3}$

which does not satisfy (5) and (6) and hence discarded.

Case 2: If $\lambda_1 \ne 0, \lambda_2 = 0$ In this case, from (1) and (2), we get

$$200 - 4x_1 = 2\lambda_1 \text{ and } 500 - 6x_2 = \lambda_1$$

$\Rightarrow$ $\qquad x_1 - 3x_2 + 200 = 0 \qquad \text{...(8)}$

If $\lambda_1 \ne 0$, then from (3), $2x_1 + x_2 - 140 = 0 \qquad \text{...(9)}$

On solving (8) and (9) we get $x_1 = \dfrac{220}{7}$, $x_2 = \dfrac{540}{7}$

which does not satisfy (6) and hence discarded.

Case 3: If $\lambda_1 = 0, \lambda_2 \ne 0$ In this case, from (1) and (2), we get

$$200 - 4x_1 = 2\lambda_2 \text{ and } 500 - 6x_2 = 3\lambda_2$$

$\Rightarrow$ $3x_1 - 3x_2 + 100 = 0 \qquad \text{...(10)}$

If $\lambda_2 \ne 0$, then from (4),

$$2x_1 + 3x_2 - 180 = 0 \qquad \text{...(11)}$$

On solving (10) and (11), we get

$$x_1 = 16, \ x_2 = \frac{148}{3}$$

$\Rightarrow \qquad \lambda_2 = 68 > 0$

Clearly, these values of x_1 and x_2 satisfy (5)

$\Rightarrow \qquad x_1 = 16, \ x_2 = \dfrac{148}{3}$ is stationary point at which

$$\text{Max.} f(X) = \frac{60160}{3}$$

Case 4: If $\lambda_1 \neq 0, \lambda_2 \neq \mathbf{0}$ In this case from (3) and (4), we get

$$2x_1 + x_2 - 140 = 0 \text{ and } 2x_1 + 3x_2 - 180 = 0$$

$\Rightarrow \qquad x_1 = 60, \ x_2 = 20$

Then, from (1) and (2), we get

$$\lambda_1 + \lambda_2 = -20 \text{ and } \lambda_1 + 3\lambda_2 = 380$$

On solving we get $\lambda_1 = -220 < 0$ and $\lambda_2 = 200 > 0$

For these values (7), is not satisfied and hence discarded.

Hence, the optimal solution is given by

$$x_1 = 16, x_2 = \frac{148}{3} \text{ and } \text{Max.} Z = \frac{60160}{3}$$

EXERCISE 3.4

Solve the following non-linear programming problems using the method of Lagrangian multipliers.

1. Minimize $Z = 3e^{2x_1+1} + 2e^{x_2+5}$

   subject to the constraints
   $$x_1 + x_2 = 7$$
   and $\quad x_1, x_2 \geq 0$

2. Min. $Z = 2x_1^2 - 24x_1 + 2x_2^2 - 8x_2 + 2x_3^2$
   $$-12x_3 + 200$$
   subject to the constraints
   $$x_1 + x_2 + x_3 = 11$$
   and $\qquad x_1, x_2, x_3 \geq 0$

3. Min. $Z = x_1^2 + x_2^2 + x_3^2$

   subject to the constraints
   $$x_1 + x_2 + 3x_3 = 2$$
   $$5x_1 + 2x_2 + x_3 = 5$$
   and $\qquad x_1, x_2, x_3 \geq 0$

4. Min. $Z = 6x_1^2 + 5x_2^2$
   subject to the constraints
   $$x_1 + 5x_2 = 3$$

   and $\quad x_1, x_2 \geq 0$

5. Max. $Z = 4x_1 + 6x_2 - 2x_1^2 - 2x_1x_2 - 2x_2^2$

   subject to the constraints
   $$x_1 + 2x_2 = 2$$
   and $\quad x_1, x_2 \geq 0$

6. Max. $Z = x_1^2 + 4x_1x_2 + x_2^2$

   subject to the constraints
   $$x_1^2 + x_2^2 = 1$$
   and $\quad x_1, x_2 \geq 0$

7. Min. $Z = 4x_1^2 + 2x_2^2 + x_3^2 - 4x_1x_2$
   subject to the constraints
   $$x_1 + x_2 + x_3 = 15$$
   $$2x_1 - x_2 + 2x_3 = 20$$
   and $\qquad x_1, x_2, x_3 \geq 0$

8. Max. $Z = 7x_1 - 0.3x_1^2 + 8x_2 - 0.4x_2^2$
   subject to the constraints
   $$4x_1 + 5x_2 = 100$$
   and $\qquad x_1, x_2 \geq 0$

Solve the following non-linear programming problemS, by using Kuhn-Tucker conditions.

9. Min.$Z = \left(x_1 - \dfrac{9}{4}\right)^2 + (x_2 - 2)^2$

subject to the constraints

$$x_2 - x_1^2 \geq 0$$
$$x_1 + x_2 \leq 6$$

and $x_1, x_2 \geq 0$

10. Min.$Z = (x_1 - 1)^2 + (x_2 - 5)^2$

subject to the constraints

$$-x_1^2 + x_2 \leq 4$$
$$-(x_1 - 2)^2 + x_2 \leq 4$$

and $x_1, x_2 \geq 0$

11. Min.$Z = -\log x_1 - \log x_2$

subject to the constraints
$$x_1 + x_2 \leq 2$$

and $x_1, x_2 \geq 0$

12. Max.$Z = 2x_1^2 + 12x_1x_2 - 7x_2^2$

subject to the constraints
$$2x_1 + 5x_2 \leq 98$$

and $x_1, x_2 \geq 0$

13. Max.$Z = 3x_1 + x_2$

subject to the constraints

$$x_1^2 + x_2^2 \leq 5$$
$$x_1 - x_2 \leq 1$$

and $x_1, x_2 \geq 0$

14. Max.$Z = 10x_1 - x_1^2 + 10x_2 - x_2^2$

subject to the constraints

$$x_1 + x_2 \leq 14$$
$$-x_1 + x_2 \leq 6$$

and $x_1, x_2 \geq 0$

15. Max.$Z = 7x_1^2 - 6x_1 + 5x_2^2$

subject to the constraints

$$x_1 + 2x_2 \leq 10$$
$$x_1 - 3x_2 \leq 9$$

and $x_1, x_2 \geq 0$

16. Max $Z. = 2x_1 - x_1^2 + x_2$

subject to the constraints
$$2x_1 + 3x_2 \leq 6$$
$$2x_1 + x_2 \leq 4$$

and $x_1, x_2 \geq 0$

ANSWERS

1. $x_1 = \dfrac{1}{3}(11 - \log 3), x_2 = \dfrac{1}{3}(10 + \log 3)$ **2.** $x_1 = 6, x_2 = 2, x_3 = 3$, Min.$Z = 102$

3. $x_1 = 0.81, x_2 = 0.35, x_3 = 0.928$, Min. $Z = 1.625$ **4.** $x_1 = \dfrac{3}{31}, x_2 = \dfrac{18}{31}$, Min.$Z = \dfrac{54}{31}$

5. $x_1 = \dfrac{1}{3}, x_2 = \dfrac{5}{6}$, Max.$Z = \dfrac{25}{6}$ **6.** $x_1 = \dfrac{1}{\sqrt{2}}, x_2 = \dfrac{1}{\sqrt{2}}$, Max.$Z = 3$

7. $x_1 = \dfrac{11}{3}, x_2 = \dfrac{10}{3}, x_3 = 8$, Min.$Z = \dfrac{820}{9}$ **8.** $x_1 = 12.06, x_2 = 10.35$, Max. $Z = 80.73$

9. $x_1 = \dfrac{3}{2}, x_2 = \dfrac{9}{4}$, Min.$Z = \dfrac{\sqrt{10}}{16}$ **10.** $x_1 = 1, x_2 = 5$, Min. $Z = 10$

11. $x_1 = 1, x_2 = 1$, Min.$Z = 0$ **12.** $x_1 = 44, x_2 = 2$, Max.$Z = 4900$

13. $x_1 = 1.43, x_2 = 0.48$, Max.$Z = 4.77$ **14.** $x_1 = 5, x_2 = 5$, Max.$Z = 50$

15. $x_1 = \dfrac{48}{5}, x_2 = \dfrac{1}{5}$, Max.$Z = 587.72$ **16.** $x_1 = \dfrac{2}{3}, x_2 = \dfrac{14}{9}$, Max.$Z = \dfrac{22}{9}$

4 Non-linear Programming : Formulation and Graphical Solutions

4.1 INTRODUCTION

It is a well known fact that the term non-linear programming problem (NLPP) refers to the problem in which the objective function becomes non-linear or one or more of the constraints have non-linear relationship or both. In this chapter we shall discuss the formulation of non-linear programming problem and their graphical solution.

4.2 GENERAL NON-LINEAR PROGRAMMING PROBLEMS (GNLPP)

The general form of a non-linear programming problem is defined as follows,

Find $x_1, x_2, ..., x_n$

which optimize the objective function

$$Z = f(x_1, x_2, ..., x_n)$$

subject to the constraints

$$g_1(x_1, x_2, ..., x_n) \ (\leq, = \text{ or} \geq) \ b_1$$
$$g_2(x_1, x_2, ..., x_n) \ (\leq, = \text{ or} \geq) \ b_2$$
$$\vdots \qquad \vdots \qquad \vdots$$
$$\vdots \qquad \vdots \qquad \vdots$$
$$g_m(x_1, x_2, ..., x_n) \ (\leq, = \text{ or} \geq) \ b_m$$

and non-negative restrictions

$$x_j \geq 0 \ \ \forall \, j = 1, 2, .., n.$$

where, either $f(x_1, x_2, ..., x_n)$ or some $g_i(x_1, x_2, ..., x_n)$; $i = 1, 2, ..., m$ or both are non-linear.

In matrix notations, the above NLPP can be written as below:

Determine $X^T = (x_1, x_2, ..., x_n)^T$

which optimize the objective function $Z = f(X)$

subject to the constraints

$$g_i(X) \ (\leq, = \text{ or} \geq) \ \ b_i \, ; i = 1, 2, ..., m$$

and $\qquad x_i \geq 0$

4.3 CANONICAL FORM OF NON-LINEAR PROGRAMMING PROBLEM

The canonical form of a non-linear programming problem can be written as follows,

Max. $Z = C(x_1, x_2, ..., x_n)$

subject to the constraints

$$a(x_1, x_2, ..., x_n) \leq 0; \ \ i = 1, 2, ..., m$$

and $\qquad x_j \geq 0, \ j = 1, 2, ..., n$

where at least one of the function $C(x_1, x_2, ..., x_n)$ and $a_i(x_1, x_2, ..., x_n)$ or both is non-linear.

☛ REMARK
• Both the function $C(x)$ and $a(x)$ defined above should be non-linear.

4.4 MATHEMATICAL FORMULATION OF NON-LINEAR PROGRAMMING PROBLEMS

Following examples give the basic idea of the formation of general non-linear programming problem.

SOLVED EXAMPLES

EXAMPLE 1. *A company manufactures two products A and B. It takes 30 minutes to process one unit of product A and 15 minutes for product B and the maximum machine time available is 35 hours per week. Product A and B require 2 kgs and 3 kgs of raw material per unit respectively. The available quantity of raw material is 180 kgs per week. The product A and B which have unlimited market potential sell for ₹ 200 and ₹ 500 per unit respectively. If the manufacturing costs for product A and B are $2x_1{}^2$ and $3x_2{}^2$ respectively, formulate the non-linear programming problem, where*

$$x_1 = quantity\ of\ product\ A\ to\ be\ produced$$
$$x_2 = quantity\ of\ product\ B\ to\ be\ produced$$

SOLUTION. According to the question,

The processing time of products A and B are 30 minutes *i.e.*, 0.5 hour and 15 minutes, *i.e.*, 0.25 hour respectively. So, total processing time for x_1 and x_2 units of two products produces per week is $0.5\,x_1 + 0.25\,x_2$ which should be less than or equal to the total available time of 35 hours. Thus, one constraint is

$$0.5\,x_1 + 0.25\,x_2 \le 35$$
$$\Rightarrow \qquad 2\,x_1 + x_2 \le 140$$

Further, since the products requires 2 kgs and 3 kgs of raw material per unit respectively, therefore the total material required per week is $2x_1 + 3x_2$ which should be less than or equal to the total quantity of raw material available per week which is 180 kg.

Therefore, second constraint is given by

$$2\,x_1 + 3x_2 \le 180$$

If Z is the total profit of the company per week, then

$$Z = (200\,x_1 - 2x_1{}^2) + (500\,x_2 - 3x_2{}^2)$$

Hence, the non-linear programming problem can be written as,

$$\text{Maximize } Z = (200\,x_1 - 2x_1{}^2) + (500\,x_2 - 3x_2{}^2)$$

subject to the constraints

$$2x_1 + x_2 \le 140$$
$$2x_1 + 3x_2 \le 180$$

and

$$x_1, x_2 \ge 0$$

EXAMPLE 2. *An engineering company has received a rush order for a maximum no. of two types of items that can be produced and transported during a two week period. The profit in thousand rupees on this order is related to the number of each type of item manufactured by company and is given by*

$$12x_1 + 10x_2 - x_1^2 - x_2^2 + 61$$

where x_1 is the no. of units (in thousands) of type -I item and x_2 is the number of units (in thousands) of type-II item. Because of other commitment over the next two weeks, the company has available only 60 hours in the shifting and packing department. It is assumed that every thousand unit of type-I and II items will requires 20 hours and 30 hours respectively in the shifting and packing departments. Given the above information, formulate the non-linear programming problem.

SOLUTION. Proceed same as in example-1, we have the following non-linear programming problem

$$\text{Maximize. } Z = 12x_1 + 10x_2 - x_1^2 - x_2^2 + 61$$

subject to the constraints

$$20 x_1 + 30 x_2 \leq 60$$

and

$$x_1, x_2 \geq 0$$

EXAMPLE 3. *(Production Allocation Problem)*

A manufacturing company produces a product consisting of two raw material say A and B. The production function is estimated as

$$Z = f(x_1, x_2) = 3.6\,x_1 - 0.4\,x_1^2 + 1.6\,x_2 - 0.2\,x_2^2$$

where Z represents the quantity (in tones) of the product produced and x_1, x_2 design the input amount of raw material A and B. The company has ₹ 50000 to spend on these two raw materials. The unit price of A is ₹10,000 and of B is ₹ 5000. Formulate the problem.

SOLUTION. Let x_1, x_2 be the input amounts of the material A and B respectively. Then the company will have to spend ₹ 10,000 x_1 + 5000 x_2 on the two materials. Since company has ₹ 50000 to spend on the two raw materials A and B, therefore, we have

$$10000 x_1 + 5000 x_2 \leq 50000$$

$$\Rightarrow \qquad 2x_1 + x_2 \leq 10$$

Hence, the NLPP is given as follows:

$$\text{Maximize } Z = f(x_1, x_2) = 3.6\,x_1 - 0.4\,x_1^2 + 1.6\,x_2 - 0.2x_2^2$$

subject to the constraint

$$2x_1 + x_2 \leq 10$$

and

$$x_1, x_2 \geq 0$$

EXAMPLE 4. *A manufacturing company produces two products Radios and T.V. sets. Sales price relationship for these two products are given below*

Product	Quantity demanded	Unit price
Radios	$1500 - 5p_1$	p_1
T.V.	$3800 - 10\,p_2$	p_2

The total cost functions for these two products are given by $200x_1 + 0.1x_1^2$ and $300x_2 + 0.1x_2^2$ respectively.

The production takes place on two assembly lines. Radio sets are assembled on assembly line-1 and T.V. sets are assembled line-2. Because of the limitations of the line capacity, the daily production is limited to no more than 80 radio sets and 60 T.V. sets. The product of both types of products requires electronic components. The production of each of these sets requires 5 units and 6 units of electronic equipments component respectively. The electronic components are supplied by another manufacturer and the supply is limited to 600 units per day. The production of one unit of radio set requires 1 man-day of labour, whereas 2 men-days of labour are required for a T.V. set. How many units of radios and T.V. sets should the company produce in order to maximize the total profit. Formulate the problem as a NLPP. (AGRA-2002)

SOLUTION. Let us assume that the company manufacture x_1 and x_2 units of radio and T.V. sets respectively.

So, $$x_1 = 1500 - 5p_1 \quad \text{and} \quad x_2 = 3800 - 10 p_2$$
$$\Rightarrow \quad p_1 = 300 - 0.2 x_1 \text{ and } p_2 = 380 - 0.1 x_2$$

Now, total revenue received by the company is given by
$$R = p_1 x_1 + p_2 x_2 = (300 - 0.2 x_1)x_1 + (380 - 0.1 x_2)x_2$$

and total cost of production

= cost of production of radio sets + cost of production of TV sets

$$= (200 x_1 + 0.1 x_1^2) + (300 x_2 + 0.1 x_2^2)$$

Now, if Z is the total profit of the company then

Z = total revenue – total cost of production

$$= [(300 - 0.2 x_1)x_1 + (380 - 0.1 x_2)x_2] - [(200 x_1 + 0.1 x_1^2) + (300 x_2 + 0.1 x_2^2)]$$

$$= 100 x_1 + 80 x_2 - 0.3 x_1^2 - 0.2 x_2^2$$

Further, since the capacity of total production of the assembly line is limited to no more than 80 radio sets and 60 T.V. sets, thus the constraints on x_1 and x_2 are given by
$$x_1 \le 80 \text{ and } x_2 \le 60$$

Also, since 5 and 6 units of electronic equipments respectively are required for the production of radio sets and T.V. sets whereas the supply of these equipments is limited to 600 units per day.

So, total no. of electronic equipments required for production
$$= 5x_1 + 6x_2 \le 600$$

and, since the production of one unit of radio set and one unit of T.V. set require 1 man-day and 2 man-day labour respectively whereas the labour supply is limited to 160 man-days.

$\Rightarrow$ Total man hour of labour required $= 1. x_1 + 2x_2 \le 160$

Also $x_1 \ge 0, x_2 \ge 0$ (because negative no. of sets cannot be produced)

Hence, the required NLPP is given as follows :

$$\text{Max}. Z = 100 x_1 + 80 x_2 - 0.3 x_1^2 - 0.2 x_2^2$$

subject to the constraints

$$5 x_1 + 6 x_2 \le 600$$
$$x_1 + 2x_1 \le 160$$
$$x_1 \le 80$$
$$x_2 \le 60$$

and $$x_1, x_2 \ge 0$$

4.5 GRAPHICAL SOLUTION OF A NON-LINEAR PROGRAMMING PROBLEM

In a linear programming problem, the optimal solution was usually obtained at one of the extremities of the convex region generated by the constraints and the objective function of the problem. But it is not necessary to find the solution at a corner or edge of the feasible region of non-linear programming problem.

The following examples will make the method clear.

SOLVED EXAMPLES

EXAMPLE 1.   *Solve the following NLPP by graphical method*

$$\text{Maximize } Z = x_1 + 2x_2$$

subject to the constraints

$$x_1^2 + x_2^2 \le 1$$
$$2x_1 + x_2 \le 2$$

and $$x_1, x_2 \ge 0$$

SOLUTION.   Clearly the constraints of the given non-linear programming problem are given by

$$x_1^2 + x_2^2 \le 1$$
$$2x_1 + x_2 \le 2$$

and $$x_1, x_2 \ge 0$$

Considering these constraints as equalities, we get

$$x_1^2 + x_2^2 = 1 \text{ (Circle)}$$

and $$2x_1 + x_2 = 2 \text{ (Straight line)}$$

Also $$Z = x_1 + 2x_2 = 0 \Rightarrow \frac{x_1}{x_2} = \frac{-2}{1}$$

Drawing these in the plane, we get the following figure

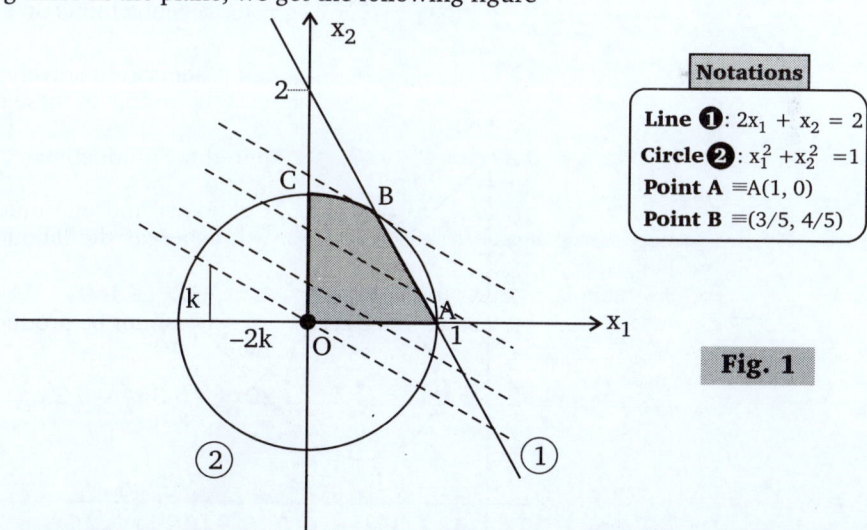

Notations

Line ❶: $2x_1 + x_2 = 2$
Circle ❷: $x_1^2 + x_2^2 = 1$
Point A $\equiv A(1, 0)$
Point B $\equiv (3/5, 4/5)$

Fig. 1

In the above figure, drawing the line $Z = x_1 + 2x_2 = 0$ through the origin and draw parallel to this line till we reach the extremities B of the permissible region B is the most distance point of the permissible region through with the line parallel to $Z = 0$ passes.

Here, B is the point of intersection of the circle

$$x_1^2 + x_2^2 = 1 \text{ and the line } 2x_1 + x_2 = 0.$$

Therefore, at B

$$x_1 = \frac{3}{5} = 0.6$$

$$x_2 = \frac{4}{5} = 0.8$$

and

$$Z = \frac{11}{5} = 2.2$$

Hence the optimal solution is given by

$$x_1 = 0.6, x_2 = 0.8 \text{ and Max. } Z = 2.2$$

EXAMPLE 2. *Solve the following non-linear programming problem graphically. Also verify the Kuhn-Tucker necessary condition for maximum of the function*

$$\textbf{Max.} \, Z = 8x_1 - x_1^2 + 8x_2 - x_2^2$$

subject to the constraints

$$x_1 + x_2 \leq 12$$
$$x_1 - x_2 \geq 4$$

and $x_1, x_2 \geq 0$ [AGRA-2009]

SOLUTION.   Here, the constraints of the given NLPP are

$$x_1 + x_2 \leq 12$$
$$x_1 - x_2 \geq 4$$
$$x_1, x_2 \geq 0$$

Convert the above inequations into equalities and drawing these lines in the plane, we get the permissible region DABD.

Also, the objective function

$$Z = 8x_1 - x_1^2 + 8x_2 - x_2^2 \text{ is a circle with centre } (4, 4)$$

Notations
Line ❶: $x_1 + x_2 = 12$
Line ❷: $x_1 - x_2 = 4$
Point A $\equiv$ A(12, 0)
Point B $\equiv$ B(4, 0)
Point C $\equiv$ C(4, 4)
Point P $\equiv$ P(6, 6)
Point Q $\equiv$ Q(6, 2)

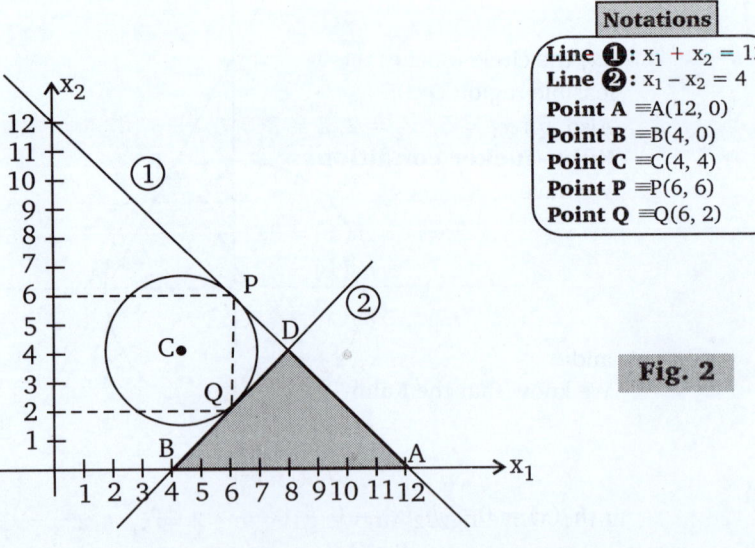

Fig. 2

Now the point giving the maximum value of Z in the point at which the feasible region is tangent to the circle given by the objective function

$$Z = 8x_1 - x_1^2 + 8x_2 - x_2^2 \qquad \qquad ...(1)$$

The centre of this circle $C = C(4, 4)$

Differentiating (1) w.r.t. x_1, we get

$$0 = 8 - 2x_1 + 8\frac{dx_2}{dx_1} - 2x_2\frac{dx_2}{dx_1}$$

$$\Rightarrow \qquad \frac{dx_2}{dx_1} = \frac{2x_1 - 8}{8 - 2x_2} = \frac{x_1 - 4}{4 - x_2} = m_1 \text{ (say)}$$

Now, for the line $x_1 + x_2 = 12$, we have

$$\frac{dx_2}{dx_1} = -1 = m_2 \text{ (say)}$$

Clearly the circle touch this line, when

$$m_1 = m_2$$

$$\Rightarrow \qquad \frac{x_1 - 4}{4 - x_2} = -1 \Rightarrow x_2 = x_1$$

Putting this value in $x_1 + x_2 = 12$, we get $x_1 = x_2 = 6$

$\Rightarrow$ the circle touches the line $x_1 + x_2 = 12$ at the point $P(6, 6)$

But the point P is not in the feasible region $DABD$

Again for the line $x_1 - x_2 = 4 \Rightarrow \dfrac{dx_2}{dx_1} = 1 = m_3$ (say)

The circle touches this line at the point where

$$m_1 = m_3 \Rightarrow \frac{x_1 - 4}{4 - x_2} = 1 \Rightarrow x_2 = 8 - x_1$$

Using this in the line $x_1 - x_2 = 4$, we get

$$x_1 = 6 \Rightarrow x_2 = 2$$

$\Rightarrow$ the circle touches the line $x_1 - x_2 = 4$ at the point $Q(6, 2)$ which is the point in the feasible region DABD.

Also, for $x_1 = 6$, $x_2 = 2$, $Z = 24$, which is the required solution of given NLPP.

Kuhn-Tucker conditions:

We have

$$f(x) = 8x_1 - x_1^2 + 8x_2 - x_2^2$$

$$h_1(x) = x_1 + x_2 - 12 \leq 0$$

$$h_2(x) = 4 - x_1 + x_2 \leq 0$$

and $\qquad x_1, x_2 \geq 0$

We know that the Kuhn-Tucker condition for max. $f(x)$ are

$$\frac{\partial f(x)}{\partial x_j} = \sum_{i=1}^{2} \lambda_i \frac{\partial h_1(x)}{\partial x_j} \qquad j = 1, 2$$

$\lambda_1 h_1(x) = 0$, $\lambda_2 h_2(x) = 0$, $h_1(x) \leq 0$, $h_2(x) \leq 0$ and $\lambda_1, \lambda_2 \geq 0$

$$8 - 2x_1 = \lambda_1 - \lambda_2 \qquad \qquad ...(2)$$

$$8 - 2x_2 = \lambda_1 + \lambda_2 \qquad \qquad ...(3)$$

$$\lambda_1(x_1 + x_2 - 12) = 0 \qquad \qquad \ldots(4)$$
$$\lambda_2(4 - x_1 + x_2) = 0 \qquad \qquad \ldots(5)$$
$$x_1 + x_2 - 12 \leq 0 \qquad \qquad \ldots(6)$$
$$4 - x_1 + x_2 \leq 0 \qquad \qquad \ldots(7)$$
and $\qquad \qquad \lambda_1, \lambda_2 \geq 0 \qquad \qquad \ldots(8)$

Clearly the point (6, 2) satisfies conditions (6) and (7).

For these values from (2) and (3), we have
$$\lambda_1 - \lambda_2 = -4$$
and $\qquad \lambda_1 + \lambda_2 = 4$

On solving we get, $\qquad \qquad \lambda_1 = 0, \lambda_2 = 4$

$\Rightarrow$ Condition (4), (5) and (8) are also satisfied.

Hence, the optimal solution $x_1 = 6, x_2 = 2$ obtained by graphical method satisfies all the Kuhn-Tucker condition for maximum of $f(x)$.

EXAMPLE 3. *Using graphical method, solve the following NLPP*

$$\textbf{Min. Z} = \textbf{x}_1^2 + \textbf{x}_2^2$$

subject to the constraints
$$\textbf{x}_1 + \textbf{x}_2 \geq \textbf{4}$$
$$\textbf{2x}_1 + \textbf{x}_2 \geq \textbf{5}$$
and $\qquad \qquad \textbf{x}_1, \textbf{x}_2 \geq \textbf{0}$

SOLUTION. Convert the above inquation into equalities , we get
$$x_1 + x_2 = 4 \text{ and } 2x_1 + x_2 = 5$$

Plot the above lines on the plane.

Since, $x_1, x_2 \geq 0$, the feasible region will lie in the first quadrant only.

Clearly, the constraints $x_1 + x_2 \geq 4$ is satisfied by all points lying in the region shaded by the vertical lines

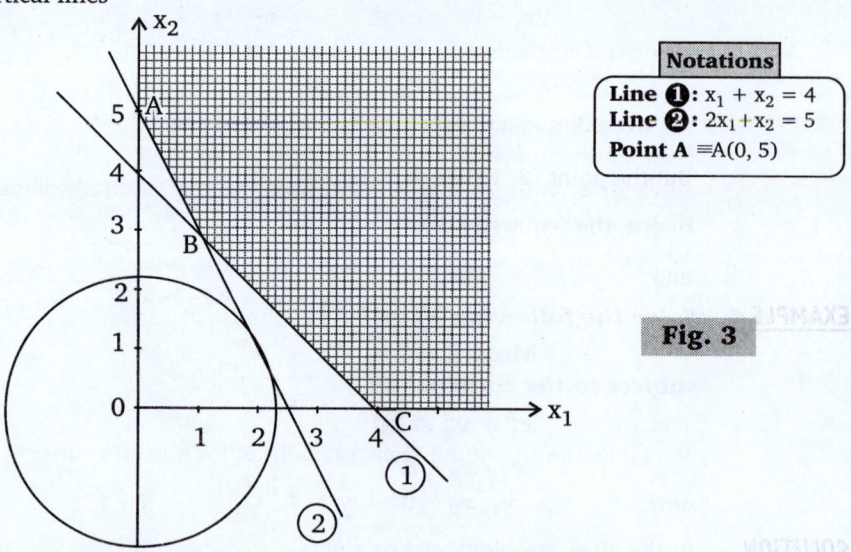

Notations

Line ❶: $x_1 + x_2 = 4$
Line ❷: $2x_1 + x_2 = 5$
Point A $\equiv$ A(0, 5)

Fig. 3

While the constraints $2x_1 + x_2 \geq 5$ is satisfied by all the points lying in the region shaded by horizontal lines only.

The shaded region by both the vertical and horizontal lines is unbounded convex feasible region x_2ABCx_1.

But we want to find a point (x_1, x_2) which gives a minimum value of $x_1^2 + x_2^2$ and lies in the convex region.

The desired point will be a point of the region at which a side of the convex region is tangent to the circle.

Now, the gradient of the circle $x_1^2 + x_2^2 = k$ (let $z = k$) is given by

$$2x_1 + 2x_2 \frac{dx_2}{dx_1} = 0 \Rightarrow \frac{dx_2}{dx_1} = -\frac{x_1}{x_2} \qquad \text{...(1)}$$

Also, gradient of the line $x_1 + x_2 = 4$ is -1 and the gradient of the line $2x_1 + x_2 = 5$ is -2.

If the line $x_1 + x_2 = 4$ is the tangent to the circle $x_1^2 + x_2^2 = k$, then

$$\frac{dx_2}{dx_1} = -\frac{x_1}{x_2} = -1 \Rightarrow x_1 = x_2 \qquad \text{...(2)}$$

and if the line $2x_1 + x_2 = 5$ is the tangent to the circle $x_1^2 + x_2^2 = k$, then

$$\frac{dx_2}{dx_1} = -\frac{x_1}{x_2} = -2 \Rightarrow x_1 = 2x_2 \qquad \text{...(3)}$$

$\Rightarrow$ The point at which the line $x_1 + x_2 = 4$ is tangent to the circle is obtained by solving the equations $x_1 + x_2 = 4$ and $x_1 = x_2$ which gives $x_1 = 2, x_2 = 2$.

In a similar way, the point at which the line $2x_1 + x_2 = 5$ touches the circle is obtained by solving the equations

$$2x_1 + x_2 = 5 \text{ and } x_1 = 2x_2 \text{ to give us } x_1 = 2, x_2 = 1.$$

$\Rightarrow$ The line $x_1 + x_2 = 4$ touches the circle $x_1^2 + x_2^2 = k$ at the point $(2, 2)$ and the line $2x_1 + x_2 = 5$ touches the circle $x_1^2 + x_2^2 = k$ at the point $(2, 1)$.

But the point $(2, 1)$ lies outside the convex region and hence it is not the desired point.

Hence, the required point $= (2, 2)$

and $\qquad$ Min. $Z = 2^2 + 2^2 = 8$

EXAMPLE 4. *Solve the following NLPP by graphical method:*

$$\textbf{Max . } Z = 2x_1 + 3x_2$$

subject to the constraints

$$x_1^2 + x_2^2 \leq 20$$

$$x_1 \cdot x_2 \leq 8$$

and $\qquad x_1, x_2 \geq 0$

SOLUTION. In the given problem, clearly the objective function is linear and constraints are non-linear.

Apply the usual method to plot the given constraints on the graph.

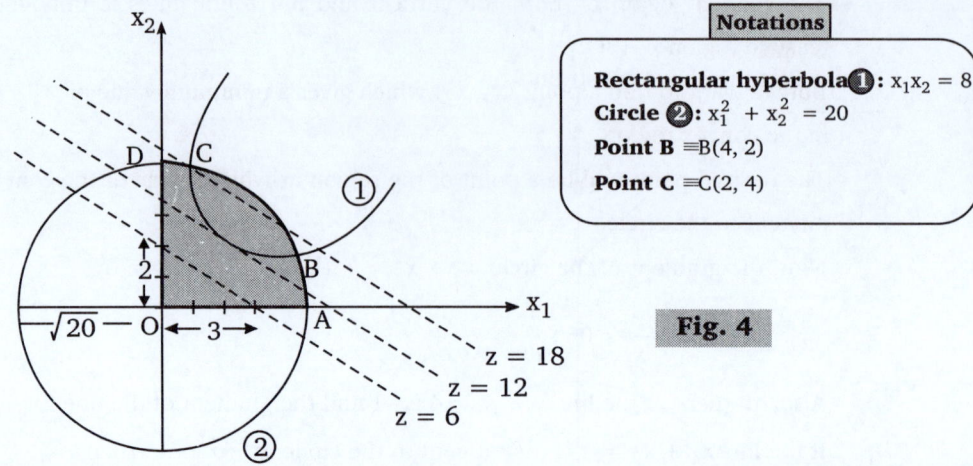

Notations

Rectangular hyperbola ①: $x_1x_2 = 8$
Circle ②: $x_1^2 + x_2^2 = 20$
Point B ≡ B(4, 2)
Point C ≡ C(2, 4)

Fig. 4

Clearly the constraints $x_1^2 + x_2^2 = 20$ represents a circle with radius $\sqrt{20}$, centre $(h, k) = (0, 0)$ and $x_1x_2 = 8$ represents a rectangular hyperbola, whose asymptotes are represented by x_1 and x_2 axis.

On solving

$$x_1^2 + x_2^2 = 20$$

and

$$x_1 x_2 = 8$$

we get

$$(x_1, x_2) = (4, 2) \text{ and } (2, 4)$$

These solution points which also satisfy both the constraints may be obtained within the shaded non-convex region *OABCDO*, which is the feasible region.

Now, we are looking for such a point (x_1, x_2) within the region *OABCDO* where the value of the given objective function $Z = 2x_1 + 3x_2$ is maximum and this point lies in the convex part of the region. We can find such point by iso-profit method.

In this method, we saw parallel objective function $2x_1 + 3x_2 = k$ lines for different value of k and stop the process when a line touches the extreme boundary point of the feasible region for some value of k. Starting with $k = 6$ and so on, we find the iso-profit line at $k = 16$ touches the extreme boundary point $C(2, 4)$ where Z is maximum.

Hence, the optimum solution is given by

$$x_1 = 2, x_2 = 4 \text{ and Max. } Z = 16$$

EXAMPLE 5. *Use graphical method to minimize the distance of the origin from the concave region bounded by the following constraints*

$$x_1 + x_2 \geq 4;$$
$$2x_1 + x_2 \geq 5$$

and $\quad\quad x_1, x_2 \geq 0$

Also, verify that Kuhn-Tucker necessary conditions holds at the point of minimum distance.

SOLUTION. Since the problem is of minimizing distance of solution point from origin, we have to minimize the radius of the circle touching the convex region bounded by the given constraints. Therefore, non-linear programming becomes

$$\text{Min.} Z(= r^2) = x_1^2 + x_2^2$$

subject to the constraints

$$x_1 + x_2 \geq 4 \quad\quad\quad\quad ...(1)$$

$$2x_1 + x_2 \geq 5 \qquad \qquad \ldots(2)$$

and $\qquad x_1, x_2 \geq 0$

Plot the above constraints on the plane graph by following the usual procedure

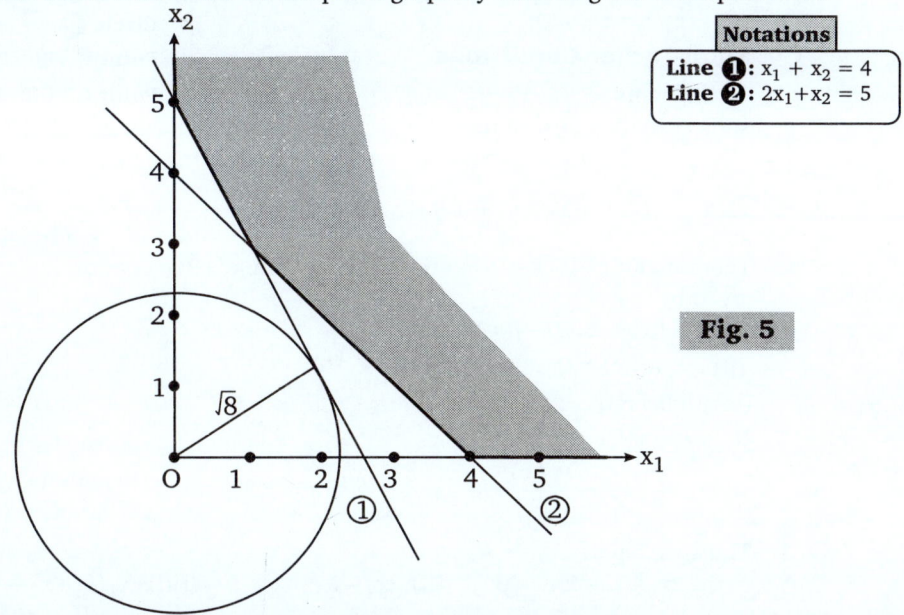

Notations
Line **❶**: $x_1 + x_2 = 4$
Line **❷**: $2x_1 + x_2 = 5$

Fig. 5

Here, the circle which represents the objective function will touch one of the sides of the convex space, convex region should be tangent to the circle.

Differentiate the equation of the circle $x_1^2 + x_2^2 = k$

We get

$$2x_1 dx_1 + 2x_2 dx_2 = 0 \Rightarrow \frac{dx_2}{dx_1} = -\frac{x_1}{x_2} \qquad \qquad \ldots(3)$$

Also from (1) and (2), we get

$$2dx_1 + dx_2 = 0 \Rightarrow \frac{dx_2}{dx_1} = -2 \qquad \qquad \ldots(4)$$

and $\qquad dx_1 + dx_2 = 0 \Rightarrow \frac{dx_2}{dx_1} = -1 \qquad \qquad \ldots(5)$

Solving (3) and (4) with constraints (2), we get

$$-\frac{x_1}{x_2} = \frac{dx_2}{dx_1} = -2$$

$\Rightarrow \qquad \qquad x_1 = 2x_2$

$\Rightarrow \qquad \qquad (x_1, x_2) = (2, 1)$

Clearly this point does not satisfy the constraint $x_1 + x_2 \geq 4$.

Now solving (3) and (5) with constraint (1), we get

$$-\frac{x_1}{x_2} = \frac{dx_2}{dx_1} = -1 \Rightarrow x_1 = x_2$$

$\Rightarrow \qquad \qquad (x_1, x_2) = (2, 2)$

This point also satisfies the constraints and hence the optimal point for the given problem.

Therefore, the minimum distance of solution from the origin is the radius $r = \sqrt{x_1^2 + x_2^2} = \sqrt{8}$.

Kuhn-Tucker Conditions

Minimize $f(x) = x_1^2 + x_2^2$

subject to the constraints

$$h_1(x) = x_1 + x_2 - 4$$

$$h_2(x) = 2x_1 + x_2 - 5$$

and $x \geq 0$

Then the Kuhn-Tucker conditions for the minimization of non-linear programming are given by

(i) $f_j(x) - \Sigma \lambda_i h_{i,j}(x) = 0$

(ii) $\lambda_i h_i(x) = 0$

(iii) $h_i(x) \geq 0$

(iv) $\lambda_i \geq 0$

where $f_j(x) = \dfrac{\partial f(x)}{\partial x_j}$, $h_{i,j} = \dfrac{\partial h_i(x)}{\partial x_j}$ $(j = 1, 2)$

Thus, we have

(i) $2x_1 - \lambda_1 - 2\lambda_2 = 0$ (ii) $2x_2 - \lambda_1 - \lambda_2 = 0$ (iii) $\lambda_1(x_1 + x_2 - 4) = 0$

(iv) $\lambda_2(2x_1 + x_2 - 5) = 0$ (v) $x_1 + x_2 - 4 \geq 0$ (vi) $2x_1 + x_2 - 5 \geq 0$

and $\lambda_1, \lambda_2 \geq 0$

Now, since the optimal solution obtained above is $x_1 = x_2 = 2$

Putting these values in constraints (i) and (ii), we get

$$\lambda_1 + 2\lambda_2 = 4 \text{ and } \lambda_1 + \lambda_2 = 4$$

On solving, we get

$$\lambda_1 = 4, \lambda_2 = 0$$

The solution $x_1 = 2$, $x_2 = 2$ and $\lambda_1 = 4$ and $\lambda_2 = 0$ satisfies all the equations (i) to (vi) and hence satisfies the Kuhn-Tucker condition.

EXAMPLE 6. *Solve graphically, the following non-linear programming problem.*

$$\textbf{Max} \textbf{.} \boldsymbol{Z = 10x_1 - x_1^2 + 10x_2 - x_2^2}$$

subject to the constraints

$$x_1 + x_2 \leq 12$$

$$x_1 - x_2 \leq 6$$

and $x_1, x_2 \geq 0$

SOLUTION. Here, the objective function is non-linear and constraints are linear.

Firstly plot the constraints on the graph following the usual procedure.

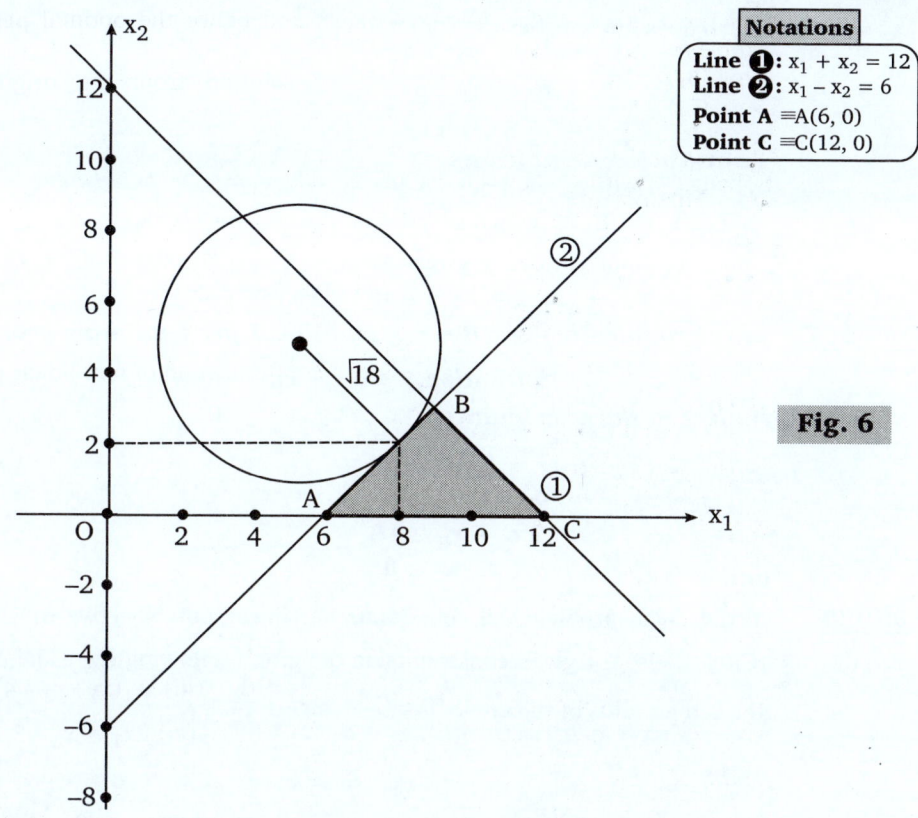

Notations
Line ❶: $x_1 + x_2 = 12$
Line ❷: $x_1 - x_2 = 6$
Point A ≡A(6, 0)
Point C ≡C(12, 0)

Fig. 6

Since the objective function is a circle, the optimum solution point (x_1, x_2) should be a part at which the side of the convex region is tangent to the circle.

Now
$$Z = 10x_1 - x_1^2 + 10x_2 - x_2^2$$

$$\Rightarrow \qquad 10 - 2x_1 + 10\frac{dx_2}{dx_1} - 2x_2\frac{dx_2}{dx_1} = 0$$

$$\Rightarrow \qquad \frac{dx_2}{dx_1} = \frac{2x_1 - 10}{10 - 2x_2} \qquad \qquad ...(1)$$

$$x_1 + x_2 = 12 \Rightarrow 1 + \frac{dx_2}{dx_1} = 0 \Rightarrow \frac{dx_2}{dx_1} = -1 \qquad \qquad ...(2)$$

and $\quad x_1 - x_2 = 6 \Rightarrow 1 - \frac{dx_2}{dx_1} = 0 \Rightarrow \frac{dx_2}{dx_1} = 1 \qquad \qquad ...(3)$

If the line $x_1 + x_2 = 12$ is tangent to the circle, then put $\frac{dx_2}{dx_1} = -1$, from (2) in (1), we get

$$\frac{dx_2}{dx_1} = \frac{2x_1 - 10}{10 - 2x_2} = -1$$

$$\Rightarrow \qquad \qquad x_2 = x_1$$

Using $x_1 = x_2$ in the equation $x_1 + x_2 = 12$, we get

$$(x_1, x_2) = (6, 6)$$

which does not satisfies the constraint $x_1 - x_2 \geq 6$ and hence not feasible.

Now (3) $\Rightarrow \quad \dfrac{dx_2}{dx_1} = \dfrac{2x_1 - 10}{10 - 2x_2} = 1$

$\Rightarrow \qquad\qquad x_1 + x_2 = 10$ ...(4)

Solving (4) with $x_1 - x_2 = 6$, we get $x_1 = 8$, $x_2 = 2$.

$\Rightarrow (x_1, x_2) = (8, 2)$, which satisfies all the constraints also.

Hence, optimal solution is given by

$\qquad\qquad x_1 = 8$, $x_2 = 2$ and Max. $Z = 32$.

EXAMPLE 7.   *Solve graphically the following non-linear programming problem.*

$$\textit{Minimize } Z = x_1^2 + x_2^2$$

subject to the constraints

$$x_1 + x_2 \geq 8$$
$$x_1 + 2x_2 \geq 10$$
$$2x_1 + x_2 \geq 10$$

and $\qquad\qquad x_1, x_2 \geq 0$

SOLUTION.   In the given problem, all constraints are linear but objective function is not linear (Circle). Plot the given constraints on the graph following the usual procedure

If r is the radius of the circle $Z = (r^2) = x_1^2 + x_2^2$

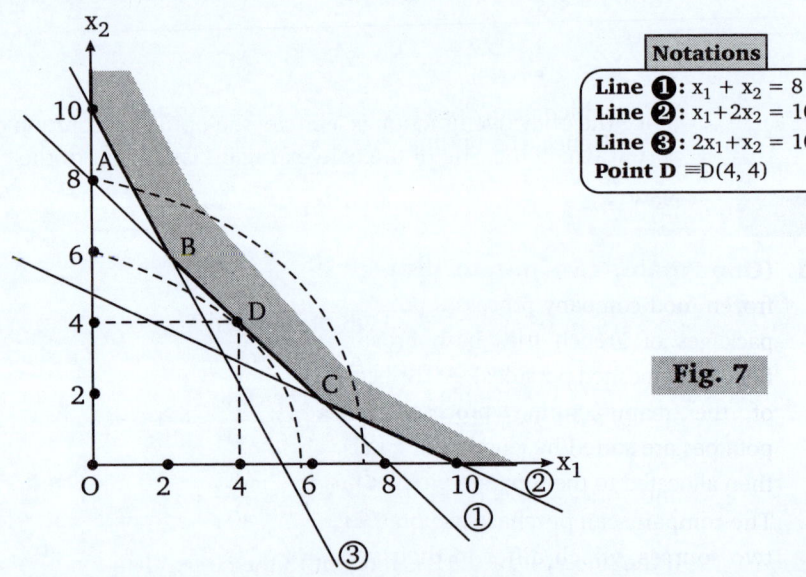

Notations
Line ❶: $x_1 + x_2 = 8$
Line ❷: $x_1 + 2x_2 = 10$
Line ❸: $2x_1 + x_2 = 10$
Point D $\equiv D(4, 4)$

Fig. 7

Then the objective function is to determine the minimum value of r so that the circle with centre $(0, 0)$ and radius r touches the solution space.

The solution point $D(4, 4)$ lies on the line $x_1 + x_2 = 8$ and line is tangent to the circle at D.

Now, since the circle touches one of the sides of the convex region, one of the side of the convex solution space would be tangent to the circle and therefore the solution can also be obtained by differentiating $Z = x_1^2 + x_2^2$ w.r.t. x_1 i.e.,

$$2x_1 + 2x_2\frac{dx_2}{dx_1} = 0 \Rightarrow \frac{dx_2}{dx_1} = -\frac{x_1}{x_2} \qquad ...(1)$$

Also, differentiating the constraints equation, we get

$$dx_1 + dx_2 = 0 \Rightarrow \frac{dx_2}{dx_1} = -1 \qquad ...(2)$$

$$dx_1 + 2dx_2 = 0 \Rightarrow \frac{dx_2}{dx_1} = -\frac{1}{2} \qquad ...(3)$$

and $\qquad 2dx_1 + dx_2 = 0 \Rightarrow \frac{dx_2}{dx_1} = -2 \qquad ...(4)$

Here, three alternative solution which can now be obtained are

(i) From (1) and (2) with $x_1 + x_2 = 8$, we have

$$\frac{dx_2}{dx_1} = -\frac{x_1}{x_2} = -1 \Rightarrow x_1 = x_2 \Rightarrow (x_1, x_2) = (4, 4)$$

This solution satisfies all the constraints so it is feasible.

(ii) Using (1) and (3) with constraint $x_1 + 2x_2 = 10$, we have

$$\frac{dx_2}{dx_1} = -\frac{x_1}{x_2} = -\frac{1}{2} \Rightarrow (x_1, x_2) = (2, 4)$$

which does not satisfies the constraints.

(iii) Taking (1) and (4) with the constraints $2x_1 - x_2 = 10$, we have

$$\frac{dx_2}{dx_1} = -\frac{x_1}{x_2} = -2 \Rightarrow x_2 = \frac{x_1}{2} \Rightarrow (x_1, x_2) = (4, 2)$$

which also, does not satisfies the constraints.

Hence, the optimal solution is given by
$$x_1 = 4, x_2 = 4 \text{ and Min. } Z = 32$$

EXERCISE 4.1

1. **(One-Potato, two potato problem):** A frozen food company processes potatoes into packages of French fries hash browns and flakes (for meshed potatoes). At the beginning of the manufacturing process, the raw potatoes are sorted by length and quality and then allocated to the separate product lines.

The company can purchase its potatoes from two sources, which differ in their yields of various sizes and quality. Each source yields different fractions of the product. French fries, has brown and flakes. Suppose that it is possible at different costs to alter these yield some what. Let f_1, f_2 and f_3 be the fractional yield per unit of weight of source-1 potatoes made into the three products.

Similarly, let g_1, g_2 and g_3 be the yield for source-2.

Suppose that each f_i and g_i can vary within $\pm 10\%$ of the yield shown below :

Product	Source-1	Source-2	Purchase limitation
French fries	0.2	0.3	1.8
Mash browns	0.2	0.1	1.2
Flakes	0.3	0.3	2.4
Relative profit	5.0	6.0	

Let $C_1(f_1, f_2, f_3)$ and $C_2(g_1, g_2, g_3)$ be the expense associated with obtaining these yields.

The problem is to determine how many

potatoes should the company purchase from each source? Formulate the problem as NLPP.

2. A manufacturing concern operates its two available machines to polish it metal products. The two machines are equally efficient although their maintenance costs are different. The daily maintenance and operation cost of the machine is given in rupees as the non-linear function

$$f(x_1,x_2)=100-1.2x_1-1.5x_2+0.3x_1^2+0.5x_2^2$$

where x_1 and x_2 are the numbers of hour of operation of machine-I and machine-II respectively.

The past records of the firm indicate that the combined operating hours of two machines should be at least 35 hours, a day in order to perform a satisfactory jobs. However, the production manager wishes to operate machine-I at least 6 hours more than machine-II because of the higher repair cost of the later. Find the optimal hours of operating the two machines and the minimum daily cost. Formulate the problem as NLPP.

Solve the following NLPP by graphical method

3. Maximize. $Z = 100x_1 - x_1^2 + 100x_2 - x_2^2$
subject to the constraints

$$x_1 + x_2 \geq 80$$
$$x_1 + 2x_2 \leq 100$$
and $$x_1, x_2 \geq 0$$

4. Minimize $f(Z) = (x_1-1)^2 + (x_2-2)^2$
subject to the constraints
$$0 \leq x_1 \leq 2, \ 0 \leq x_1 \leq 1$$

5. Min. $Z = 4(x_1 - 6)^2 + 6(x_2 - 2)^2$
subject to the constraints
$$0.5\,x_1 + x_2 \leq 4; \ 3x_1 + x_2 \leq 15; \ x_1 + x_2 \geq 1$$
and $$x_1, x_2 \geq 0$$

6. Min. $Z = (x_1 - 2)^2 + (x_2 - 1)^2$
subject to the constraints
$$-x_1^2 + x_2 \geq 0; \ -x_1 - x_2 + 2 \geq 0 \text{ and } x_1, x_2 \geq 0$$
Solve the following problems graphically and show that Kuhn-tucker necessary conditions for a maxima do not hold. What do you conclude?

7. Max. $Z = x_1$
subject o the constraints
$$(1-x_1)^2 - x_2 \geq 0$$
$$x_1, x_2 \geq 0$$

8. Max. $Z = x_1$
subject to the constraints
$$(3-x_1)^3 - (x_2 - 2) \geq 0$$
$$(3-x_1)^3 - (x_2 - 2) \geq 0$$
and $$x_1, x_2 \geq 0$$

ANSWERS

1. Max. $Z = 5(f_1, f_2, f_3) + 6(g_1, g_2, g_3)$ subject to the constraints
    $$0.2\,f_1 + 0.3\,g_1 \leq 1.8; \ 0.2\,f_2 + 0.1\,g_2 \leq 1.2;$$
    $$0.3\,f_3 + 0.3\,g_3 \leq 2.4; \ f_1, f_2, f_3$$

2. Min. $Z = 100 - 1.2\,x_1 - 1.5\,x_2 + 0.3x_1^2 + 0.5x_2^2$
    subject to the constraints $6x_1 + x_2 \geq 35; \ x_1 \geq 0; \ x_2 \geq 0$

3. $x_1 = 4, x_2 = 2, x_3 = 1$, Min $Z = -35$

4. $x_1 = 0, x_2 = 1$, Min $Z = 2$ 5. $x_1 = \dfrac{129}{29}, \ x_2 = \dfrac{48}{29}$, Min. $Z = \dfrac{7800}{841}$

6. $x_1 = 1, x_2 = 1$, Min $Z = 1$

7. $x_1 = 0, x_2 = 0$, Max. $Z = 1$, Constraint qualification is not satisfied.

8. $x_1 = 3, x_2 = 2$, Max. $Z = 3$, Constraint qualification is not satisfied.

5 Quadratic and Separable Programming

5.1 INTRODUCTION

In linear programming problem, because of linearity, we are able to develop a very efficient algorithm called the simplex method but in non-linear programming problem no such general algorithm exists for the solution. However for problems with certain suitable structures, efficient algorithm have been developed. To solve non-linear programming problem, we use quadratic and separable programming for which specific computational method have been designed. In a mathematical programming, when the objective function is quadratic and constraints are linear, we apply quadratic programming techniques. Separable programming is also a simplex based method and is applicable if the objective function and constraints are separable function.

5.2 QUADRATIC PROGRAMMING

An non-linear programming problem in which the objective function is quadratic and the constraints are linear is called quadratic programming problem.

The general structure of quadratic programming problem is given as below:

$$\text{Optimize } Z = \sum_{j=1}^{n} C_j x_j + \frac{1}{2} \sum_{j=1}^{n} \sum_{k=1}^{n} x_j d_{jk} x_k$$

subject to the constraints

$$\sum_{j=1}^{n} a_{ij} x_j \le b_i$$

and
$$x_j \ge 0, \quad i = 1, 2, \ldots, m; j = 1, 2, \ldots, n$$

where
$$d_{jk} = d_{kj} \quad \forall \, j, k \text{ and } b_i \ge 0$$

In matrix form the above quadratic programming problem (QPP) can be written as

$$\text{Optimize } Z = \mathbf{C}X + \frac{1}{2} X^T \mathbf{D} X$$

subject to the constraints

$$\mathbf{A}X \le \mathbf{b} \qquad \text{and} \qquad X \ge 0$$

where
$$X = (x_1, x_2, \ldots, x_n)^T; \qquad \mathbf{C} = (C_1, C_2, \ldots, C_n)$$
$$\mathbf{b} = (b_1, b_2, \ldots, b_m)^T$$
$$\mathbf{D} = [d_{jk}] \text{ is an } n \times n \text{ symmetric matrix.}$$
$$\mathbf{A} = [a_{ij}] \text{ is } m \times n \text{ matrix.}$$

☛ REMARKS

- The matrix D defined above is symmetric and positive definite.
- Objective function of the quadratic programming problem is strictly convex in X for minimization and concave in X for maximization.
- If the matrix D is a null matrix, the QPP reduced to the standard LPP.

5.3 KUHN-TUCKER CONDITIONS FOR QUADRATIC PROGRAMMING PROBLEMS

Consider the following QPP

$$\text{Max } f(X) = Z = \sum_{j=1}^{n} C_j x_j + \frac{1}{2} \sum_{j=1}^{n} d_{jk} x_j x_k$$

subject to the constraints

$$\sum_{j=1}^{n} a_{ij} x_j \le b_i \quad i = 1, 2, ..., m; j = 1, 2, ..., n$$

and $$x_j \ge 0$$

where $$d_{jk} = d_{kj} \text{ and } b_i \ge 0 \quad \forall i$$

Introducing slack variables s_i^2 and r_j^2, then above problem becomes

$$\text{max } Z = \sum_{j=1}^{n} C_j x_j + \frac{1}{2} \sum_{j=1}^{n} \sum_{k=1}^{n} d_{jk} x_j x_k$$

subject to the constraints

$$a_i'X + s_i^2 = b_i \quad i = 1,2,...,m$$

$$-x_j + r_j^2 = 0 \quad j = 1,2,...,n$$

Now, define the Lagrangian function such that

$$L(x,s,\mu,\lambda,r) = f(X) - \sum_{i=1}^{m} \lambda_i (a_i'X + s_i^2 - b_i) - \sum_{j=1}^{n} \mu_j (-x_j + r_j^2)$$

Forming the necessary condition, we obtain

$$\frac{\partial L}{\partial x_j} = \frac{\partial f}{\partial x_j} - \sum_{i=1}^{m} \lambda_i a_{ij} + \mu_j = 0, \quad j = 1,2,...,n \qquad ...(1)$$

$$\lambda_i (a_i X - b_i) = 0 \qquad ...(2)$$

$$\mu_i x_i = 0 \qquad ...(3)$$

$$AX \le b \qquad ...(4)$$

and $$X, \lambda, \mu \ge 0$$

Equation (1) can be rewritten as

$$\frac{\partial L}{\partial x_j} = \left[C_j + \frac{1}{2} \left(2 \sum_{k=1}^{n} C_{jk} x_k \right) \right] - \sum_{i=1}^{m} \lambda_i a_{ij} + \mu_j = 0, \quad j = 1,2,...,n$$

Letting $s_i^2 = q_i \ge 0$ then above equation becomes,

$$-\mu_j + C_j + \sum_{k=1}^{n} d_{jk} x_k - \sum_{i=1}^{m} \lambda_i a_{ij} = 0, \quad j = 1,2,...,n \qquad ...(5)$$

$$AX + Iq = b, \quad X \ge 0, q \ge 0, \lambda, \mu \ge 0$$

and $$\lambda_i q_i = 0, \ i = 1,2,...,m, \ \mu_j x_j = 0, \ j = 1,2,...,n$$

It must be noted that except for the final condition $\lambda_i q_i = 0 = \mu_j x_j$, the remaining equations are linear in X, λ, μ and q. The problem thus becomes equivalent to finding the solution to a set of linear equations which also satisfies the additional conditions $\lambda_i q_i = 0 = \mu_j x_j$. Since, $f(X)$ is strictly concave and the solution space is convex the feasible solution satisfying all these conditions must give the optimum solution directly.

5.4 WOLFE'S MODIFIED SIMPLEX METHOD

Consider the following quadratic programming problem

$$\text{Max. } Z = f(X) = \sum_{j=1}^{n} C_j x_j + \frac{1}{2} \sum_{j,k=1}^{n} x_j d_{jk} x_k$$

subject to the constraints

$$\sum_{j=1}^{n} a_{ij}x_j \le b_i$$

and $\qquad x_j \ge 0, \; i = 1, 2, ..., m; j = 1, 2, ..., n$

where $\qquad d_{jk} = d_{kj}$ and $b_i \ge 0$

Wolfe suggested a solution procedure for the above QPP using the ordinary simplex method with one slight modification as given in the following working procedure.

WORKING PROCEDURE

STEP 1. Write the given QPP into standard maximization form (*i.e.*, if it is a problem of minimization, then first convert it into maximization form).

STEP 2. Convert the inequality constraints into equations by introducing the slack variable s_i^2 in the ith constraints ($i = 1, 2, ..., m$) and the slack variables s_{m+j}^2 in the ith non-negative constraints ($j = 1, 2, ..., n$).

STEP 3. Construct the Lagrangian function

$$L(x,s,\lambda) = f(X) - \sum_{i=1}^{m} \lambda_i \left[\sum_{j=1}^{n} a_{ij}x_j - b_i + s_i^2 \right] - \sum_{j=1}^{n} \lambda_{m+j}(-x_j + s_{m+j}^2)$$

STEP 4. Construct the Kuhn-Tucker conditions by differentiating $L(x, s, \lambda)$ partially w.r.t. x, s and λ and put the first order partial derivatives equal to zero.

STEP 5. Introduce non-negative artificial variables $A_j : j = 1, 2, ..., n$ in the Kuhn-Tucker conditions and get the following equations

$$C_j + \sum_{k=1}^{n} d_{jk}x_k - \sum_{j=1}^{m} \lambda_i a_{ij} + \lambda_{m+j} = 0, \quad j = 1,2,...,m$$

STEP 6. Construct the objective function Max. $Z = -A_1 - A_2 - ... - A_n$.

STEP 7. Apply the simplex method to obtain initial basic feasible solution of the following LPP
$$\text{Max. } Z = -A_1 - A_2 - ... - A_n$$
subject to the constraints

$$\sum_{k=1}^{n} d_{jk}x_k - \sum_{i=1}^{m} \lambda_i a_{ij} + \lambda_{m+j} + A_j = -C_j$$

$$\sum_{j=1}^{n} a_{ij}x_j + x_{n+i} = b_i$$

and $\; A_j, \lambda_i, \lambda_{m+j}, x_j \ge 0$

where $x_{n+1} = s_i^2, \; i = 1,2,...,n$ and satisfying the complementary slackness conditions given by

$$\sum_{j=1}^{n} \lambda_{m+j}x_j + \sum_{j=1}^{m} x_{n+i}\lambda_i = 0$$

STEP 8. Use Phase-I of the artificial variable techniques (two phase method) to find the optimum solution to the LPP of step 7, which satisfy the complementary slackness conditions.

☛ REMARK

- The solution of the above system is obtained by using Phase-I of simplex method. Since, our aim is to obtain a feasible solution, the solution does not require the consideration of Phase-II. The only necessary thing is to maintain the conditions $\lambda_i s_i = 0 = \mu_j x_j$ all the time. This implies that if λ_i is in the basic solution with positive values, then x_i can not be basic with positive value. Similarly μ_j and x_j can not be positive simultaneously.

Phase-I will end in the usual manner with the sum of all artificial variables equal to zero only if the feasible solution to the problem exists.

SOLVED EXAMPLES

EXAMPLE I.   *Solve the following QPP using Wolfe's modified simplex method.*

$$\text{Max. } Z = 4x_1 + 6x_2 - 2x_1^2 - 2x_1 x_2 - 2x_2^2$$

subject to the constraints

$$x_1 + 2x_2 \leq 2$$

and $x_1, x_2 \geq 0$ [IAS–1994; GUWAHATI(MCA)–1992; DELHI(MBA)–2012]

SOLUTION.   Using the slack variables s_1^2, s_2^2, s_3^2, the modified problem can be written as

$$\text{Max. } Z = 4x_1 + 6x_2 - 2x_1^2 - 2x_1 x_2 - 2x_2^2$$

subject to the constraints

$$x_1 + 2x_2 + s_1^2 = 2$$

$$-x_1 + s_2^2 = 0$$

$$-x_2 + s_3^2 = 0$$

Let us define the Lagrangian function as given below

$$L(x_1, x_2; s_1, s_2, s_3; \lambda_1, \lambda_2, \lambda_3) = (4x_1 + 6x_2 - 2x_1^2 - 2x_1 x_2 - 2x_2^2)$$

$$- \lambda_1 (x_1 + 2x_2 + s_1^2 - 2) - \lambda_2 (-x_1 + s_2^2)$$

$$- \lambda_3 (-x_2 + s_3^2)$$

$$= 0$$

For the necessary condition of maxima, we must have

$$\frac{\partial L}{\partial x_1} = 0 \implies 4 - 4x_1 - 2x_2 - \lambda_1 + \lambda_2 = 0$$

$$\frac{\partial L}{\partial x_2} = 0 \implies 6 - 2x_1 - 4x_2 - 2\lambda_1 + \lambda_3 = 0$$

$$\frac{\partial L}{\partial s_1} = 0 \implies -2\lambda_1 s_1 = 0; \frac{\partial L}{\partial s_2} = 0 \implies -2\lambda_2 s_2 = 0$$

$$\frac{\partial L}{\partial s_3} = 0 \implies -2\lambda_3 s_3 = 0; \frac{\partial L}{\partial \lambda_1} = 0 \implies x_1 + 2x_2 + s_1^2 - 2 = 0$$

$$\frac{\partial L}{\partial \lambda_2} = 0 \implies -x_1 + s_2^2 = 0$$

and $\frac{\partial L}{\partial \lambda_3} = 0 \implies -x_2 + s_3^2 = 0$

After some simplification, we have the following system of equations

$$\left.\begin{aligned} 4x_1 + 2x_2 + \lambda_1 - \lambda_2 &= 4 \\ 2x_1 + 4x_2 - 2\lambda_1 - \lambda_3 &= 6 \\ x_1 + 2x_2 + s_1^2 &= 2 \end{aligned}\right\} \quad \dots(1)$$

and $\left.\begin{aligned} \lambda_1 s_1^2 + \lambda_2 s_2^2 + \lambda_3 s_3^2 = 0 \Rightarrow \lambda_1 s_1^2 + \lambda_2 x_1 + \lambda_3 x_2 = 0 \\ x_1, x_2, s_1^2, \lambda_i \geq 0, \quad i = 1, 2, 3 \end{aligned}\right\} \quad \dots(2)$

Now, to obtain the solution to the above simultaneous equations (1), introduce the artificial variables $A_1, A_2 (\geq 0)$ in the first two constraints of (1). Then given problem can be written as

$$\text{Max. } Z = -A_1 - A_2$$

subject to the constraints

$$4x_1 + 2x_2 + \lambda_1 - \lambda_2 + A_1 = 4$$
$$2x_1 + 4x_2 + 2\lambda_1 - \lambda_3 + A_2 = 6$$
$$x_1 + 2x_2 + x_3 = 2$$

and $\quad x_1, x_2, x_3 \geq 0, A_1, A_2, \lambda_i \geq 0$

We note that here we have replaced s_1^2 by x_3 and satisfied complementary slackness conditions $\Sigma\lambda_i x_i = 0$. Now, apply the procedure of phase-1 of two phase method we have the following simplex table

Simplex Table-1

B.V.	C_B	X_B	x_1	x_2	x_3	λ_1	λ_2	λ_3	A_1	A_2	Min. Ratio $X_B \vert x_1$
	C_j		0	0	0	0	0	0	-1	-1	
A_1	-1	4	④	2	0	1	-1	0	1	0	1 (min) $\rightarrow$
A_2	-1	6	2	4	0	2	0	-1	0	1	3
x_3	0	2	1	2	1	0	0	0	0	0	2
	Δ_j		6 $\uparrow$	6	0	3	-1	-1	0	0	

In the above table, clearly we have x_1 and x_2 are the variables with the most positive entry. Let x_1 enter in the basis and A_1 will leave. Then next simplex table is given as below.

Simplex Table-2

B.V.	C_B	X_B	x_1	x_2	x_3	λ_1	λ_2	λ_3	A_2	Min. Ratio $X_B \vert x_2$
	C_j		0	0	0	0	0	0	-1	
x_1	0	1	1	1/2	0	1/4	$-1/4$	0	0	2
A_2	-1	4	0	3	0	3/2	1/2	-1	1	4/3
x_3	0	1	0	③/②	1	$-1/4$	1/4	0	0	2/3(min) $\rightarrow$
	Δ_j		0	3 $\uparrow$	0	3/2	1/2	-1	0	

In the above table, x_2 is the variable with the most positive entry (i.e., 3) in the Δ_j-row so x_2 enter in the basis and x_3 will leave. Then we have the following simplex table.

Simplex Table-3

B.V.	C_B	X_B	x_1	x_2	x_3	λ_1	λ_2	λ_3	A_2	Min. Ratio $X_B \vert \lambda_1$
	C_j		0	0	0	0	0	0	-1	
x_1	0	2/3	1	0	$-1/3$	1/3	$-1/3$	0	0	2
A_2	-1	2	0	0	-2	②	0	-1	1	1(min) $\rightarrow$
x_2	0	2/3	0	1	2/3	$-1/6$	1/6	0	0	—
	Δ_j		0	0	-2	2 $\uparrow$	0	-1	0	

Here, λ_1 is the variable with the most positive entry 2 in Δ_j-row so λ_1 enter in the basis and A_2 will leave. Then we have the following simplex table.

Simplex Table-4

	C_j		0	0	0	0	0	0
B.V.	C_B	X_B	x_1	x_2	x_3	A_1	A_2	A_3
x_1	0	1/3	1	0	0	0	–1/3	1/6
λ_1	0	1	0	0	–1	1	0	–1/6
x_2	0	5/6	0	1	1/2	0	1/6	–1/6
	Δ_j		0	0	0	0	0	0

In the above table, all $\Delta_j \le 0 \Rightarrow$ solution is optimal and is given by
$$x_1 = 1/3, x_2 = 5/6 \text{ and}$$

$$\text{Max.} Z = 4\left(\frac{1}{3}\right) + 6\left(\frac{5}{6}\right) - 2\left(\frac{1}{3}\right)^2 - 2\left(\frac{1}{3}\right)\left(\frac{5}{6}\right) - 2\left(\frac{5}{6}\right)^2 = \frac{25}{6}$$

EXAMPLE 2. *Solve the following QPP using Wolfe's method:*

$$\textbf{Max. } \boldsymbol{Z = 2x_1 + x_2 - x_1^2}$$

subject to the constraints

$$\boldsymbol{2x_1 + 3x_2 \le 6}$$
$$\boldsymbol{2x_1 + x_2 \le 4}$$

and $\quad \boldsymbol{x_1, x_2 \ge 0}$ $\qquad$ [ANDHRA(BE)–1996; MADRAS–1990; MEERUT(M.PHIL)–2012]

SOLUTION. Considering non-negativity constraints conditions $x_1 \ge 0, x_2 \ge 0$ as inequality constraints and adding the slack variables s_1^2, s_2^2, s_3^2 and s_4^2 to all the inequalities, the above problem can be written as

$$\text{Max. } Z = 2x_1 + x_2 - x_1^2$$

subject to the constraints

$$2x_1 + 3x_2 + s_1^2 = 6$$
$$2x_1 + x_2 + s_2^2 = 4$$
$$-x_1 + s_3^2 = 0$$
$$-x_2 + s_4^2 = 0$$

Let us define the Lagrangian function as given below

$$L(X, S, \lambda) = (2x_1 + x_2 - x_1^2) - \lambda_1(2x_1 + 3x_2 + s_1^2 - 6) - \lambda_2(2x_1 + x_2 + s_2^2 - 4)$$
$$- \lambda_3(-x_1 + s_3^2) - \lambda_4(-x_2 + s_4^2)$$

Now, the necessary condition for the maximum of L (and hence for Z) we must have

$$\frac{\partial L}{\partial x_1} = 0 \qquad \Rightarrow \qquad -2 - 2x_1 - 2\lambda_1 - 2\lambda_2 + \lambda_3 = 0$$

$$\frac{\partial L}{\partial x_2} = 0 \qquad \Rightarrow \qquad 1 - 3\lambda_1 - \lambda_2 + \lambda_4 = 0$$

$$\frac{\partial L}{\partial s_1} = 0 \quad \Rightarrow \quad -2\lambda_1 s_1 = 0 \qquad \frac{\partial L}{\partial s_2} = 0 \quad \Rightarrow \quad -2\lambda_2 s_2 = 0$$

$$\frac{\partial L}{\partial s_3} = 0 \quad \Rightarrow \quad -2\lambda_3 s_3 = 0 \qquad \frac{\partial L}{\partial s_4} = 0 \quad \Rightarrow \quad -2\lambda_4 s_4 = 0$$

$$\frac{\partial}{\partial \lambda} = 0 \quad \Rightarrow \quad 2x_1 + 3x_2 + s_1 - 6 = 0$$

$$\frac{\partial}{\partial \lambda} = 0 \quad \Rightarrow \quad 2x_1 + x_2 + s_2 - 4 = 0$$

$$\frac{\partial}{\partial \lambda} = 0 \quad \Rightarrow \quad -x_1 + s_3 = 0$$

$$\frac{\partial}{\partial \lambda} = 0 \quad \Rightarrow \quad -x_2 + s_4 = 0$$

On simplification, above equations can be written as

$$2x_1 + 2\lambda_1 + 2\lambda_2 - \lambda_3 = 2$$
$$3\lambda_1 + \lambda_2 - \lambda_4 = 1$$
$$2x_1 + 3x_2 + s_1^2 = 6$$
$$2x_1 + x_2 + s_2^2 = 4$$
$$\lambda_1 s_1 = \lambda_2 s_2 = 0$$
$$\lambda_3 x_1 = \lambda_4 x_2 = 0$$

and $\quad x_1, x_2; \lambda_1, \lambda_2, \lambda_3, \lambda_4; s_1, s_2 \geq 0$

Now, introducing the artificial variables A_1 and A_2, the modified QPP becomes

Max. $Z = -A_1 - A_2$

subject to the constraints

$$2x_1 + 2\lambda_1 + 2\lambda_2 - \lambda_3 + A_1 = 2$$
$$3\lambda_1 + \lambda_2 - \lambda_4 + A_2 = 1$$
$$2x_1 + 3x_2 + s_1^2 = 6$$
$$2x_1 + x_2 + s_2^2 = 4$$

$$\lambda_1 s_1 = \lambda_2 s_2 = 0$$
$$\lambda_3 x_1 = \lambda_4 x_2 = 0$$
$$x_i, s_j, \lambda_k, A_l \geq 0$$

Now, apply the phase-1 of two phase method, we have the following simplex table

Simplex Table-1

B.V.	C_j		0	0	0	0	0	0	0	0	-1	-1	Min. Ratio
	C_B	X_B	x_1	x_2	λ_1	λ_2	λ_3	λ_4	s_1	s_2	A_1	A_2	
A_1	1	2	②	0	2	2	-1	0	0	0	1	0	1 (min) $\rightarrow$
A_2	1	1	0	0	3	1	0	-1	0	0	0	1	—
s_1	0	6	2	3	0	0	0	0	1	0	0	0	3
s_2	0	4	2	1	0	0	0	0	0	1	0	0	2
	Δ_j		2	0	5	3	-1	-1	0	0	0	0	
			$\uparrow$										

In the above table, the largest value in Δ_j-row is 5, but we can not enter λ_1 (or λ_2) in the basis because complementry slackness conditions $\lambda_1 s_1 = \lambda_2 s_2 = 0$. Since, $\lambda_3 = 0$, x_1 can be entered into the basis with A_1 as leaving variable.

Then, new simplex table is given as under

Simplex Table-2

	C_j		0	0	0	0	0	0	0	0	-1	Min. Ratio
B.V.	C_B	X_B	x_1	x_2	λ_1	λ_2	λ_3	λ_4	s_1	s_2	A_2	
x_1	0	1	1	0	1	1	1/2	0	0	0	0	—
A_2	1	1	0	0	3	1	0	-1	0	0	1	—
s_1	0	4	0	③	-2	-2	1	0	1	0	0	4/3 (min)→
s_2	0	2	0	1	-2	-2	1	0	0	1	0	2
	Δ_j		0	0	3	1	0	-1	0	0	0	
				↑								

In the above table, again we can not enter λ_1, λ_2, and λ_3 in the basis because s_1, s_2 and x_1 are already in the basis. Therefore, enter x_2 into the basis with s_1 as the leaving variable because $\lambda_4 = 0$. The next simplex table is given as follows:

Simplex Table-3

	C_j		0	0	0	0	0	0	0	0	-1	Min. Ratio
B.V.	C_B	X_B	x_1	x_2	λ_1	λ_2	λ_3	λ_4	s_1	s_2	A_2	
x_1	0	1	1	0	1	1	$-1/2$	0	0	0	0	1
A_2	1	1	0	0	③	1	0	-1	0	0	1	1/3 (min)→
x_2	0	4/3	0	1	$-2/4$	$-2/3$	1/3	0	1/3	0	0	
s_2	0	2/3	0	0	$-4/3$	$-4/3$	2/3	0	$-1/3$	1	0	
	Δ_j		0	0	3	1	0	-1	0	0	-1	
					↑							

In the above table, since $s_1 = 0$, λ_1 can be entered into the basis and A_2 will leave the basis. Then we have the following simplex table

Simplex Table-4

	C_j		0	0	0	0	0	0	0	0
B.V.	C_B	X_B	x_1	x_2	λ_1	λ_2	λ_3	λ_4	s_1	s_2
x_1	0	2/3	1	0	0	2/3	$-1/2$	1/3	0	0
λ_1	0	1/3	0	0	1	1/3	0	$-1/3$	0	0
x_2	0	14/9	0	1	0	$-4/9$	1/3	$-2/9$	1/3	0
s_2	0	10/9	0	0	0	$-8/9$	2/3	$-4/9$	$-1/3$	1
	Δ_j		0	0	0	0	0	0	0	0

Here, clearly all $\Delta_j \leq 0 \Rightarrow$ solution is optimal and is given by

$$x_1 = \frac{2}{3}, x_2 = \frac{14}{9}, \lambda_1 = \frac{1}{3}, \lambda_2 = \lambda_3 = \lambda_4 = 0, s_1 = 0, s_2 = \frac{10}{9}$$

This solution also satisfy the complementary slackness conditions

$$\lambda_1 s_1 = \lambda_2 s_2 = 0, \lambda_3 x_1 = \lambda_4 x_2 = 0$$

Also, $Z^* = 0$, so current solution is also feasible. Hence, the maximum value of the objective function of the given QPP is given by

$$\text{Max}.Z = 2\left(\frac{2}{3}\right) + \left(\frac{14}{9}\right) - \left(\frac{2}{3}\right)^2 = \frac{22}{9}$$

EXAMPLE 3. *Solve the following quadratic programming problem by using Wolfe's modified simplex method.*

$$\textbf{Max. } Z = 2x_1 + 3x_2 - 2x_1^2$$

subject to the constraints
$$x_1 + 4x_2 \leq 4$$
$$x_1 + x_2 \leq 2$$
and $\quad x_1, x_2 \geq 0$

SOLUTION. Using the slack variables s_1^2, s_2^2, s_3^2 and s_4^2 in the given QPP, we get the following modified form.

$$\text{Max. } Z = 2x_1 + 3x_2 - 2x_1^2$$
subject to the constraints
$$x_1 + 4x_2 + s_1^2 = 4$$
$$x_1 + x_2 + s_2^2 = 2$$
$$-x_1 + s_3^2 = 0$$
$$-x_2 + s_4^2 = 0$$

Now, the Lagrangian function is given by
$$L(x_1, x_2, s_1, s_2, s_3, s_4, \lambda_1, \lambda_2, \lambda_3, \lambda_4)$$
$$= (2x_1 + 3x_2 - 2x_1^2) - \lambda_1(x_1 + 4x_2 + s_1^2 - 4)$$
$$- \lambda_2(x_1 + x_2 + s_2^2 - 2) - \lambda_3(-x_1 + s_3^2) - \lambda_4(-x_2 + s_4^2)$$

Now, the necessary conditions of the minimum of L (and hence Z) we must have

$$\frac{\partial L}{\partial x_1} = 0 \Rightarrow 2 - 4x_1 - \lambda_1 - \lambda_2 + \lambda_3 = 0$$

$$\frac{\partial L}{\partial x_2} = 0 \Rightarrow 3 - 4\lambda_1 - \lambda_2 + \lambda_4 = 0$$

$$\frac{\partial L}{\partial s_1} = 0 \Rightarrow -2\lambda_1 s_1 = 0; \frac{\partial L}{\partial s_2} = 0 \Rightarrow -2\lambda_2 s_2 = 0$$

$$\frac{\partial L}{\partial s_3} = 0 \Rightarrow -2\lambda_3 s_3 = 0; \frac{\partial L}{\partial s_4} = 0 \Rightarrow -2\lambda_4 s_4 = 0$$

$$\frac{\partial L}{\partial \lambda_1} = 0 \Rightarrow x_1 + 4x_2 + s_1^2 - 4 = 0$$

$$\frac{\partial L}{\partial \lambda_2} = 0 \Rightarrow x_1 + x_2 + s_2^2 - 2 = 0$$

$$\frac{\partial L}{\partial \lambda_3} = 0 \Rightarrow -x_1 + s_3^2 = 0$$

and $\quad \frac{\partial L}{\partial \lambda_4} = 0 \Rightarrow -x_2 + s_4^2 = 0$

After some simplification, we get
$$\left. \begin{array}{l} 4x_1 + \lambda_1 + \lambda_2 - \lambda_3 = 2 \\ 4\lambda_1 + \lambda_2 - \lambda_4 = 3 \end{array} \right] \qquad \qquad \dots(1)$$

and
$$x_1 + 4x_2 + s_1^2 = 4$$
$$x_1 + x_2 + s_2^2 = 2$$
$$\lambda_1 s_1^2 + \lambda_2 s_2^2 + x_1 \lambda_3 + x_2 \lambda_4 = 0$$
...(2)

$$x_1, x_2, s_1, s_2, s_3, s_4, \lambda_1, \lambda_2, \lambda_3, \lambda_4 \geq 0$$

Now, proceed same as in previous questions, introduce the artificial variables A_1, A_2 (≥ 0) in the equations of (1) we get the following form

$$\text{Max. } Z = -A_1 - A_2$$

subject to the constraints

$$4x_1 + \lambda_1 + \lambda_2 - \lambda_3 + A_1 = 2$$
$$4\lambda_1 + \lambda_2 - \lambda_4 + A_2 = 3$$
$$x_1 + 4x_2 + x_3 = 4$$
$$x_1 + x_2 + x_4 = 2$$

and $x_i, s_i, A_i, \lambda_i \geq 0$

Here, we have replaced s_1^2 by x_3 and s_2^2 by x_4 and satisfied the complementry slackness conditions $\Sigma \lambda_i x_i = 0$. Now, the optimum solution of the above problem shall be obtained using phase-1 of two phase method. So, we have the following initial simplex table.

Simplex Table-1

B.V.	C_B	X_B	x_1	x_2	x_3	x_4	λ_1	λ_2	λ_3	λ_4	A_1	A_2	Min. Ratio $X_B \vert x_1$
	C_j		0	0	0	0	0	0	0	0	-1	-1	
A_1	-1	2	④	0	0	0	1	1	-1	0	1	0	1/2(min) →
A_2	-1	3	0	0	0	0	4	1	0	-1	0	1	—
x_3	0	4	1	4	1	0	0	0	0	0	0	0	4
x_4	0	2	1	1	0	1	0	0	0	0	0	0	2
	Δ_j		4	0	0	0	5	2	-1	-1	0	0	
			↑										

In the above table we observe that λ_1 is the variable with the most positive value (*i.e.*, 5) and so it should enter the basis, but it will not because x_3 is in the basis (complementary slackness condition $\lambda_1 x_3 = 0$). The next most positive entry is 4 for the x_1 column in Δ_j-row. Hence, x_1 enter in the basis and A_1 will leave. Then we have the following simplex table.

Simplex Table-2

B.V.	C_B	X_B	x_1	x_2	x_3	x_4	λ_1	λ_2	λ_3	λ_4	A_2	Min. Ratio $X_B \vert x_2$
	C_j		0	0	0	0	0	0	0	0	-1	
x_1	0	1/2	1	0	0	0	1/4	1/4	$-1/4$	0	0	—
A_2	-1	3	0	0	0	0	4	1	0	-1	1	—
x_3	0	7/2	0	④	1	0	$-1/4$	$-1/4$	1/4	0	0	7/8(min)→
x_4	0	3/2	0	1	0	1	$-1/4$	$-1/4$	1/4	0	0	3/2
	Δ_j		0	0	0	0	4	1	0	-1	0	
				↑								

Here, we see that either λ_1 or λ_2 will enter in the basis, but since x_3 and x_4 are still in the basis they can not enter the basis because of the complementary slackness conditions $\lambda_1 x_3 = 0$ and $\lambda_2 x_4 = 0$. Here, x_2 is eligible to enter in the basis because λ_4 is not in the basis ($\lambda_4 x_2 = 0$). Then we have the following simplex table.

Simplex Table-3

	C_j		0	0	0	0	0	0	0	0	-1	Min. Ratio
B.V.	C_B	X_B	x_1	x_2	x_3	x_4	λ_1	λ_2	λ_3	λ_4	A_2	X_B / λ_2
x_1	0	1/2	1	0	0	0	1/4	$-1/4$	$-1/4$	0	0	2
A_2	-1	3	0	0	0	0	④	1	0	-1	1	3/8(min)$\rightarrow$
x_2	0	7/8	0	1	1/4	0	$-1/16$	$-1/16$	1/16	0	0	—
x_4	0	5/8	0	0	$-1/4$	1	$-3/16$	$-3/16$	3/16	0	0	—
	Δ_j		0	0	0	0	4	1	0	-1	0	
							$\uparrow$					

In the above table, we observe that either λ_1 or λ_2 will enter in the basis but x_4 is still in the basis so λ_2 can not enter because of the complementary slackness conditions $\lambda_2 x_4 = 0$. So, λ_1 enter in the basis.

Hence, we have the following simplex table

Simplex Table-4

	C_j		0	0	0	0	0	0	0	0
B.V.	C_B	X_B	x_1	x_2	x_3	x_4	λ_1	λ_2	λ_3	λ_4
x_1	0	5/16	1	0	0	0	0	$-5/16$	$-1/4$	1/16
λ_2	0	3/4	0	0	0	0	1	1/4	0	$-1/4$
x_2	0	59/64	0	1	1/4	0	0	$-3/64$	1/16	$-1/64$
x_4	0	49/64	0	0	$-1/4$	1	0	$-9/64$	3/16	$-3/64$
	Δ_j		0	0	0	0	0	0	0	0

In the above table, we observe that all $\Delta_j \leq 0$.

$\Rightarrow$ Solution is optimum and is given by

$$x_1 = \frac{5}{16}, x_2 = \frac{59}{64} \text{ and Max. } Z = 3.19$$

EXAMPLE 4. *Solve the following QPP by using Wolfe's method*

$$\text{Max. } Z = 6x_1 + 3x_2 - 4x_1 x_2 - 2x_1^2 - 3x_2^2$$

subject to the constraints

$$x_1 + x_2 \leq 1$$
$$2x_1 + 3x_2 \leq 4$$

and $$x_1, x_2 \geq 0$$

SOLUTION. Using the slack variables s_1^2, s_2^2, s_3^2 and s_4^2 we have the modified form of the given problem as follows:

$$\text{Max. } Z = 6x_1 + 3x_2 - 4x_1 x_2 - 2x_1^2 - 3x_2^2$$

s.t. $$x_1 + x_2 + s_1^2 = 1$$

$$2x_1 + 3x_2 + s_2^2 = 4$$

$$-x_1 + s_3^2 = 0$$

$$-x_2 + s_4^2 = 0$$

Now, construct the Lagrangian function as given below

$$L(x_1, x_2; s_1, s_2, s_3, s_4; \lambda_1, \lambda_2, \lambda_3, \lambda_4)$$

$$= (6x_1 + 3x_2 - 4x_1x_2 - 2x_1^2 - 3x_2^2)$$

$$- \lambda_1(x_1 + x_2 + s_1^2 - 1) - \lambda_2(2x_1 + 3x_2 + s_2^2 - 4)$$

$$- \lambda_3(-x_1 + s_3^2) - \lambda_4(-x_2 + s_4^2)$$

For the maxima of L (and hence Z) we must have

$$\frac{\partial L}{\partial x_1} = 0 \quad \Rightarrow \quad 6 - 4x_2 - 4x_1 - \lambda_1 - 2\lambda_2 + \lambda_3 = 0$$

$$\frac{\partial L}{\partial x_2} = 0 \quad \Rightarrow \quad 3 - 4x_1 - 6x_2 - \lambda_1 - 3\lambda_2 + \lambda_4 = 0$$

$$\frac{\partial L}{\partial s_1} = 0 \quad \Rightarrow \quad -2\lambda_1 s_1 = 0; \frac{\partial L}{\partial s_2} = 0 \quad \Rightarrow \quad -2\lambda_2 s_2 = 0$$

$$\frac{\partial L}{\partial s_3} = 0 \quad \Rightarrow \quad -2\lambda_3 s_3 = 0; \frac{\partial L}{\partial s_4} = 0 \quad \Rightarrow \quad -2\lambda_4 s_4 = 0$$

$$\frac{\partial L}{\partial \lambda_1} = 0 \quad \Rightarrow \quad x_1 + x_2 + s_1^2 - 1 = 0$$

$$\frac{\partial L}{\partial \lambda_2} = 0 \quad \Rightarrow \quad 2x_1 + 3x_2 + s_2^2 - 4 = 0$$

$$\frac{\partial L}{\partial \lambda_3} = 0 \quad \Rightarrow \quad -x_1 + s_3^2 = 0$$

and $$\frac{\partial L}{\partial \lambda_4} = 0 \quad \Rightarrow \quad -x_2 + s_4^2 = 0$$

After some simplification, we get

$$\left.\begin{array}{r} 4x_1 + 4x_2 + \lambda_1 + 2\lambda_2 - \lambda_3 = 6 \\ 4x_1 + 6x_2 + \lambda_1 + 3\lambda_2 - \lambda_4 = 3 \\ x_1 + x_2 + s_1^2 = 1 \end{array}\right] \qquad \dots(1)$$

and

$$\left.\begin{array}{r} 2x_1 + 3x_2 + s_2^2 = 4 \\ \lambda_1 s_1^2 + \lambda_2 s_2^2 + x_1\lambda_3 + x_2\lambda_4 = 0 \\ x_1, x_2, s_1^2, s_2^2, \lambda_i \geq 0 \end{array}\right] \qquad \dots(2)$$

Proceed same as in previous questions, let us introduce artificial variables A_1 and A_2 (≥ 0) in the first two constraints of (1), we have the following problem

$$\text{Max } Z = -A_1 - A_2$$

subject to the constraints

$$4x_1 + 4x_2 + \lambda_1 + 2\lambda_2 - \lambda_3 + A_1 = 6$$

$$4x_1 + 6x_2 + \lambda_1 + 3\lambda_2 - \lambda_4 + A_2 = 3$$

$$x_1 + x_2 + x_3 = 1$$

$$2x_1 + 3x_2 + x_4 = 4$$

and $\quad x_1, x_2, x_3, x_4, A_1, A_2, \lambda_i \geq 0 \ \forall \ i = 1, 2, 3, 4$

Here, we have replaced s_1^2 by x_3 and s_2^2 by x_4 and satisfied the complementary slackness conditions $\Sigma \lambda_i x_i = 0$. The optimum solution to the above LPP will now be obtained by phase-1 of two phase method. Thus, we have the following starting table,

Simplex Table-1

B.V.	C_B	X_B	C_j 0 x_1	0 x_2	0 x_3	0 x_4	0 λ_1	0 λ_2	0 λ_3	0 λ_4	-1 A_1	-1 A_2	Min. Ratio $X_B\|x_2$
A_1	-1	6	4	4	0	0	1	2	-1	0	1	0	3/2
A_2	-1	3	4	⑥	0	0	1	3	0	-1	0	1	1/2(min) →
x_3	0	1	1	1	1	0	0	0	0	0	0	0	1
x_4	0	4	2	3	0	1	0	0	0	0	0	0	4/3
	Δ_j		8	10 ↑	0	0	2	5	-1	-1	0	0	

In the above table, we observe that x_2 is the variable with the most positive entry 10 in the Δ_j-row and hence will enter the basis and A_2 will leave. Then we have the following simplex table.

Simplex Table-2

B.V.	C_B	X_B	C_j 0 x_1	0 x_2	0 x_3	0 x_4	0 λ_1	0 λ_2	0 λ_3	0 λ_4	-1 A_1	Min. Ratio $X_B\|x_1$
A_1	-1	4	4/3	0	0	0	1/3	0	-1	2/3	1	3
x_2	0	1/2	②/③	1	0	0	1/6	1/2	0	$-1/6$	0	3/4(min) →
x_3	0	1/2	1/3	0	1	0	$-1/6$	$-1/2$	0	1/6	0	3/2
x_4	0	5/2	0	0	0	1	$-1/2$	$-3/2$	0	1/2	0	—
	Δ_j		4/3 ↑	0	0	0	1/3	0	-1	2/3	0	

In the above table, we see that x_1 is the variable with the most positive entry 4/3 in the Δ_j-row and hence will enter the basis and x_2 will leave. Then we have the following simplex table.

Simplex Table-3

B.V.	C_B	X_B	C_j 0 x_1	0 x_2	0 x_3	0 x_4	0 λ_1	0 λ_2	0 λ_3	0 λ_4	-1 A_1	Min. Ratio
A_1	-1	3	0	-2	0	0	0	-1	-1	1	1	3
x_1	0	3/4	1	3/2	0	0	1/4	3/4	0	$-1/4$	0	—
x_3	0	1/4	0	$-1/2$	1	0	$-1/4$	$-3/4$	0	①/④	0	4/3(min) →
x_4	0	5/2	0	0	0	1	$-1/2$	$-3/2$	0	1/2	0	5
	Δ_j		0	-2	0	0	0	-1	-1	1 ↑	0	

Here, λ_4 is the variable with the most positive entry 1 in Δ_j-row and hence will enter the basis and x_3 will leave. Then the next simplex table is given as below:

Simplex Table-4

	C_j		0	0	0	0	0	0	0	0	– 1	Min. Ratio
B.V.	C_B	X_B	x_1	x_2	x_3	x_4	λ_1	λ_2	λ_3	λ_4	A_1	X_B / λ_1
A_1	–1	6	4	4	0	0	①	2	–1	0	1	6 →
x_1	0	1	1	1	1	0	0	0	0	0	0	—
λ_4	0	1	0	–2	4	0	–1	–3	0	1	0	—
x_4	0	2	0	1	–2	1	0	0	0	0	0	—
	Δ_j		0	4	0	0	1	2	–1	0	0	
							↑					

In the above table, either x_2 or λ_1 will enter the basis but λ_4 is still in the basis, so x_2 can not enter in the basis because of the complementary slackness condition $\lambda_2 x_4 = 0$. Hence, λ_1 enter in the basis and A_1 leaves. Therefore, we have the following simplex table.

Simplex Table-5

	C_j		0	0	0	0	0	0	0	0
B.V.	C_B	X_B	x_1	x_2	x_3	x_4	λ_1	λ_2	λ_3	λ_4
λ_1	0	6	4	4	0	0	1	2	–1	0
x_1	0	1	1	1	1	0	0	0	0	0
λ_4	0	7	4	2	4	0	0	–1	–1	1
x_4	0	2	0	1	–2	1	0	0	0	0
	Δ_j		0	0	0	0	0	0	0	0

In the above table, we observe that all $\Delta_j \leq 0$.

$\Rightarrow$ solution is optimum and is given by

$$x_1 = 1, x_2 = 0$$

and   Max. $Z = 6(1) + 3(0) - 4(1)(0) - 2(1)^2 - 3(0)^2 = 4$

5.5 BEALE'S METHOD

In Wolfe's method, we use the Kuhn-Tucker conditions, but in this method instead of Kuhn-Tucker conditions, we used the results based on calculus.

Consider the quadratic programming problem given by

$$\text{Max. } Z = f(X) = C^T \cdot X + \frac{1}{2} X^T Q X$$

subject to the constraints

$$AX (\leq, \geq \text{ or } =) b$$

and $X \geq 0$, where $X = (x_1, x_2, ..., x_n) \in R^n$

A is a $m \times n$, b is $m \times 1$, C is $n \times 1$ and Q is an $n \times n$ symmetric matrix.

To solve the above QPP by Beale's method, we use the following procedure,

WORKING PROCEDURE

STEP 1. Convert the given QPP into the standard form of maximization.

STEP 2. Set the given QPP in standard form using slack and surplus variables.

STEP 3. Choose arbitrarily any m variables as the basic variables such that remaining $(n-m)$ variables becomes non-basic.

STEP 4. Write each basic variable X_B in terms of non-basic variables X_{NB} and u_i if any using the given constraints.

STEP 5. Write the objective function $f(X)$ in terms of non-basic variables X_{NB} (and u_i if any)

STEP 6. Obtain the partial derivaties of $f(X)$ w.r.t. the non-basic variables at the point $X_{NB} = 0$ (or $u = 0$). Then we have the following three cases:

If $\left(\dfrac{\partial f}{\partial X_{NB_k}}\right)_{\substack{X_{NB}=0 \\ u=0}} = 0$ for each $k = 1, 2, ..., n - m$ and $\left(\dfrac{\partial f}{\partial u_i}\right)_{\substack{X_{NB}=0 \\ u=0}} = 0$ then the current basic solution is optimal. Then go to step 9.

If $\left(\dfrac{\partial f}{\partial X_{NB_k}}\right)_{\substack{X_{NB}=0 \\ u=0}} > 0$ for at least one k, then select the most positive one. Then the corresponding non-basic variables will enter the basis. Then go to step 7.

If $\left(\dfrac{\partial f}{\partial X_{NB_k}}\right)_{\substack{X_{NB}=0 \\ u=0}} = 0$ for each $k = 1, 2, ..., n - m$ and $\left(\dfrac{\partial f}{\partial u_i}\right)_{\substack{X_{NB_k}=0 \\ u=0}} \neq 0$ for some

$i = r$, then introduce a new non-basic variable $u_j = \dfrac{1}{2}\left(\dfrac{\partial f}{\partial u_r}\right)$ and treat u_r as a basic variable and go to step 4.

STEP 7. Consider $X_{NB_i} = x_k$ to be the entering variable identified in the above step 6. Find the minimum ratio, i.e., $\min\left\{\dfrac{\alpha_{h_0}}{|\alpha_{hk}|}, \dfrac{\gamma_{k_0}}{|\gamma_{kk}|}\right\}$ for all basic variables x_h, where α_{h_0} is the constant term and α_{hk} is the coefficient of x_k in the expression of basic variables x_k when expressed in terms of the non-basic ones and γ_{k_0} is the constant term and γ_{kk} is the coefficient of x_k in $\dfrac{\partial f}{\partial x_k}$. Here, we have the following cases:

CASE 1 If the minimum ratio occur for some $\dfrac{\alpha_{h_0}}{|\alpha_{hk}|}$, the corresponding basic variables x_h will be the outgoing vector, i.e., x_h will leave the basis.

CASE 2 If the minimum ratio occurs for some $\dfrac{\gamma_{k_0}}{|\gamma_{kk}|}$, there exist criterion corresponding to a non-basic variables. Then we introduce an additional non-basic variable (called free variable) defined by

$$u_i = \frac{1}{2}\frac{\partial f}{\partial x_k}, u_i \text{ is unrestricted}$$

STEP 8. Go to step 4 and repeat the same procedure until an optimum basic solution is obtained.

STEP 9. Find the optimum value of X_B and $f(X)$ by setting $X_{NB} = 0$ in the expression obtained in steps 4 and 5.

☞ **REMARKS**
- The free variable u_j is introduced in the set of constraints only for computational purpose and its value is zero at the next feasible solution.
- Since u_j is unrestricted in sign, so while evaluating $\dfrac{\partial f}{\partial u_j}$, both increase and decrease must be checked.

> At any iteration, if a free variable becomes a basic variable and is non-zero then drop the new constraints containing it. This should be done because it is a free variable and therefore will neither be chosen to leave the basis nor will appear in the selection of leaving variable.

SOLVED EXAMPLES

EXAMPLE 1. *Solve the following quadratic programming problem by Beale's method*

$$\textbf{Max. } Z = 2x_1 + 3x_2 - x_1^2$$

subject to the constraints

$$x_1 + 2x_2 \leq 4$$

and $\qquad x_1, x_2 \geq 0$

SOLUTION. Using the slack variable s_1, the above QPP can be written as

$$\text{Max. } Z = 2x_1 + 3x_2 - x_1^2$$

subject to the constraints

$$x_1 + 2x_2 + s_1 = 4$$

and $\qquad x_1, x_2, s_1 \geq 0$

Treat s_1 as a basic variable and x_1, x_2 as non-basic.

i.e., $\qquad X_B = (s_1)$ and $X_{NB} = (x_1, x_2)$

Then we have

$$s_1 = 4 - x_1 - 2x_2 \qquad\qquad\qquad \text{...(1)}$$

and $\qquad f(X) = 2x_1 + 3x_2 - x_1^2 \qquad\qquad \text{...(2)}$

Differentiating (2) partially w.r.t. x_1 and x_2 we get

$$\left(\frac{\partial f}{\partial x_1}\right)_{X_{NB}=0} = (2 - 2x_1)_{x_1, x_2 = 0} = 2$$

and $\qquad \left(\dfrac{\partial f}{\partial x_2}\right)_{X_{NB}=0} = (3)_{x_1, x_2 = 0} = 3$

Hence, the most positive is α_2, therefore x_2 will enter in the basis.

Now, $\qquad \beta_1 = \min\left\{\dfrac{\alpha_{30}}{|\alpha_{32}|}, \dfrac{\gamma_{20}}{|\gamma_{22}|}\right\} = \min\left\{\dfrac{4}{|-2|}, \dfrac{3}{0}\right\} = 2$

$\Rightarrow \quad s_1$ is the outgoing vector and hence leave the basis.

Thus, we have $\quad X_B = (x_2), X_{NB} = (x_1, s_1)$

Then $\qquad\qquad x_2 = \dfrac{1}{2}(4 - x_1 - s_1) = 2 - \dfrac{1}{2}x_1 - \dfrac{1}{2}s_1$

$\Rightarrow \qquad\qquad f = 2x_1 + 3\left(2 - \dfrac{1}{2}x_1 - \dfrac{1}{2}x_3\right) - x_1^2$

$$= 2x_1 + 6 - \frac{3}{2}x_1 - \frac{3}{2}s_1 - x_1^2$$

$$= 6 + \frac{1}{2}x_1 - x_1^2 - \frac{3}{2}s_1$$

$$\Rightarrow \quad \left(\frac{\partial f}{\partial x_1}\right)_{X_{NB}=0} = \left(\frac{1}{2} - 2x_1\right)_{x_1, x_2=0} = \frac{1}{2}$$

and $\quad \left(\dfrac{\partial f}{\partial x_2}\right)_{X_{NB}=0} = \left(-\dfrac{3}{2}\right)_{x_1, x_2=0} = -\dfrac{3}{2}$

Here, x_1 enter in the basis.

Now, $\qquad \beta_2 = \min\left\{\dfrac{\alpha_{20}}{|\alpha_{21}|}, \dfrac{\gamma_{10}}{|\gamma_{11}|}\right\} = \min\left\{\dfrac{2}{\left|-\dfrac{1}{2}\right|}, \dfrac{1/2}{(-2)}\right\} = \dfrac{1}{4}$

which is corresponds to $\dfrac{\gamma_{10}}{|\gamma_{11}|}$ so x_2 does not enter in the basis.

Thus, we have to introduce a non-basic variable u_1 such that

$$u_1 = \frac{1}{2}\frac{\partial f}{\partial x_1} = \frac{1}{2}\left(\frac{1}{2} - 2x_1\right) = \frac{1}{4} - x_1 \quad \Rightarrow \quad x_1 = \frac{1}{4} - u_1$$

$$X_B = (x_1, x_2), X_{NB} = (u_1, s_1)$$

$$x_2 = 2 - \frac{1}{2}x_1 - \frac{1}{2}s_1 = 2 - \frac{1}{2}\left(\frac{1}{4} - u_1\right) - \frac{1}{2}s_1$$

$$= 2 - \frac{1}{8} + \frac{1}{2}u_1 - \frac{1}{2}s_1 = \frac{15}{8} + \frac{1}{2}u_1 - \frac{1}{2}s_1$$

and $\qquad f = 6 + x_1\left(\dfrac{1}{2} - x_1\right) - \dfrac{3}{2}s_1 = 6 + \left(\dfrac{1}{4} - u_1\right)\left(\dfrac{1}{2} - \dfrac{1}{4} + u_1\right) - \dfrac{3}{2}s_1$

$$= 6 + \left(\frac{1}{4} - u_1\right)\left(\frac{1}{4} + u_1\right) - \frac{3}{2}s_1 = 6 + \frac{1}{16} - u_1^2 - \frac{3}{2}s_1$$

$$= \frac{97}{16} - u_1^2 - \frac{3}{2}s_1$$

Differentiating the f defined above partially w.r.t. s_1 and u_1 we get

$$\left(\frac{\partial f}{\partial s_1}\right)_{X_{NB}=0} = \left(-\frac{3}{2}\right)_{x_1, x_2=0} = \frac{-3}{2}$$

and $\quad \left(\dfrac{\partial f}{\partial u_1}\right)_{X_{NB}=0} = (-2u_1)_{x_1, x_2=0} = 0$

which gives the optimal solution and is given by

$$x_1 = \frac{1}{4}, x_2 = \frac{15}{8} \text{ and Max } Z = \frac{97}{16}$$

EXAMPLE 2. *Solve the following quadratic programming problem by Beale's method.*

$$\textbf{Max. } Z = 2x_1 + 3x_2 - 2x_2^2$$

subject to the constraints

$$x_1 + 4x_2 \le 4$$

$$x_1 + x_2 \leq 2$$

and $\qquad x_1, x_2 \geq 0$ $\hfill$ [BHARATHIAR–1992; SAMBHALPUR–2006; RAJ.–2013]

SOLUTION. Introducing slack variables s_1 and s_2 the given QPP can be written as

$$\text{Max. } Z = 2x_1 + 3x_2 - 2x_2^2$$

subject to the constraints

$$x_1 + 4x_2 + s_1 = 4 \qquad \qquad \text{...(1)}$$
$$x_1 + x_2 + s_2 = 4 \qquad \qquad \text{...(2)}$$

and $\qquad x_1, x_2, s_1, s_2 \geq 0$

Making s_1, s_2 as basic variables and x_1, x_2 non-basic, then

$$X_B = (s_1, s_2) \text{ and } X_{NB} = (s_1, s_2)$$

Now, from (1), $\qquad s_1 = 4 + 1(-x_1) + 4(-x_2)$

and from (2), $\qquad s_2 = 2 + 1(-x_1) + 2(-x_2)$

We choose the initial basic feasible solution $x_1 = x_2 = 0$, $s_1 = 4$ and $s_2 = 2$.

Then, the initial value of the objective function is $Z = 0$.

also $\qquad\qquad X_B = (s_1, s_2) = (4, 2)$

and $\qquad\qquad X_{NB} = (x_1, x_2) = (0, 0)$

Expressing Z in terms of non-basic variables x_1 and x_2, we get

$$f = Z = 2x_1 + 3x_2 - 2x_2^2$$

$\Rightarrow \qquad\qquad \dfrac{\partial f}{\partial x_1} = 2, \dfrac{\partial f}{\partial x_2} = 3 - 4x_2$

$\Rightarrow \qquad \left(\dfrac{\partial f}{\partial x_1}\right)_{X_{NB}=0} = (2)_{x_1, x_2 = 0}$ and $\left(\dfrac{\partial f}{\partial x_2}\right)_{X_{NB}=0} = (3 - 4x_2)_{x_1, x_2 = 0} = 3$

Here, the most positive is α_2.

$\Rightarrow \quad x_2$ will enter in the basis.

Now, critical value β_1 of x_2 is given by

$$\beta_1 = \min\left\{\frac{4}{4}, \frac{2}{1}\right\} = 1$$

The partial derivative $\dfrac{\partial f}{\partial x_2}$ becomes zero at $x_2 = \dfrac{3}{4}$ $(x_1 = 0)$ therefore,

$$\beta_2 = \frac{|\alpha_2|}{2\gamma_{22}} = \frac{|3|}{2(2)} = \frac{3}{4}$$

The new value of the entering variable x_2 is given by

$$x_2 = \min\{\beta_1, \beta_2\} = \left\{1, \frac{3}{4}\right\} = \frac{3}{4}$$

$\Rightarrow \quad x_2$ does not enter in the basis.

Therefore, introduce a non-basic variable u_1 such that

$$u_1 = \frac{\partial f}{\partial x_2} = 3 - 4x_2 \Rightarrow 4x_2 + u_1 = 3$$

$\therefore \qquad\qquad \text{New } X_B = (s_1, s_2, u_1)$

and $\qquad\qquad X_{NB} = (x_1, x_2)$

Now, introduce x_2 into the basis and remove u_1 from the basis as in the following table :

Simplex Table-1

B.V.	X_B	x_1	x_2	s_1	s_2	u_1
s_1	1	1	0	1	0	1
s_2	5/4	1	0	0	1	1/4
x_2	3/4	0	1	0	0	−1/4

Here, $X_B = (s_1, s_2, x_2) = (1, 5/4, 3/4)$ and $X_{NB} = (x_1, u_1) = (0, 0)$

Now, expressing basic variables into non basic variables as given below :

$$x_2 = \frac{3}{4} - \frac{1}{4}u_1, \quad s_1 = 1 - x_1 - u_1, \quad s_2 = \frac{5}{4} - x_1 - \frac{1}{4}u_1$$

Eliminating the basic variable x_2 from the objective function and expressing it in terms of non-basic variables x_1 and u_1, we get

$$f = 2x_1 + 3\left(\frac{3}{4} - \frac{u_1}{4}\right) - 2\left(\frac{3}{4} - \frac{u_1}{4}\right)^2 = \frac{9}{8} + 2x_1 - \frac{u_1^2}{8}$$

$$\Rightarrow \qquad \frac{\partial f}{\partial x_1} = 2, \quad \frac{\partial f}{\partial u_1} = -\frac{u_1}{4}$$

$$\Rightarrow \qquad \left(\frac{\partial f}{\partial x_1}\right)_{X_{NB}=0} = \left(2\right)_{\substack{x_1=0 \\ u_1=0}} = 2 \text{ and } \left(\frac{\partial f}{\partial u_1}\right)_{X_{NB}=0} = \left(-\frac{u_1}{4}\right)_{(0,0)}$$

Here, $\alpha_1 = 2$, $\alpha_2 = 0$. Choosing x_1 to enter in the basis and using the above table, we get

(i) The largest value of x_1 without deriving any basic variables x_1, s_2 and x_2 to zero.

Since $\quad x_2 = \frac{3}{4} - \frac{1}{4}u_1, \quad s_1 = 1 - x_1 - u_1, \quad s_2 = \frac{5}{4} - x_1 - \frac{1}{4}u_1$

$$\Rightarrow \qquad \beta = \min\left\{\frac{1}{1}, \frac{(5/4)}{1}\right\} = 1$$

(ii) Since $\frac{\partial f}{\partial x_1} \neq 0 \Rightarrow \beta_2 = 2$

Therefore,

$$x_1 = \min\{\beta_1, \beta_2\} = 1$$

This value of x_1 corresponds to β_1. Thus the new optimal solution is given in the following table:

Simplex Table-2

B.V.	X_B	x_1	x_2	s_1	s_2	u_1
x_1	1	1	0	1	0	1
s_2	1/4	0	0	−1	1	−3/4
x_2	3/4	0	1	0	0	−1/4

In the above table,

$$X_B = (x_1, s_2, x_2) = (1, 1/4, 3/4)$$

and $\qquad X_{NB} = (s_1, u_1) = (0, 0)$

Now, expressing basic variables x_1, x_2, s_2 in terms of non-basic variables s_1, u_1, we get

$$x_1 = 1 - s_1 - u_1, \quad s_2 = \frac{1}{4} + s_1 + \frac{3}{4}u_1, \quad x_2 = \frac{3}{4} + \frac{1}{4}u_1$$

Also, expressing objective function Z in terms of non-basic variables s_1 and u_1, we get

$$f = \frac{9}{8} + 2(1 - s_1 - u_1) - \frac{1}{8}u_1^2 = \frac{25}{8} - 2s_1 - 2u_1 - \frac{1}{8}u_1^2$$

$$\frac{\partial f}{\partial s_1} = -2 \Rightarrow \left(\frac{\partial f}{\partial s_1}\right)_{X_{NB}=0} = (-2)_{(s_1,u_1)=(0,0)} = -2 < 0$$

and

$$\frac{\partial f}{\partial u_1} = -2 - \frac{1}{4}u_1 \Rightarrow \left(\frac{\partial f}{\partial u_1}\right)_{X_{NB}=0} = \left(-2 - \frac{1}{4}u_1\right)_{(0,0)} = -2 < 0$$

$\Rightarrow$ solution is optimal and is given by

$$x_1 = 1, \quad x_2 = \frac{3}{4} \quad \text{and Max. } Z = \frac{25}{8}$$

EXAMPLE 3. *Solve the following QPP by using Beale's method*

$$\textbf{Min.} \, \boldsymbol{Z = 6 - 6x_1 + 2x_1^2 - 2x_1x_2 + 2x_2^2}$$

subject to the constraints

$$\boldsymbol{x_1 + x_2 \leq 2}$$

and $\qquad\qquad\qquad \boldsymbol{x_1, x_2 \geq 0}$

SOLUTION. Convert the given minimization QPP as maximization, we get

$$\text{Max}(-Z) = f(x) = -6 + 6x_1 - 2x_1^2 + 2x_1x_2 - 2x_2^2$$

Now, using slack variable $s_1 \, (\geq 0)$ the given QPP can be written as follows :

$$\text{Max}(-Z) = f = -6 + 6x_1 - 2x_1^2 + 2x_1x_2 - 2x_2^2$$

subject to the constraints
$$x_1 + x_2 + s_1 = 2$$
$$x_1, x_2, s_1 \geq 0$$

Choosing s_1 as basic variable and x_1, x_2 as non-basic.

Therefore, $\qquad\qquad X_B = (s_1)$ and $X_{NB} = (x_1, x_2)$

Now expressing basic variables in f in terms of non-basic variables, we get

$$s_1 = 2 - x_1 - x_2$$

and $\qquad\qquad f = -6 + 6x_1 - 2x_1^2 + 2x_1x_2 - 2x_2^2$

$$\Rightarrow \left(\frac{\partial f}{\partial x_1}\right)_{X_{NB}=0} = (6 - 4x_1 + 2x_2)_{\substack{x_1=0 \\ x_2=0}} = 6 \text{ and } \left(\frac{\partial f}{\partial x_2}\right)_{X_{NB}=0} = (-2x_1 + 4x_2)_{\substack{x_1=0 \\ x_2=0}} = 0$$

Here α_1 is most positive, so x_1 will enter the basis.

Now, $\quad \min\left\{\dfrac{\alpha_{30}}{|\alpha_{31}|}, \dfrac{\gamma_{10}}{|\gamma_{11}|}\right\} = \min\left\{\dfrac{2}{|-1|}, \dfrac{6}{|-4|}\right\} = \dfrac{3}{2}$

which corresponds to $\dfrac{\gamma_{10}}{|\gamma_{11}|}$

$\Rightarrow s_1$ does not leave the basis.

Therefore, we introduce a new non-basic variable u_1, such that

$$u_1 = \frac{1}{2}\frac{\partial f}{\partial x_1} = 3 - 2x_1 + x_2$$

$$\therefore \qquad X_B = (s_1, x_1), \; X_{NB} = (x_2, u_1)$$

$$\Rightarrow \qquad x_1 = \frac{3}{2} - \frac{1}{2}u_1 + \frac{1}{2}x_2$$

$$s_1 = 2 - x_1 - x_2 = 2 - \left(\frac{3}{2} - \frac{1}{2}u_1 + \frac{1}{2}x_2\right) - x_2$$

$$= \frac{1}{2} + \frac{1}{2}u_1 - \frac{3}{2}x_2$$

and

$$f = -6 + 6x_1 - 2x_1^2 + 2x_1x_2 - 2x_2^2 = -6 + 2x_1(3 - x_1 + x_2) - 2x_2^2$$

$$= -6 + 2\left(\frac{3}{2} - \frac{1}{2}u_1 + \frac{1}{2}x_2\right)\left(3 - \frac{3}{2} + \frac{1}{2}u_1 - \frac{1}{2}x_2 + x_2\right) - 2x_2^2$$

$$= -6 + (3 - u_1 + x_2)\left(\frac{3}{2} + \frac{1}{2}u_1 + \frac{1}{2}x_2\right) - 2x_2^2$$

$$= -6 + \frac{1}{2}(3 + x_2 - u_1)(3 + x_2 + u_1) - 2x_2^2$$

$$= -6 + \frac{1}{2}\left\{(3 + x_2)^2\right\} - 2x_2^2$$

$$= -6 + \frac{1}{2}\left(9 + x_2 - 6x_2 - u_1^2\right) - 2x_2^2$$

$$= -\frac{3}{2} - \frac{u_1^2}{2} - 3x_2 - \frac{3}{2}x_2^2$$

$$\Rightarrow \qquad \left(\frac{\partial f}{\partial x_2}\right)_{X_{NB}=0} = (3 - 3x_2)_{\substack{x_1=0 \\ u_1=0}} = 3$$

and

$$\left(\frac{\partial f}{\partial u_1}\right)_{X_{NB}=0} = (-u_1)_{\substack{x_1=0 \\ u_1=0}} = 0$$

$\Rightarrow x_2$ enter in the basis.

Now, $\min\left\{\dfrac{\alpha_{30}}{|\alpha_{32}|}, \dfrac{\alpha_{10}}{|\alpha_{12}|}, \dfrac{\gamma_{20}}{|\gamma_{22}|}\right\} = \min\left\{\dfrac{1/2}{|-3/2|}, \dfrac{3/2}{|1/2|}, \dfrac{3}{|-3|}\right\} = \dfrac{1}{3}$

$\Rightarrow s_1$ will leave the basis and thus now

$$X_B = (x_1, x_2) \quad \text{and} \quad X_{NB} = (u_1, s_1)$$

Now, $\qquad x_2 = \dfrac{1}{3} + \dfrac{1}{3}u_1 - \dfrac{2}{3}s_1$

$$x_1 = \frac{3}{2} - \frac{1}{2}u_1 + \frac{1}{2}x_2 = \frac{3}{2} - \frac{1}{2}u_1 + \frac{1}{2}\left(\frac{1}{3} + \frac{1}{3}u_1 - \frac{2}{3}s_1\right)$$

$$= \frac{3}{2} - \frac{1}{2}u_1 + \frac{1}{6} + \frac{1}{6}u_1 - \frac{1}{3}s_1$$

$$= \frac{5}{3} - \frac{1}{3}u_1 + \frac{1}{3}s_1$$

and

$$f = -\frac{3}{2} - \frac{u_1^2}{2} + \frac{3}{2}x_2(2 - x_2)$$

$$= -\frac{3}{2} - \frac{u_1^2}{2} + \frac{3}{2}\left(\frac{1}{3} + \frac{1}{3}u_1 - \frac{2}{3}s_1\right)\left(2 - \frac{1}{3} - \frac{1}{3}u_1 + \frac{2}{3}s_1\right)$$

$$= -\frac{3}{2} - \frac{u_1^2}{2} + \frac{1}{6}(1 + u_1 - 2s_1)(5 - u_1 + 2s_1)$$

$$= -\frac{3}{2} - \frac{u_1^2}{2} + \frac{1}{6}\left(5 - u_1 + 2s_1 + 5u_1 - u_1^2 + 2s_1u_1 - 10s_1 + 2s_1u_1 - 4s_1^2\right)$$

$$= -\frac{3}{2} - \frac{u_1^2}{2} + \frac{5}{6} + \frac{4u_1}{6} - \frac{8}{6}s_1 - \frac{u_1^2}{6} + \frac{4}{6}s_1u_1 - \frac{4}{6}s_1^2$$

$$= -\frac{4}{6} - \frac{4}{6}u_1^2 + \frac{4}{6}u_1 - \frac{4}{3}s_1 - \frac{4}{6}s_1^2 + \frac{4}{6}s_1u_1$$

$$= -\frac{2}{3} + \frac{2}{3}u_1 - \frac{2}{3}u_1^2 - \frac{4}{3}s_1 + \frac{2}{3}s_1u_1 - \frac{2}{3}s_1^2$$

which implies

$$\left(\frac{\partial f}{\partial s_1}\right)_{X_{NB}=0} = \left(-\frac{4}{3} - \frac{2}{3}u_1 - \frac{4}{3}s_1\right)_{\substack{x_1=0 \\ u_1=0}} = -\frac{4}{3}$$

$$\left(\frac{\partial f}{\partial u_1}\right)_{X_{NB}=0} = \left(\frac{2}{3} - \frac{4}{3}u_1 + \frac{2}{3}s_1\right)_{\substack{s_1=0 \\ u_1=0}} = \frac{2}{3}$$

Since $\dfrac{\partial f}{\partial u_1} \neq 0$, this solution can be improved further.

Let x_1 does not enter in the basis, so we introduce another non-basic free variable u_2 such that

$$u_2 = \frac{1}{2}\frac{\partial f}{\partial u_1} = \frac{1}{3} - \frac{1}{3}u_1 + \frac{1}{3}s_1$$

Now $X_B = (x_1, x_2, u_1)$ and $X_{NB} = (s_1, u_2)$

$$u_1 = \frac{1}{2} - \frac{3}{2}u_2 + \frac{1}{2}s_1$$

$$x_2 = \frac{1}{3} + \frac{1}{3}u_1 - \frac{2}{3}s_1 = \frac{1}{3} + \frac{1}{3}\left(\frac{1}{2} - \frac{3}{2}u_2 + \frac{1}{2}s_1\right) - \frac{2}{3}s_1$$

$$= \frac{1}{2} - \frac{1}{2}u_2 - \frac{1}{2}s_1$$

$$x_1 = \frac{5}{3} - \frac{1}{3}u_1 - \frac{1}{3}s_1 = \frac{5}{3} - \frac{1}{3}\left(\frac{1}{2} - \frac{3}{2}u_2 + \frac{1}{2}s_1\right) - \frac{1}{3}s_1$$

$$= \frac{5}{3} - \frac{1}{6} + \frac{1}{2}u_2 - \frac{1}{6}s_1 - \frac{1}{3}s_1$$

$$= \frac{3}{2} + \frac{1}{2}u_2 - \frac{1}{2}s_1$$

and $$f = -\frac{2}{3} + \frac{2}{3}u_1 - \frac{4}{3}s_1 + \frac{2}{3}s_1u_1 - \frac{2}{3}u_1^2 - \frac{2}{3}s_1^2$$

$$= -\frac{2}{3} + \frac{2}{3}u_1(1 + s_1 - u_1) - \frac{4}{3}s_1 - \frac{2}{3}s_1^2$$

$$= -\frac{2}{3} + \frac{2}{3}\left(\frac{1}{2} - \frac{3}{2}u_2 + \frac{1}{2}s_1\right)\left(1 + s_1 - \frac{1}{2} + \frac{3}{2}u_2 - \frac{1}{2}s_1\right) - \frac{4}{3}s_1 - \frac{2}{3}s_1^2$$

$$= -\frac{2}{3} + \frac{1}{6}(1 - 3u_2 + s_1)(2 + 2s_1 - 1 + 3u_2 - s_1) - \frac{4}{3}s_1 - \frac{2}{3}s_1^2$$

$$= -\frac{2}{3} + \frac{1}{6}(1 - 3u_2 + s_1)(1 + s_1 + 3u_2) - \frac{4}{3}s_1 - \frac{2}{3}s_1^2$$

$$= -\frac{2}{3} + \frac{1}{6}\{(1 + s_1)^2 - 9u_2^2\} - \frac{4}{3}s_1 - \frac{2}{3}s_1^2$$

$$= -\frac{2}{3} + \frac{1}{6}[1 + s_1^2 + 2s_1 - 9u_2^2] - \frac{4}{3}s_1 - \frac{2}{3}s_1^2$$

$$= -\frac{2}{3} + \frac{1}{6} + \frac{s_1^2}{6} + \frac{1}{3}s_1 - \frac{3}{2}u_2^2 - \frac{4}{3}s_1 - \frac{2}{3}s_1^2$$

$$= -\frac{1}{2} - \frac{1}{2}s_1^2 - s_1 - \frac{3}{2}u_2^3$$

which implies

$$\left(\frac{\partial f}{\partial s_1}\right)_{X_{NB}=0} = (-s_1 - 1)_{\substack{s_1=0 \\ u_2=0}} = -1 \text{ and } \left(\frac{\partial f}{\partial u_2}\right)_{X_{NB}=0} = \left(-\frac{3}{2} \cdot 2u_2\right)_{\substack{s_1=0 \\ u_2=0}} = 0$$

which gives an optimum solution. Now ignoring the free variable u_1, the optimal solution is given by

$$x_1 = \frac{3}{2},\ x_2 = \frac{1}{2},\ Z = -f = -\left(-\frac{1}{2}\right) = \frac{1}{2}$$

EXAMPLE 4. *Solve the following QPP by Beale's method*

$$Min.\ Z = -4x_1 + x_2^2 - 2x_1x_2 + 2x_2^2$$

subject to the constraints

$$2x_1 + x_2 \geq 6$$
$$x_1 - 4x_2 \geq 0$$

and $$x_1, x_2 \geq 0$$

SOLUTION. Introducing the surplus variables s_1 and s_2 and convert the problem of minimization into maximization, we can write

$$Max.\ Z = 4x_1 - x_1^2 + 2x_1x_2 - 2x_2^2$$

s.t. $$2x_1 + x_2 - s_1 = 6 \Rightarrow s_1 = -6 + 2x_1 + x_2$$
$$x_1 - 4x_2 - s_2 = 0 \Rightarrow s_2 = x_1 - 4x_2$$

and $$x_1, x_2, s_1, s_2 \geq 0.$$

Let s_1, s_2 be the basic and x_1, x_2 be the non-basic variables.

Now $$\alpha_1 = \left(\frac{\partial f}{\partial x_1}\right)_{X_{NB}=0} = (4 - 2x_1 + 2x_2)_{\substack{x_1=0 \\ x_2=0}} = 4 > 0$$

$$\alpha_2 = \left(\frac{\partial f}{\partial x_2}\right)_{X_{NB}=0} = (2x_1 + 4x_2)_{\substack{x_1=0 \\ x_2=0}} = 0$$

$$X_B = (s_1, s_2) = (-6, 0),\ X_{NB} = (x_1, x_2) = (0, 0)$$

Here $\qquad \alpha_1 = 4, \alpha_2 = 0$

Clearly both are positive, so choose

$\qquad x_1$ (most positive value of α_1) to enter into the basis

Then critical value β_1 of x_1 is given by

$$\beta_1 = \min.\left\{-\frac{6}{|2|}, \frac{0}{|1|}\right\} = -3$$

$\Rightarrow s_1$ will leave the basis. Expressing the new basic variables x_1, s_2 and f in terms of new non-basic variables x_2 and s_1 such that

$$x_1 = 3 - \frac{1}{2}x_2 + \frac{1}{2}s_1, \quad s_2 = 3 - \frac{3}{2}x_2 + \frac{1}{2}s_1$$

$$f = 4\left(3 - \frac{1}{2}x_2 + \frac{1}{2}s_1\right) - \left(3 - \frac{1}{2}x_2 + \frac{1}{2}s_1\right)^2 + 2\left(3 - \frac{1}{2}x_2 + \frac{1}{2}s_1\right)x_2 - 2x_2^2$$

$$= 9 + x_2 - s_1 + \frac{3}{2} + \frac{3}{2}x_2 s_1 - \frac{13}{4}x_2^2 - \frac{1}{4}s_1^2$$

which implies

$$\left(\frac{\partial f}{\partial x_2}\right)_{X_{NB}=0} = \left(1 + \frac{3}{2}s_1 - \frac{13}{2}x_2\right)_{\substack{x_2=0 \\ s_1=0}} = 1$$

$$\Rightarrow \qquad \left(\frac{\partial f}{\partial s_1}\right)_{X_{NB}=0} = \left(-1 + \frac{3}{2}x_2 - \frac{1}{2}s_1\right)_{\substack{x_2=0 \\ s_1=0}} = -1$$

$\Rightarrow x_2$ will enter the basis.

Now compute $\min\left\{\dfrac{3}{|-1/2|}, \dfrac{3}{|-3/2|}\right\} = 2$

Since, the minimum ratio corresponds to β_2, we introduce a non-basic free variable defined by

$$u_1 = \frac{1}{2}\frac{\partial f}{\partial x_2} = \frac{1}{2} + \frac{3}{4}s_1 - \frac{13}{4}x_2$$

Therefore, $\qquad X_B = (x_1, s_2, x_2)$ and $X_{NB} = (s_1, u_1)$

Expressing basic variables and f in terms of non-basic variables, we have

$$x_1 = \frac{38}{13} - \frac{3}{26}s_1 + \frac{2}{13}u_1, \quad x_2 = \frac{2}{13} + \frac{3}{13}s_1 - \frac{4}{13}u_1, \quad s_2 = \frac{30}{13} - \frac{27}{26}s_1 + \frac{18}{13}u_1$$

$$f = 9 + \frac{1}{13}(2 + 3s_1 - 4u_1) - s_1 + \frac{3}{26}s_1(2 + 3s_1 - 4u_1) - \frac{1}{52}(2 + 3s_1 - u_1)^2 - \frac{1}{4}s_1^2$$

which implies

$$\left(\frac{\partial f}{\partial s_1}\right)_{X_{NB}=0} = \left(\frac{3}{13} - 1 + \frac{3}{26}(2 - u_1) + \frac{18}{26}s_1 - \frac{6}{52}(2 + 3s_1 - 4u_1) - \frac{1}{2}s_1\right)_{\substack{s_1=0 \\ u_1=0}}$$

$$= -\frac{9}{13} < 0$$

and $\qquad \left(\dfrac{\partial f}{\partial x_1}\right)_{X_{NB}=0} = \left(-\dfrac{4}{13} - \dfrac{12}{26}s_1 + \dfrac{8}{52}(2 + 3s_1 - 4u_1)\right)_{\substack{s_1=0 \\ u_1=0}} = 0$

Clearly both the above values ≤ 0. So the optimal value of Z is obtained by setting

$u_1 = 0$, $s_1 = 0$ in the current objective function given by

$$Z = 9 + \frac{2}{13} - \frac{2}{52} = \frac{474}{52}$$

Hence, the optimum solution of the given QPP is

$$x_1 = \frac{38}{13}, \ x_2 = \frac{2}{13} \ \text{and} \ \text{Min.} Z = \frac{474}{52}$$

5.6 SEPARABLE PROGRAMMING

We know that a separable programming is an indirect method to solve non-linear programming problem. It is useful for solving those non-linear programming problem in which the objective function and constraints are separable (A function of n variables $f(x_1, x_2, ..., x_n)$ is said to be separable if it can be written as the sum of n functions $f_1(x_1), f_2(x_2), ..., f_n(x_n)$).

If function is not separable, then it can be made separable by using simplified approximations.

5.7 SOME GENERAL DEFINITIONS RELATED TO SEPARABLE PROGRAMMING

(I) SEPARABLE PROGRAMMING PROBLEM

In a non-linear programming problem if the objective function can be expressed as a linear combinations of several different single-variable functions, of which some or all are non-linear. Such type of NLPP is called separable programming problem.

(II) SEPARABLE CONVEX PROGRAMMING

A separable programming in which separate functions are convex, also the non-linear function $f(x)$ is convex in case of minimization and concave in case of maximization is called separable convex programming.

5.8 PIECEWISE LINEAR APPROXIMATION OF NON-LINEAR FUNCTIONS

Consider

$$\text{Optimize } Z = \sum_{j=1}^{n} f_j(x_j)$$

subject to the constraints

$$\sum_{j=1}^{n} a_{ij}x_j = b_i, \ i = 1, 2, ..., m$$

and

$$x_j \geq 0$$

where $f_j(x_j)$ is the jth separable function which is to be approximated over a defined interval.

Let $(a_k, b_k) \ \forall \ k = 1, 2, ..., p$ be the kth breaking point joining a linear segment which approximated $f(x)$. Further let us define the weight function w_k such that

$$\sum_{k=1}^{p} w_k = 1$$

Further, add an additional constraint, (if necessary) w_{k+1} such that w_k and w_{k+1} are equated to zero to find the weighted average of breaking point. Therefore, $f(x)$ is approximated as given below

$$f(x) = \sum_{k=1}^{p} b_k w_k, \ x = \sum_{k=1}^{p} a_k w_k$$

provided

$$0 \le w_1 \le y_1$$

$$0 \le w_2 \le y_1 + y_2$$

$$0 \le w_3 \le y_2 + y_3$$

$$::$$

$$0 \le w_{k-1} \le y_{k-2} + y_{k-1}$$

$$0 \le w_k \le y_{k-1}$$

and

$$\sum_{k=1}^{P} w_k = 1, \quad \sum_{k=1}^{P-1} y_k = 1$$

$$y_k = 0 \text{ or } 1 \; \forall \; k$$

The variables for approximation are now w_k and y_k.

5.9 MIXED INTEGER APPROXIMATION OF SEPARABLE PROBLEMS

We can approximate a single variable non-linear separable function by a piecewise linear function using mixed integer programming.

Let us suppose the no. of breaking point for j^{th} variable x_j be equal to p_j and g_{ik} be the kth breaking values. If w_{jk} be the weight associated with the kth breaking point of jth variable x_j. Then we have the following mixed integer programming

$$\text{Optimize . } Z = \sum_{j=1}^{n} \sum_{k=1}^{P_j} f_j(a_{jk})w_{jk}$$

subject to the constraints

$$Z = \sum_{j=1}^{n} \sum_{k=1}^{P_j} g_{ij} \, | \, a_{jk} \, | \, w_{jk} \le b_i, \; i = 1, 2, ..., m$$

$$0 \le w_{j1} \le y_{j1}$$

$$\vdots \qquad \vdots \; \vdots \qquad \vdots \; \vdots \qquad \vdots$$

$$0 = w_{jk} \le y_{jk-1} + y_{jk} \; , \; k = 2, 3, ..., p_j^{-1}$$

and

$$\sum_{k=1}^{P_j} w_{jk} = 1, \quad \sum_{k=1}^{P_j-1} y_{jk} = 1,$$

$$y_{jk} = 0 \text{ or } 1 \; \forall \; j \text{ and } k.$$

5.10 VALIDITY OF MIXED INTEGER APPROXIMATION

The mixed integer approximation is valid under the following two conditions:

(i) For each j, no more than two w_{jk} should appear in the basis

(ii) Two w_{jk} can be positive only if they are adjacent

☛ REMARKS
- The restricted basic method gives only a local optimum, while the mixed integer programming method gives the global optimum.
- The solution obtained above may not be feasible for the original non-linear programming.

5.11 METHOD OF SOLUTION OF SEPARABLE PROGRAMMING PROBLEMS

To solve the given separable programming, use the following procedure:

WORKING PROCEDURE

We use the following steps :

STEP 1. Convert the minimization problem into maximization.

STEP 2. Examine the concavity (convexity) conditions of the function $f_j(x_j)$ and $g_{ij}\, x_j$. If condition is satisfied, go to the next step otherwise stop.

STEP 3. Divide the given interval $0 \le x_j \le t_j$ into a no. of breaking points a_{jk} ($k = 1, 2, ..., p_j$) so that

$$a_{j1} = 0, a_{j1} < a_{j2} < ... < a_{jp_j}$$

STEP 4. For each point a_{jk} obtain piecewise linear approximation $f_j(x_j)$ and $g_{ij}(x_j)$ $\forall$ i and j.

STEP 5. Solve the resulting LPP by two-phase method by assuming w_{i1} ($i = 1, 2, ..., m$) as artificial variable.

STEP 6. Find the optimum solution of the original non-linear programming problem by using

$$x_j^* = \sum_{k=1}^{p_j} a_{jk} w_{jk}, \qquad j = 1, 2, ..., n$$

REMARKS

- The solution space of the approximate problem may have additional extreme points which do not exist in the solution space of the original problem.
- Separable programming gives the approximate solution of the problem.
- To find the better approximation, have the greater no. of breaking points.

SOLVED EXAMPLES

EXAMPLE I. *Solve the following NLPP by using separable programming algorithm.*

$$\text{Max. } Z = x_1 + x_2^4$$

subject to the constraints

$$3x_1 + 2x_2^2 \le 9$$

and $\qquad x_1, x_2 \ge 0$

SOLUTION. Here, we have the following separable functions.

$$f_1(x_1) = x_1, \qquad f_2(x_2) = x_2^2$$

$$g_{11}(x_1) = 3x_1, \qquad g_{12}(x_2) = 2\, x_2^2$$

Clearly the function $f_1(x_1)$ and $g_{11}(x_1)$ satisfy concavity (convexity) conditions.

From the constraints of the given problem, we have

$$x_1 \le 3 \text{ and } x_2 \le \sqrt{9/2} = 2.13 < 3$$

So, 3 is the upper limit for x_1 and x_2 both.

$\therefore$ Divide the interval [0, 3] into four breaking points a_{jk} of equal intervals such that

$$a_{j1} = 0, \; a_{j1} < a_{j2} < a_{j3} < a_{j4} = 3$$

$\therefore$ The piecewise linear approximation for $f_2(x_2)$ and $g_{12}(x_2)$ are given in the following table:

k	a_{jk}	$f_2(x_2=a_{jk})$	$g_{12}(x_2=a_{jk})$
1	0	0	0
2	1	1	2
3	2	16	8
4	3	81	18

So,
$$f_2(x_2) = w_{21}f_2(a_{21})+w_{22}f_2(a_{22})+w_{23}f_2(a_{23}) + w_{24}f_2(a_{24})$$
$$= w_{21}(0) + w_{22}(1) + w_{23}(16) + w_{24}(81)$$
$$= w_{22}+ 16\,w_{23} +81\,w_{24}$$

and
$$g_{12}(x_2) = w_{21}g_{12}(a_{21})+w_{22}g_{12}(a_{22})+w_{23}f_{12}(a_{23}) +w_{24}g_{12}(a_{24})$$
$$= w_{21}(0) + w_{22}(2) + w_{23}(8) + w_{24}(18)$$
$$= 2w_{22} + 8\,w_{23} +18\,w_{24}$$

Therefore, the approximated linear programming problem is given below
$$\text{Max. } f(x) = x_1 + w_{22} + 16\,w_{23} + 81w_{24}$$
subject to the constraints
$$3x_1+ 2w_{22} + 8w_{23} +18\,w_{24} \le 9$$
$$w_{21}+ w_{22} + w_{23} + w_{24} = 1$$
and
$$x_1, w_{21}, w_{22}, w_{23}, w_{24} \ge 0$$

with the following two conditions:

(i) For each j, no more than two w_{jk} are positive

and (ii) if two w_{jk} are positive, they must correspond to adjacent point.

Now, treating w_{21} as the artificial variable, solve the given LPP by two-phase simplex method as follows :

The simplex tables of phase-2 are given as below :

Simplex Table-1

B.V.	C_j		1	1	16	81	0	0	Min. Ratio
	C_B	X_B	x_1	w_{22}	w_{23}	w_{24}	s_1	w_{21}	
s_1	0	9	3	2	8	18	1	0	9/8
w_{21}	0	1	0	1	①	1	0	1	1/1(min.) →
	$\Delta_j = C_j - Z_j$		1	1	16 ↑	81	—	—	

From the above table we observed that w_{24} should enter the basis. Since w_{21} is artificial basic variable, it must be dropped before w_{21} enter the basis. By minimum ratio rule, s_1 is the leaving variable. So w_{24} can not enter the basis.

So, consider the next best entering variable w_{23}.

Again the artificial variable w_{21} must be dropped.

Then, we have the following table.

Simplex Table-2

	C_j		1	1	16	81	0	Min. Ratio
B.V.	C_B	X_B	x_1	w_{22}	w_{23}	w_{24}	s_1	
s_1	0	1	3	−6	0	⑩	1	1/10 (min.) →
w_{23}	16	1	0	1	1	1	0	1/1
	Δ_j		1	−15	−	65 ↑	−	

Clearly, in the above table w_{24} is the entering variable.
Also, s_1 is the leaving variable. Then we have the following table.

Simplex Table-3

	C_j		1	1	16	81	0	Min. Ratio
B.V.	C_B	X_B	x_1	w_{22}	w_{23}	w_{24}	s_1	
w_{24}	81	1/6	3/10	−6/10	0	1	1/10	
w_{23}	16	9/10	−3/10	16/10	1	0	−1/10	
	Δ_j		−37/2	24	−	−	13/2	

From the above table, we observe that w_{22} should enter in the basis which is not possible because w_{24} can't be dropped from the current solution. So procedure terminated.
Hence, the optimum solution of the approximate LP is given by

$$w_{23} = \frac{9}{10}, \ w_{24} = \frac{1}{10} \quad \text{Max.} f(x) = 22.5$$

Hence, the optimum solution of the original NLPP is given by

$$x_j = \sum_{k=1}^{4} a_{jk} w_{jk}, \ j = 1, 2$$

$$\Rightarrow \quad x_2 = a_{21} \omega_{21} + a_{22} \omega_{22} + a_{23} \omega_{23} + a_{24} w_{24}$$

$$= 0.0 + 1.0 + 2.\frac{9}{10} + 3\left(\frac{1}{10}\right) = 2.1$$

$$x_1 = 0$$

and Max. $f(x) = 22.5$

EXERCISE 5.1

Use Wolfe's method, solve the following quadratic programming problem:

1. Min. $Z = 6 - 6x_1 + 2x_1^2 - 2x_1x_2 + 2x_2$
subject to the constraints
$$x_1 + x_2 \le 2$$
and $x_1, x_2 \ge 0$

2. Min. $Z = x_1^2 + x_2^2 + x_3^2$
subject to the constraints
$$x_1 + x_2 + 3x_3 = 2$$
$$5x_1 + 2x_2 + x_3 = 5$$
and $\quad x_1, x_2, x_3 \ge 0$

3. Max. $Z = 8x_1 + 10x_2 - x_1^2 - x_2^2$
subject to the constraints
$$3x_1 + 2x_2 \le 6$$
and $\quad x_1, x_2 \ge 0$

Use the Beale's method solve the following QPP:

4. Max. $Z = 10x_1 + 25x_2 - 10x_1^2 - x_2^2 - 4x_1x_2$
subject to the constraints
$$x_1 + 2x_2 \le 10$$
$$x_1 + x_2 \le 9$$
and $\quad x_1, x_2 \ge 0$

5. Max. $Z = 2x_1x_2 - 5x_1 - 13x_2 + 3x_2^2 - 10$

subject to the constraints

$$x_1 + x_2 \le 1$$
$$4x_1 + x_2 \ge 2$$

and $\qquad x_1, x_2 \ge 0$

6. Max. $Z = 4x_1 + 6x_2 - x_1^2 - 3x_2^2$

subject to the constraints

$$x_1 + x_2 \le 4$$

and $\qquad x_1, x_2 \ge 0$

Solve the following NLPP using separable programming algorithm.

7. Max. $Z = 3x_1 + 2x_2$

subject to the constraints

$$4x_1^2 + x_2^2 \le 16$$

and $\qquad x_1, x_2 \ge 0$

8. Max. $Z = 16 - 2(x_1 - 3)^2 - (x_2 - 7)^2$

subject to the constraints

$$x_1^2 + x_2 \le 16$$

and $\qquad x_1, x_2 \ge 0$

ANSWERS

1. $x_1 = \dfrac{3}{2}, x_2 = \dfrac{1}{2}, \text{Max } Z = \dfrac{1}{2}$

2. $x_1 = \dfrac{81}{100}, x_2 = \dfrac{7}{20}, x_3 = \dfrac{7}{20}, \text{Max.} Z = \dfrac{17}{20}$

3. $x_1 = \dfrac{4}{13}, x_2 = \dfrac{33}{13}, \text{Max.} Z = \dfrac{267}{13}$

4. $x_1 = 0, x_2 = 5, \text{Max. } Z = 100$

5. $x_1 = 1, x_2 = 0, \text{Max.} Z = -15$

6. $x_1 = 2, x_2 = 1, \text{Max. } Z = 7$

7. $x_1 = 1, x_2 = \dfrac{24}{7}, \text{Max.} f = \dfrac{69}{7}$

8. $x_1 = 3, x_2 = 7, \text{Max.} f = 16$

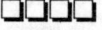

6 Geometric Programming

The geometric programming derives its name based on some geometric concepts such as orthogonality and arithmetic-geometric inequalities. This technique is initially derived from inequalities rather than the calculus and its extensions. It is developed by R. Duffin, E. L. Peterson and C. Zener in 1964, which is used to finds the solution of the problem considering an associated dual problem. If differ from other optimization techniques in the emphasis it places on the relative magnitude of the terms of the objective function rather than the variables. In this technique we first find the optimal value of the objective function instead of the value of decision variables.

In this chapter we shall discuss the constrained and unconstrained case of geometric programming and to do this we shall derive the inequality using the classical optimization theorem and then using the inequality we shall indicate how these relationship may be used to find optimal solutions to non-linear programming problems.

> The geometric programming approach instead solving a non-linear programming problem, first find the optimal value of objective function by solving its dual problem and then determines an optimal solution to the given non-linear programming problem from the optimal soluion of the dual.

6.2 GENERAL FORM OF GEOMETRIC PROGRAMMING

The general mathematical form of geometric programming is given below.

$$\text{Min } f(\boldsymbol{x}) = \sum_{j=1}^{n} C_j \cdot u_j(x)$$

with $C_j > 0$ and $u_j(x)$ has the form

$$u_j(x) = \prod_{i=1}^{m} (x_i)^{a_{ij}}$$

where a_{ij} may be any real number.

☛ REMARK

- Here u_j, because of the positive coefficients and variable and real exponents are called **posynomial**.

6.3 FORMULATION OF GEOMETRIC PROGRAMMING PROBLEM

The objective and constraints functions in the geometric programming can have the following form

$$\text{maximize } z = f(\boldsymbol{x}) = \prod_{j=1}^{n} x_j \qquad \qquad \dots(1)$$

subject to
$$\sum_{j=1}^{n} x_j = C < \infty$$

and
$$x_j \geq 0, j = 1,2,\dots,n$$

Since, the maximum will not occur where any of the $x_i = 0$. Since $f(x)$ is also zero at this point, so for the moment we ignore these inequalities and solve the following simpler problem.

$$\max f(x) = \prod_{j=1}^{n} x_j \qquad \dots(2)$$

subject to
$$\sum_{j=1}^{n} x_j = C < \infty$$

6.4 GEOMETRIC-ARITHMETIC MEAN INEQUALITY

Consider the problem

$$z = \max f(\boldsymbol{x}) = \prod_{j=1}^{n} x_j$$

subject to
$$\prod_{j=1}^{n} x_j = C < \infty$$

The Lagrangian function is given by

$$L(\boldsymbol{x},\lambda) = f(\boldsymbol{x}) + \lambda\left(\sum_{j=1}^{n} x_j - c\right) \qquad \dots(1)$$

The necessary conditions are

$$\frac{\partial L}{\partial x_i} = \prod_{j=1, j\neq i}^{n} (x_j) + \lambda = 0, \quad i = 1,2,\dots,n \qquad \dots(2)$$

On solving for λ, we get

$$\lambda = -\prod_{j=1, j\neq i}^{n} x_j \qquad \dots(3)$$

Since, i can be any integer from 1 to n, therefore we can write

$$\lambda = -\prod_{j=1, j\neq k}^{n} x_j \qquad \dots(4)$$

Now, equating (3) and (4) we get

$$x_1 x_2 \dots x_{i-1} x_{i+1} \dots x_n = x_1 x_2 \dots x_{k-1} x_{k+1} \dots x_n$$

Since, $x_i \neq 0$, we get

$$x_k = x_i = a \ \ \forall i,k = 1,2,\dots,n \text{ , for some constant } a$$

$\Rightarrow$
$$\sum_{j=1}^{n} x_j = c = \sum_{j=1}^{n} a = na$$

$\Rightarrow$
$$x_i^{\circ} = a = \frac{c}{n}, \quad i = 1,2,\dots,n$$

and
$$f(x^{\circ}) = \prod_{j=1}^{n} \frac{c}{n} = \left(\frac{c}{n}\right)^n$$

Therefore,
$$\max. f(\boldsymbol{x}) = \left(\frac{c}{n}\right)^n$$

$\Rightarrow$
$$f(\boldsymbol{x}) \leq \left(\frac{c}{n}\right)^n \qquad \dots(5)$$

where $c = \sum\limits_{j=1}^{n} x_j$

Thus, $\qquad f(x) = \sum\limits_{j=1}^{n} x_j \le \left(\dfrac{\sum\limits_{j=1}^{n} x_j}{n}\right)^{n}$ ...(6)

Taking n^{th} roots of both the sides of (6) we get

$$\left(\sum\limits_{j=1}^{n} x_j\right)^{1/n} \le \frac{1}{n}\sum\limits_{j=1}^{n} x_j \qquad ...(7)$$

with equality only when $x_j = \dfrac{c}{n}$

This is called geometric-arithmetic mean inequality.

☞ **REMARK**
- The above inequality for lower bound can be written as

$$\frac{1}{n}\sum\limits_{j=1}^{n} x_j \ge \left(\prod\limits_{j=1}^{n} x_j\right)^{1/n}$$

6.5 MORE GENERAL FORMULATION OF GEOMETRIC PROGRAMMING PROBLEM

From the above remark, we have

$$\frac{1}{n}\sum\limits_{j=1}^{n} x_j \ge \left(\prod\limits_{j=1}^{n} x_j\right)^{1/n}$$

which can also be written in general form as

$$\sum y_j \cdot x_j \ge \prod (x_j) \qquad ...(1)$$

where $y_j's$ are non-negative weights whose sum is unity. Using these results, we can derive the geometric programming relationship.

Consider, $\qquad \min. f(x) = \sum\limits_{j=1}^{n} u_j(x)$

where $u_j(x) = C_j \prod\limits_{i=1}^{n} (x_i)^{a_{ij}}$, $j = 1, 2, ..., n$

Here, all $C_j > 0$ and n is finite. The exponent a_{ij} are real but unrestricted in sign.

> The function $f(x)$ takes the form of a polynomial except that the exponents a_{ij} may be negative. For the reason all $C_j > 0$ and being closely related to polynomials, Duffin and Zener have given $f(x)$ the name posynomial.

☞ **REMARK**
- The above problem is called as the primal problem. The variable x_i are assumed to be strictly positive so that the region $x_i \le 0$ represents the infeasible space.

6.6 NECESSARY CONDITION FOR OPTIMALITY

Consider the following general mathematical form of geometric programming

$$\min f(x) = \sum\limits_{j=1}^{n} C_j u_j(x)$$

with $C_j > 0$ and $u_j(x)$ has the form

$$u_j(x) = \prod_{i=1}^{m} (x_i)^{a_{ij}}$$

Now, the necessary condition for the optimality is given by

$$\frac{\partial f(x)}{\partial x_r} = 0, \sum_{j=1}^{n} C_j \frac{\partial u_j(x)}{\partial x_r} = 0 \qquad \qquad ...(1)$$

But,

$$\frac{\partial u_j(x)}{\partial x_r} = \frac{a_{rj}}{x_r} u_j(x) \qquad \qquad ...(2)$$

Using (2) in (1) we get

$$\frac{\partial f(x)}{\partial x_r} = \frac{1}{x_r} \sum_{j=1}^{n} a_{rj} C_j u_j = 0$$

Let $f^*(x)$ be the minimum value of $f(x)$.

Now, since each x_r and C_r is positive so, $f^*(x)$ will be also positive.

Therefore, dividing $\dfrac{\partial f(x)}{\partial x_r}$ by $f^*(x)$ we get

$$\sum_{j=1}^{n} \frac{a_{rj} C_j u_j}{f^*(x)} = 0 \qquad \qquad ...(3)$$

Let us have the transformation given by

$$y_j = \frac{C_j u_j}{f^*(x)}, \quad j = 1, 2, ..., n \qquad \qquad ...(4)$$

Now using (3) and the necessary condition for local minima we can find

$$\sum_{j=1}^{n} a_{rj} y_j = 0, \quad r = 1, 2, ..., m \qquad \qquad ...(5)$$

Then using the definition of y_j we may get

$$\sum_{j=1}^{n} y_j = \frac{1}{f^*(x)} \sum_{j=1}^{n} C_j \cdot u_j = 1 \qquad \qquad ...(6)$$

at the optimal solution.

The above conditions (5) and (6) are the necessary condition for optimality of a non-linear function and are known as orthogonality and normality conditions respectively. These conditions have unique value of y_j for $m + 1 = n$ and all equations are independent. But if $n > m + 1$ then values of y_j no longer remains independent.

MATRIX NOTATIONS

Let us define

$$A = \begin{bmatrix} 1 & 1 & \cdots & 1 \\ a_{11} & a_{12} & \cdots & a_{1n} \\ \vdots & & & \\ a_{m1} & a_{m2} & \cdots & a_{mn} \end{bmatrix}; \quad y = \begin{bmatrix} y_1 \\ y_2 \\ \vdots \\ y_m \end{bmatrix}; \quad b = \begin{bmatrix} 1 \\ 0 \\ \vdots \\ 0 \end{bmatrix}$$

Now to form the normality and orthogonality conditions we have

$$Ay = b$$

Thus, the original non-linear programming problem is reduced to one of finding the correct set of values of y which satisfy these linear non-homogeneous equations.

Hence to find the unique value of y_j we use the following results.

(i) If Rank(A, b) > Rank(A) then there will be no solution.

Here, $(A, b) = \begin{bmatrix} 1 & 1 & \cdots & 1 & b_0 \\ a_{11} & a_{12} & \cdots & a_{1n} & b_1 \\ \vdots & & & \vdots & \vdots \\ a_{m1} & a_{m2} & \cdots & a_{mn} & b_m \end{bmatrix}$

(ii) There will be unique solution if A is a square matrix and Rank (A, b) = Rank(A).

(iii) There will be an infinite no. of solution if rank$(A) < n$ *i.e.*, $n > m + 1$.

Now, we conclude that

(a) When condition (i) exists, there is no vector $x > 0$ for which $f(x)$ obtain a minimum.

(b) There is a unique minimum when condition (ii) is satisfies. In this case we simply solve for y by

$$y = A^{-1}b$$

and hence, the optimal solution is obtained in terms of y by carrying out simple algebraic calculations.

Expression for minimum $f(x)$: We know that,

At the optimal solution

$$f^*(x) = \frac{C_j \cdot u_j}{y_j} = \frac{1}{y_j} C_j \prod_{i=1}^{m} (x_j)^{a_{ij}}$$

Now, raising both sides to power y_j and taking the product, we get

$$\prod_{j=1}^{n} \{f^*(x)\}^{y_j} = \prod_{j=1}^{n} \left\{ \frac{1}{y_j} C_j \sum_{j=1}^{m} (x_i)^{a_{ij}} \right\}^{y_j}$$

Using $\Sigma y_i = 1$ and $\prod_{j=1}^{n} \{f^*(x)\}^{y_j} = [f^*(x)]^{\Sigma y_i} = f^*(x)$ in the RHS of the above equation, we get

$$\prod_{j=1}^{n} \left[\left(\frac{C_j}{y_j} \right) \prod_{i=1}^{m} (x_i)^{a_{ij}} \right]^{y_j} = \prod_{j=1}^{n} \left(\frac{C_j}{y_j} \right)^{y_j} \prod_{j=1}^{m} \left\{ \prod_{j=1}^{m} (x_i)^{a_{ij}} \right\}^{y_j} = \prod_{j=1}^{n} \left(\frac{C_j}{y_j} \right)^{y_j} \prod_{i=1}^{m} (x_i)^{\Sigma a_{ij} y_j}$$

$$= \prod_{j=1}^{n} \left(\frac{C_j}{y_j} \right)^{y_j} \prod_{i=1}^{m} x_i^* = \prod_{j=1}^{n} \left(\frac{C_j}{y_j} \right)^{y_j}$$

Therefore, $\min. f(x) = f^*(x) = \prod_{j=1}^{n} \left(\frac{C_j}{y_j} \right)^{y_j}$ and hence $f(x) \geq \prod_{j=1}^{n} \left(\frac{C_j}{y_j} \right)^{y_j}$ where y_j must satisfy the

orthogonality and normality conditions given by (5) and (6).

For the given value of $f^*(x)$ and unique value of y_j the solution to a set of equations can be obtained except for calculating the values of x_j from

$$C_j \cdot \prod_{i=1}^{m} (x_i)^{a_{ij}} = y_j f^*(x)$$

Further if $n > m + 1$ *i.e.* condition (iii) is satisfied, then we have

$$\max. f(x) = \prod_{j=1}^{n} \left(\frac{C_j}{y_j} \right)^{y_j}$$

subject to the constraints

$$Ay = b \qquad\qquad (\because \min. f(x) = \max. f(x) = \prod_{j=1}^{n} \left(\frac{C_j}{y_j} \right)^{y_j})$$

This procedure shows that the solution to the original polynomial $f(x)$ can be transformed into the solution of a set of linear equations in y_j. All $y_j's$ are determined from the necessary conditions for minimum. However these conditions are also the sufficient conditions.

Let us consider the primal form of non-linear programming problem given by

$$f(x) = \sum_{j=1}^{n} y_j \left(\frac{u_j}{y_j} \right)$$

where $y_j's$ are defines as dual variables.

Now, define the function

$$f(y) = \prod_{j=1}^{n} \left(\frac{u_j}{y_j} \right)^{y_j} = \prod_{j=1}^{n} \left(\frac{C_j}{y_j} \right)^{y_j}$$

Since, $\Sigma y_i = 1$ and $y_j > 0$ then by Cauchy's inequality we have, $f(y) \le f(x)$

The function $f(y)$ with its variables $y_1, y_2, ..., y_n$ defines the dual non-linear programming problem to its primal problem. Finally, by duality theorem we have

$$\max. f(y) = \min. f(x)$$

SOLVED EXAMPLES

EXAMPLE 1.   **When $n > m + 1$, solve the following non-linear programming problem by geometric programming method**

$$\min f(x) = 7x_1 x_2^{-1} + 3x_2 x_3^{-2} + 5x_1^{-3} x_2 x_3 + x_1 x_2 x_3$$

such that $x_1, x_2, x_3 \ge 0$

SOLUTION.   We can write the given function as

$$f(x) = 7x_1^1 x_2^{-1} x_3^0 + 3x_1^0 x_2^1 x_3^{-2} + 5x_1^{-3} x_2^1 x_3^1 + x_1^1 x_2^1 x_3^1$$

so that

$$(C_1, C_2, C_3, C_4) = (7, 3, 5, 1)$$

and

$$\begin{bmatrix} a_{11} & a_{12} & a_{13} \\ a_{21} & a_{22} & a_{23} \\ a_{31} & a_{32} & a_{33} \end{bmatrix} = \begin{bmatrix} 1 & 0 & -3 & 1 \\ -1 & 1 & 1 & 1 \\ 0 & -2 & 1 & 1 \end{bmatrix}$$

Now, the orthogonality and normality conditions are given by

$$\begin{bmatrix} 1 & 0 & -3 & 1 \\ -1 & 1 & 1 & 1 \\ 0 & -2 & 1 & 1 \\ 1 & 1 & 1 & 1 \end{bmatrix} \begin{bmatrix} y_1 \\ y_2 \\ y_3 \\ y_4 \end{bmatrix} = \begin{bmatrix} 0 \\ 0 \\ 0 \\ 1 \end{bmatrix}$$

$$\Rightarrow \qquad y_1 + 0y_2 - 3y_3 + y_4 = 0$$

$$-y_1 + y_2 + y_3 + y_4 = 0$$

$$0y_1 - 2y_2 + y_3 + y_4 = 0$$

and $\qquad y_1 + y_2 + y_3 + y_4 = 0$

On solving the above equations, we get

$$y_1^* = \frac{1}{2}, y_2^* = \frac{1}{6}, y_3^* = \frac{5}{24} \text{ and } y_4^* = \frac{3}{24}$$

$$\Rightarrow \qquad y_1^* = \frac{12}{24}, y_2^* = \frac{4}{24}, y_3^* = \frac{5}{24} \text{ and } y_4^* = \frac{3}{24}$$

Therefore, $f(x) = \left(\dfrac{7}{12/24}\right)^{12/24} \cdot \left(\dfrac{3}{4/24}\right)^{4/24} \left(\dfrac{5}{5/24}\right)^{5/24} \left(\dfrac{1}{3/24}\right)^{3/24}$

$\qquad = \dfrac{761}{50} = 15.22$

Now, from the substitution $u_j = y_j f^*(x);\ f^*(x) = \min f(x) = \dfrac{u_j}{y_j}$

We have

$$7x_1 x_2^{-1} = u_1 = \frac{1}{2}(15.22) = 7.61$$

$$3x_2 x_3^{-2} = u_2 = \frac{1}{6}(15.22) = 2.54$$

$$5x_1^{-1}x_2 x_3 = u_3 = \frac{5}{24}(15.22) = 3.17$$

$$x_1 x_2 x_3 = u_4 = \frac{1}{8}(15.22) = 1.90$$

The solution of these equations are given by
$$x_1^* = 1.315, x_2^* = 1.21 \text{ and } x_3^* = 1.20$$
which is the required optimal solution of the given primal problem.

EXAMPLE 2. **When $n > m + 1$, solve the following non-linear programming problem by geometric programming method**
$$\min.f(x) = 5x_1 x_2^{-1} + 2x_1^{-1}x_2 + 5x_1 + x_2^{-1}$$

SOLUTION. We can write the given function as
$$f(x) = 5x_1^1 x_2^{-1} + 2x_1^{-1}x_2^1 + 5x_1^1 x_2^0 + x_1^0 x_2^{-1}$$
so that $(C_1, C_2, C_3, C_4) = (5, 2, 5, 1)$
The orthogonality and normality conditions are given by

$$\begin{bmatrix} 1 & -1 & 1 & 0 \\ -1 & 1 & 0 & -1 \\ 1 & 1 & 1 & 1 \end{bmatrix}\begin{bmatrix} y_1 \\ y_2 \\ y_3 \\ y_4 \end{bmatrix} = \begin{bmatrix} 0 \\ 0 \\ 1 \end{bmatrix}$$

Since $n > m + 1$, these equations do not give the required y_j directly. Therefore, solving for y_1, y_2 and y_3 in terms of y_4, we get

$$\begin{bmatrix} 1 & -1 & 1 \\ -1 & 1 & 0 \\ 1 & 1 & 1 \end{bmatrix}\begin{bmatrix} y_1 \\ y_2 \\ y_3 \end{bmatrix} = \begin{bmatrix} 0 \\ y_4 \\ 1-y_4 \end{bmatrix}$$

$\Rightarrow \qquad y_1 = \dfrac{1-3y_4}{2} = 0.5(1-3y_4)$

$\qquad\qquad y_2 = \dfrac{1-y_4}{2} = 0.5(1-y_4)$

and $\qquad y_3 = y_4$

Now, the corresponding dual problem can be written as

$$\min.f(y) = \left[\dfrac{5}{0.5(1-3y_4)}\right]^{0.5(1-3y_4)}\left[\dfrac{2}{0.5(1-y_4)}\right]^{0.5(1-y_4)}\left[\dfrac{5}{y_4}\right]^{y_4}\left[\dfrac{1}{y_4}\right]^{y_4}$$

which becomes a problem of maxima of one variable only. So, we apply the principle of maxima and minima of differential calculus.

Taking log of both the sides, we get

$$\log f(y) = 0.5(1-3y_4)\{\log 10 - \log(1-3y_4)\} + 0.5(1-y_4)\{\log 4 - \log(1-y_4)\}$$
$$+ y_4\{\log 5 - \log y_4\} + y_4\{\log 1 - \log y_4\} \qquad \ldots(1)$$

Differentiating w.r.t. y_4, we get

$$\frac{\partial}{\partial y_4}\log f(y) = -\frac{3}{2}\log 10 - \left\{-\frac{3}{2} + \left(-\frac{3}{2}\right)\log(1-3y_4)\right\} - \frac{1}{2}\log 4 - \left\{-\frac{1}{2} + \left(-\frac{1}{2}\right)\log(1-y_4)\right\}$$
$$+ \log 5 - \{1 + \log y_4\} + \log 1 - \{1 + \log y_4\}$$

For the condition of maxima and minima we must have $\dfrac{\partial}{\partial y_4}(\log f(y)) = 0$

$$\Rightarrow \quad -\log\left\{\frac{2\times 10^{3/2}}{}\right\} + \log\left\{\frac{(1-3y_4)^{3/2}(1-y_4)^{1/2}}{}\right\} = 0$$

$$\Rightarrow \qquad \sqrt{\frac{\{(1-3y_4)^3(1-y_4)\}}{y_4^2}} = 12.6$$

$$\Rightarrow \qquad y_4^* = 0.16$$

Hence, $\quad y_3^* = 0.16, y_2^* = 0.42$ and $y_1^* = 0.26$

Then $f^*(y) = f^*(x) = \left(\dfrac{5}{0.26}\right)^{0.26}\left(\dfrac{2}{0.42}\right)^{0.42}\left(\dfrac{5}{0.16}\right)^{0.16} = 9.661$

Therefore, $u_3 = 0.16(9.661) = 1.546 = 5x_1$

$$u_4 = 0.16(9.661) = 0.1546 = x_2^{-1}$$

Hence, $x_1^* = 0.309, x_2^* = 0.647$

EXAMPLE 3.   ***Solve the following non-linear programming problem.***

$$f(x) = f(x_1, x_2) = 100x_1 + 50x_1^2 + 20x_2^{-1} + 300x_1^{-5}x_2^2$$

SOLUTION.   The given function can be written as

$$f(\mathbf{x}) = f(x_1, x_2) = 100x_1^1 x_2^0 + 50x_1^2 x_2^0 + 20x_1^0 x_2^{-1} + 300x_1^{-5}x_2^2 \qquad \ldots(1)$$

So, that $(C_1, C_2, C_3, C_4) = (100, 50, 20, 300)$

and $\begin{bmatrix} a_{11} & a_{12} & a_{13} & a_{14} \\ a_{21} & a_{22} & a_{23} & a_{24} \end{bmatrix} = \begin{bmatrix} 1 & 2 & 0 & -5 \\ 0 & 0 & -1 & 2 \end{bmatrix}$

Also, the orthogonality and normality conditions are given by

$$\begin{bmatrix} 1 & 2 & 0 & -5 \\ 0 & 0 & -1 & 2 \\ 1 & 1 & 1 & 1 \end{bmatrix}\begin{bmatrix} y_1 \\ y_2 \\ y_3 \\ y_4 \end{bmatrix} = \begin{bmatrix} 0 \\ 0 \\ 1 \end{bmatrix}$$

We observe that here $N > (n + 1)$ these equations do not give the required $y_j : j = 1, 2, 3, 4$ directly. Thus, we can find the values of other variables in term of a single variable. Now we solve y_1, y_2, y_3 in terms of y_4 we get

$$y_1 = 2 - 11y_4, \qquad y_2 = 8y_4 - 1, \qquad y_3 = 2y_4$$

Now the dual problem can be written as follows:

$$\max. f(\mathbf{y}) = \max. f(y_1, y_2, y_3, y_4)$$

$$= \left(\frac{C_1}{y_1}\right)^{y_1}\left(\frac{C_2}{y_2}\right)^{y_2}\left(\frac{C_3}{y_3}\right)^{y_3}\left(\frac{C_4}{y_4}\right)^{y_4}$$

$$= \left(\frac{100}{2-11y_4}\right)^{2-11y_4}\left(\frac{50}{8y_4-1}\right)^{8y_4-1}\left(\frac{20}{2y_4}\right)^{2y_4}\left(\frac{300}{y_4}\right)^{y_4}$$

Taking log of both the sides of above equations, we get

$$\log f(\mathbf{y}) = F(\mathbf{y}) = (2-11y_4)[\log 100 - \log(2-11y_4)]$$
$$+ (8y_4 - 1)[\log 50 - \log(8y_4 - 1)]$$
$$+ 2y_4[\log 20 - \log(2y_4)] + y_4[\log 300 - \log(y_4)]$$

For the maximum of F we must have $\dfrac{\partial F}{\partial y_4} = 0$

$$\Rightarrow \quad -11[\log 100 - \log(2-11y_4)] + (2-11y_4)\frac{11}{2-11y_4}$$

$$+ 8[\log 50 - \log(8y_4 - 1)] + (8y_4 - 11)\left(-\frac{8}{8y_4 - 1}\right)$$

$$+ 2[\log 20 - \log(2y_4)] + 2y_4\left(\frac{-2}{2y_4}\right) + 1[\log 300 - \log(y_4)] + y_4\left(\frac{-1}{y_4}\right) = 0$$

$$\Rightarrow \quad \frac{(2-11\log y_4)^{11}}{(8y_4 - 1)^8 (2y_4)^2 y_4} = \frac{(100)^{11}}{(50)^8 (20)^2 (300)} = 2130$$

$$\Rightarrow \quad y_4 \approx 0.147$$

So, $\quad y_1^* = 2 - 11y_4^* = 0.385; y_2^* = 8y_4^* - 1 = 0.175$ and $y_3^* = 2y_4^* = 0.294$

and optimal value of the objective function is given by

$$y^* = f^* = \left(\frac{100}{0.385}\right)^{0.385}\left(\frac{50}{0.175}\right)^{0.175}\left(\frac{20}{0.294}\right)^{0.294}\left(\frac{300}{0.147}\right)^{0.147}$$

$$= 8.5 \times 2.69 \times 3.46 \times 3.06 = 242$$

Also, $\quad u_1 = y_1^* f^* = 0.385 \times 242 = 92.2$

$$u_2 = y_2^* f^* = 0.175 \times 242 = 42.4$$

$$u_3 = y_3^* f^* = 0.294 \times 242 = 71.1$$

and $\quad u_4 = y_4^* f^* = 0.147 \times 242 = 35.6$

Finally on solving the above equations, we get

$$x_1^* = 0.922 \text{ and } x_2^* = 0.281$$

6.7 PRIMAL GEOMETRIC PROGRAMMING PROBLEM WITH EQUALITY CONSTRAINTS

Consider the case of minimization as given follows:

$$\min. Z = f(\mathbf{x})$$

subject to the constraints

$$g_j(\mathbf{x}) = \sum_{r=1}^{P(i)} C_{ir} u_{ir}(\mathbf{x}), \quad i = 1, 2, \dots, m$$

where $P(i)$ is the no. of terms in the ith constraints and

$$u_{ir} = \prod_{j=1}^{n} (x_j)^{\alpha_{ir_n}}$$

Now, first we form the Lagrangian function

$$L(\boldsymbol{x}, \lambda) = f(\boldsymbol{x}) + \sum_{i=1}^{m} \lambda_i [g_i(\boldsymbol{x}) - 1]$$

such that

(i) $\quad \dfrac{\partial L}{\partial x_i} = \dfrac{\partial f(\boldsymbol{x})}{\partial x_t} + \sum_{i=1}^{m} \dfrac{\partial g_i(\boldsymbol{x})}{\partial x_t} = 0, \ t = 1, 2, \ldots, n$

(ii) $\quad \dfrac{\partial L}{\partial \lambda_i} = g_i(\boldsymbol{x}) - 1 = 0; \ i = 1, 2, \ldots, m$

Now, (i) $\Rightarrow \dfrac{\partial L}{\partial x_t} = \sum_{j=1}^{n} \dfrac{C_j a_{ij} u_j(\boldsymbol{x})}{x_t} + \sum_{i=1}^{m} \lambda_i \left[\sum_{r=1}^{P(i)} \dfrac{C_{ir} a_{irt} u_{ir}(\boldsymbol{x})}{x_t} \right]$

We may again introduce variables y_j and y_{ir} as follows:

$$y_j = \frac{C_j u_j}{f^*(\boldsymbol{x})} \text{ and } y_{ir} = \frac{\lambda_i C_{ir} u_{ir}}{f^*(\boldsymbol{x})}$$

Putting these values of y_j and y_{ir} in $\dfrac{\partial L}{\partial x_t} = 0$ we get

$$\sum_{j=1}^{n} a_{ij} y_j + \sum_{i=1}^{m} \sum_{r=1}^{P(i)} a_{irt} y_{ir} = 0; t = 1, 2, \ldots n \qquad \text{(Orthogonality condition)}$$

and $$\sum_{j=1}^{n} y_j = 0 \qquad \text{(Normality Condition)}$$

Here, all y_j were positive. However, in the equality constraints case y_j are again positive but y_{ir} may be negative because λ_i need not be non-negative. It is required to have all $y_{ir} > 0$ to construct a dual function. If we reverse the order of constructing the Lagrangian function, the sign of the Lagrangian multipliers will change. Hence, in any problem if one of the λ_{ir} is negative then we can reverse its sign simply by writting that term in the Lagrangian functions as

$$\lambda_q [1 - g_q(\boldsymbol{x})]$$

Then the condition of normality and orthogonality can be derived by solving a system of linear equtions

$$\sum_{j=1}^{n} y_j = 1 \qquad \text{(Normality condition)}$$

and $\quad \sum_{j=1}^{n} a_{ij} y_j + \sum_{i=1}^{m} \sum_{r=1}^{P(i)} a_{irt} y_{ir} = 0 \qquad \text{(Orthogonality Condition)}$

If these equations have a unique solution, the optimal of the original problem can be obtained from the definitions of y_j and y_{ir} in terms of $f^*(\boldsymbol{x})$ and x. In this case when there are infinite no. of solutions we maximize the dual function given by

$$f(y) = \prod_{j=1}^{n} \left(\frac{C_j}{y_j} \right)^{y_j} \prod_{i=1}^{m} \left[\prod_{r=1}^{P(i)} \left(\frac{C_{rj}}{y_{rj}} \right)^{y_{rj}} \right] \prod_{i=1}^{m} (v_i)^{v_i} \text{ where } v_r = \sum_{r=1}^{P_y} y_{ir}$$

subject to the normality and orthogonality conditions.

SOLVED EXAMPLES

EXAMPLE 1. *Solve the following non-linear programming problem:*

$$\text{Min} . f(x) = 2x_1 x_2^{-3} + 4x_1^{-1} x_2^{-2} + \frac{32}{3} x_1 x_2$$

subject to the constraints

$$10 x_1^{-1} x_2^2 = 1$$

and $\qquad x_1, x_2 \geq 0$

SOLUTION. The dual of the NLPP can be written as

$$\text{max} . f(y) = \left(\frac{2}{y_1}\right)^{y_1} \left(\frac{4}{y_2}\right)^{y_2} \left(\frac{32}{3y_3}\right)^{y_3} \left(\frac{0.1}{y_4}\right)^{y_4}$$

subject to the constraints

$$y_1 + y_2 + y_3 =$$
$$y_1 - y_2 + y_3 - y_4 =$$
$$-3y_1 - 2y_2 + y_3 + 2y_4 = 0$$

Now, express each term of the objective function in terms of y_1 we get

$$\text{max} . f(y_1) = \left(\frac{2}{y_1}\right)^{y_1} \left(\frac{4}{1 - \left(\frac{4}{3}\right) y_1}\right)^{1 - \frac{4}{3} y_1} \left(\frac{32}{y_1}\right)^{\frac{1}{3} y_1} (0.1)^{\frac{8}{3} y_1 - 1}$$

where $y_2 = 1 - \frac{4}{3} y_1; y_3 = \frac{y_1}{3}$ and $y_4 = \frac{8}{3} y_1 - 1$

Taking log on both the sides and then differentiating w.r.t. y_1 we get

$$F(y) = \log\{f(y_1)\} = y_1 \log\left(\frac{2}{y_1}\right) + \left\{1 - \frac{4}{3} y_1\right\} \log\left\{\frac{4}{1 - \frac{4}{3} y_1}\right\}$$

$$+ \frac{y_1}{3} \log\left(\frac{32}{y_1}\right) + \left(\frac{8}{3} y_1 - 1\right) \log(0.1)$$

$$= y_1 \{\log 2 - \log y_1\} + \left\{1 - \frac{4}{3} y_1\right\} + \left\{\log 4 - \log\left(1 - \frac{4}{3} y_1\right)\right\}$$

$$+ \frac{y_1}{3} \{\log 32 - \log y_1\} + \left(\frac{8}{3} y_1 - 1\right) \log(0.1)$$

$$\Rightarrow \qquad \frac{dF}{dy_1} = \log\left(\frac{2}{y_1}\right) + 2 - \left(\frac{16}{3}\right) y_1 + \log\left(\frac{32}{y_1}\right) + \frac{8}{3} \log(0.1)$$

For the maxima and minima of F, we must have $\dfrac{dF}{dy_1} = 0$

$$\Rightarrow \qquad y_1 = 0.662$$

Therefore, values of other variables becomes

$$y_2 = 0.217, y_3 = 0.221 \text{ and } y_4 = 0.766$$

Now, using $f^*(x) = \dfrac{C_j \cdot u_j}{y_j}$

We can find the following values

$$y_1 = \frac{C_1 u_1}{f^*(\boldsymbol{x})} = \frac{2x_1 x_2^{-1}}{f^*(\boldsymbol{x})}, y_2 = \frac{C_2 u_2}{f^*(\boldsymbol{x})} = \frac{4x_1^{-1} x_2^{-1}}{f^*(\boldsymbol{x})}$$

$$y_3 = \frac{C_3 u_3}{f^*(\boldsymbol{x})} = \frac{32x_1 x_2}{3f^*(\boldsymbol{x})}, y_4 = \frac{C_4 u_4}{f^*(\boldsymbol{x})} = \frac{10x_1^{-1} x_2^2}{f^*(\boldsymbol{x})}$$

On simplification, we get

$$x_1 = 2.5 \text{ and } x_2 = 0.5$$

6.8 CONSTRAINED OPTIMIZATION PROBLEM

Consider the following problem

$$f(\boldsymbol{x}) = \sum_i f_i(\boldsymbol{x}) \qquad \qquad \dots(1)$$

subject to the constraints

$$g_k(\boldsymbol{x}) = \sum_{i \in j(k)} f_i(\boldsymbol{x}) \le 1, \quad k = 1, 2, \dots, p \qquad \dots(2)$$

and $\qquad\qquad \boldsymbol{x} \ge 0 \qquad\qquad\qquad\qquad\qquad\qquad\qquad\qquad \dots(3)$

where $f(\boldsymbol{x})$ and each $g_k(\boldsymbol{x})$ are posynomials.

and

$$f_i(\boldsymbol{x}) = C_1 x_1^{a_{i1}} \cdot x_2^{a_{i2}} \cdot \dots \cdot x_n^{a_{in}}, \; C_i > 0, x_i > 0$$
$$J(k) = \{m_k, m_k + 1, m_k + 2, \dots, n_k\}, \quad k = 0, 1, \dots, p$$
$$m_0 = 1, m_1 = n_0 + 1, \dots, m_p = n_{p-1} + 1, n_p = s$$

Let us denote n_0 by m and let $y_1, y_2, \dots, y_m$ be the positive weights such that

$$y = y_1 + y_2 + \dots + y_m$$

Let $u_i = \dfrac{y_i}{y}, \; i = 1, 2, \dots, m$ $\qquad\qquad\qquad\qquad\qquad\qquad \dots(4)$

Then clearly u_i are positive weights whose sum is unity.

Now let $F_i = \dfrac{f_i}{y_i}, \; i = 1, 2, \dots m$ $\qquad\qquad\qquad\qquad\qquad\qquad \dots(5)$

Then, we have

$$u_1 F_1 + u_2 F_2 + \dots u_m F_m \ge F_1^{u_1} \cdot F_2^{u_2} \cdot \dots \cdot F_m^{u_m}$$

$$(f_1 + f_2 + \dots + f_m)^{y_1} \ge y^y \left\{ \left(\frac{f_1}{y_1}\right)^{y_1} \left(\frac{f_2}{y_2}\right)^{y_2} \dots \left(\frac{f_m}{y_m}\right)^{y_m} \right\} = A \qquad \dots(6)$$

$\Rightarrow \qquad\qquad \{f(\boldsymbol{x})\}^y \ge A \qquad\qquad\qquad\qquad\qquad\qquad\qquad \dots(7)$

Again using (6) for g_i with positive weights $z_1, z_2, \dots, z_{n_1 - n_0}$ such that

$$z_1 + z_2 + \dots + z_{n_1 - n_0} = z$$

We have

$$1 \ge \{g_i(\boldsymbol{x})\}^z \ge z^z \left\{ \left(\frac{f_{m+1}}{z_1}\right)^{z_1} \dots \left(\frac{f_{n_1}}{z_{n_1 - n_0}}\right)^{z_{n_1 - n_0}} \right\} \qquad \dots(8)$$

From (7) and (8) we have

$$\{f(\boldsymbol{x})\}^y \ge y^y z^z \left\{ \left(\frac{f_1}{y_1}\right)^{y_1} \dots \left(\frac{f_m}{y_m}\right)^{y_m} \left(\frac{f_{m+1}}{z_1}\right)^{z_1} \dots \left(\frac{f_{n_1}}{z_{n_1 - n_0}}\right)^{z_{n_1 - n_0}} \right\} \qquad \dots(9)$$

without loss of any generality, we may assume that $y = 1$

Then $(q) \Rightarrow$ $\qquad$ $f(\mathbf{x}) \geq z^z \left\{ \left(\dfrac{f_1}{y_1} \right)^{y_1} \cdots \left(\dfrac{f_m}{y_m} \right)^{y_m} \left(\dfrac{f_{m+1}}{z_1} \right)^{z_1} \cdots \left(\dfrac{f_{n_1}}{z_{n_1 - n_0}} \right)^{z_{n_1 - n_0}} \right\}$ $\qquad$ ...(10)

where $y_1 + y_2 + \ldots + y_m = 1, \quad y_i > 0$

$z_1 + z_2 + \ldots + z_{n_1 - m} = z, \quad z_i > 0$

SOLVED EXAMPLES

EXAMPLE 1. *Solve the following non-linear programming problem*

$$\text{Min.} f(x) = 40x_1^{-1}x_2^{-1}x_3^{-1} + 40x_1 x_2$$

subject to

$$g_i(x) = \frac{1}{2}x_1 x_2 + \frac{1}{4}x_2 x_3 \leq 1$$

and $x_1, x_2, x_3 \geq 0$

SOLUTION. The dual of the above given NLPP is obtained from

$$F(u) = \left(\frac{40x_1^{-1}x_2^{-1}x_3^{-1}}{u_1} \right)^{u_1} \left(\frac{40x_1 x_3}{u_2} \right)^{u_2} \left(\frac{x_1 x_2}{2u_3} \right)^{u_3} \left(\frac{x_2 x_3}{4u_4} \right)^{u_4}$$

and is given by

$$\phi(u) = \left(\frac{10}{u_1} \right)^{u_1} \left(\frac{40}{u_2} \right)^{u_2} \left(\frac{1}{2u_3} \right)^{u_3} \left(\frac{1}{4u_4} \right)^{u_4} \cdot (u_3 + u_4)^{u_3 + u_4}$$

subject to

$$-u_1 + u_2 + u_3 = 0$$
$$-u_1 + u_3 + u_4 = 0$$
$$-u_1 + u_2 + u_4 = 0$$
$$u_1 + u_2 = 1$$

and $u_1, u_2, u_3, u_4 \geq 0$

On solving these equations, we get

$$u_1 = \frac{2}{3}, u_2 = u_3 = u_4 = \frac{1}{3}$$

$\Rightarrow$ dual has a unique solution with $u > 0$. Hence, it is the optimal solution u^*.

Thus, $\qquad f(x^*) = \phi(u^*) = \left(\frac{40}{2/3} \right)^{2/3} \left(\frac{40}{1/3} \right)^{1/3} \left(\frac{1}{2/3} \right)^{1/3} \left(\frac{1}{4/3} \right)^{1/3} \left(\frac{2}{3} \right)^{2/3} = 60$

$\Rightarrow \qquad \min f(\mathbf{x}) = 60$

Since, we know that if u^* is the optimal dual solution, every optimal solution $\mathbf{x}^*$ of primal problem satisfies

$$C_1 x_1^{y_{i1}} x_2^{y_{i2}} \ldots x_n^{y_{in}} = u_i^* \phi(u^*) \text{ and } C_1 x_1^{a_{i1}} \cdot x_2^{a_{i2}} \ldots x_n^{a_{in}} = \frac{u_i^*}{z_k^*}$$

where k ranges over all positive integers for which $z_k^* = \Sigma u_i^* > 0$

Therefore, we have

$$40x_1^{-1}x_2^{-1}x_3^{-1} = u_1^* \phi(u^*) = \frac{2}{3} \times 60$$

$$40x_1 x_3 = u_2^* \phi(u^*) = \frac{1}{3} \times 60$$

and $\qquad \dfrac{1}{2}x_1 x_2 = \dfrac{\dfrac{1}{3}}{\dfrac{1}{3}+\dfrac{1}{3}} = \dfrac{1}{2}$

$$\dfrac{1}{4}x_2 x_3 = \dfrac{1}{2}$$

Solving the above equtions by logarithmic method, we get

$$-t_1 - t_2 - t_3 = \log 1$$

$$t_1 + t_3 = \log\left(\dfrac{1}{2}\right)$$

$$t_1 + t_2 = \log 1$$

$$t_2 + t_3 = \log 3$$

where $t_i = \log x_i$

On solving we get the optimal solution given by

$$x_1^* = \dfrac{1}{2}, x_2^* = 2, x_3^* = 1$$

6.9 DEGREE OF DIFFICULTY

We can define the degree of difficulty as follows:

Degree of difficulty = (Total no. of terms in $f(x)$ and in all $g_i(x)$)

$$- \text{(Total no. of variables)} - 1$$

$$= s - n - 1$$

Now, we have the following observations:

1. If degree of difficulty is one then we will get the solution of dual constraints in terms of one arbitrary y_i.

2. If degree of difficulty is 2 then we get $y_i's$ in terms of two arbitrary y_i and so on.

☛ REMARK

 • Geometric programming is not applicable if degree of difficulty is negative.

SOLVED EXAMPLES

EXAMPLE 1.   **Solve the following NLPP**

$$\textbf{Min } f(x) = 2x_1 x_2 + 2x_1 x_2^{-1} x_3 + 4x_1^{-1} x_2^2 x_3^{-1/2}$$

subject to the constraints

$$\sqrt{3}x_2^{-1} + 3x_1^{-1} x_3^{-1/2} \le 1$$

and $\qquad x_1, x_2, x_3 \ge 0$

SOLUTION.   Clearly the degree of difficulty is one. Now, the dual problem is given by

$$\phi(y) = \left(\dfrac{2}{y_1}\right)^{y_1}\left(\dfrac{2}{y_2}\right)^{y_2}\left(\dfrac{4}{y_3}\right)^{y_3}\left(\dfrac{\sqrt{3}}{y_4}\right)^{y_4}\left(\dfrac{3}{y_5}\right)^{y_5}(y_4 + y_5)^{y_4 + y_5}$$

subject to the constraints

$$y_1 + y_2 - y_3 - y_5 = 0$$

$$y_1 - y_2 + 2y_3 - y_4 = 0$$

$$y_2 - \dfrac{1}{2}y_3 - \dfrac{1}{2}y_5 = 0$$

$$y_1 + y_2 + y_3 = 0$$

and $\quad y_1, y_2, y_3 \geq 0$

Solving the above system by row operations we get

$$y_1 = y_2 = \frac{1 + y_5}{4}, y_3 = \frac{1 - y_5}{2}, y_4 = 1 - y_5$$

Therefore,

$$\phi(y) = \left(\frac{8}{1 + y_5}\right)^{\frac{1+y_5}{4}} \left(\frac{8}{1 + y_5}\right)^{\frac{1+y_5}{4}} \left(\frac{8}{1 - y_5}\right)^{\frac{1-y_5}{2}} \left(\frac{\sqrt{3}}{1 - y_5}\right)^{1-y_5} \left(\frac{3}{y_5}\right)^{y_5}$$

Taking log of both the sides, we get

$$\log \phi(y) = \log 8 - \frac{1 + y_5}{2} \log(1 + y_5) - \frac{3}{2}(1 - y_5)\log(1 - y_5)$$

$$- y_5 \log y_5 + \frac{1}{2}(1 + y_5)\log 3$$

Now, for maxima of $\phi(y)$ we have

$$\frac{d}{dy_5} \log \phi(y) = 0$$

$$\Rightarrow \quad -\frac{1}{2}\log(1 + y_5) + \frac{3}{2}\log(1 - y_5) - \log y_5 + \frac{1}{2}\log 3 = 0$$

$$\Rightarrow \quad\quad\quad\quad\quad\quad\quad 4y_5^3 - 8y_5^2 + 9y_5 - 3 = 0$$

$$\Rightarrow \quad\quad\quad\quad y_5 = \frac{1}{2}$$

We can easily verify that the second derivative of $\log \phi(y)$ w.r.t. y_5 at $y_5 = \frac{1}{2}$ is negative.

Therefore, $\quad\quad y_1 = y_2 = \frac{3}{8}, y_3 = \frac{1}{4}, y_4 = \frac{1}{2}, y_5 = \frac{1}{2}$

and $\quad\quad \max. \phi(y) = 32 = \min. f(x)$

and thus

$$2x_1 x_2 = 12; \quad 2x_1 x_2^{-1} x_3 = 12$$

$$4x_1^{-1} x_2^2 x_3^{-1/2} = 8; \quad \sqrt{3} x_2^{-1} = \frac{1}{2}$$

$$3x_1^{-1} x_3^{-1/2} = \frac{1}{2}$$

On solving we get

$$x_1^* = \sqrt{3}, x_2^* = 2\sqrt{3}, x_3^* = 12$$

EXERCISE 6.1

Using Geometric programming solve the following non-linear programming problem.

1. Min. $z = 40x_1^{-1} x_2^{-1} x_3^{-1} + 40x_2 x_3 + 20x_1 x_2$
$\quad\quad + 10x_1 x_3$,
$\quad\quad x_1, x_2, x_3 \geq 0$

2. Min. $z = 5x_1 x_2^{-1} x_3^2 + x_1^{-2} x_2^{-1} + 10x_2^2$
$\quad\quad + 2x_1^{-1} x_2 x_3^{-2}$,

$\quad\quad x_1, x_2, x_3 \geq 0$

3. Min. $f(x) = 2x_1 + 4x_2 + 10x_1^{-1} x_2^{-1}$, $\quad x_1, x_2 \geq 0$

4. Min. $f(x) = C_1 x_1^{-1} x_2^{-1} x_3^{-1} + C_2 x_2 x_3 + C_3 x_1 x_3$
$\quad\quad + C_4 x_1 x_2$
where $C_i > 0$, $x_j > 0$, $i = 1, 2, 3, 4; j = 1, 2, 3$

5. Min. $f(x) = 2x_1^2 x_2^3 + 2x_1^{-3} x_2^{-2}$
subject to $x_1 x_2^{-1} \leq \frac{1}{4}, x_1, x_2 > 0$

6. $\text{Min.} f(\ ) = 10x_1^{-1}x_2^{-1}x_3^{-1} + 10x_2x_3$
   subject to $2x_1x_3 + x_1x_2 = 4$
   and $x_1, x_2, x_3 > 0$

7. $\text{Min.} f(\boldsymbol{x}) = 4x_1x_2^2x_3^3 + x_1^{-2}x_3^2$

subject to $6x_1^{-1}x_2^{-2}x_3^{-1} + 4x_2^{-3}x_3^{-3} \leq 15$
$x_1, x_2, x_3 > 0$

8. $\text{Min.} f(\boldsymbol{x}) = x_1^2 + x_2^2$
   subject to $x_1x_2 \geq 1, \quad x_1, x_2 > 0$

ANSWERS

1. $x_1 = 2, x_2 = 1, x_3 = \dfrac{1}{2}, \text{min.} z = 100$

2. $x_1 = 1.26, x_2 = 0.41, x_3 = 0.59, \text{min.} f(\boldsymbol{x}) = 10.28$

3. $x_1 = 14.1, x_2 = 23, \text{min.} f(\boldsymbol{x}) = 112.9$

4. $\text{min.} f(\boldsymbol{x}) = \left(\dfrac{5}{2C_1}\right)^{2/5} (5C_2)^{1/5}(5C_3)^{1/5}(5C_4)^{1/5}$

5. $x_1 = \dfrac{1}{2}, x_2 = 2$

6. $x_1 = x_2 = 1$

7. $x_1 = 1.125, x_2 = 0.713, x_3 = 1.07, \text{min.} f = 5\left(\dfrac{2}{3}\right)^{2/3}$

□□□□

7 Fractional Programming

INTRODUCTION

The fractional programming technique is used to solve the problem of maximizing the ratio of two linear functions subject to a set of linear equalities and the non-negative restrictions. Such type of problem can be directly solved by starting with a basic feasible solution.

In this chapter we shall discuss the fractional programming in details.

7.2 FRACTIONAL PROGRAMMING

This is a newly developed important techniques of mathematical programming namely fractional programming. It is also known as 'Linear fractional programming'. In this technique we solve the problem of maximizing the ratio of two linear functions subject to the set of linear constraints and non-negative restrictions. This problem can be solved directly by starting with a basic feasible solution and showing the conditions for imroving the current basic feasible solution. To test the optimality of the solution, we use the optimality criterion.

7.3 MATHEMATICAL FORMULATION OF LINEAR FRACTIONAL PROGRAMMING

The linear fractional programming problem can be formulated mathematically as follows:

$$\max. Z = \frac{C_1' \boldsymbol{x} + \alpha}{C_2' \boldsymbol{x} + \beta} \qquad \qquad ...(1)$$

subject to the constraints

$$\left. \begin{array}{r} A\boldsymbol{x} = \boldsymbol{b} \\ \boldsymbol{x} \geq 0 \end{array} \right\} \qquad \qquad ...(2)$$

and

where

  (i) $\boldsymbol{x}$, C_1 and C_2 are $n \times 1$ column vectors.

 (ii) A is a $m \times n$ matrix

(iii) $\boldsymbol{b}$ is a $m \times 1$ column vector

 (iv) the desh ($'$) denote the transpose

  (v) $\alpha, \beta \in \boldsymbol{R}$

It should be noted that the constraints set

$$S = \{\boldsymbol{x} : A\boldsymbol{x} = \boldsymbol{b}, \ \boldsymbol{x} \geq 0\}$$

is non-empty and bounded.

7.4 LINEAR FRACTIONAL PROGRAMMING ALGORITHM

Consider

$$\max. Z = \frac{(C_1' x + \alpha)}{(C_2' x + \beta)}$$

subject to the constraints

$$A x = b, \quad x \geq 0$$

such that the denominator is positive for all feasible solution.
If x_B be the starting basic feasible solution such that

$$B \cdot x_B = b$$
$$\Rightarrow \qquad x_B = B^{-1} b \qquad x_B \geq 0$$

where $B = (\beta_1, \beta_2, ..., \beta_m)$

Now, let $Z^{(1)} = C_{1B}' x_B + \alpha$ and $Z^{(2)} = C_{2B}' x_B + \beta$

where C_{1B}' and C_{2B}' are the vectors having their components as the coefficients associated with the basic variables in the numerator and the denominator of the objective function respectively.

Further, for the basic feasible solution, we assume that

$$x_j = B^{-1} a_j \cdot Z_j^{(1)} = C_{1B}' x_j Z_j^{(2)} = C_{2B}' x_j$$

are determinable for every column a_j belong to A but not to B.

To improve the IBFS

Let $\hat{x}_B$ be the new basic feasible solution. Then

$$\hat{x}_B = \hat{B}^{-1} b \text{ where } \hat{B} = \{\hat{\beta}_1, \hat{\beta}_2, ..., \hat{\beta}_n\}$$

So, we have to obtained new non-singular matrix from B by replacing β_r by a_j. Therefore,

$$\hat{\beta}_i = \beta_i \ (i \neq r), \quad \hat{\beta}_r = a_j$$

Then we find the following values

$$\hat{x}_{B_i} = x_{B_i} - \frac{x_{ij}}{x_{rj}} x_{B_r} \ (i \neq r)$$

$$\hat{x}_{B_r} = \frac{x_{B_r}}{x_{rj}} = \theta \ (\text{say})$$

where $a_j = \sum_{i=1}^{m} x_{ij} \cdot \beta_i$

Now we have to justify whether Z is improved

Let the value of the objective fuction for the original basic feasible solution is $Z = \frac{Z^{(1)}}{Z^{(2)}}$

Suppose the new value of the objective function be $\hat{Z} = \frac{\hat{Z}^{(1)}}{\hat{Z}^{(2)}}$.

Then we have

$$\hat{Z}^{(1)} = Z^{(1)} - \theta(\hat{Z}_j^{(1)} - C_{1j})$$

$$\hat{Z}^{(2)} = Z^{(2)} - \theta(\hat{Z}_j^{(2)} - C_{2j})$$

The value of the new objective function will improve if

$$\frac{Z^{(1)} - \theta(Z_j^{(1)} - C_{1j})}{Z^{(2)} - \theta(Z_j^{(2)} - C_{2j})} > \frac{Z^{(1)}}{Z^{(2)}} \quad i.e. \quad \frac{Z^{(1)} - \theta(Z_j^{(1)} - C_{1j})}{Z^{(2)} - \theta(Z_j^{(2)} - C_{2j})} - \frac{Z^{(1)}}{Z^{(2)}} > 0$$

or $\quad Z^{(2)}[Z^{(1)} - \theta(Z_j^{(1)} - C_{1j})] - Z^{(1)}[Z^{(2)} - \theta(Z_j^{(2)} - C_{2j})] > 0$

or $\quad -Z^{(2)}[Z^{(1)} - \theta(Z_j^{(1)} - C_{1j})] + Z^{(1)}[Z^{(2)} - \theta(Z_j^{(2)} - C_{2j})] < 0$

Let us denote

$$\Delta_j = Z^{(1)}[Z^{(2)} - \theta(Z_j^{(2)} - C_{2j})] - Z^{(2)}[Z^{(1)} - \theta(Z_j^{(1)} - C_{1j})]$$

Now, for $\Delta_j > 0$, we have the following cases:

(i) If $Z_j^{(2)} - C_{2j} > 0$ then $\dfrac{(Z_j^{(1)} - C_{1j})}{(Z_j^{(2)} - C_{2j})} < \dfrac{Z^{(1)}}{Z^{(2)}}$

(ii) If $Z_j^{(2)} - C_{2j} < 0$ then $\dfrac{(Z_j^{(1)} - C_{1j})}{(Z_j^{(2)} - C_{2j})} > \dfrac{Z^{(1)}}{Z^{(2)}}$

(iii) If $Z_j^{(2)} - C_{2j} = 0$ then $Z_j^{(1)} - C_{1j} > 0$

Then we have the following result

"Given a basic feasible solution $x_B = B^{-1}b$, if for any column vector a_j in A but not in B, $\Delta_j > 0$ holds and if at least one $x_{ij} > 0$ $(i = 1, 2, ..., m)$ then it is possible to find a new basic feasible solution by replacing one of the column in B by a_j and the new value of the objective function satisfies $\hat{Z} \geq Z$."

> If the problem
>
> $$\text{max. } Z = \frac{(C_1'x + \alpha)}{(C_2'x + \beta)}$$
>
> subject to $Ax = b$, $x \geq 0$
> has a basic feasible solution $x_B = B^{-1}b$ with
>
> $$Z^* = \frac{(C_{1B}'x_B + \alpha)}{(C_{2B}'x_B + \beta)}$$
>
> such that $\Delta_j \geq 0$ for every column a_j in A then Z^* will be maximum value of Z and the basic solution will be an optimum solution.

SOLVED EXAMPLES

EXAMPLE 1. *Solve the fractional linear programming problem*

$$\text{max.} Z = \frac{5x_1 + 3x_2}{5x_1 + 2x_2 + 1}$$

subject to the constraints

$$3x_1 + 5x_2 \leq 15$$
$$5x_1 + 2x_2 \leq 10$$

and $x_1, x_2 \geq 0$

SOLUTION. Introducing the slack variables s_1 and s_2 the above problem can be written as follows:

$$\text{max.} Z = \frac{5x_1 + 3x_2}{5x_1 + 2x_2 + 1} = \frac{Z^{(1)}}{Z^{(2)}}$$

subject to the constraints
$$3x_1 + 5x_2 + s_1 = 15$$
$$5x_1 + 2x_2 + s_2 = 10$$
$$x_1, x_2, s_1, s_2 \geq 0$$

Now we have the following table:

Starting table

| | | | $C_1 \rightarrow$ | 5 | 3 | 0 | 0 | |
| | | | $C_2 \rightarrow$ | 5 | 2 | 0 | 0 | |

B.V.	C_{2B}	C_{1B}	x_B	x_1	x_2	s_1	s_2	$\min\left(\dfrac{x_B}{x_1}\right)$
s_1	0	0	15	3	5	1	0	15/3
s_2	0	0	10	⑤	2	0	1	10/5 →
		$Z^{(1)} = C_{1B}x_B = 0$		+5	+3	0	0	
		$Z^{(2)} = C_{2B}x_B = 1$		+5	+2	0	0	
		$Z = \dfrac{Z^{(1)}}{Z^{(2)}} = 0$		+5	+3	—	—	
				↑			↓	

In the above table we find Δ_1 is minimum. So, Z can be increased by taking x_1 in the basis. Apply the same procedure of ordering simplex method, we can easily find that s_2 will be the outgoing vector.

So, introducing x_1 and dropping s_2, we get the following table:

| | | | $C_{1j} \rightarrow$ | 5 | 3 | 0 | 0 | |
| | | | $C_{2j} \rightarrow$ | 5 | 2 | 0 | 0 | |

B.V.	C_{2B}	C_{1B}	x_B	x_1	x_2	s_1	s_2	$\min\left(\dfrac{x_B}{x_2}\right)$
s_1	0	0	9	0	⑲/5	1	–3/5	$9/(19/5) = (45/19)\leftarrow$
x_1	5	5	2	1	2/5	0	1/5	$2/(2/5) = 5$
		$Z^{(1)} = C_{1B}x_B = 10$		0	+1	0	–1	
		$Z^{(2)} = C_{2B}x_B = 11$		0	0	0	–1	
		$Z = \dfrac{Z^{(1)}}{Z^{(2)}} = \dfrac{10}{11}$		—	+11	—	–1	
				↓	↑			

Further introducing x_2 and dropping s_1 we get the following table:

$$C_{1j}: \quad 5 \quad 3 \quad 0 \quad 0$$
$$C_{2j}: \quad 5 \quad 2 \quad 0 \quad 0$$

B.V.	C_{2B}	C_{1B}	x_B	x_1	x_2	s_1	s_2	$\min\left(\dfrac{x_B}{s_2}\right)$
x_2	2	3	45/19	0	1	5/19	−3/19	
x_1	5	5	20/19	1	0	−2/19	5/19	
$Z^{(1)} = C_{1B}x_B = \dfrac{235}{19}$				0	0	−5/19	−16/19	
$Z^{(2)} = C_{2B}x_B = \dfrac{209}{19}$				0	0	0	1	
$Z = \dfrac{Z^{(1)}}{Z^{(2)}} = \dfrac{235}{209}$				—	—	−1045/361	+1121/361	
				↓			↑	

Finally introducing s_2 and dropping x_1 we get the following table:

$$C_{1j}: \quad 5 \quad 3 \quad 0. \quad 0$$
$$C_{2j}: \quad 5 \quad 2 \quad 0 \quad 0$$

B.V.	C_{2B}	C_{1B}	x_B	x_1	x_2	s_1	s_2
x_2	2	3	3	3/5	1	1/5	0
s_2	0	0	4	19/5	0	−2/5	1
$Z^{(1)} = C_{1B}x_B = 9$				+16/5	0	−3/5	0
$Z^{(2)} = C_{2B}x_B = 7$				+19/5	0	−2/5	0
$Z = \dfrac{Z^{(1)}}{Z^{(2)}} = \dfrac{9}{7}$				−59/5	—	−3/5	—

Since, all $\Delta_j \leq 0$ therefore, solution is optimum and is given by

$$x_1 = 0, x_2 = 3, s_1 = 0, s_2 = 4 \text{ and max. } Z = \frac{9}{7}$$

7.5 GRAPHICAL METHOD FOR LINEAR FRACTIONAL PROGRAMMING

We know that a programming problem is called a linear fractional programming if
  (i) the objective function is the quotient of two linear variates, such that denominator does not vanish anywhere on the set of feasible solution.
  (ii) the set of feasible solution is as in a linear programming problem.

To obtain the solution of linear fractional programming by graphical method we proceed as follows.

Let S, the set of constraints be a convex set which is the set of feasible solutions. The optimum of an objective function occurs at one of the vertices of S.

In linear fractional programming of two variables. Let

$$N = \text{Numerator of } f(\pmb{x})$$

and $\qquad D = \text{denominator of } f(\pmb{x})$

The point of intersection of $N = 0$ and $D = 0$ denoted by (x_0, y_0) called the rotation point. On the line $N = 0$, $f(\pmb{x}) = 0$ and on the line $D = 0$, $f(x) = \pm\infty$. The part of the line $D = 0$ corresponding to $f = \infty$ and the part corresponding to $f = -\infty$ will be identified by substituting two points on the either side of (x_0, y_0) in N and observing whether the value of N is positive or negative. Despite the linearity of the level curves of objective function in fractional programming the level curves are not parallel to one another as they are in linear programming.

We observe that

(i) For a minimization problem, the value of $f(x)$ increases as one rotates from $D = 0$ to $N = 0$ about (x_0, y_0) in such a direction so that one moves towards S. Clearly the line $D = 0$ will be outside S, since $D \neq 0$ on S. The first point encountered on S as one moves, is the point where minimum occurs.

(ii) For a maximization problem, the value of $f(x)$ decreases as one rotates from $D = 0$ to $N = 0$ about the point (x_0, y_0) in such a direction so that one moves towards S. The first point encountered on S, is the point where maximum occurs.

SOLVED EXAMPLES

EXAMPLE 1.   *Solve the following linear fractional programming problem by graphical method*

$$\text{Min } f(x) = \frac{-8x_1 + 9x_2 + 4}{x_1 + x_2 + 8}$$

subject to the constraints

$$5x_1 + 4x_2 \leq 40$$
$$x_1 + 2x_2 \leq 12$$
$$5x_1 + 19x_2 \leq 95$$

and $x_1, x_2 \geq 0$

SOLUTION. We know that for a minimization problem, the value of $f(x)$ increase as one rotates from $D = 0$ i.e. $f = -\infty$ to $N = 0$ about the point (x_0, y_0) in such a direction so that one moves towards the solution set S. Here we observe that the line $D = 0$ will be outside S since by assumptions $D \neq 0$ on S. So, as one move from $D = 0$ $(f = -\infty)$ to $N = 0$ in the anticlockwise direction about the point $(x_0, y_0) = (-4, -4)$ the point of minimum is encountered at $(8, 0)$ which is shown as in the following graph.

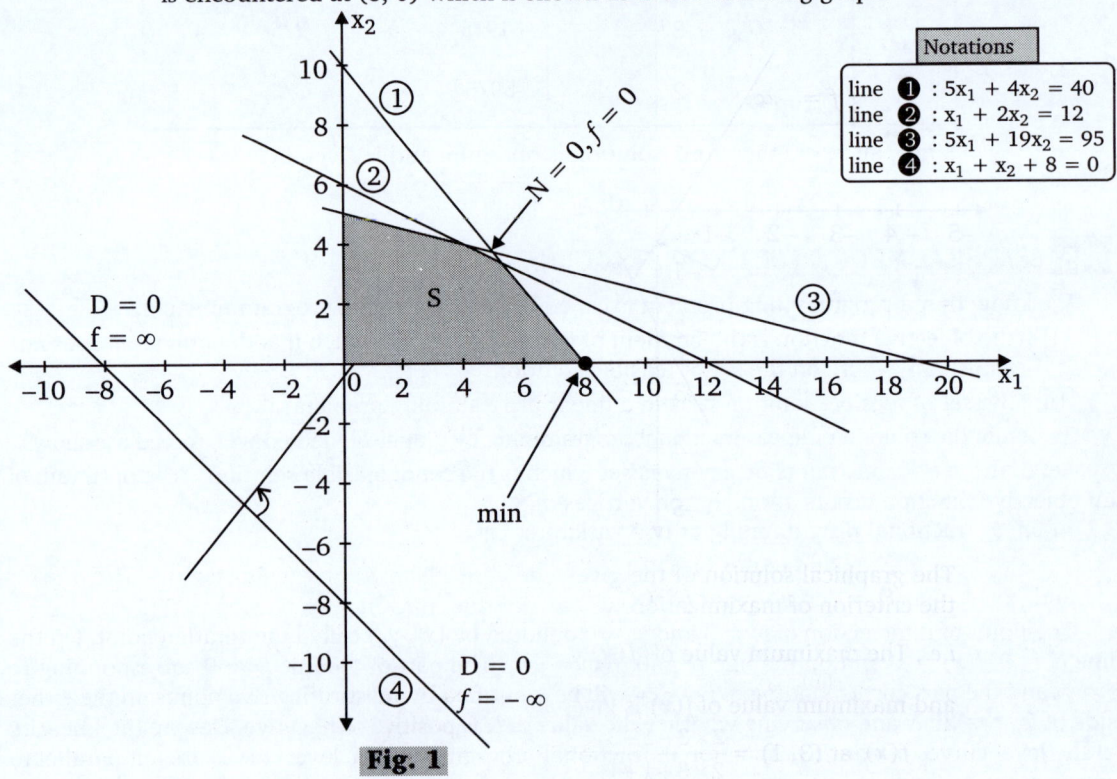

Fig. 1

Since, the point of minimum is (8, 0).

$\Rightarrow \quad x_1 = 8$ and $x_2 = 0$

Hence, $\quad \min f(x) = \dfrac{-8x_1 + 9x_2 + 4}{x_1 + x_2 + 8}$ at (8, 0)

$$= \dfrac{-8 \times 8 + 9 \times 0 + 4}{8 + 0 + 8} = \dfrac{-64 + 4}{16} = \dfrac{-15}{4}$$

EXAMPLE 2. *Solve the following linear fractional programming by graphical method.*

$$\text{max. } f(x) = \dfrac{2x_1 - x_2 - 3}{2x_1 + x_2 + 1}$$

subject to the constraints

$$x_1 + x_2 \leq 4$$

$$x_1 - 2x_2 \leq 1$$

and $\quad x_1, x_2 \geq 0$

SOLUTION. For a maximization problem, we know that the value of $f(x)$ decreases as one rotates from $D = 0$ ($f(x) = \infty$) to $N = 0$ about the point (x_0, y_0) in such a direction so that one moves towards S.

Then the first point encountered on S is the point where maximum occurs.

Here, we have the following graph.

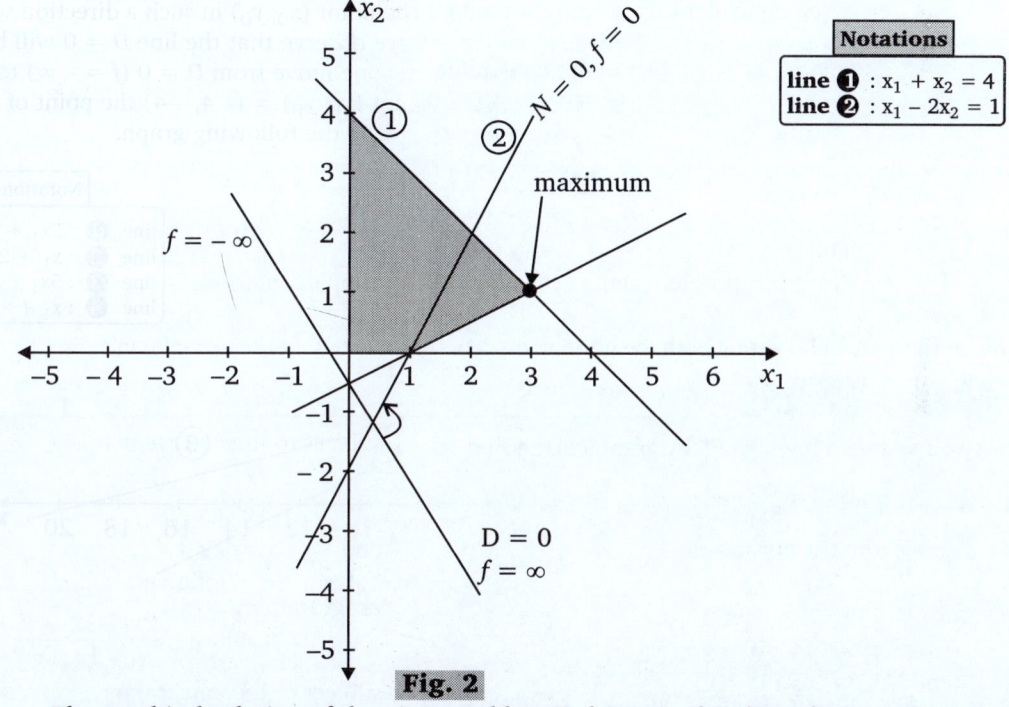

Notations

line ❶ : $x_1 + x_2 = 4$
line ❷ : $x_1 - 2x_2 = 1$

Fig. 2

The graphical solution of the given problem is shown in the above figure. Now using the criterion of maximization we can find that maximum occurs at (3, 1).

i.e., The maximum value of $f(x) = \dfrac{2x_1 - x_2 - 3}{2x_1 + x_2 + 1}$ at $(x_1, x_2) = (3, 1)$.

and maximum value of $f(x)$ is given by

$$f(x) \text{ at } (3, 1) = \dfrac{2 \times 3 - 1 - 3}{2 \times 3 + 1 + 1} = \dfrac{6 - 4}{8} = \dfrac{2}{8} = \dfrac{1}{4}$$

7.6 CHARMES AND COOPER METHOD FOR LINEAR FRACTIONAL PROGRAMMING

In this method the given linear fractional programming is converted into a linear programming problem and then can be solved by the usual methods of the solution of linear programming problem. For the understanding of this method, consider the following example.

EXAMPLE 1. **Solve Min** $f(x) = \dfrac{-8x_1 + 9x_2 + 4}{x_1 + x_2 + 8}$

subject to the constraints

$$5x_1 + 4x_2 \le 40$$
$$x_1 + 2x_2 \le 12$$
$$5x_1 + 19x_2 \le 95$$

and $x_1, x_2 \ge 0$

SOLUTION. The denominator of $f(x)$ is taken equal to v, which will be non-zero in solution set S. Further, we assume that the denominator of $f(x)$ is positive everywhere in S. If it is negative everywhere then the negative sign is taken to the numerator. The given linear fractional programming is converted into one with variables y_1, y_2 and v by using the following substitution.

$$\frac{x_1}{v} = y_1 \ge 0; \frac{x_2}{v} = y_2 \ge 0 \text{ and } \frac{1}{v} = u \ge 0$$

The substitution $x_1 + x_2 + 8 = v \ne 0$ is taken as a constraints given as $y_1 + y_2 + 84 = 1$
Then equivalent linear programming problem is given by

$$\text{Min}.y_0 = -8y_1 + 9y_2 + 4u$$
subject to the constraints
$$5y_1 + 4y_2 - 40u \le 0$$
$$y_1 + 2y_2 - 12u \le 0$$
$$5y_1 + 19y_2 - 95u \le 0$$
$$y_1 + y_2 + 8u = 1$$

and $y_1, y_2, u \ge 0$

Now, this problem can be solved easily by simplex method.

☛ REMARK

• Here, we observe that both the no. of constraints and the number of variables increases by one each.

EXERCISE 7.1

1. Solve the following LFP by graphical method
$$\text{max}.Z = \frac{x_1 + x_2 - 1}{5x_1 + x_2 - 1}$$
subject to the constraints
$$3x_1 + 2x_2 \ge 6$$
$$x_1 \le 3$$
$$x_2 \le 3$$
and $x_1, x_2 \ge 0$

2. Solve the following linear fractional programming
$$\text{max}.Z = \frac{-x_1 - 3}{x_2 + 1}$$

subject to the constraints
$$2x_1 + x_2 \ge 2$$
$$-4x_1 + x_2 \le 2$$
and $x_1, x_2 \ge 0$

3. Solve the following linear fractional programming.
$$\text{min}.Z = \frac{2x_1 - 3x_2 + x_3 + 1}{x_1 + 2x_2 + x_3 + 1}$$
subject to the constraints
$$2x_1 + x_2 + x_3 \le 1$$
$$x_1 - x_2 + 2x_3 \ge 2$$
and $x_1, x_2, x_3 \ge 0$

ANSWERS

1. $x_1 = 0, x_2 = 3$ **2.** $x_1 = -3, x_2 = -1$ **3.** $x_1 = x_2 = 0, x_3 = 1$

8 Stochastic Programming

8.1 INTRODUCTION

Stochastic programming deals with the situations where some or all of the parameters of the optimization problem are described by stochastic (or random or probabilistic) variable rather than deterministic quantities.

The general solution like simplex method for linear programming problems has been developed with random parameters. The stochastic programming approach is used for formulating linear programming problem under uncertainity so as to consider random effects on parameter explicity in the solution of the model.

8.2 APPROACHES FOR SOLVING A STOCHASTIC PROGRAMMING PROBLEM

There are following three approaches for solving a stochastic programming problem.

8.2.1 SEQUENTIAL STOCHASTIC PROGRAMMING

If the given problem involve two or more decision variables at different points in time with the condition that at least one of the later decision may be affected not only by previous decision, but also by some random (stochastic) parameters whose value will be observed before later decision are made, then we use sequential stochastic programming.

The general form of sequential stochastic problem is given below.

$$\text{Optimize (max or min) } Z = \sum_{j=1}^{k} E(C_j x_j) + \sum_{r=1}^{Q} P_r(C_{rj} x_{rj})$$

subject to the constraints

$$\sum_{j=1}^{k} a_{ij} x_j = b_i, \quad i = 1, 2, \ldots, s$$

$$\sum_{j=1}^{K} a_{rij} x_j + \sum_{j=n+1}^{L} a_{rij} x_{rj} = b_{rj}$$

and

$$x_j \geq 0 \quad i = s+1, s+2, \ldots, m$$

$$x_{rj} \geq 0 \quad \forall r \text{ and } j$$

Since this is a two-stage approach x_j ($j = 1, 2, \ldots, k \leq n$) represents level of x in the first stage for $j = k + 1, \ldots, n$ it represents level of x after all random values are known. The contraints i contains only first stage variable for a_{ij} and b_i.

Also, Q = A finite number of possible sets of values, for C_j, a_{ij} and b_i.

P_r = Probability of occurance of an event

Since the model parameters are fixed in the first stage, and after the random event occurs every time the parameters can be revised according to the decision rules, therefore this approach is also called two-stage approach.

8.2.2 NON-SEQUENTIAL STOCHASTIC PROGRAMMING

This is a one stage technique which transforms the given non-linear stochastic programming problem into a deterministic linear programming model. This approach is used to replace the cost coefficients C_j of the objective function, where problem reduces to minimization or maximization of the expected value $E(C)$ of random parameter representing cost.

Here, the optimal value of the objective function obtained is greater than the actual optimal solution for the case of minimization and less than the actual optimal value in case of maximization.

8.2.3 CHANCE-CONSTRAINED PROGRAMMING

The general form of chance-constrained programming mathematical model is given as below.

optimize (max. or min.) $Z = \sum\limits_{j=1}^{n} C_j \cdot x_j$

subject to the constraints

$$P\left[\sum_{j=1}^{n} a_{ij}x_j \leq b_i\right] \geq 1 - \alpha_i; \quad i = 1, 2, \ldots, m$$

and $x_j \geq 0$

where a_i = Risk level for constraints i and $0 \leq \alpha_i \leq 1$ for each level of b_i

WORKING PROCEDURE

STEP 1. The given chance-constrained programming is first transformed into an equivalent deterministic non-linear programming problem.

STEP 2. Solve the problem obtained in step 1 by seperable programming.

Here we have the following cases:

Case I. When only C is a random variable

Let $C_j : j = 1, 2, \ldots, n$ be the normal variate given by $N(\mu_j, \sigma^2)$ where μ_j and σ_j^2 are expected value and variance of random variance C_j. Then $\sum\limits_{j=1}^{n} C_j \cdot x_j$ has also a normal distribution with expected value.

$$E(f) = \sum_{j=1}^{n} \mu_j \cdot x_j$$

Here, $A = [a_{ij}]$ and b are deterministic.

So, there is no need of chance-constraints technique. Then, equivalent deterministic LPP is given as follows.

optimize (max. or min.) $E(f) = \sum\limits_{j=1}^{n} \mu_j x_j$

subject to $Ax \geq, \leq b$

$x \geq 0$

Case II. When only b is a random variable

Let $b_j : j = 1$ to m be the normal random variable given by $N(\mu_{b_j}, \sigma_{b_j}^2)$. Then we use chance constraint technique. So, we have

$$P\left\{b_i \geq \sum_{j=1}^{n} a_{ij}x_j\right\} \geq 1 - p_i$$

Then by central limit theorem, we can write

$$P\left\{\frac{b_i - E(b_i)}{\sqrt{v(b_i)}} \geq \frac{\sum_{j=1}^{n} a_{ij} \cdot x_j - E(b_i)}{\sqrt{v(b_i)}}\right\} \geq 1 - p_i, \text{ where } Z_i = \frac{\{b_i - E(b_i)\}}{\sqrt{v(b_i)}} \quad N(0,1)$$

Now, let $Z_i = \dfrac{\left\{\sum\limits_{j=1}^{n} a_{ij}x_j - E(b_i)\right\}}{\sqrt{v(b_i)}}$

Then we can write the chance-constraints as

$$1 - P(Z_i \leq z_i) \geq 1 - p_i \qquad \Rightarrow \qquad \phi(Z_i) \leq p_i = \phi(K_{p_i})$$

where ϕ is a cumulative distribution function of $N(0, 1)$. Since, ϕ is a non-decreasing continuous function, therefore we have

$$Z_i \leq K_{p_i}$$

$$\sum_{j=1}^{n} a_{ij} \cdot x_j \leq E(b_i) + K_{p_i}\sqrt{v(b_i)}$$

Hence, the LPP equivalent to stochastic programming is given as below.

optimize $x_0 = C'\boldsymbol{x}$

subject to

$$\sum_{j=1}^{n} a_{ij}x_j \leq E(b_i) + K_p\sqrt{v(b_i)} \quad i = 1, 2, \cdots, m$$

and $\quad x_j \geq 0; \ j = 1, 2, ..., n$

Case III. When only A is a random variable

Let a_{ij} ($i = 1$ to m, $j = 1$ to n) be the random variable of the normal variate $N(\mu_{ij}, \sigma_{ij}^2)$

Let $r_k = \sum\limits_{j=1}^{n} a_{ij} \cdot x_j; \ i = 1$ to m

Then the random variable r_i is a normal variate with expected value and variance given by

$$E(r_i) = \sum_{j=1}^{n} \mu_{ij}x_j \quad \text{and} \quad V(r_i) = x'v_i(x)$$

where $v_i = i^{\text{th}}$ covariance matrix

$$= \begin{bmatrix} V(a_{i1}) & Cov(a_{i1}, a_{i2}) & \cdots & Cov(a_{i1}, a_{in}) \\ Cov(a_{i2}, a_{i1}) & V(a_{i2}) & \cdots & Cov(a_{i2}, a_{in}) \\ \vdots & & & \\ Cov(a_{in}, a_{i1}) & Cov(a_{in}, a_{i2}) & \cdots & V(a_{in}) \end{bmatrix}$$

Now, i^{th} probabilistic constraints is given by

$$P(r_i \leq b_i) \geq 1 - p_i$$

$$P(Z_i \leq z_i) \geq 1 - p_i = q_i \qquad (\because p + q = 1)$$

and
$$z_i = \frac{\{r_i - E(r_i)\}}{\sqrt{v(r_i)}} \text{ is } N(0, 1)$$

and
$$z = \frac{\{b_i - E(r_i)\}}{\sqrt{v(r_i)}}$$

Therefore,
$$\phi(z_i) \geq q_i = \phi(K_{qi})$$

$$\Rightarrow \qquad K_{qi} \leq Z_i$$

or $\qquad E(r_i) + K_{qi}\sqrt{v(r_i)} - b_i \leq 0$

Putting the values of $E(r_i)$ and $v(r_i)$ in the above equation we get

$$\sum_{j=1}^{n} \mu_{ij} x_j + K_{qi}\sqrt{x'v_i x - b_i} \leq 0$$

Hence, the deterministic non-linear programming problem equivalent to given stochastic linear programming is given by

optimize (max. or min.) $Z = C'\boldsymbol{x}$

subject to the constraints

$$\sum_{i=1}^{n} \mu_{ij} x_j + K_{qi}\sqrt{x'v_i x} \leq b_i, \quad i = 1, 2, \ldots, m \text{ and } x_i \geq 0 \quad \forall i$$

☞ **REMARK**

- If $a'_{ij}s$ are independent variable, then their covariances are zero, hence, the deterministic equivalent problem can be reduces as given below:

  optimize $Z = C'\boldsymbol{x}$

  subject to the constraints

  $$\sum_{j=1}^{n} \mu_{ij} \cdot x_j + K_{qi} y_i \leq b_i, \quad i = 1, 2, \ldots, m$$

  $$\sum_{j=1}^{n} V(a_{ij}) x_j^2 - y_i^2 = 0, \quad i = 1, 2, \ldots, m$$

  and $\qquad x_i, y_i \geq 0 \ \forall i$

Case IV. When C, b and A are random variables

We know that C occurs only in the objective function so the deterministic objective function is

$$Z = E(f) = \sum_{j=1}^{n} \mu_j \cdot x_j \; ; \text{ where } C_j \text{ is } N(\mu_j, \sigma_j^2) \text{ for } j = 1, 2, \ldots, n$$

Consider the ith constraints

$$P\left\{ \sum_{j=1}^{n} a_{ij} x_j - b_i \leq 0 \right\} \geq 1 - p_i$$

Now, let a_{ij} be $N(\mu_{ij}, \sigma_{ij}^2) \forall i$ and j and let b_i be $N(\mu_{b_i}, \sigma_{b_i}^2)$. Also, let

$$h = \sum_{j=1}^{n} a_{ij} x_j - b_i = \sum_{j=1}^{n+1} a_{ij} \cdot x_j, \quad i = 1, 2, \ldots, m \text{ where } a_{i,n+1} = b_i \text{ and } x_{n+1} = -1. \text{ Then } h_i \text{ is a normal variate}$$

with

$$E(h_i) = \sum_{j=1}^{n} \mu_{ij} x_j - \mu_{b_i} \qquad \text{and} \qquad V(h_i) = \boldsymbol{x}'v_i \boldsymbol{x}$$

where $\boldsymbol{x} = (x_1, x_2, \ldots, x_n, x_{n+1})'$

and $v_i = \begin{bmatrix} V(a_{i1}) & Cov(a_{i1},a_{i2}) & \dots & Cov(a_{i1},a_{i,n+1}) \\ \vdots & & & \\ Cov(a_{i,n+1},a_{i1}) & Cov(a_{i,n+1},a_{i2}) & \dots & V(a_{i,n+1}) \end{bmatrix}$

Using the above values the ith stochastic constraints becomes

$$P(h_i \le 0) \ge 1 - p_i$$

$$\phi\{E(h_i) / \sqrt{v(h_i)}\} \le p_i = \phi(K_{p_i})$$

Hence, the deterministic non-linear problem equivalent to the given stochastic LPP is as follows:

$$\text{optimize (max. or min.) } E(f) = \sum_{j=1}^{n} \mu_j \cdot x_j$$

subject to the constraints

$$E(h_i) - K_{p_i}\sqrt{v(h_i)} \le 0, \quad i = 1, 2, \dots, m \text{ and } x_i \ge 0 \ \forall i$$

SOLVED EXAMPLES

EXAMPLE 1. *Convert the following Chance-constrained problem into deterministic model.*

$$\text{Max. } Z = x_1 + 4x_2 + 2x_3$$

subject to the constraints

$$P[a_{11}x_1 + a_{12}x_2 + a_{13}x_3 \le 8] \ge 0.95$$

$$P[5x_1 + x_2 + 6x_3 \le b_2] \ge 0.1$$

and $\quad x_1, x_2, x_3 \ge 0$

SOLUTION. Let us assume that parameters $a_{ij}'s$ ($j = 1, 2, 3$) are all independent and normally distributed random variables with mean and variance.

$$E(a_{11}) = 1 \qquad E(a_{12}) = 3 \qquad E(a_{13}) = 9$$
$$Var(a_{11}) = 25 \qquad Var(a_{12}) = 16 \qquad Var(a_{13}) = 4$$

Further, it is assumed that the parameter b_2 is normally distributed with mean 7 and variance 9.

Now from the standard normal variate, we have

$$K_{a_1} = K_{0.05} \approx 1.645 \quad \text{and} \qquad K_{a_2} = K_{0.10} \approx 1.285$$

Then the statement $P(h_i \le b_i) \ge 1 - \alpha_i$ is realized if and only if

$$\frac{b_i - E(h_i)}{\sqrt{Var(h_i)}} \ge K_{a_i}; \ h_i = \sum_{j=1}^{n} a_{ij}x_j$$

which gives the following non-linear constraints

$$[\sum_{j=1}^{n} E(a_{ij})x_j + K_{a_i}\sqrt{x'D_ix}] \le b_i; \quad i = 1, 2$$

where $x = [x_1, x_2, \dots, x_n]'$ and $D_i = i^{th}$ covariance matrix. Hence, equivalent deterministic constraints becomes

$$(x_1 + 3x_2 + 9x_3) + 1.645\sqrt{25x_1^2 + 16x_2^2 + 4x_3^2} \le 8$$

$$(5x_1 + x_2 + 6x_3) \le [7 + 1.285(3)] = 10.855$$

Now, let $y^2 = 25x_1^2 + 16x_2^2 + 4x_3^2$, then the given chance-constrained problem reduces to the following form.

$$\text{maximize } Z = x_1 + 4x_2 + 2x_3$$

subject to the constraints

$$x_1 + 3x_2 + 9x_3 + 1645y \le 8$$

$$25x_1^2 + 16x_2^2 + 4x_3^2 - y = 0$$

$$5x_1 + x_2 + 6x_3 \le 10.855$$

and $\qquad x_1, x_2, x_3 \geq 0$

which can be solved by seperable programming.

EXAMPLE 2. *Convert the following stochastic problem into an equivalent deterministic model.*

$$\text{max. } Z = 2x_1 - x_2 + x_3$$

subject to the constraints

$$P(a_{11}x_1 + a_{12}x_2 + a_{13}x_3 \leq 5) \geq 0.9$$

$$P(2x_1 + 3x_2 + 4x_3 \leq b_2) \geq 0.8$$

and $\qquad x_i \geq 0 \quad \forall i$

In the above problem $a_{11}, a_{12}, a_{13}, b_2$ *are independent normal variates with expected values and variances 1, 5, 8, 10 and 36, 25, 16, 9 respectively.*

SOLUTION. For the first constraints

Since, $q_1 = 0.9$, we have $K_{q_1} = 1.28$

Therefore, first probabilistic constraints is replaced by the following deterministic constraints

$$x_1 + 5x_2 + 8x_3 + 1.28y_1 \leq 5$$

$$36x_1^2 + 25x_2^2 + 16x_3^2 + y_1^2 = 0$$

Now, for the second constraints we have $p_2 = 1 - 0.8 = 0.2$ and so $K_{p_2} = -0.84$

So, the second probabilistic constraints is replaced by the following deterministic constraints

$$2x_1 + 3x_2 + 4x_3 \leq 10 + (-0.84)(3) = 7.48$$

Hence, the deterministic non-linear equivalent problem is given as below

$$\text{max. } Z = 2x_1 - x_2 + x_3$$

subject to the constraints

$$x_1 + 5x_2 + 8x_3 + 1.28y_1 \leq 5$$

$$36x_1^2 + 25x_2^2 + 16x_3^2 - y_1^2 = 0$$

$$2x_1 + 3x_2 + 4x_3 \leq 7.48$$

and $\qquad x_1, x_2, x_3, y_1 \geq 0$

EXERCISE 8.1

1. The width of a slot on a duralumin forging is normally distributed. The specification of the slot with is 0.900 ± 0.005. The parameters $\mu = 0.9$ and $\sigma = 0.003$ are known from the past experience in production process. What is the percentage of scrap forgings?

2. An automobile body is assembled using a large number of spot welds. The number of defective welds (x) closely follows the distribution

$$P(x = \phi) = \frac{e^{-2}2^4}{d!}; \quad d = 0, 1, 2, \ldots$$

Find the probability that the number of defective welds is less than or equal to 2.

3. Solve, if the objective is to maximize the variance of f.

$$\text{max. } f = 4x_1 + 2x_2 + 3x_3 + 4x_4$$

subject to

$$x_1 + x_3 + x_4 \leq 24$$

$$3x_1 + 2x_2 + 2x_3 + 4x_4 \leq 48$$

$$2x_1 + 2x_2 + 3x_3 + 2x_4 \leq 36$$

and $\qquad x_i \geq 0$

ANSWERS

1. 9.5%   2. 0.6767   3. $x^* = [0,0,0,12], f^* = 12$

9 One-dimensional Minimization Methods

9.1 INTRODUCTION

A mathematical programming problem involves determining the maxima and minima of $f(x)$ subject to the constraints $g_i(x) \leq = $ or ≥ 0 where $f(x)$ and $g(x)$ are real valued functions of $X = [x_1, x_2, ..., x_n]$ in n-dimensional space. If some or all the functions $f(x)$, $g_i(x)$ are non-linear then the mathematical programming problem is called non-linear programming problem.

In this chapter we shall discuss one-dimensional minimization methods to solve non-linear programming problem.

9.2 BASIC DEFINITIONS

(1) **Unimodal function:** A function of one variable is said to be unimodal if, given that two values of the variable are on the same side of the optimum, the one nearer the optimum gives the better functional values. Mathematically, a function $f(x)$ is said to be unimodular if

    (i) $x_1 < x_2 < x^*$ $\Rightarrow$ $f(x_2) < f(x_1)$

and   (ii) $x_2 > x_1 > x^*$ $\Rightarrow$ $f(x_1) < f(x_2)$

    where x^* is the minimum point.

☛ REMARKS
- A unimodular function is one that has only one maximum or minimum in a given interval.
- A unimodular function can be a non-differentiable or even a discontinuous function.

(2) **Multimodal function:** A function which is not unimodular is called multimodal.

☛ REMARK
- The range of multimodal function can be subdivided into several parts and the function treated as a unimodular in each part.

(3) **Strictly unimodular function:** A function is said to be strictly unimodular if it is unimodular and has no intervals of finite length in which the function is of constant value.

(4) **Global minima:** A function $f(x)$ defined on a set S attains its global minimum at a point $x^{**} \in S$ if and only if

$$f(x^{**}) \leq f(x) \ \forall \ x \in S$$

(5) **Local minima:** A function $f(x)$ defined on a set S has a local minima at a point $x^* \in S$ if and only if

$$f(x^*) \leq f(x) \ \forall \ x \text{ within a small distance from } x^*.$$

☛ REMARKS
- Under the assumption of unimodality the local minimum automatically becomes the global minimum.
- When the function is not unimodal multiple local optima are possible and global minima can be found only by locating all local optima and selecting the best one.

By reversing the direction of inequalities in the above definitions, we may get the equivalence definition of global maximum and local maximum.

(6) Interval of Uncertainty: Initially the interval $[a, b]$ in which the optimum of the objective function is needed is called the uncertainty interval and after two stages (experiments) finding $f(x_1)$ and $f(x_2)$ the uncertainty interval reduces to $]x_1, b[$, $[x_1, x_2]$ or $[a, x_2]$.

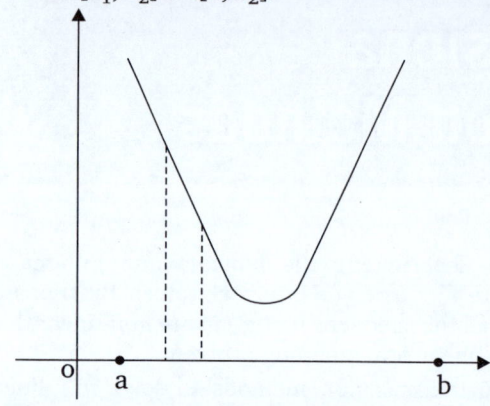

$$f(x_1) > f(x_2) \Rightarrow \text{min.} \in]x_1, b]$$

$$f(x_1) = f(x_2) \Rightarrow \text{min.} \in [x_1, x_2]$$

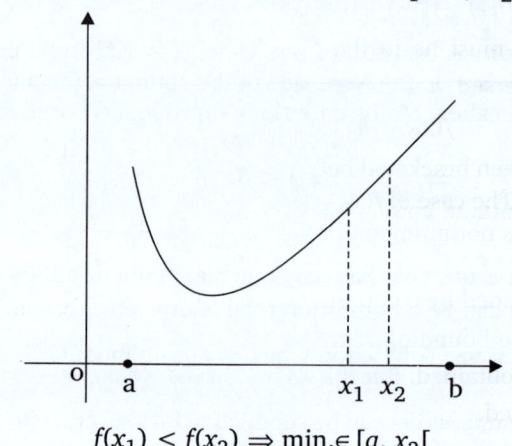

$$f(x_1) < f(x_2) \Rightarrow \text{min.} \in [a, x_2[$$

Fig. 1

(7) Measures of effectivness: Let L_0 be the initial uncertainty interval length and L_n be the uncertainty interval length after n experiments then the ratio $\dfrac{L_0}{L_n} = \alpha \leq 1$ is known as the measure of effectiveness, for comparing the measure of effectiveness for comparing different search methods.

9.3 REGION ELIMINATION METHODS

To find the optimal solution we develop a number of single variable search method for locating the optimal point in a given interval. Search methods that locates a single variable optimum by successively eliminating subintervals so as to reduce the remaining interval of search are called region elimination methods.

9.3.1 ELIMINATION PROPERTY

Let f is strictly unimodal on the interval $a \leq x \leq b$ with a minimum at x^*. Let x_1 and x_2 be two points in the interval such that $a < x_1 < x_2 < b$. Then

(i) If $f(x_1) > f(x_2)$ then the minimum of $f(x)$ does not lie in the interval $]a, x_1[$, *i.e.*, $x^* \in]x_1, b[$.

(ii) If $f(x_1) < f(x_2)$ then the minimum does not lie in the interval $]x_2, b[$, *i.e.*, $x^* \in]a, x_2[$.

☛ REMARK

- If $f(x_1) = f(x_2)$ we could eliminate both ends $]a, x_1[$ and $]x_2, b[$ and the minimum must occur in the interval $]x_1, x_2[$ provided $f(x)$ is strictly unimodal.

9.3.2 PHASES OF SEARCH METHODS

Generally search methods can be broken down into following two phases:

(i) Bounding Phase: At the initial phase, starting at some selected trial points the optimum is roughly bracketed within a finite interval by using the elimination property. This bounding search is conducted using some self made expanding pattern.

For example: Consider the recursion formula given by

$$x_{k+1} = x_k + 2^k h \text{ for } k = 0, 1, 2, \ldots$$

where x_0 is an arbitrarily selected starting point and h is a step size parameter of suitable magnitude. If

$$f(x_0 - |h|) \geq f(x_0) \geq f(x_0 + |h|)$$

then the minimum must lie to the right of x_0 and h is chosen to be positive. But if the inequalities are reversed, h is chosen to be negative if

$$f(x_0 - |h|) \geq f(x_0) \leq f(x_0 + |h|)$$

the minimum has been bracketed between $x_0 - |h|$ and $x_0 + |h|$ and then bounding search can be terminated. The case of $f(x_0 - |h|) \leq f(x_0) \geq f(x_0 + |h|)$ is ruled out and in this case the given function is not unimodal.

☛ REMARKS

- The sign of h is determined by comparing $f(x_0)$, $f(x_0 + |h|)$ and $f(x_0 - |h|)$.

- The effectiveness of the bounding search depends directly on the step size h. If h is large, a large initial interval is obtained. But if h is small many evaluations may be necessary before a bound can be established.

(ii) Interval Refinement Phase: After the bounding phase, when a bracket has been established around the optimum then we apply the interval reduction scheme to find a refined estimate of the optimum point. The amount of subinterval eliminated at each step depends on the location of the trial points x_1 and x_2 within the search interval.

9.4 ELIMINATION METHODS

(i) Interval Halving Method: This method delete exactly one half the interval at each stage, *i.e.*, exactly one half of the current interval of uncertainty is deleted in every stage. It requires three experiments in the first stage and two experiments in each subsequent stage. This is the reason that it is also called a *three point equal intervals*.

WORKING PROCEDURE

The basic steps in interval halving method for finding the minimum of a function $f(x)$ over the interval $L_0 = [a, b]$ are given below

STEP 1. Divide the initial interval of uncertainty $L_0 = [a, b]$ into four equal parts such that

$$x_0 = \frac{1}{2}(a+b), \quad L_0 = b - a$$

$$x_1 = a + \frac{1}{4}L_0, \quad x_2 = b - \frac{1}{4}L_0$$

STEP 2. Evaluate $f(x)$ at x_0, x_1 and x_2, i.e., we have to find $f(x_0)$, $f(x_1)$ and $f(x_2)$.

STEP 3. Compare $f(x_1)$ and $f(x_0)$. Then we have the following cases:
  (i) If $f(x_1) < f(x_0)$, then delete the interval (x_0, b) by setting $b = x_0$. Label x_1 and x_0 as the new x_0 and b respectively and go to step-4.
  (ii) If $f(x_1) \geq f(x_0)$ go to step-4.

STEP 4. Compare $f(x_2)$ and $f(x_0)$
  (i) If $f(x_2) < f(x_0)$, delete the interval (a, x_0) by setting $a = x_0$. Since the mid point of new interval will now be x_2, set $x_0 = x_2$ and go to step 5.
  (ii) If $f(x_2) \geq f(x_0)$, delete the interval (a, x_1) and (x_2, b). Set $a = x_1$ and $b = x_2$. Also x_0 continues to be the mid point of the new interval.
  Go to step-3.

STEP 5. Compute $L = b - a$. Test whether the new interval of uncertainty $L = b - a$ satisfies the convergence criterion, i.e., if $|L|$ is small then terminate. Otherwise return to step-2.

☛ **REMARKS**
  • In interval halving method, the function value at the middle point of the interval of uncretainity, f_0 will be available in all the stages except the first stage.
  • At each stage of the algorithm, exactly half of the length of the search interval is deleted.
  • At most two functional evaluations are necessary at each subsequent steps.
  • After n functional evaluation, the initial search interval will be reduced to $\left(\frac{1}{2}\right)^{n/2}$.
  • The interval of uncertainty remaining at the end of n experiments ($n \geq 3$ and odd) is given by

$$L_n = \left(\frac{1}{2}\right)^{\frac{n-1}{2}} \cdot L_0$$

SOLVED EXAMPLES

EXAMPLE 1. *Find the minimum of $f(x) = (100 - x)^2$ over the interval $60 \leq x \leq 150$.*

SOLUTION. Clearly, here we have
$$a = 60, b = 150, L = 150 - 60 = 90$$

Now, $$x_0 = \frac{60+150}{2} = \frac{210}{2} = 105$$

Now we proceed as follows:

Stage 1. $$x_1 = a + \frac{1}{4}L = 60 + \frac{90}{4} = 82.5$$

$$x_2 = b - \frac{1}{4}L = 150 - \frac{90}{4} = 127.5$$

Also, $f(82.5) = 306.25 > f(105) = 25$

and $f(127.5) = 756.25 > f(105) = 25$

Since, $f(x_1) > f(x_0)$ and $f(x_2) > f(x_0)$, so drop the interval (a, x_1) and (x_2, b). Hence, length of the search interval from 90 to 45.

Stage 2. Here, $a = 82.5, b = 127.5, x_0 = 105$

$$L = 127.5 - 82.5 = 45$$

$$x_1 = 82.5 + \frac{45}{4} = 93.75$$

$$x_2 = 127.5 - \frac{45}{4} = 116.25$$

Also, $f(93.75) = 39.06 > f(105) = 25$

and $f(116.25) = 264.06 > f(105) = 25$

Since, $f(x_1) > f(x_0)$ and $f(x_2) > f(x_0)$, therefore the interval of uncertainty is $(93.75, 116.25)$.

Stage 3. Here, we have

$$a = 93.75, b = 116.25, x_0 = 105$$
$$L = 116.25 - 93.75 = 22.5$$
$$x_1 = 93.375$$
$$x_2 = 110.625$$

and $f(x_1) = 0.39 < f(x_0) \ (= f(105))$

Hence, delete the interval $(105, 116.25)$. The new interval of uncertainty is now $(93.75, 105)$ and its mid-point is 99.375 (old x_1). Hence, in three stages (six functional evaluation), the initial search interval of length 90 has been reduced   exactly to

$$90\left(\frac{1}{2}\right)^3 = 11.25.$$

EXAMPLE 2. **Find the minimum of $f(x) = x(x - 1.5)$ in the interval (0, 10) to within 10% of the exact value.**

SOLUTION. Clearly, if the middle point of the final interval of uncertainty is taken as the optimal point, the specified accuracy can be achieved if

$$\frac{1}{2}L_n < \frac{L_0}{10}, i.e., \left(\frac{1}{2}\right)^{\frac{n-1}{2}} L_0 \le \frac{L_0}{5} \qquad \qquad ...(1)$$

Here, since $L_0 = 1 \ (= 1 - 0)$, then from (1)

$$2^{\frac{1}{(n-2)/2}} \le \frac{1}{5} \ \Rightarrow \ 2^{(n-1)/2} \ge 5 \qquad \qquad ...(2)$$

Since, n has to be odd, then from (2), the minimum possible value of n is 7.

Stage 1. Here, $L_0 = [a = 0, b = 1]$

$$x_1 = 0.25 \qquad \Rightarrow \qquad f(x_1) = 0.25 \ (- 1.25) = - 0.3125$$
$$x_0 = 0.50 \qquad \Rightarrow \qquad f(x_0) = 0.50 \ (-1.00) = - 0.5000$$
$$x_2 = 0.75 \qquad \Rightarrow \qquad f(x_2) = 0.75 \ (- 1.75) = - 0.5625$$

Clearly, $f(x_1) > f(x_0) > f(x_2)$, so we delete the interval $(a, x_0) = (0, 0.5)$

Stage 2. Label x_2 and x_0 as the new x_0 and a, thus, $a = 0.5, x_0 = 0.75, b = 1.0$

By dividing the new interval of uncertainty $L_3 = (0.5, 1.0)$ into four equal parts, we get

$$x_1 = 0.625 \qquad \Rightarrow \qquad f(x_1) = 0.625 \ (- 0.875) = - 0.546875$$
$$x_0 = 0.750 \qquad \Rightarrow \qquad f(x_0) = 0.750 \ (- 0.750) = - 0.562500$$
$$x_2 = 0.875 \qquad \Rightarrow \qquad f(x_2) = 0.875 \ (- 0.625) = - 0.546875$$

Clearly, $f(x_1) > f(x_0)$ and $f(x_2) > f(x_0)$ we delete both the intervals (a, x_1) and (x_2, b).

Stage 3. Label x_1, x_0 and x_2 as the new a, x_0 and b respectivly. The new interval of uncertainty will be given by

$$L_5 = (0.625, 0.875)$$

Now by dividing the new interval $(0.625, 0.875)$ into four equal parts, we get the following values

$$x_1 = 0.6875 \quad \Rightarrow \quad f(x_1) = 0.6875 \, (- 0.8125) = - 0.558594$$
$$x_0 = 0.75 \quad \Rightarrow \quad f(x_0) = 0.75 \, (- 0.75) = - 0.5625$$
$$x_2 = 0.8125 \quad \Rightarrow \quad f(x_2) = 0.8125 \, (- 0.6875) = - 0.558594$$

Clerly, since

$$f(x_1) > f(x_0) \text{ and } f(x_2) > f(x_0),$$ therefore, we delete both the interval (a, x_1) and (x_2, b)

Now, the new interval of uncetainity is given by

$$L_7 = (0.6875, 0.8125)$$

Finally, by taking the middle point of this interval L_7 as optimum, we get

$$x_{(min)} = 0.75 \text{ and } f(min) = - 0.5625$$

9.5 FIBONACCI METHOD

The Fibonacci method can be used to find the minimum of a function of one variable even if the fuction is not continuous.

ASSUMPTIONS

1. The initial interval of uncertainty say $[a, b]$ has to be known.

2. The function to be optimized to a given degree of accuracy must be unimodular.

3. The no. of functions evaluations to be used in search has to be specified before used.

Fibonacci's sequence: A Fibonacci's sequence $<F_n>$ is defined by

$$F_n = F_{n-1} + F_{n-2}$$

where n is an integer greater than 1.

and $F_0 = F_1 = 1$

For example. 1, 1, 2, 3, 5, 8, 13, 21, 34, 55, 89, 144, ... is a fibonacci sequence.

Procedure of Fibonacci Method: Let $[a, b]$ be the initial interval of uncertainty with the length $L_0 = b - a$ ...(1)

and n be the no. of experiments to be performed. Now, select two points x_1 and x_2 in the interval $[a, b]$ of length L_0 such that

$$x_1 = a + \frac{F_{n-2}}{F_n} L_0 = a + L_2^* \qquad \qquad ...(2)$$

where

$$L_2^* = \frac{F_{n-2}}{F_n} L_0$$

and

$$x_2 = b - \frac{F_{n-2}}{F_n} L_0 = b - L_2^* \qquad \qquad ...(3)$$

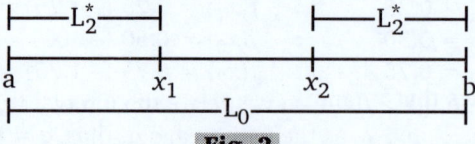

Fig. 2

Therefore,

$$b - x_2 = b - b + \frac{F_{n-2}}{F_n} L_0 \qquad \qquad \text{(Using (3))}$$

$$= x_1 - a \qquad \qquad \text{(Using (2))}$$

Clearly, here we have $x_1 - a = b - x_2$

$\Rightarrow$ x_1 and x_2 are symmetrically placed with respect to the end points a and b of the interval of length L_0.

Now from (1) and (2) we have

$$x_2 = b - \frac{F_{n-2}}{F_n} L_0 = a + L_0 - \frac{F_{n-2}}{F_n} L_0$$

$$= a + \left(\frac{F_n - F_{n-2}}{F_n} \right) L_0$$

$$= a + \frac{F_{n-1}}{F_n} L_0 \qquad \text{(Using (1))} \qquad ...(4)$$

Further, let $y = f(x)$ be a unimodel function having minimum value in $[a, b]$ with two points x_1 and x_2 with the following cases:

CASE I. If $x_1 < x_2$ and $f(x_1) < f(x_2)$ then using the assumption of unimodal, we may conclude that the minimum value does not lie on the right of x_2 and hence drop the interval $(x_2, b]$ and then next interval of uncertainty is $[a, x_2)$.

CASE II. If $x_1 < x_2$ and $f(x_1) > f(x_2)$, drop $[a, x_1)$. The next uncertainty interval will be $[x_1, b)$.

Now, using (i) and (ii) there is a small interval of uncertainty L_2 given by

$$L_2 = L_0 - L_2^* = L_0 - \frac{F_{n-2}}{F_n} L_0 = L_0 \left(\frac{F_n - F_{n-2}}{F_n} \right) = \frac{F_{n-1}}{F_n} L_0 \qquad ...(5)$$

Here, for the above case (i) the interval of further search will be $[a, x_2)$ in which one observations is at x_1

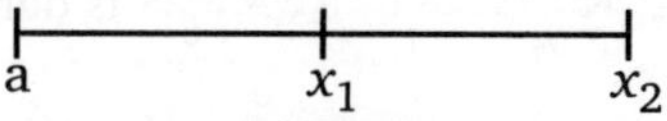

Fig. 3

Also, in case (ii) the interval of further search will be $(x_1, b]$ in which one observations is at x_2.

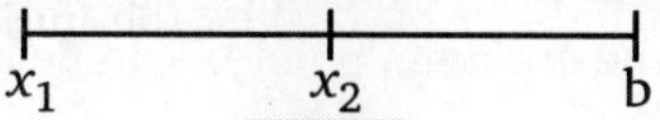

Fig. 4

The next iteration is on the interval of uncertainty $[a, x_2)$ of length L_2. Let the interval $[a, x_2)$ be renamed as $[a_1, b_1]$.

Further choose x_3 and x_4 on the inerval $[a, x_2]$ of length L_2 such that the conditions given below

$$x_3 = a + \frac{F_{n-3}}{F_n} L_2 \quad \text{and} \quad x_4 = x_2 - \frac{F_{n-2}}{F_{n-1}} L_2 \qquad ...(6)$$

are satisfied.

Now, $\qquad x_3 - a = \frac{F_{n-3}}{F_{n-1}} L_2 = x_2 - x_4 \qquad ...(7)$

Now, suppose that $\dfrac{F_{n-3}}{F_{n-1}} L_2 = L_3^* \qquad ...(8)$

Clearly, equation (7) implies that x_3 and x_4 selected according to (6) are symmetric w.r.t. to the end points of L_2.

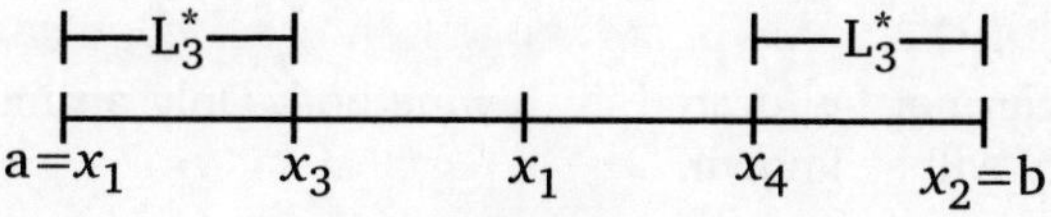

Fig. 5

Now,
$$x_2 - x_4 = \frac{F_{n-3}}{F_{n-1}} L_2 = \frac{F_{n-3}}{F_n} L_0 \qquad \text{(Using (5) and (7))} \qquad \ldots(9)$$

and
$$x_2 - x_4 = a + \frac{F_{n-1}}{F_n} L_0 - a - \frac{F_{n-2}}{F_n} L_0 \qquad \text{(Using (2) and (4))}$$

$$= \frac{F_{n-1} - F_{n-2}}{F_n} L_0 = \frac{F_{n-3}}{F_n} L_0 \qquad \ldots(10)$$

So,
$$x_2 - x_1 = x_2 - x_4 = \frac{F_{n-3}}{F_n} L_0 = L_3^* \qquad \text{(By (9) and (10))}$$

which implies that x_4 coincides with x_1.

$\Rightarrow$ In the second iteration (and also in subsequent iteration) we need to evaluate the function value at only one new point (i.e., x_3 in this case).

Similarly, if for case (ii) the new search interval $[x_1, b]$ is retained.

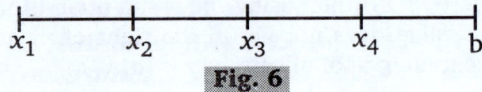

Fig. 6

Then we take
$$x_3 = x_1 + \frac{F_{n-3}}{F_n} L_2$$

and
$$x_4 = b - \frac{F_{n-3}}{F_{n-1}} L_2$$

Here $x_2 = x_3$, i.e., x_2 and new point x_3 will be coincide.

The length of the uncertainty interval after the third experiment is given by
$$L_3 = L_2 - L_3^* = \frac{F_{n-1}}{F_n} L_0 - \frac{F_{n-3}}{F_{n-1}} L_2 = \frac{F_{n-1}}{F_n} L_0 - \frac{F_{n-3}}{F_n} L_0 = \left(\frac{F_{n-1} - F_{n-3}}{F_n} \right) L_0$$

$\Rightarrow$
$$L_3 = \frac{F_{n-2}}{F_n} L_0 \qquad \ldots(11)$$

Repeat the above process till the last two experiments are equidistant from the end points of the uncertainty interval $[a_n, b_n]$. Then the optimum point x^* can be approximately taken as
$$x^* = \frac{a_n + b_n}{2}$$

Finally, the distance of jth experiments from one end of the interval of uncertainty is given by
$$L_j^* = \frac{F_{n-j}}{F_n} L_0 \qquad \ldots(12)$$

The length of the uncertainty interval at this stage can be derived as given below
$$L_j = \frac{F_{n-(j-1)}}{F_n} L_0, \qquad j = 2, 3, \ldots \qquad \ldots(13)$$

Therefore, for $j = n$, $\dfrac{L_n}{L_0} = \dfrac{F_1}{F_n} = \dfrac{1}{F_n}$

Finally the ratio $\dfrac{L_n}{L_0} = \dfrac{1}{F_n}$ permits the required no. of experiments n that is necessary to achieve the desired accuracy in locating the optimal point.

☛ **REMARKS**
- The exact optimum can not be located in this method. Only an interval known as the final interval of uncertainty will be known.
- The final interval of uncertainty can be made as small as desired by using more computations.

After conducting $n - 1$ experiments and drop the appropriate interval in each step, the remaining interval will contain one experiment precisely at its middle point. However, the nth (final) experiment is also to be placed at the centre of the present interval of uncertainty, *i.e.,* the position of the nth experiment will be same as that of $(n - 1)^{th}$ one and this is true for any chosen value of n.

An Important Table

Value of n	Fibonacci no, F_n	Value of n	Fibonacci no, F_n
0	1	11	144
1	1	12	233
2	2	13	377
3	3	14	610
4	5	15	987
5	8	16	1597
6	13	17	2584
7	21	18	4181
8	34	19	6765
9	55	20	10946
10	89		

SOLVED EXAMPLES

EXAMPLE I. *Using the Fibonacci method, determine the minimum of $x^2 - 3x + 5$ in the interval [1, 2.6] taking $n = 6$.*

SOLUTION. Let $f(x) = x^2 - 3x + 5$

Here, $L_0 = 2.6 - 1 = 1.6$

$n = 6$

and initial interval $= [1, 2.6]$

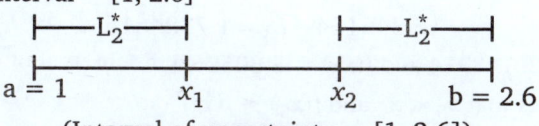

(Interval of uncertainty $= [1, 2.6]$)

Fig. 7

First and second experiments

Let x_1 and x_2 be any two points in the interval $[1, 2.6]$ at a distance

$$L_2^* = \frac{F_{n-2}}{F_n} L_0 = \frac{F_4}{F_6} L_0 = \frac{5}{13} \times 1.6 \qquad (\because F_4 = 5, F_6 = 13)$$

$$= 0.6153846$$

From the end points 1 and 2.6 respectively.

$\therefore \qquad x_1 = a + L_2^* = 1.6153846$

and $\qquad x_2 = b - L_2^* = 2.6 - 0.6153846 = 1.9846154$

and $\qquad f(x_1) = 2.763314, f(x_2) = 2.984852$

which shows that

$$f(x_1) < f(x_2), \qquad x_1 < x_2$$

So, drop the interval $]1.9846154, 2.6]$ and the new uncertainty interval becomes $[1, 1.9846154]$.

Third experiment

Let us suppose x_3 and x_4 be two points in the interval $]1, 1.9846154]$ at a distance

$$L_3^* = \frac{F_{n-3}}{F_n} L_0 = \frac{F_3}{F_6} L_0$$

$$= \frac{3}{13} \times 1.6 \qquad\qquad (\because F_3 = 3, F_6 = 13)$$

$$= 0.36923$$

$$\Rightarrow \qquad x_3 = 1 + L_3^* = 1.3692308$$

$$x_4 = 1.9846154 - L_3^* = 1.6153846$$

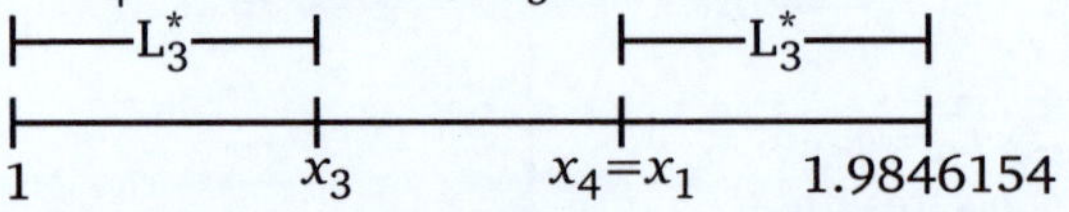

(Interval of uncertainty $= [1, 1.9846154]$)

Fig. 8

Now, $\qquad f(x_3) = 2.7671006$ and $f(x_4) = 2.763314$

which implies that

$$x_3 < x_4 \text{ and } f(x_3) > f(x_4)$$

So, drop the interval $[1, 1.3692308[$ and the new uncertainty interval becomes $[1.3692308, 1.9846154]$.

Fourth Experiment

Let us suppose x_5 and x_6 be two points in the interval $[1.3692308, 1.9846154]$ at a distance

$$L_4^* = \frac{F_{n-4}}{F_n} L_0 = \frac{F_2}{F_6} L_0 = \frac{2}{13} \times 1.6 \qquad\qquad (\because F_2 = 2, F_6 = 13)$$

$$= 0.2461538$$

Now, $\qquad x_5 = 1.3692308 + L_4^* = 1.6153846$

$$x_6 = 1.9846154 - L_4^* = 1.7384616$$

and $f(x_5) = 2.763314$ and $f(x_6) = 2.8068639$

which implies that $x_5 < x_6$ and $f(x_5) < f(x_6)$

$\Rightarrow$ Drop the interval $[1.7384616, 1.9846154]$

So, the next interval of uncertainty is given by $[1.3692308, 1.7384616]$

Fifth Experiment

Let us suppose that x_7 and x_8 be the two point in the interval of uncertainty $[1.3692308, 1.7384616]$ at a distance

$$L_5^* = \frac{F_{n-5}}{F_n} L_0 = \frac{F_1}{F_6} L_0 = \frac{1}{13} \times 1.6 = 0.1230769$$

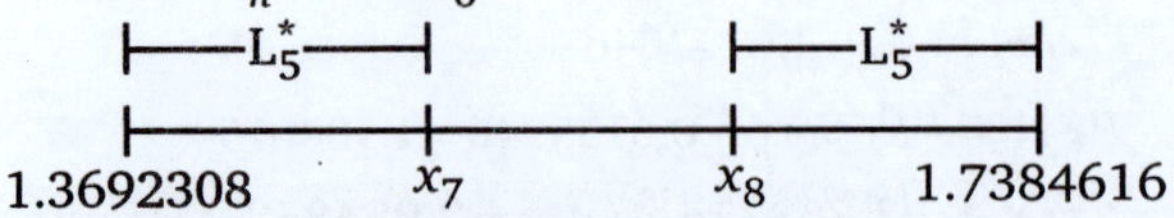

(Interval of uncertainty $= [1.3692308, 1.7384616]$)

Fig. 9

Now, $\qquad x_7 = 1.3692308 + L_5^* = 1.4923077$

$\qquad\qquad x_8 = 1.7384616 - L_5^* = 1.6153846$

and $\qquad f(x_7) = 2.750059$ and $f(x_8) = 2.7633136$

Clearly we have $x_7 < x_8$ and $f(x_7) < f(x_8)$

Then again using concept of unimoduality, drop the interval

$\qquad\qquad$ [1.6153846, 1.7384616]

Therefore, the next interval of uncertainty is given by [1.3692308, 1.6153846[

Sixth Experiment

Here, we have

$$L_6^* = \frac{F_{n-6}}{F_n} L_0 = \frac{F_0}{F_6} L_0 = \frac{1}{13} \times 1.6 = 0.1230769 = L_5^*$$

$\Rightarrow \qquad\qquad L_6^* = L_5^*$

$\Rightarrow \qquad$ There is no fresh point.

So, the final uncertainty interval is given by (1.3692308, 1.6153846)

whose middle point is given by

$$x^* = \frac{1.3692308 + 1.6153846}{2} = 1.4923077$$

Hence, minimum value of $f(x) = 2.750059$ at $x = 1.4923077$.

EXAMPLE 2. **Find the minimum of $x^2 - 2x$, $0 \le x \le 1.5$ taking $n = 4$ using the Fibonacci's method.**

SOLUTION. We have $\qquad f(x) = x^2 - 2x$

$\qquad\qquad\qquad L_0 = 1.5 - 0 = 1.5$

$\qquad\qquad\qquad n = 4$

Clearly, $f(x)$ is unimodal in the given interval [0, 1.5]. Now, we proceed as follows:

First and Second Experiment

Let x_1 and x_2 be two points in the given intervals at a distance $L_2^* = \dfrac{F_{n-2}}{F_n} L_0$

such that $\qquad x_1 = 0 + L_2^* = 0 + \dfrac{F_{n-2}}{F_n} L_0 = \dfrac{2}{5} \times 1.5 = 0.6$

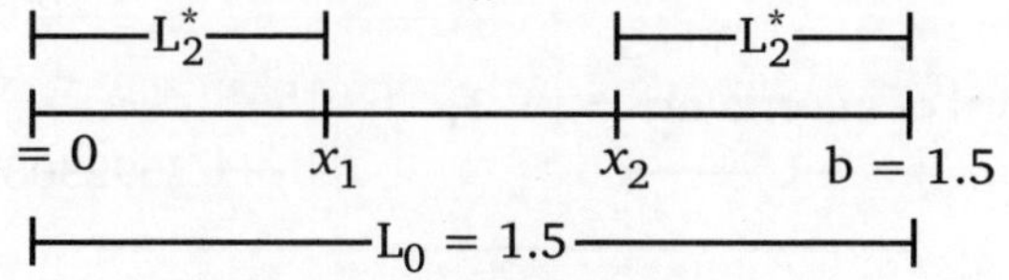

(Interval of uncertainty = [0, 1.5])

Fig. 10

Since, x_1 and x_2 are symmetrically placed, x_2 will be at a distance 0.6 from $b = 1.5$

$\therefore \qquad\qquad x_2 = 0.9$

Also, $\qquad f(x_1) = f(0.6) = -0.84$

$\qquad\qquad f(x_2) = f(0.9) = -0.99$

Clearly, $f(0.6) > f(0.9)$

$\Rightarrow \qquad\qquad f(x_1) > f(x_2)$

∴ Using unimodality, drop the interval $]0, 0.6[$. Then new interval of uncertainty $\leq [0.6, 1.5]$.

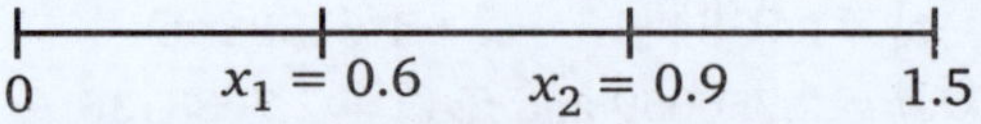

(Interval of uncertainty = [0.6, 1.5])

Fig. 11

Third Experiment

Let us take two point $x_3 (=0.9)$ and x_4 in the new uncertainty interval $[0.6, 1.5]$.

Here, x_3 is the same as x_2 which is at a distance $L_3^* = 0.3$ from 0.6

∴ $x_4 = 1.2$, being symmetrical at a distance 0.3 from 1.5

So, $x_3 = 0.9$ and $x_4 = 1.2$

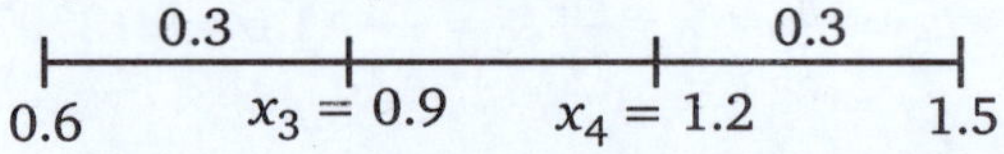

(Interval of uncertainty = [0.6, 1.2])

Fig. 12

$$f(x_3) = f(0.9) = -0.99 \text{ and } f(x_4) = f(1.2) = -0.96$$

⇒ $f(x_4) > f(x_3)$

Then using unimodality, drop the interval $(1.2, 1.5)$.

Therefore, the new interval of uncertainty is $[0.6, 1.2]$

Here, 0.9 is the middle point of the interval of uncertainty, so we do not get a new point x_4 as $x_4 = x_3 = 0.9 = x^*$. Hence, minimum of $f(x) = -0.99$ at $x = 0.9$. The corrcect minimum is $f(x) = 1$ at $x = 1$.

EXAMPLE 3. *Minimize* $f(x) = 0.65 - \left[\dfrac{0.75}{(1+x^2)}\right] - 0.65x \tan^{-1}\dfrac{1}{x}$ *in the interval*

$[0, 3]$ *by the Fibonacci's method using* $n = 6$.

SOLUTION. Hence, we have

$$f(x) = 0.65 - \left[\frac{0.75}{(1+x^2)}\right] - 0.65x \tan^{-1}\frac{1}{x}$$

$n = 6$
$L_0 = 3 - 0 = 3$

Initial interval of uncertainty = $[0, 3]$

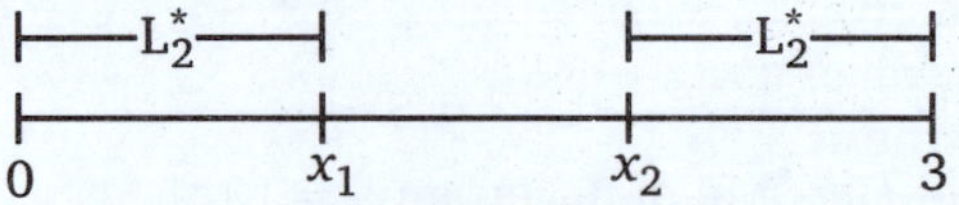

(Interval of uncertainty = [0, 3])

Fig. 13

First and second experiments

Let x_1 and x_2 be any two points in the interval $[0, 3]$ at a distance

$$L_2^* = \frac{F_{n-2}}{F_n} L_0 = \frac{5}{13}(3.0) = 1.153846$$

from the end points 0 and 3 respectively.

Therefore, $x_1 = 0 + L_2^* = 1.153846$

$$x_2 = 3 - L_2^* = 3.0 - 1.153846 = 1.846154$$

and
$$f(x_1) = -0.207270$$
$$f(x_2) = -0.115843$$

which shows that $x_1 < x_2$ and $f(x_1) < f(x_2)$

So, by unimodality, drop the interval [1.846154, 3.0] and the new uncertainty interval becomes

[0, 1.846154]

Third Experiment

Let x_3 and x_4 be two positions in the interval [0, 1.846154] at a distance L_3^* given by

$$L_3^* = \frac{F_{n-3}}{F_n} \cdot L_0 = \frac{F_3}{F_6} L_0 = \frac{3}{13} \times 3 = 0.692308$$

$\Rightarrow$
$$x_3 = 0 + L_3 = 0 + 0.692308 = 0.692308$$
$$x_4 = 1.846154 - 0.692308 = 1.153846$$

and
$$f(x_3) = -0.291364$$

and
$$f(x_4) = -0.207270$$

which shows that $x_3 < x_4$ and $f(x_3) < f(x_4)$

So, drop the interval [1.153846, 1.846154]

$\therefore$ the new uncertainty interval is given by [0.692308, 1.153846]

Fourth Experiment

The fourth experiment is located at
$$x_5 = 0 + (1.153846 - 0.692308) = 0.461538$$

and
$$x_6 = 0.692308$$

$\Rightarrow$
$$f(x_5) = -0.309811 \text{ and } f(x_6) = -0.291364$$

Clearly, $f(x_5) < f(x_6)$ so using the unimodality, drop the interval

[0.692308, 1.153846]

$\Rightarrow$ the next interval of uncertainty = [0, 0.692308]

Fifth Experiment

The location of the fifth experiment can be obtained by following the usual procedure.

Here,
$$x_7 = 0.692308 - 0.461538 = 0.230770$$
$$x_8 = 0.461538$$

and
$$f(x_7) = -0.263678, f(x_8) = -0.230770$$

Clearly,
$$f(x_7) > f(x_8)$$

So, drop the interval [0, 0.230770]

$\Rightarrow$ next interval of uncertainty = [0.230770, 0.692308]

Sixth Experiment

The final experiment is positioned at
$$x_9 = 0.230770 + [0.692308 - 0.461538] = 0.461540$$

and
$$x_{10} = 0.461538$$

$\Rightarrow$
$$f(x_9) = -0.309810, f(x_{10}) = -0.309811$$

We observed that $f(x_9) \approx f(x_{10})$

$\Rightarrow$ There is no fresh point.

So, final interval of uncertainty is given by [0.230770, 0.461540]

whose middle point is given by $x^* = \dfrac{0.230770 + 0.461540}{2} = 0.346155$

which is the required optimal point.

☞ **REMARK**

- The obtained minimum and correct mimimum should be same, however it is slightly different due to round-off- error.

9.6 GOLDEN SECTION METHOD

In Fibonacci's method, the total no. of experiments to be conducted has to be specified before begining the calculations. But in Golden section method this is not required and total no. of experiments to be conducted are unlimited. In this method, total no. of experiments can be decided during the computation. Also, in this method we start with the assumption that we are going to conduct a large no. of experiments.

Let n be the total no. of experiments to be conducted and is large enough to satisfy the limit

$$\lim_{n \to \infty} \frac{F_{n-1}}{F_n} = \lim_{n \to \infty} \frac{F_{n-2}}{F_{n-1}} = \lim_{n \to \infty} \frac{F_{n-3}}{F_{n-2}} = \frac{1}{\lambda} \quad \text{(say)}$$

Then we have

$$L_2 = \lim_{n \to \infty} \frac{F_{n-1}}{F_n} L_0 = \frac{L_0}{\lambda}$$

$$L_3 = \lim_{n \to \infty} \frac{F_{n-2}}{F_n} L_0 = \lim_{n \to \infty} \frac{F_{n-2}}{F_{n-1}} \cdot \frac{F_{n-1}}{F_n} L_0 = \frac{L_0}{\lambda^2} \quad \text{and so on.}$$

In general $\quad L_k = \frac{L_0}{\lambda^{k-1}} = \left(\frac{1}{\lambda}\right)^{k-1} \cdot L_0$

Now by definition of Fibonacci's sequence $<F_n>$ we can write

$$F_n = F_{n-1} + F_{n-2}, \quad n \text{ is an integer greter than 1.}$$

Therefore, $\quad \dfrac{F_n}{F_{n-1}} = 1 + \dfrac{F_{n-2}}{F_{n-1}}$

On taking the limit as $n \to \infty$ we get

$$\lim_{n \to \infty} \frac{F_n}{F_{n-1}} = \lim_{n \to \infty} \left(1 + \frac{F_{n-2}}{F_{n-1}}\right)$$

$\Rightarrow \qquad\qquad \lambda = 1 + \dfrac{1}{\lambda} \qquad\qquad \Rightarrow \qquad\qquad \lambda^2 - \lambda - 1 = 0$

On solving we get $\lambda = \dfrac{1 + \sqrt{5}}{2} = 1.618033988$

$\Rightarrow \qquad\qquad \dfrac{1}{\lambda} = 0.618$

WORKING PROCEDURE

STEP 1.   Let us take two points in the initial interval of uncertainty $[a, b]$ which are symmetrically placed from both the end points a and b at a distance

$$L_2^* = \lim_{n \to \infty} \frac{F_{n-2}}{F_n} L_0 = \lim_{n \to \infty} \frac{F_{n-2}}{F_{n-1}} \cdot \frac{F_{n-1}}{F_n} L_0 = \frac{L_0}{\lambda^2} = (0.618)^2 \cdot L_0$$

        where $\quad L_0 = b - a$

STEP 2.   In the kth-iteration $L_k^* = (0.618)^k L_0$

STEP 3.   Repeat step 2 until the stage where the interval of uncertainty is as small as desired.

STEP 4.   Find the optimal point by taking the middle point of the final uncertainty interval.

☞ REMARKS
- The procedure of Golden section is same as the Fibonacci's method except that the location of first two experiments is defined by

$$L_2^* = \frac{F_{n-2}}{F_n} L_0 = \frac{F_{n-2}}{F_{n-1}} \cdot \frac{F_{n-1}}{F_n} L_0 = \frac{L_0}{\lambda^2} = 0.382 L_0$$

- The Golden section search should require the evaluation of only one new point at each step.
- The Golden section iterations can be terminated either by specifying a limit on the number of evaluation (and hence the accuracy in the variable) or by specifying a relative accuracy in the function value.

An Important Table

Value of n	2	3	4	5	6	7	8	9	10	∞
Ratio $\dfrac{F_{n-1}}{F_n}$	0.5	0.667	0.6	0.625	0.6136	0.619	0.6177	0.6181	0.6184	0.618

In Euclid geometry, the division of a line segment into two equal parts so that the ratio of the whole to the larger part is equal to the ratio of the larger to the smaller $\left(i.e., \dfrac{d+b}{b} = \dfrac{d}{b} = \lambda \right)$ is known Golden section or Golden mean. That is why this method is known as Golden section method.

SOLVED EXAMPLES

EXAMPLE 1. *Using Golden section method, determine the maximum of the unimodal function $f(x) = x(5 - x)$ in the interval [0, 8] wherein the maximum lies.*

SOLUTION. We have $f(x) = x[5 - x]$

$$[a, b] = [0, 8] \quad \Rightarrow \quad L_0 = 8 - 0 = 8$$

First and second Experiments

Let x_1 and x_2 be two points between [0, 8] at a distance

$$L_2^* = (0.618)^2 L_0 = (0.618)^2 \cdot 8 = 3.055392$$

(Interval of uncertainty = [0, 8])

Fig. 14

Now, $x_1 = 0 + L_2^* = 3.055392$ $\Rightarrow$ $f(x_1) = 5.941975$

and $x_2 = 8 - L_2^* = 8 - 3.055392 = 4.944$ $\Rightarrow$ $f(x_2) = 0.276864$

Clearly $x_1 < x_2$ and $f(x_1) > f(x_2)$

Then by unimodality of $f(x)$ we can say that the maximum points do not lie on the right of x_2 and drop the interval]4.944, 8].

So, the next interval of uncertainty is [0, 4.944] with length $L_2 = (0.628)^{2-1} \cdot L_0 = 4.944$.

Third Experiment

Let us take two points x_3 and x_4 in [0, 4.944] at a distance $L_3^* = (0.618)^3$ and $L_0 = (0.618)^3 \times 8 = 1.888$.

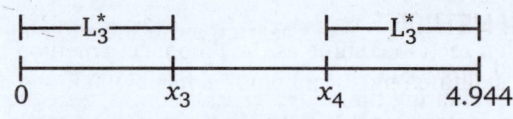

(Interval of uncertainty = [0, 4.944])

Fig. 15

Now, $x_3 = 0 + L_3^* = 1.888$ $\Rightarrow$ $f(x_3) = 5.875456$

and $x_4 = 4.9444 - L_3^* = 3.056$ $\Rightarrow$ $f(x_4) = 5.940864$

which shows that $x_3 < x_4$ and $f(x_3) < f(x_4)$.

Then by unimodality, the maximum point do not lie on the left of x_3, so drop the interval [0, 1.888] and next interval of uncertainty is [1.888, 4.944].

Fourth Experiment

Let us take two points x_5 and x_6 in the interval [1.888, 4.944] at a distance

$$L_4^* = (0.618)^4 L_0 = 1.1669$$

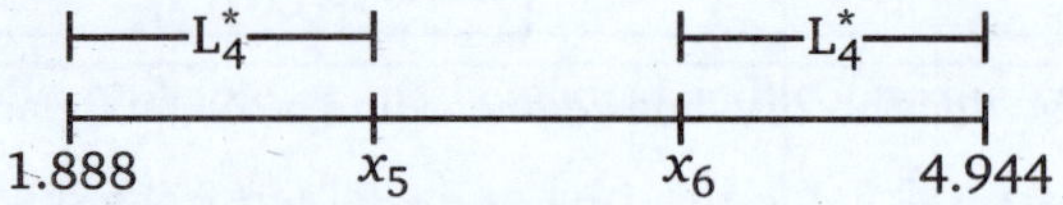

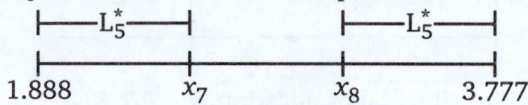

(Interval of uncertainty = [1.888, 4.944])

Fig. 16

Now, $x_5 = 1.888 + L_4^* = 3.055$ $\Rightarrow$ $f(x_5) = 5.941975$

and $x_6 = 4.944 - L_4^* = 3.777$ $\Rightarrow$ $f(x_6) = 4.6190$

which shows that

$$x_5 < x_6 \text{ and } f(x_5) > f(x_6)$$

Then again by unimodality of $f(x)$, the maximum point do not lie on the right of x_6 and therefore drop the interval]3.777, 4.944] and next interval of uncertainty is given by [1.888, 3.777].

Fifth Experiment

Let us take two points x_7 and x_8 in the interval [1.888, 3.777] at a distance $L_5^* = (0.618)^5 L_0 = 0.721$ from the two end points.

$$\vdash\!\!-L_5^*\!\!-\!\!\dashv \qquad\qquad \vdash\!\!-L_5^*\!\!-\!\!\dashv$$

$$\begin{array}{cccc} | & + & + & | \\ 1.888 & x_7 & x_8 & 3.777 \end{array}$$

Fig. 17

Now, $x_7 = 1.888 + L_5^* = 2.609$ $\Rightarrow$ $f(x_7) = 6.238597$

and $x_8 = 3.777 - L_5^* = 3.056$ $\Rightarrow$ $f(x_8) = 5.940864$

Clearly $x_7 > x_8$ and $f(x_7) > f(x_8)$

Therefore, by the unimodality of $f(x)$ the maximum points do not lie on the right of x_8.

Drop the the interval]3.056, 3.777].

The next uncertainty interval is [1.888, 3.056].

We observe that after the 5th experiment the uncertainty interval is sufficiently small (*i.e.*, 1.16).

Hence, the maximum point, $x^* = \dfrac{1.888 + 3.056}{2} = 2.472$

and $f(x^*) = \max f(x) = 6.249216$

9.7 EXHAUSTIVE SEARCH METHOD

This method can be used to solve problems where the interval in which the optimal is known to lie is finite. It consists of evaluating the objective function at a predetermined number of equally spaced points in the interval (x_s, x_e) where x_s and x_e denote respectively the starting and end points of the interval of uncertainty. Then by using the assumption of unimodality, reduce the interval of uncertainty.

Let $f(x)$ be a function defined on the interval $[x_s, x_e]$ and let be evaluated at eight equally spaced interior points x_1 to x_8. If the function values as shown in the following figure.

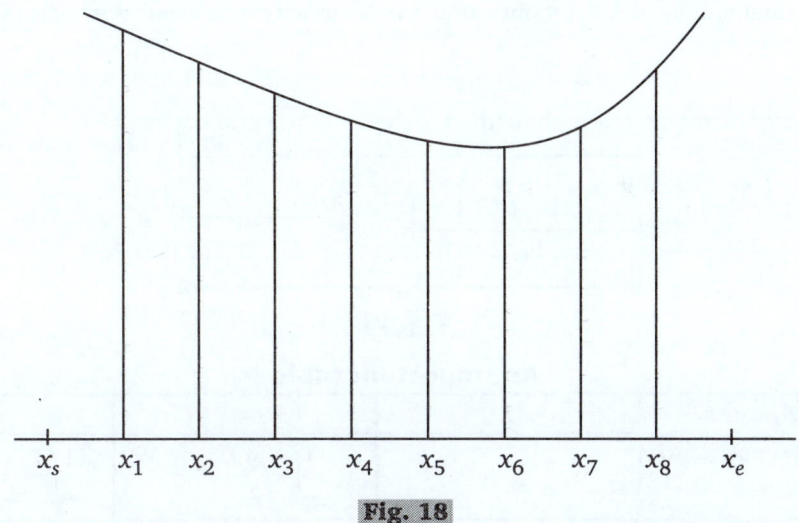

Fig. 18

In the above figure the minimum point must lie according the unimodality between x_5 and x_7. So the interval (x_5, x_7) may be considered as the final interval of uncertainty.

In general, the final uncertainty interval is given by $L_n = x_{j+1} - x_{j-1} = \dfrac{2}{n+1} L_0$

☞ **REMARK**
- This method is also called 'simultaneous search method' because the function is evaluated at all n points simultaneously.

An Important Table

No. of trials	2	3	4	5	6	...	n
L_n/L_0	2/3	2/4	2/5	2/6	2/7	...	$2/(n+1)$

9.8 DICHOTOMOUS SEARCH METHOD

We know that the exhaustive search method is a simultaneously search method, *i.e.*, all the experiments are conducted before knowing the location of the optimal point. But Dichotomous search method is a sequented search method in which the result of any experiments influence of the location of the subsequent experiment.

In this method, we place two experiments, very close to the centre of uncertainty and almost half of the interval of uncertainty is eliminated on the basis of the relative values of the objective function at two points.

Let $f(x)$ be a function defined on an interval $[a, b]$ and $L_0 = b - a$. Further let the position of the two experiments be given by two points x_1 and x_2 such that

$$x_1 = \frac{L_0}{2} - \frac{\delta}{2} \quad \text{and} \quad x_2 = \frac{L_0}{2} + \frac{\delta}{2}$$

where δ is a small positive number which can be chosen such that two experiments gives the different results.

Then new interval of uncertainty is given by $\left(\frac{L_0}{2} + \frac{\delta}{2}\right)$. This search consists of conducting a pair of experiments at the centre of the current interval of uncertainty. Thus, the next pair of experiments is conducted at the centre of remaining interval of uncertainty.

In general, final interval of uncertainty after n experiments (n is even) is given as below

$$L_n = \frac{L_0}{2^{n/2}} + \delta\left(1 - \frac{1}{2^{n/2}}\right)$$

The working of dicotomous search method is shown in the graph given below

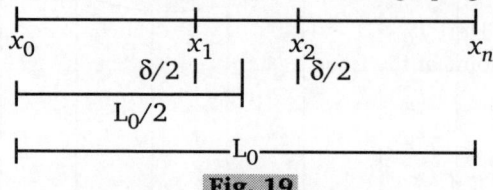

Fig. 19

An Important table

No. of Experiments	2	4	6
Final interval of uncertainty	$\frac{1}{2}(L_0 + \delta)$	$\frac{1}{2}\left(\frac{L_0 + \delta}{2}\right) + \frac{\delta}{2}$	$\frac{1}{2}\left(\frac{L_0 + \delta}{4} + \frac{\delta}{2}\right) + \frac{\delta}{2}$

SOLVED EXAMPLES

EXAMPLE 1. *Using exhaustive search method, find the minimum of $f(x) = x(x - 1.5)$ in the interval (0.0, 1.00) to within 10% of the exact value.*

SOLUTION. Let us take the middle point of the final interval of uncertainty as the approximate optimum point.

The maximum deviation $= \dfrac{1}{n+1}$ times the initial interval of uncertainty

Therefore, to find the optimum within 10% of the exact value, we must have

$$\frac{1}{n+1} \le \frac{1}{10}, i.e., n \ge 9$$

Now, we have the following calculation table

i	x_i	$f(x_i) = x_i(x_i - 1.5)$
1	0.1	-0.14
2	0.2	-0.26
3	0.3	-0.36
4	0.4	-0.44
5	0.5	-0.50
6	0.6	-0.54
7	0.7	-0.56
8	0.8	-0.56
9	0.9	-0.54

From the above table, we observe that $x_7 = x_8$. Therefore, by the assumption of unimodality, the final interval of uncertainty is $L_9 = (0.7, 0.8)$.

Therefore, the optimum point $x^* = \dfrac{0.7 + 0.8}{2} = 0.75$

and
$$\begin{aligned}
\min f(x) &= 0.75(0.75 - 1.5) \\
&= 0.75(-0.75) \\
&= -0.5625
\end{aligned}$$

EXAMPLE 2. *Using Dichotomous search method, find the minimum of f(x) = x(x – 1.5) in the interval (0.0, 1.00) to within 10% of the exact value.*

SOLUTION. We know that the ratio of final to initial intervals of uncertainty is given by

$$\frac{L_n}{L_0} = \frac{1}{2^{n/2}} + \frac{\delta}{L_0}\left(1 - \frac{1}{2^{n/2}}\right)$$

Let $\delta = 0.001$. Here $L_0 = 1 - 0 = 1$
If the middle point of the final interval is taken as the optimum point, then

$$\frac{1}{2}\frac{L_n}{L_0} \le \frac{1}{10}$$

$$\Rightarrow \qquad \frac{1}{2^{n/2}} + -\left(-\frac{1}{2^{n/2}} \right) \le \frac{1}{1}$$

$$\Rightarrow \qquad \frac{1}{2^{n/2}} + \frac{1}{1000}\left(1 - \frac{1}{2^{n/2}}\right) \le \frac{1}{5} \qquad\qquad (\because\ L_0 = 1, \delta = 0.001)$$

$$\Rightarrow \qquad \frac{999}{1000}\cdot\frac{1}{2^{n/2}} \le \frac{995}{5000}$$

$$\Rightarrow \qquad 2^{n/2} \ge \frac{999}{199} = 5.0$$

Since n should be even so we have to take $n = 6$
Now, we apply the Dichotomous search method.

First and Second Experiments

We have $\qquad x_1 = \dfrac{L_0}{2} - \dfrac{\delta}{2} = 0.5 - 0.0005 = 0.4995$

and $\qquad x_2 = \dfrac{L_0}{2} + \dfrac{\delta}{2} = 0.5 + 0.0005 = 0.5005$

which gives $\qquad f(x_1) = 0.4995(-1.0005) = -0.49975$

$$f(x_2) = 0.5005(-0.9995) = -0.50025$$

Clearly, $f(x_2) < f(x_1)$, therefore by unimodality of the function, the new interval of uncertainty is given by $[0.4995, 1.0)$.

Third Experiment

Let us take two points x_3 and x_4 in the new interval of uncertainty.

Then $\qquad x_3 = \left(0.4995 + \dfrac{1.0 - 0.4995}{2}\right) - 0.0005 = 0.74925$

$$x_4 = \left(0.4995 + \frac{1.0 - 0.4995}{2}\right) + 0.0005 = 0.75025$$

which gives $\qquad f(x_3) = 0.74925(-0.75075) = -0.5624994375$

$$f(x_4) = 0.75025(-0.74975) = -0.5624999375$$

Clearly, $f(x_3) > f(x_4)$. Then by unimodality of $f(x)$, drop the interval $(0.4995, x_3)$. Then

the new interval of uncertainty is given by
$$(x_3, 1.0) = (0.74925, 1.0)$$

Final Experiment

Let us take two points x_5 and x_6 in the new interval of uncertainty such that

$$x_5 = \left(0.74925 + \frac{1.0 - 0.74925}{2}\right) - 0.0005 = 0.874125$$

$$x_6 = \left(0.74925 + \frac{1.0 - 0.74925}{2}\right) + 0.0005 = 0.875125$$

which gives $f(x_5) = 0.874125(-0.625875) = -0.5470929844$

and $f(x_6) = 0.875125(-0.624875) = -0.5468437342$

Clearly, $f(x_5) < f(x_6)$, then again by unimodality of $f(x)$ the new interval of uncertainty is given by
$$(x_3, x_6) = (0.74925, 0.875125)$$

So, the optimum point $x^* = \dfrac{0.74925 + 0.87512}{2} = 0.8121875$.

and optimum value of $f(x) = -0.5586327148$

An Important table
(Final Interval of Uncertainty)

S.No.	Name of method	Final interval of uncertainty (L_n)
1.	Exhaustive search	$\dfrac{2}{n+1}L_0$
2.	Dichotomous search	$\dfrac{L_0}{2^{n/2}} + \delta\left(1 - \dfrac{1}{2^{n/2}}\right)$, n is even
3.	Internal halving method	$\left(\dfrac{1}{2}\right)^{\frac{n-1}{2}} \cdot L_0$, $n \geq 3$ and odd
4.	Fibonacci method	$\dfrac{1}{F_n} \cdot L_n$
5.	Golden section method	$(0.618)^{n-1} \cdot L_0$

9.9 INTERPOLATION METHODS

The interpolation methods were originally developed as one-dimensional searches within multivariate optimization technique. In this section we will discuss the following three interpolation methods:

   1. Quadratic Interpolation method   **2.** Cubic Interpolation method
   3. Direct Roots method

9.9.1 QUADRATIC INTERPOLATION MEHTOD

We know that the aim of all the one dimensional minimization methods is to find λ^*, the smallest non-negative value of λ for which the function
$$f(\lambda) = f(x + \lambda s) \qquad \qquad ...(1)$$
attain a local minimum, where s is any specified vector. In this method we have the following stages:

Stage 1. Let s_i be the ith component of the vector s then find
$$\Delta = \max |s_i|$$
and then divide each component of s by Δ.

or find $\Delta = (s_1^2 + s_2^2 + \ldots + s_n^2)^{1/2}$ and divide each component of s by Δ.

Stage 2. Define a quadratic function $h(\lambda)$ such that
$$h(\lambda) = a + b\lambda + c\lambda^2 \qquad \qquad ...(1)$$

For minima of $h(\lambda)$, we must have

$$\frac{dh}{d\lambda} = 0$$

$\Rightarrow \qquad b + 2c\lambda = 0$

$\Rightarrow \qquad \lambda = -\frac{b}{2c} = \bar{\lambda}^* \text{ (say)}$ ...(2)

Again, by sufficient condition of minima of $h(\lambda)$ we have

$$\frac{d^2h}{d\lambda^2} > 0 \text{ at } \lambda = \bar{\lambda}^*$$

$\Rightarrow \qquad 2c > 0 \qquad \Rightarrow \qquad c > 0$ ...(3)

Now, we have to find the constants a, b and c in (1). For this we have to evaluate $f(\lambda)$ at three points say $\lambda = A$, $\lambda = B$ and $\lambda = C$ such that f_A, f_B and f_C are the corresponding function values of $f(\lambda)$ respectively. Therefore,

$$\left.\begin{array}{l} f_A = a + bA + cA^2 \\ f_B = a + bB + cB^2 \\ f_C = a + bC + cC^2 \end{array}\right\}$$ ...(4)

On solving (4) for a, b and c we get

$$a = \frac{f_A BC(C-B) + f_B CA(A-C) + f_c AB(B-A)}{(A-B)(B-C)(C-A)}$$ ...(5)

$$b = \frac{f_A(B^2 - C^2) + f_B(C^2 - A^2) + f_c(A^2 - B^2)}{(A-B)(B-C)(C-A)}$$ ...(6)

$$c = -\frac{f_A(B-C) + f_B(C-A) + f_c(A-B)}{(A-B)(B-C)(C-A)}$$ ...(7)

Using (5), (6) and (7) in (2) we get

$$\bar{\lambda}^* = -\frac{b}{2c} = \frac{f_A(B^2 - C^2) + f_B(C^2 - A^2) + f_C(A^2 - B^2)}{2[f_A(B-C) + f_B(C-A) + f_C(A-B)]}$$ ...(8)

provided that $c > 0$.

Now, for simplicity, let us take $A = 0$, $B = t$ and $C = 2t$ then from (5) to (8) we get

$$a = f_A$$ ...(9)

$$b = \frac{4f_B - 3f_A - f_C}{2t}$$ ...(10)

$$c = \frac{f_C + f_A - 2f_B}{2t^2}$$ ...(11)

and $\qquad \bar{\lambda}^* = \frac{4f_B - 3f_A - f_C}{4f_B - 2f_C - 2f_A} \cdot t$ ...(12)

provided that $c > 0$, i.e., $\dfrac{f_C + f_A - 2f_B}{2t^2} > 0$ ...(13)

From (13) $\qquad \dfrac{f_A + f_C}{2} > f_B$

$\Rightarrow$ The fundamental value f_B should be smaller than the average value of f_A and f_C.

Now, we have the following procedure.

STEP 1. If $f_A = f(\lambda = 0)$ and the initial step size t_0 is known then evaluate the function at $\lambda = t_0$ to get $f_1 = \lambda(\lambda = t_0)$.

STEP 2. If $f_1 > f_A$ then set $f_C = f_1$ and evaluate the function f at $\lambda = \dfrac{t_0}{2}$ and $\bar{\lambda}^*$ at $t = \dfrac{t_0}{2}$.

STEP 3.   If $f_1 \le f_A$, set $f_B = f_1$ and evaluate the function f at $\lambda = 2t_0$ to find $f_2 = f(\lambda = 2t_0)$.

STEP 4.   If $f_2 > f_1$, set $f_C = f_2$ and evaluate $\bar{\lambda}^*$ at $t = t_0$.

STEP 5.   If $f_2 < f_1$, set $f_1 = f_2$ and $t_0 = 2t_0$ and repeat the step 2 to 4 until we find $\bar{\lambda}^*$.

Stage 3. At this stage, we want to make sure that $\bar{\lambda}^*$ (obtained at stage 2) is sufficiently closed to the true minimum λ^* of $f(\lambda)$ by using any one of the following convergence criteria.

(i) $\left| \dfrac{h(\bar{\lambda}^*) - f(\bar{\lambda}^*)}{f(\bar{\lambda}^*)} \right| \le \varepsilon_1$ ...(14)

(ii) $\left| \dfrac{f(\bar{\lambda}^* + \Delta\bar{\lambda}^*) - f(\bar{\lambda}^* - \Delta\bar{\lambda}^*)}{2\Delta\bar{\lambda}^*} \right| \le \varepsilon_2$ ...(15)

where ε_1, ε_2 are small positive numbers to be specified depending on the accuracy desired. If criteria (14) and (15) are not satisfied, then we used a new quadratic fuction

$$h'(\lambda) = a' + b' + c'\lambda^2$$

and then evaluate the constants a', b' and c' by using $f_A = f(\lambda = 0)$, $f_B = f(\lambda = t_0)$, $f_C = f(\lambda = 2t_0)$ and $\bar{f} = f(\lambda = \bar{\lambda}^*)$.

☛ REMARKS

- Stage one can be dropped is the given problem is not a multivariable optimization problem. The process of tring to fit another polynomial to obtain a better approximation to $\bar{\lambda}^*$ is known as 'refitting the polynomial'.
- For refitting the polynomial we consider all possible situations and select the best three points of the present A, B, C and $\bar{\lambda}^*$.

An Important table (For refitting)

Case	Characteristics	New points for Refiting	
		New	**Old**
1.	$\bar{\lambda}^* > B$ $\bar{f} < f_B$	A B C Neglect old A	B $\bar{\lambda}^*$ C
2.	$\bar{\lambda}^* > B$ $\bar{f} > f_B$	A B C Neglect old C	A B $\bar{\lambda}^*$
3.	$\bar{\lambda}^* < B$ $\bar{f} < f_B$	A B C Neglect old C	A $\bar{\lambda}^*$ B
4.	$\bar{\lambda}^* < B$ $\bar{f} > f_B$	A B C Neglect old A	$\bar{\lambda}^*$ B C

9.9.2 CUBIC INTERPOLATION METHOD

This method is used to find the minimizing step length λ^* in following four stages:

STAGE 1. Calculate $\Delta = \max\{s_i\}$ (where $|s_i|$ is the absolute value of ith component of the vector s) and divide each component of s by Δ or alternatively, find

$$\Delta = (s_1^2 + s_2^2 + \ldots + s_n^2)^{1/2}$$

and divide each comonent of s by Δ.

STAGE 2. We have to find two points A and B at which the slope $\dfrac{df}{d\lambda}$ has different sign.

For this, initially we can take $A = 0$ and then try to find a point $\lambda = B$ at which $\dfrac{df}{d\lambda} > 0$.

Point B can be taken as the first value of t_0, $2t_0$, $4t_0$, ... at which $f' \geq 0$ (t_0 is the preassigned initial step size). Then λ^* is bounded in the interval $A < \lambda^* \leq \Delta$.

STAGE 3. Let

$$h(\lambda) = a + b\lambda + c\lambda^2 + d\lambda^3 \qquad \ldots(1)$$

We used (1) to approximate the function $f(\lambda)$ between A and B by finding the values

$$f_A = f(\lambda = A)$$

$$f_A' = \frac{df}{d\lambda}(\lambda = A)$$

$$f_B = f(\lambda = B)$$

and $$f_B' = \frac{df}{d\lambda}(\lambda = B)$$

in order to evaluate the constants a, b, c and d in (1).

$\therefore$ From (1) we have

$$\left. \begin{aligned} f_A &= a + bA + cA^2 + dA^3 \\ f_B &= a + bB + cB^2 + dB^2 \\ f_A' &= b + 2cA + 3dA^2 \\ f_B' &= b + 2cB + 3dB^2 \end{aligned} \right| \qquad \ldots(2)$$

On solving system (2) for a, b, c and d we get

$$a = f_A - bA - cA^2 - dA^3 \qquad \ldots(3)$$

$$b = \frac{1}{(A-B)^2}(B^2 f_A' + A^2 f_B' + 2ABP) \qquad \ldots(4)$$

$$c = -\frac{1}{(A-B)^2}[(A+B)P + Bf_A' + Af_B'] \qquad \ldots(5)$$

and $$d = \frac{1}{3(A-B)^2}(2P + f_A' + f_B') \qquad \ldots(6)$$

where $$P = \frac{3(f_A - f_B)}{B - A} + f_A' + f_B' \qquad \ldots(7)$$

Now, for the minima of $h(\lambda)$, the necessary condition is given by

$$\frac{dh}{d\lambda} = b + 2c\lambda + 3d\lambda^2 = 0$$

$\Rightarrow$ $$\bar{\lambda}^* = \frac{-c \pm (c^2 - 3bd)^{1/2}}{3d} \qquad \ldots(8)$$

Also, by sufficient condition for minima of $h(\lambda)$

$$\left.\frac{d^2h}{d\lambda^2}\right|_{d\lambda=\bar{\lambda}^*} = 2c + 6d\bar{\lambda}^* > 0 \qquad \ldots(9)$$

Using (5) and (6) in (9) we get

$$\bar{\lambda}^* = A + \frac{f_A' + P \pm Q}{f_A' + f_B' + 2P}(B - A) \qquad \ldots(10)$$

where

$$Q = (P^2 - f_A f_B')^{1/2} \qquad \ldots(11)$$

$$2(B-A)(2P + f_A' + f_B')(f_A' + P \pm Q) - 2(B-A)(f_A'^2 + Pf_B' + 3Pf_A' + 2P^2)$$

$$- 2(B+A)f_A' \cdot f_B' > 0 \qquad \ldots(12)$$

In particular, for $A = 0$, we have

$$\left.\begin{aligned}
a &= f_A \\
b &= f_A' \\
c &= -\frac{1}{B}(P + f_A') \\
d &= \frac{1}{3B^2}(2P + f_A' + f_B') \\
\bar{\lambda}^* &= B\frac{f_A' + P \pm Q}{f_A' + f_B' + 2P} \\
Q &= (P^2 - f_A' f_B')^{1/2} > 0
\end{aligned}\right\} \qquad \ldots(13)$$

where

$$P = \frac{3(f_A - f_B)}{B} + f_A' + f_B' \qquad \ldots(14)$$

Clearly, there are two values of $\bar{\lambda}^*$ given by (10) and (13) which gives two possibilities of $h'(\lambda) = 0$.

To avoid imaginary values of Q we must have

$$P^2 - f_A' f_B' \geq 0$$

which is satisfied automatically because A and B are selected such that $f_A' < 0$ and $f_B' \geq 0$
.

STAGE 4. At this stage, we want to make sure that $\bar{\lambda}^*$ (obtained at satge 3) is sufficiently closed to the minimum λ^* of $f(\lambda)$ by using any one of the following convergence criteria

$$\left|\frac{h(\bar{\lambda}^*) - f(\bar{\lambda}^*)}{f(\bar{\lambda}^*)}\right| \leq \varepsilon_1 \qquad \ldots(15)$$

$$\left|\frac{df}{d\lambda}\right|_{\bar{\lambda}^*} = |s^T \nabla f|_{\bar{\lambda}^*} \leq \varepsilon_2 \qquad \ldots(16)$$

or

$$\left|\frac{s^T \nabla f}{\|s\| \|\nabla f\|}\right|_{\bar{\lambda}^*} \leq \varepsilon_2 \qquad \ldots(17)$$

where ε_1 and ε_2 are very small positive numbers whose value depends on the desired accuracy.
If (15) and (17) are not satisfied, then used a new cubic equation.

$$h'(\lambda) = a' + b'\lambda + c'\lambda^2 + d'\lambda^3$$

Then proceed same as above, the constants a', b', c', d' can be calculated using the function and derivaties at the best two points out of three A, B and $\bar{\lambda}^*$. Then use (10) for finding the optimal size $\bar{\lambda}^*$. The new points are taken as $\bar{\lambda}^*$ and B respectively, otherwise (i.e., if $f'(\bar{\lambda}^*) > 0$) the new point A and B are taken as A and $\bar{\lambda}^*$. Then again test the convergence criterion given by (15) and (16). If convergence is achieved, stop the process.

SOLVED EXAMPLES

EXAMPLE 1. *Using Cubic interpolation method, find the minimum of* $f = \lambda^5 - 5\lambda^3 - 20\lambda + 5$.

SOLUTION. Here, we have $f = \lambda^5 - 5\lambda^3 - 20\lambda + 5$

$$\Rightarrow \quad \frac{df}{d\lambda} = 5\lambda^4 - 15\lambda^2 - 20$$

So, $\frac{df}{d\lambda}$ at $\lambda = 0 = -20 < 0$

Now, start with $t_0 = 0.4$ and find the value of the derivation at t_0, $2t_0$, $4t_0$, ... etc. We get

$$f'(t_0 = 0.4) = 5(0.4)^4 - 15(0.4)^2 - 20.0 = -22.272$$

$$f'(2t_0 = 0.8) = 5(0.8)^4 - 15(0.8)^2 - 20.0 = -27.552$$

$$f'(4t_0 = 1.6) = 5(1.6)^4 - 15(1.6)^2 - 20.0 = -25.632$$

and $$f'(8t_0 = 3.2) = 5(3.2)^4 - 15(3.2)^2 - 20.0 = 350.688$$

Using these value in the equations (2) of cubic interpolation formula, we get

$$A = 0, f_A = 5.0, f'_A = -20.0$$

$$B = 3.2, f_B = 113.0, f'_B = 350.688$$

$$A < \lambda^* < B$$

Now, we have to find the value of $\bar{\lambda}^*$.

Firstly, we shall compute P and Q as given below, we know that

$$P = \frac{3(f_A - f_B)}{B - A} + f'_A + f'_B$$

$$= \frac{3(5.0 - 113.0)}{3.2} - 20.0 + 350.688$$

$$= 229.588$$

and $$Q = (P^2 - f_A f'_B)^{1/2}$$

$$= [(229.588)^2 + (20.0)(350.688)]^{1/2} = 244.0$$

Here, $$\bar{\lambda}^* = A + \frac{f'_A + P \pm Q}{f'_A + f'_B + 2P}(B - A)$$

$$= (3.2)\left(\frac{-20.0 + 229.588 \pm 244}{-20 + 350.688 + 459.176}\right) = 1.84 \text{ or } -0.1396$$

On discarding the negative sign, we get

$$\bar{\lambda}^* = 1.84$$

Now, we have to check the convergene criterion.

We know that if $\bar{\lambda}^*$ is losed to the true minimum λ^* then $f'(\bar{\lambda}^*) = \frac{df(\bar{\lambda}^*)}{d\lambda}$ should be zero.

Here $f' = 5\lambda^4 - 15\lambda^2 - 20$

$$\Rightarrow \qquad f'(\bar{\lambda}^*) = f'(1.84) = 5(1.84)^4 - 15(1.84)^2 - 20$$
$$= -13.0$$

which is not a small number

so we should go to the next iteration

Since $f'(\bar{\lambda}^*) < 0$ we take $A = \bar{\lambda}^*$

and
$$f_A = f(\bar{\lambda}^*) = (1.84)^5 - 5(1.84)^3 - 20(1.84) + 5$$
$$= -41.70$$

Therefore, we have

$$A = 1.84, \ f_A = -41.70 \text{ and } f_A' = -13.0$$

$$B = 3.2, \ f_B = 113.0, \ f_B' = 350.688$$

$$\Rightarrow \qquad A < \bar{\lambda}^* < B$$

Now, we proceed to the next iteration as follows.

$$P = \frac{3(f_A - f_B)}{B - A} + f_A' + f_B'$$

$$= \frac{3(-41.7 - 113.0)}{3.20 - 1.84} - 13.0 + 350.688$$

$$= -3.312$$

and
$$Q = (P^2 - f_A' f_B')^{1/2}$$

$$= [(-3.312)^2 + (13.0)(350.688)]^{1/2} = 67.5$$

so,
$$\bar{\lambda}^* = A + \frac{f_A' + P \pm Q}{f_A' + f_B' + 2P}(B - A)$$

$$= 1.84 + \frac{(-13 - 3.312 \pm 67.5)}{(-13 + 350.688 - 6.624)}(3.2 - 1.84)$$

$$= 2.05$$

Again, we have to check the convergence criteron as follows:

$$f'(\bar{\lambda}^*) = 5(2.05)^4 - 15.0(2.05)^2 - 20.0$$
$$= 5.35 > 0$$

which is a large value

Again we go the next iteration with $B = \bar{\lambda}^* = 2.05$ $\qquad (\because f'(\bar{\lambda}^*) \geq 0)$

and therefore, $\qquad f_B = (2.05)^5 - 5.0(2.05)^3 - 20(2.05) + 5$
$$= -42.90$$

So, $\quad A = 1.84, f_A = -41.70, f_A' = -13.0$

$$B = 2.05, f_B = -42.90, f_B' = 5.35$$

$$A < \lambda^* < B$$

Now, for the next iteration, we have the following values.

$$P = \frac{3(f_A - f_B)}{B - A} + f_A' + f_B'$$

$$= \frac{3(-41.70 + 42.90)}{(2.05 - 1.84)} - 13 + 5.35 = 9.49$$

and
$$Q = (P^2 - f_A' f_B')^{1/2}$$

$$= [(9.49)^2 + (13)(5.35)]^{1/2} = 12.61$$

So, $$\bar{\lambda}^* = A + \frac{f_A' + P \pm Q}{f_A' + f_B' + 2P}(B - A)$$

$$= 1.84 + \frac{(-13) + 9.49 \pm 12.61}{(-13) + 5.35 + 18.98}(2.05 - 1.84)$$

$$= 2.0086$$

Finally, we again check the criterion of convergence.
We have,

$$f'(\bar{\lambda}^*) = (5.0)(2.0086)^4 - 15.0(2.0086)^2 - 20 = 0.855$$

So, assuming that this value is close to zero we can stop the process.

Hence, $\lambda^* = \bar{\lambda}^* = 2.0086$

EXMPLE 2. *Using quadratic interpolation method, find the minimum of $f = \lambda^5 + -5\lambda^3 - 20\lambda + 5$*

SOLUTION. Let $$f = \lambda^5 - 5\lambda^3 - 20\lambda + 5$$

Since this is not a multivariable optimization problem, we will proceed directly to stage 2.

Let us take $t_0 = 0.5$ and $A = 0$

Then, $$f_A = f(A) = f(\lambda = 0) = 5$$

$$f_1 = f(\lambda = t_0) = 0.03125 - 5(0.125) - 20(0.5) + 5 = -5.59375$$

Clearly, $f_1 < f_A$, so set $f_B = f_1 = -5.59375$ and then we find

$$f_2 = f(\lambda = 2t_0 = 1.0) = -19$$

Since, $f_2 > f_1$, we set new $t_0 = 1$ and $f_1 = -19$. Again we find $f_B < f_A$ and therefore set $f_B = f_1 = -19.0$ and $f_2 = f(\lambda = 2t_0 = 2) = -43$

Again since, $f_2 < f_1$ we set $t_0 = 2$ and $f_1 = -43$

$\Rightarrow$ $\qquad\qquad f_1 < f_A$

Thus, set $f_B = f_1 = -43$

and $$f_2 = f(\lambda = 2t_0 = 4) = 629$$

$\Rightarrow$ $\qquad\qquad f_2 > f_1$

$\therefore$ $\qquad\qquad f_c = f_2 = 629$

Now, $$\bar{\lambda}^* = \frac{4f_B - 3f_A - f_c}{4f_B - 2f_c - 2f_A} \cdot t$$

$$= \frac{4(-43) - 3(5) - 629}{4(-43) - 2(629) - 2(5)}(2) = 1.135$$

Now, we will check the convergence criterion

Here, $A = 0$, $f_A = 5$, $B = 2$, $f_B = -43$, $C = 4$ and $f_C = 629$

Then proceed same as in the derivation of quadratic interpolation formula, the value of a, b and c are given by

$$a = f_A = 5$$

$$b = \frac{4f_B - 3f_A - f_C}{2t} = -204$$

$$c = \frac{f_C + f_A - 2f_B}{2t^2} = 90$$

and $$h(\bar{\lambda}^*) = h(1.135) = 5 - 204(1.135) - 90(1.135)^2 = -110.9$$

Now, since

$$\bar{f} = f(\bar{\lambda}^*) = (1.135)^5 - 5(1.135)^3 - 20(1.135) + 5.0$$
$$= -23.127$$

Now, $\left| \dfrac{h(\bar{\lambda}^*) - f(\bar{\lambda}^*)}{f(\bar{\lambda}^*)} \right| = \left| \dfrac{-116.5 + 23.127}{-23.127} \right| = 3.8$

Which is a large quantity

$\Rightarrow$ Convergence is not achieved.

So, we have to go to the next iteration.

$\because \quad \bar{\lambda}^* = B$ and $\bar{f} > f_B$, we take the new values of A, B and C as given below

$$A = 1.135 \qquad \Rightarrow \qquad f_A = -23.127$$
$$B = 2.0 \qquad \Rightarrow \qquad f_B = -43.0$$
$$C = 4.0 \qquad \Rightarrow \qquad f_C = 629.0$$

and new $\bar{\lambda}^*$ is given by

$$\bar{\lambda}^* = \frac{f_A(B^2 - C^2) + f_B(C^2 - A^2) + f_C(A^2 - B^2)}{2[f_A(B - C) + f_B(C - A) + f_C(A - B)]}$$

$$= \frac{[(-23.127)(4.0 - 16.0) + (-43.0)(16.0 - 1.29) + (629.0)(1.29 - 4.0)]}{2[(-23.127)(2 - 4) + (-43)(4 - 1.135) + (629)(1.135 - 2)]}$$

$$= 1.661$$

Again we check the Convergence creiterion.

We have, $a = 288$, $b = -417$ and $c = 125.3$

and $\quad h(\bar{\lambda}^*) = h(1.661) = 288 - 417(1.661) + 125.3(1.661)^2$
$$= -59.7$$

and $\qquad \tilde{f} = f(\bar{\lambda}^*) = 12.8 - 5(4.59) - 20(1.661) + 5$
$$= -38.37$$

Hence, $\left| \dfrac{h(\bar{\lambda}^*) - f(\bar{\lambda}^*)}{f(\bar{\lambda}^*)} \right| = \left| \dfrac{-59.70 + 38.37}{-38.37} \right| = 0.556$

To make this quantity, sufficiently small, we do the more iteration.

9.10 DIRECT ROOT METHOD

We know that, the necessary condition for $f(\lambda)$ to have a minimum of λ^* is given by

$$f'(\lambda^*) = 0 \qquad \qquad \text{(By the principle of maxima and minima)}$$

In direct roots methods, we have to find the root of the equation $f'(\lambda) = 0$

In this section we shall discuss the following direct methods.

(i) Newton Raphson method

(ii) Bisection method

(iii) Secant method

(1) NEWTON-RAPHSON METHOD

The Newton-Raphson method requires that the function $f(x)$ be twice differentiable. It begins with a point x_1 that is the initial estimate to the stationary point or root of the equation $f'(x) = 0$. A linear approximation to the function $f'(x)$ at x_1 is consricted and the point of which the linear approximation variables is taken as the next approximation. Expand the function $f(x)$ by Taylor's series as follows

$$f(x) = f\left(x_i + (x - x_i)\right) = f(x_i) + (x - x_i)f'(x_i) + \frac{1}{2}(x - x_i)^2 f''(x_i) \qquad ...(1)$$

(using second order approximateion)

Now by setting the derivative of equation (1) equal to zero, for the minimum of $f(x)$, we get

$$f'(x) = f'(x_i) + f''(x_i)(x - x_i) = 0 \qquad ...(2)$$

If x_i is an approximation to the minimum of $f(x)$, equation (2) can be rearranged to obtain an improved approximation given by

$$x_{i+1} = x_i - \frac{f'(x_i)}{f''(x_i)} \qquad ...(3)$$

Which is the required Newton's method

☛ REMARKS

- Newton-Raphon method has a quadratic convergence *i.e.*, if $f''(x_i) = 0$, this method has a fastest convergence property.

(2) BISECTION METHOD

Let $f(x)$ be a unimodel function defined over the given search interval. Then by the principle of maxima and minima, the optimal point can be obtained by $f'(x) = 0$.

If at a point t, $f'(t) < 0$ then assuming that the function is unimodal, the minimum can not lie to the left of t. In this case the interval $x \le t$ can be dropped. But if $f'(t) > 0$ then minimum can not lie to the right of t and in this case the interval $x \ge t$ can be dropped.

Let x_1 and x_2 be two points such that

$$f'(x_1) < 0 \text{ and } f'(x_2) > 0$$

Then stationary point will lies between x_1 and x_2

Thus we have to find the derivative of $f(x)$ at the mid-point

$$t = \frac{x_1 + x_2}{2}$$

If $f'(t) > 0$ then the interval (t, x_2) can be eliminated from the search. But if $f'(t) < 0$ then the interval (x_1, t) can be eliminated.

WORKING PROCEDURE

STEP 1. Find two points x_1 and x_2 such that $f'(x_1) < 0$ and $f'(x_2) > 0$

STEP 2. Calculate mid-point $t = \frac{x_1 + x_2}{2}$ and $f'(t)$

STEP 3. If $|f'(t)| < \varepsilon$, terminate the process, otherwise.

If $f'(t) < 0$ set, $x_1 = t$ go to step 2.

If $f'(t) > 0$ set, $x_2 = t$ go to step 2.

☛ REMARKS

- This method is conducted using just a single point rather than a pair of points, to identify a point at which $f'(x) = 0$
- The search logic of Bisection method is based purely on the sign of the derivative and does not use its magnitude.
- The Bisection method is also known as 'Bolzano method'.

(III) SECANT METHOD

In the secant method we use both the sign of derivatives and its magnitude.

Let $f(x)$ be a function defined on an interval $[a, b]$. We have to find the roots of $f'(x) = 0$ in

(a, b), if exists.

Let x_1 and x_2 be two points in (a, b) such that

$$f'(x_1) f'(x_2) < 0$$

i.e., $f'(x_1)$ and $f'(x_2)$ have opposite sign

The secant alogrithm then approximate the function $f'(x)$ and determine the next point where the secant line of $f'(x)$ is zero.

Thus, the next approximation to the stationary point x^* is given by

$$t = x_2 - \frac{f'(x_2)(x_2 - x_1)}{[f'(x_2) - f'(x_1)]}$$

or

$$x_{i+1} = x_i - \frac{f'(x_{i+1})(x_{i+1} - x_i)}{[f'(x_{i+1}) - f'(x_i)]}$$

Now, if $|f'(t)| \leq \varepsilon$, a small positive quantity terminate the above algorithm, otherwise select t and one of the point x_1 or x_2 such that their derivatives are of opposite sign and repeat the above secant step.

☛ REMARKS

• A straight line between the two points is called secant line.
• Second-method can eliminate more than half the interval in some instances.

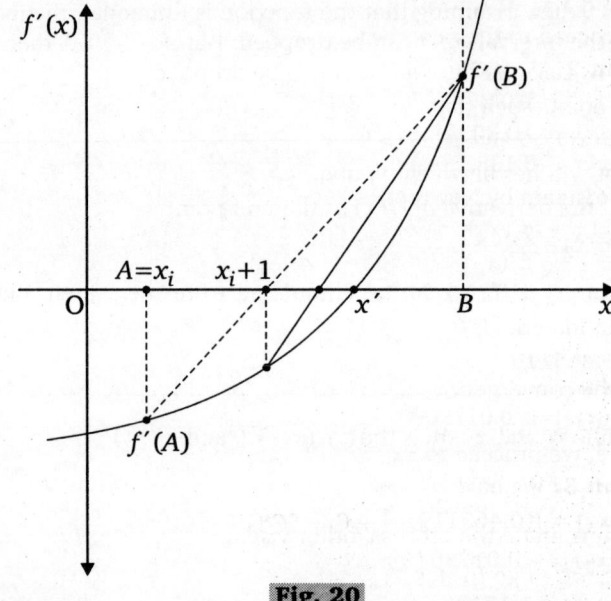

Fig. 20

SOLVED EXAMPLES

EXAMPLE 1.   *Using Newton-Raphson method, find the minimum of the function*

$$f(x) = 0.65 - \frac{0.75}{1 + x^2} - 0.65x \tan^{-1} \frac{1}{x} \text{ with the starting point } x_1 = 0.1. \text{ Use}$$

$$\varepsilon = 0.01 \text{ for checking the convergence.}$$

SOLUTION.   We have,

$$f(x) = 0.65 - \frac{0.75}{1 + x^2} - 0.65x \tan^{-1} \frac{1}{x}$$

$$\Rightarrow \quad f'(x) = \frac{1.5x}{(1+x^2)^2} + \frac{0.65x}{1+x^2} - 0.65\tan^{-1}\frac{1}{x}$$

and $f''(x) = \dfrac{1.5(1-3x^2)}{(1+x^2)^3} + \dfrac{0.65(1-x^2)}{(1+x^2)^2} + \dfrac{0.65}{1+x^2}$

$$= \frac{2.8 - 3.2x^2}{(1+x^2)^3}$$

Now, we proceed as follows:

Iteration 1: We have

$$x_1 = 0.1$$

$$\Rightarrow \quad f(x_1) = f(0.01) = -0.188197$$

$$f'(x_1) = -0.744832$$

and $f''(x_1) = 2.68659$

Now by Newton-Raphson method

$$x_2 = x_1 - \frac{f'(x_1)}{f''(x_1)} = 0.1 - \frac{(-0.744832)}{2.68659} = 0.377241$$

Also, for the convergene

$$|f'(x_2)| = |-0.138230| > \varepsilon$$

so, we proceed to the next iteration.

Iteration 2: We have

$$f(x_2) \quad = f(0.377241)$$
$$= -0.303279$$

$$f'(x_2) = -0.138230 \text{ and } f''(x_2) = 1.57296$$

Therefore, again by Newton-Raphson method

$$x_3 = x_2 - \frac{f'(x_2)}{f''(x_2)}$$

$$\Rightarrow \quad = 0.377241 - \frac{f'(0.377241)}{f''(0.377241)}$$

$$= 0.465119$$

and for the convergence

$$|f'(x_3)| = |-0.0179078| > \varepsilon$$

Therefore, we proceed to the next iteration

Iteration 3: We have

$$f(x_3) = f(0.465119) = -0.309881$$

$$f'(x_3) = -0.0179078$$

$$f''(x_3) = 1.17126$$

Then by Newton-Raphson method, we have

$$x_4 = x_3 - \frac{f'(x_3)}{f''(x_3)}$$

$$= 0.465119 - \frac{(-0.0179078)}{1.17126} = 0.480409$$

and for the convergenece

$$|f'(x_1)| = |-0.0005003| \approx \varepsilon$$

Hence, the optimum solution is given by

$$x^* = x_4 = 0.480409$$

EXAMPLE 2. *Find the minimum of the function* $f(x) = 2x^2 + \dfrac{16}{x}$ *over the interval* **[1, 5]** *using secant method.*

SOLUTION. We have

$$f(x) = 2x^2 + \frac{16}{x}$$

$$\Rightarrow \quad f'(x) = 4x - \frac{16}{x^2}$$

Iteration 1: Let us take $A = 1$

Then $B = 5$

$$f'(5) = 19.36$$

$$f'(1) = -12$$

Then by secant method

$$t = 5 - \frac{19.36 \times 4}{19.36 + 12} = 2.53$$

and $f'(t) = 7.62 > 0$

Therefore set $B = 2.53$

Iteration 2: $t = 2.53 - \dfrac{7.62}{(7.62 + 12)} \times 1.53 = 1.94$

and $f'(t) = 3.51 > 0$

set $B = 1.94$

Proceed same as above until we get $|f'(t)| < \varepsilon$

EXAMPLE 3. *Using secant method, find the minimum of*

$$f(x) = 0.65 - \frac{0.75}{1 + x^2} - 0.63x \tan^{-1}\frac{1}{x}$$

with step size $t_0 = 0.1$, $x_1 = 0.0$ *and* $\varepsilon = 0.01$

SOLUTION. We have

$$x_1 = A = 0, t_0 = 0.1, f'(A) = -1.02102$$

$$B = A + t_0 = 0.1$$

$$f'(B) = -0.744832$$

$\because$ $f'(B) < 0$, we set new A = 0.1, therefore $f'(A) = -0.744832$

$$t_0 = 2(0.1) = 0.2$$

$$B = x_1 + t_0 = 0.2$$

Then, $f(B) = -0.490343$

Clearly, $f'(B) < 0$

Therefore, we set

$$A = 0.2$$

$$f'(A) = -0.490343$$

$$t_0 = 2(0.2) = 0.4$$

$$B = x_1 + t_0 = 0.4$$

$$f'(B) = -0.103652$$

$\Rightarrow \quad f'(B) < 0$

Then, set $A = 0.4$

$\Rightarrow \quad f'(A) = -0.103652$

$$t_0 = 2(0.4) = 0.8$$

$$B = x_1 + t_0 = 0.8$$

Then, $f'(B) = 0.180800 > 0$

Now, we proceed to find the value of x_2.

Iteration 1:

$\because \qquad A > x_1 = 0.4, \; f'(A) = -0.103652$

$$B = 0.8, \; f'(B) = 0.180800$$

Then by secant method

$$x_2 = A - \frac{f'(A)(B - A)}{f'(B) - f'(A)}$$

$$= 0.4 - \frac{(-0.103652)(0.8 - 0.4)}{0.180800 + 0.103652} = 0.545757$$

for the convergence criterion

$$|f'(x_2)| = |-0.0105789| > \varepsilon$$

Iteration 2:

Here, $f'(x_2) = 0.0105789 > 0$

Set $A = 0.4 \Rightarrow f'(A) = -0.103652$

$$B = x_2 = 0.545757 \Rightarrow f'(B) = f'(x_2) = 0.0105789$$

Then, again by secant method

$$x_3 = A - \frac{f'(A)(B - A)}{f'(B) - f'(A)}$$

$$= 0.4 - \frac{(-0.103652)(0.545757 - 0.4)}{0.0105789 + 0.103652}$$

$$= 0.490632$$

Finally, for the convergence criterion

$$|f'(x_3)| = |0.00151235| < \varepsilon$$

$\Rightarrow$ Process is converges

Hence optimal solution is given by

$$x^* = \lambda_3 = 0.490632$$

HOW TO APPLY?

1. If initial interval of uncertainty is known then apply fibonacci method.
2. If initial interval of uncertainly is not know, apply quadratic interpretation method.
3. If the first derivations of the function being minimized are available, apply cubic interpetation method or secant method.
4. If both the first and second derivatives of the functions are available, apply Newton-Raphson method.

EXERCISE 9.1

1. Find the minimum of the function.

$$f(x) = 0.65 - \frac{0.75}{1+x^2} - 0.65x \tan^{-1}\frac{1}{x}$$

   Using quadratic interpeation formula with an initial size of 0.1.

2. Find the minimum of the function $f(x)$ given in ques. 1 using the cubic interpolation method

3. Find the maximum of the function given by

$$f(x) = \frac{0.5}{\sqrt{1+x^2}} - \sqrt{1+x^2}\left(1 - \frac{0.5}{1+x^2}\right) + x$$

   using the quadratic intepolation method

   with an initial step length of 0.1

4. Find the maximum of the function given in the above question using the following method.

   (i) Newton method with the initial point 0.6

   (ii) Secant method

5. Minimize the function

$$f = x_1 - x_2 + 2x_1^2 + 2x_1x_2 + x_2^2$$ starting with

$$x_1 = \begin{bmatrix} 0 \\ 0 \end{bmatrix}$$ along the dimension $$s = \begin{bmatrix} -1 \\ 0 \end{bmatrix}$$ using

   the quadratic interpolation method with an initial step length 0.1.

ANSWERS

1. 0.484 2. 0.481 3. 0.7817   4. (i) 0.78615   (ii) 0.789192   5. 0.25

10 Unconstrained Optimization Techniques

In this chapter we shall examine the fundamental concepts and useful methods for finding minima of constrained and unconstrained functions. The study of unconstrained minimization techniques provide the basic understanding necessary for the study of constrained minimization method. The unconstrained minimization method can be used to solve certain complex engineering analysis problems.

The general form of unconstrained minimization problem can be written as follows:

Find $\qquad X = \begin{bmatrix} x_1 \\ x_2 \\ \vdots \\ x_n \end{bmatrix}$ which minimizes $f(X)$.

GENERAL DEFINITIONS

(1) Rate of convergence: An optimization method is said to have convergence of order p if

$$\frac{\left\| X_{i+1} - X^* \right\|}{\left\| X_i - X^* \right\|^p} \le k, \quad k \ge 0, p \ge 1$$

(2) Linearly convergent method: If the rate of convergence of a method is 1, *i.e.*, $p = 1$, $0 \le k \le 1$ then this method is said to be linearly convergent. (slow convergence)

(3) Quadratically convergent method: If the rate of convergence of a method is 2, *i.e.*, $p = 2$, then this method is said to be quadratically convergence. (fast convergence)

(4) Super linear convergence method: An optimization method is said to have super linear convergence if

$$\lim_{i \to \infty} \frac{\left\| X_{i+1} - X^* \right\|}{\left\| X_i - X^* \right\|^p} \to 0 \quad \text{(Fast convergence)}$$

☛ REMARK
• The different iterative methods have different rate of convergence.

10.2 UNCONSTRAINED MINIMIZATION METHODS

The unconstrained minimization methods can be classified into following two categories:

1. Direct search method   **2.** Descent method (Indirect search method)

In direct search method, there is a requirement of only the objective function values but not the partial derivatives of the function for finding the minimum. On the other hand the descent method requires in addition to the function values, the first and second order partial derivatives of the objective function.

☛ REMARKS

- The direct search methods are also known as non-gradient methods or Zeroth order methods.
- The descent methods are also known as gradient methods.
- Among the gradient methods those requiring only first order derivative are called 'First order method', those requiring both first and second order partial derivatives are known as 'Second order method'.

> Direct search methods are more suitable for simple problem involving a relatively small number of variables and generally less efficient than the descent methods.

10.3 SOME DIRECT METHOD

10.3.1 UNIVARIATE METHOD

In univariate method, we change only one variable at a time and search is carried out in the axial directions covering all the direction one by one to get a sequence of improved approximation to the minimum point. In this method we start at a base point x_i in the ith iteration, we fix the value of $(n-1)$ variables and vary the remaining variable. Then problem becomes a one dimensional minimization problem ($\because$ only one variable is change), which can be solved easily.

Now, change any one of the $(n-1)$ variables that were fixed in previous iteration and continue this process by take each coordinates. Finally, n directions are searched sequentially. Continue this process until no further improvement is possible in the objective function in any of the n directions of a cycle.

WORKING PROCEDURE

STEP 1. Choose a fixed point $X_1(x_1, y_1)$ and a search direction $u_1 = (1, 0)^T$.

STEP 2. Take a step of size λ_1 in this direction and find the optimum point $\lambda_1 = \lambda_1^*$ for which $f(X_1 + \lambda_1^* u_1)$ is minimum.

Set a new point $X_2 = X_1 + \lambda_1^* u_1$.

STEP 3. Repeat the above steps by taking the optimum step length λ_2^* in the direction $u_2 = (0, 1)^T$ from the point X_2, to arrive at the next point X_3 given by

$$X_3 = X_2 + \lambda_2^* u_2$$

It complete the first iteration.

STEP 4. Repeat the process for the completion of the next cycle of iterations until no significant change is achieved in the value of the objective function, *i.e.*, until the variation in the value of the function is negligible.

☛ REMARKS

- In the above procedure, we consider the function of two variables x_1 and x_2 only, although, this procedure can be extended for function of more than two variables as well.
- Univariate method will not converge rapidly to the optimum solution as it has a tendency to oscillate with steadily decreasing progress toward the optimum.
- The univariate method can be applied to find the minimum of any function that possesses continuous derivatives.

> For n-dimensional case, the search directions are given as below:
> $$u_i^T = \begin{cases} (1,0,0,\ldots,0), \text{for } i = 1, n+1, 2n+1, \ldots \\ (0,1,0,\ldots,0), \text{for } i = 2, n+2, 2n+2, \ldots \\ (0,0,1,\ldots,0), \text{for } i = 3, n+3, 2n+3, \ldots \\ \vdots \\ (0,0,0,\ldots,1), \text{for } i = n, 2n, 3n, \ldots \end{cases}$$

SOLVED EXAMPLES

EXAMPLE 1. *Minimize* $f(X) = f(x_1, x_2) = 2x_1^2 + 3x_2^2 - x_1x_2$ *using univariate method by taking* $X_1 = \begin{bmatrix} 1 \\ 1 \end{bmatrix}$ *as a starting point.*

SOLUTION. Let $f(x_1, x_2) = 2x_1^2 + 3x_2^2 - x_1x_2$ ...(1)

Iteration-1. Consider the unit vectors $u_1 = (1, 0)^T$ and $u_2 = (0, 1)^T$ in the directions of x_1 and x_2 respectively.

$$\because \qquad X_1 = \begin{pmatrix} 1 \\ 1 \end{pmatrix}$$

Then $f(x_1) = 2 \cdot 1^2 + 3 \cdot 1^2 - 1 \cdot 1 = 4$

Now take a step size λ_1 in the x_1-direction to get X_2 such that

$$X_2 = X_1 + \lambda_1 \begin{pmatrix} 1 \\ 0 \end{pmatrix} = \begin{pmatrix} 1 \\ 1 \end{pmatrix} + \begin{pmatrix} \lambda_1 \\ 0 \end{pmatrix} = \begin{pmatrix} 1 + \lambda \\ 1 \end{pmatrix}$$

Now, to obtain the optimum size λ_1, we find

$$\text{Min.} f\begin{pmatrix} 1 + \lambda_1 \\ 1 \end{pmatrix} = 2(1 + \lambda_1)^2 + 3 - (1 + \lambda_1)$$

Using the principle of maxima and minima, for the minimum of f, we have

$$\frac{df}{d\lambda_1} = 0 \implies 4(1 + \lambda_1) - 1 = 0 \implies \lambda_1 = -\frac{3}{4}$$

and $$\frac{d^2 f}{d\lambda_1^2} = 4 > 0$$

$\implies$ f is minimum for $\lambda_1 = -\dfrac{3}{4}$

So, new point $X_2 = \begin{pmatrix} 1 + \lambda_1 \\ 1 \end{pmatrix} = \begin{pmatrix} 1 - 3/4 \\ 1 \end{pmatrix} = \begin{pmatrix} 1/4 \\ 1 \end{pmatrix}$

and $$f(X_2) = 2\left(\frac{1}{4}\right)^2 + 3(1)^2 - \frac{1}{4}(1) = \frac{23}{8} = 2.875$$

Further, to find the optimum step size λ_2 in x_2-direction

Minimize $\quad f(X_2 + \lambda_2(0,1)^T) = f\left(\begin{pmatrix} 1/4 \\ 1 \end{pmatrix} + \begin{pmatrix} 0 \\ \lambda_2 \end{pmatrix}\right)$

$$= f\left(\begin{pmatrix} 1/4 \\ 1 + \lambda_2 \end{pmatrix}\right) = \frac{1}{8} + 3(1 + \lambda_2)^2 - \frac{1}{4}(1 + \lambda_2)$$

Again, for minimum of f,

$$\frac{df}{d\lambda_2} = 0 \implies 6(1 + \lambda_2) - \frac{1}{4} = 0 \implies \lambda_2 = -\frac{23}{24} \quad \text{and} \quad \frac{d^2 f}{d\lambda_2^2} = 6 > 0$$

$\implies$ f is minimum for $\lambda_2 = -\dfrac{23}{24}$.

$\therefore$ New point $X_3 = X_2 + \lambda_2(0,1)^T = \begin{pmatrix} 1/4 \\ 1 + \lambda_2 \end{pmatrix} = \begin{pmatrix} 1/4 \\ 1/24 \end{pmatrix}$

and $$f(X_3) = 2\left(\frac{1}{4}\right)^2 + 3\left(\frac{1}{24}\right)^2 - \left(\frac{1}{4}\right)\left(\frac{1}{24}\right) = 0.11979$$

Iteration-2. Consider X_3 as the new base point. Now, to find optimum step size λ_3 in x_3-direction, we

Minimize $f(X_3 + \lambda_3(1, 0)^T)$

$$= f\left(\begin{pmatrix} 1/4 \\ 1/24 \end{pmatrix} + \begin{pmatrix} \lambda_3 \\ 0 \end{pmatrix}\right) = f\begin{pmatrix} 1/4 + \lambda_3 \\ 1/24 \end{pmatrix}$$

$$= 2\left(\frac{1}{4} + \lambda_3\right)^2 + 3\left(\frac{1}{24}\right)^2 - \frac{1}{24}\left(\frac{1}{4} + \lambda_3\right) \qquad \text{(By (1))}$$

Again, for minimum of f, $\dfrac{df}{d\lambda_3} = 0$

$$\Rightarrow \qquad 4\left(\frac{1}{4} + \lambda_3\right) - \frac{1}{24} = 0$$

$$\Rightarrow \qquad \lambda_3 = -\frac{23}{96}$$

and $\qquad \dfrac{d^2 f}{d\lambda_3^2} = 4 > 0$

$$\Rightarrow \quad f \text{ is minimum for } \lambda_3 = -\frac{23}{96}$$

So, new point $X_4 = \begin{pmatrix} (1/4) + (-23/96) \\ 1/24 \end{pmatrix} = \begin{pmatrix} 1/96 \\ 1/24 \end{pmatrix}$

and $\qquad f(X_4) = 2\left(\dfrac{1}{96}\right)^2 + 3\left(\dfrac{1}{24}\right)^2 - \dfrac{1}{96}\left(\dfrac{1}{24}\right) = \dfrac{23}{4608} = 0.00499$

Now considering X_4 as the new base point, to find optimum step size λ_4 in x_2-direction.

Minimize $f(X_4 + \lambda_4(0, 1)^T)$

$$= f\left(\begin{pmatrix} 1/96 \\ 1/24 \end{pmatrix} + \begin{pmatrix} 0 \\ \lambda_4 \end{pmatrix}\right) = f\begin{pmatrix} \dfrac{1}{96} \\ \dfrac{1}{24} + \lambda_4 \end{pmatrix}$$

$$= 2\left(\frac{1}{96}\right)^2 + 3\left(\frac{1}{24} + \lambda_4\right)^2 - \frac{1}{96}\left(\frac{1}{24} + \lambda_4\right)$$

Again for minimum of f, $\dfrac{df}{d\lambda_4} = 0$

$$\Rightarrow \qquad 6\left(\frac{1}{24} + \lambda_4\right) - \frac{1}{96} = 0$$

$$\Rightarrow \qquad \lambda_4 = -\frac{23}{576}$$

and $\qquad \dfrac{d^2 f}{d\lambda_4^2} = 6 > 0$

$$\Rightarrow \quad f \text{ is minimum for } \lambda_4 = -\frac{23}{576}$$

$$\therefore \quad \text{the new point } X_5 = \left(\begin{pmatrix} 1/96 \\ \dfrac{1}{24} \end{pmatrix} - \begin{pmatrix} \dfrac{23}{576} \end{pmatrix}\right) = \begin{pmatrix} 1/96 \\ 1/576 \end{pmatrix}$$

and $$f(X_5) = 2\left(\frac{1}{96}\right)^2 + 3\left(\frac{1}{576}\right)^2 - \left(\frac{1}{96}\right)\left(\frac{1}{576}\right) \qquad \text{(By (1))}$$

$$= 0.00020797$$

Hence, at this stage

Minimum of $f(X) = 0.00020797$ when $x_1 = \dfrac{1}{96}$ and $x_2 = \dfrac{1}{576}$

EXAMPLE 2. *Using univariate search method,*
Minimize $f(X) = f(x_1, x_2) = x_1^2 - x_1 x_2 + 3x_2^2$
by taking starting point as $X_1 = (1, 1)^T.$

SOLUTION. We have $f(X) = f(x_1, x_2) = x_1^2 - x_1 x_2 + 3x_2^2$ ...(1)

Iteration-1.

STEP-1. Consider the unit vectors $u_1 = (1, 0)^T$ and $u_2 = (0, 1)^T$ in x_1 and x_2-directions

respectively. Given $X_1 = \begin{pmatrix} 1 \\ 1 \end{pmatrix}$

$$\therefore \qquad f(X_1) = 1^2 - 1 \cdot 1 + 3 \cdot 1^2 = 3$$

Now, we take a step size λ_1 in the x_1-direction to arrive at the value of X_1.

Now proceed same as in example-1, we have to minimize $f(1 + \lambda_1, 1)$

From (1)

$$f(1 + \lambda, 1) = (1 + \lambda_1)^2 - (1 + \lambda_1) + 3$$

For the minimum of f,

$$\frac{df}{d\lambda_1} = 0 \qquad \Rightarrow \qquad 2(1 + \lambda_1) - 1 = 0 \quad \Rightarrow \quad \lambda_1 = -1/2$$

and $$\frac{d^2 f}{d\lambda_1^2} = 2 > 0$$

$$\Rightarrow \qquad f \text{ is minimum when } \lambda_1 = -\frac{1}{2}$$

So, the new point is given by

$$X_2 = \begin{pmatrix} 1 + \lambda_1 \\ 1 \end{pmatrix} = \begin{pmatrix} 1 - 1/2 \\ 1 \end{pmatrix} = \left(\frac{1}{2}, 1\right)$$

STEP-2. Now find step size λ_2 along x_2 by minimizing $f\left(\frac{1}{2}, 1 + \lambda_2\right)$

Now, $$f\left(\frac{1}{2}, 1 + \lambda_2\right) = \frac{1}{4} - \frac{1}{2}(1 + \lambda_2) + 3(1 + \lambda_2)^2 \qquad \text{(By (1))}$$

Now for minimum of f

$$\frac{df}{d\lambda_2} = 0 \Rightarrow -\frac{1}{2} + 6(1 + \lambda_2) = 0 \Rightarrow \lambda_2 = -\frac{11}{12}$$

and $$\frac{d^2 f}{d\lambda_2^2} = 6 > 0$$

$$\Rightarrow \qquad f \text{ is minimum at } \lambda_2 = -\frac{11}{12}$$

So, the new point is

$$X_3 = \left(\frac{1}{2}, 1 + \lambda_2\right) = \left(\frac{1}{2}, \frac{1}{12}\right)$$

Now, repeat the above process with starting point X_3. Go on repeating the process till the quantities $|\lambda_k|$ are less than some prefixed tolerance.

EXAMPLE 3. *Minimize* $f(X) = f(x_1, x_2) = x_1 - x_2 + 2x_1^2 + 2x_1x_2 + x_2^2$ *with the starting point* $X_1 = [0, 0]^T$ *by Univariate method.*

SOLUTION. We have

$$f(X) = f(x_1, x_2) = x_1 - x_2 + 2x_1^2 + 2x_1x_2 + x_2^2 \qquad \qquad \text{...(1)}$$

We can minimize $f(X)$ by using the following steps:

STEP 1. Consider the unit vector $u_1 = (1, 0)^T$ and $u_2 = (0, 1)^T$ in x_1 and x_2 direction respectively.

Given $\qquad X_1 = \begin{pmatrix} 0 \\ 0 \end{pmatrix}$

$\Rightarrow \qquad f(X_1) = 0 - 0 + 2 \cdot 0^2 + 2 \cdot 0 \cdot 0 + 0^2 = 0$

We take a step size λ_1 in x_1-direction to get the value of X_2

$\therefore \qquad X_2 = X_1 + \lambda_1 \begin{pmatrix} 1 \\ 0 \end{pmatrix} = \begin{pmatrix} 0 \\ 0 \end{pmatrix} + \begin{pmatrix} \lambda_1 \\ 0 \end{pmatrix} = \begin{pmatrix} \lambda_1 \\ 0 \end{pmatrix}$

(We can easily verify that $-u_1$ is the correct direction for minimizing f from X_1)

So, $\qquad X_2 = X_1 - \lambda_1 \begin{pmatrix} 1 \\ 0 \end{pmatrix} = \begin{pmatrix} -\lambda_1 \\ 0 \end{pmatrix}$

To obtain the optimum size λ, we minimize

$$f \begin{pmatrix} -\lambda_1 \\ 0 \end{pmatrix} = (-\lambda_1) - 0 + 2(-\lambda_1)^2 + 0 + 0 = 2\lambda_1^2 - \lambda_1$$

Now, for the minimum of f,

$$\frac{df}{d\lambda_1} = 0 \qquad \Rightarrow \qquad 4\lambda_1 - 1 = 0$$

$$\Rightarrow \qquad \lambda_1 = 1/4$$

and $\dfrac{d^2 f}{d\lambda_1^2} = 4 > 0$

$\Rightarrow \qquad f$ is minimum at $\lambda_1 = 1/4$

Now, set the new point

$$X_3 = X_1 - \lambda_1 u_1 = \begin{pmatrix} 0 \\ 0 \end{pmatrix} - \frac{1}{4} \begin{pmatrix} 1 \\ 0 \end{pmatrix} = \begin{pmatrix} -1/4 \\ 0 \end{pmatrix}$$

and $\quad f(X_3) = f\left(-\dfrac{1}{4}, 0\right) = -\dfrac{1}{8}$ \hfill (By (1))

STEP 2. Now considering X_3 as the new base point in $u_2 = (0, 1)^T$ direction. Then we proceed same as in example-1.

We have to minimize $f(X_3 + \lambda_2 u_2)$

Now, $\quad f(X_3 + \lambda_2 u_2) = f(X_3 + \lambda_2(0, 1)^T)$

$\qquad \qquad = f(-0.25, \lambda_2)$

$\qquad \qquad = -0.25 - \lambda_2 + 2(0.25)^2 - 2(0.25)(\lambda_2) + \lambda_2^2$ \hfill (By (1))

$\qquad \qquad = \lambda_2^2 - 1.5\lambda_2 - 0.125$

Again by the principle of maxima and minima

$$\frac{df}{d\lambda_2} = 0 \Rightarrow 2\lambda_2 - 1.5 = 0, \text{ i.e., } \lambda_2 = 0.75$$

and $\quad \dfrac{d^2 f}{d\lambda_2^2} = 2 > 0$

$\Rightarrow \qquad f$ is minimum at $\lambda 2 = 0.75$

Now, set the new point

$$X_4 = X_3 + \lambda_2 u_2 = \begin{pmatrix} -0.25 \\ 0 \end{pmatrix} + 0.75 \begin{pmatrix} 0 \\ 1 \end{pmatrix} = \begin{pmatrix} -0.25 \\ 0.75 \end{pmatrix}$$

Then by (1), $f(X_4) = -0.6875$

Continuing the above process until the optimum solution

$$X^* = \begin{pmatrix} -1.0 \\ 1.5 \end{pmatrix} \text{ with } f(X^*) = -1.25 \text{ is found.}$$

10.3.2 PATTERN SEARCH METHOD

In univariate method, we search for the minimum of the given functions along directions parallel to the coordinate axes (axial directions) only. It can be observed that this method may not converge in some cases, *i.e.*, the rate of convergence in univariate method to arrive at the minimal point is very slow. This problem can be avoided by changing the directions of search in a favourable manner instead of retaining them always parallel to the coordinate axes.

    (i) Pattern Directions: The line joining the alternate points of the search in the general direction of the minimum are known as pattern directions.

☛ REMARKS
- If the objective function is a quadratic in two variables, all pattern directions pass through the minimum.
- The above property is not valid for multivariable functions even when they are quadratic.

    (ii) Pattern search method: The methods that use pattern directions as search directions are called pattern search methods.

    There are following two pattern search methods.
        (i)   Hooke and Jeeve's method
        (ii) Powel's method

(1) HOOKE AND JEEVE'S METHOD

It is a sequential technique in which each step consist of the following two types of move

    (i) Exploratory move

    In exploratory moves, given a specified step size, which may be different for each coordinate direction and change during the search, the explanation proceeds from an initial point by specified step size in each coordinate direction. If the function value does not increase, the step is considered successful, otherwise the step is retracted and replaced by a step in the opposite direction which in turn is retained depending upon whether it succeeds or fails. Finally, the exploratory move is completed when all coordinates have been investigated. Here, the resulting point is known as base point.

    (ii) Pattern move

    A pattern move consists of a single step from the present base point along the line from the previous to the current base point.

The general procedure of Hooke and Jeeve's method is given as below:

WORKING PROCEDURE

STEP 1. Start with a fixed point say $X_1 = \begin{bmatrix} x_1 \\ x_2 \\ \vdots \\ x_n \end{bmatrix}$ and prescribed step size is considered to be

constants (*i.e.*, $\Delta x_1 = \Delta x_2 = \dots = \Delta x_n = \text{constant}$)

STEP 2. Made the search in each direction $u_i : i = 1, 2, ..., n$ where u_i is the unit vector along the direction x_i whose i^{th} component is 1 and all other components are zero.

(The search is made first in the positive direction and then if necessary in the negative direction of each axis to arrive at a temporary base point)

STEP 3. Define the temporary base point Y_k, obtained from X_k by slightly changing (perturbed) the ith components of X_k $(Y_{k_0} = X_k)$ as given below

$$Y_{ki} = \begin{cases} Y_{k,i-1} + \Delta x_i u_i; \text{if } f^+ = f(Y_{k,i-1} + \Delta x_i u_i) < f = f(Y_{k,i-1}) \\ Y_{k,i-1} - \Delta x_i u_i; \text{if } f^- = f(Y_{k,i-1} - \Delta x_i u_i) < f = f(Y_{k,i-1}) \\ Y_{k,i-1}; \text{if } f(Y_{k,i-1}) = f \le \min(f^+, f^-) \end{cases}$$

STEP 4. Continue this process of finding a new base point for $i = 1, 2, 3, ..., n$ until all directions are covered.

STEP 5. If the point $Y_{k,n}$ remains same as X_k reduce the step length Δx_i (say by a factor of 2) set $i = 1$ and go to step 3.

But if $Y_{k,n}$ is different from X_k, obtain the new base point as

$$X_{k+1} = Y_{k,n}$$

and go to the next step.

STEP 6. Establish a pattern direction u as

$$u = X_{k+1} - X_k$$

and find a point $Y_{k+1,0} = X_{k+1} + \lambda u$

where λ is the step length (for simplicity, it can be taken as 1)

STEP 7. Repeat the steps until a desired accuracy is achieved or the change in the value of the function satisfies the given conditions.

SOLVED EXAMPLES

EXAMPLE 1. *Using Hooke and Jeeve's method, minimize the function*

$$f(X) = 2x_1^2 + x_2^2 + 2x_1x_2 + x_1 - x_2$$

taking the starting base point as $X_1 = (0, 0)^T$ and $\Delta x_1 = \Delta x_2 = 0.8$

SOLUTION. We have

$$f(X) = f(x_1, x_2) = 2x_1^2 + x_2^2 + 2x_1x_2 + x_1 - x_2 \qquad ...(1)$$

Iteration-1.

STEP 1. Take the starting base point $X_1 = \begin{bmatrix} 0 \\ 0 \end{bmatrix}$ and step length $\Delta x_1 = 0.8, \Delta x_2 = 0.8$

$$\therefore \qquad f(X_1) = f(0, 0) = 0 \qquad \text{(By (1))}$$

Now, moving in direction u_1 from the base point X_1. So, obtain the temporary base point as given below:

$$f^+(X_1 + \Delta x_1 \cdot u_1) = f^+[(0,0) + 0.8(1,0)] = f^+(0.8, 0) = 2.08 \nleq f(X_1)$$

$$f^-(X_1 - \Delta x_1 \cdot u_1) = f^+(-0.8, 0) = 0.48 \nleq f(X_1)$$

Now, since f^+, f^- are not less than $f(X_1)$, movement in u_1-direction is not beneficial and hence discarded.

So, take $Y_{11} = X_1$ and $f(Y_{11}) = f(X_1) = 0$.

STEP 2. Now resuming movement in u_2-direction as follows:

$$f^+(X_1 + \Delta x_2 \cdot u_2) = f^+[(0,0) + 0.8(0,1)] = f^+(0, 0.8) = -0.16 < f(X_1)$$

Since, $Y_{12} = X_1 + \Delta x_2 u_2 = (0, 0.8)$ is different from X_1, the new base point is taken as

$$Y_{12} = X_2 = (0, 0.8) \text{ and } f(X_2) = -0.16 \qquad \text{(By (1))}$$

STEP 3. In step 1 and 2, the movements have been made in both the directions (u_1 and u_2) so third movement is to be made in the first pattern direction.
$$S_{p1} = X_2 - X_1 = (0, 0.8) - (0, 0) = (0, 0.8)$$
through step length λ from X_2 so that $f(X_2 + \lambda S_{p1})$ is minimum.

Hence, $f(X_2 + \lambda S_{p1}) = f[X_2 + \lambda(X_2 - X_1)] = f\{0, 0.8(1 + \lambda)\}$
$$= [(0.8)^2(1 + \lambda)^2 - 0.8(1 + \lambda)]$$

Using the principle of extrema, for the minima of f, we must have
$$\frac{df}{d\lambda} = 0 \Rightarrow 2 \times 0.64(1 + \lambda) - 0.8 = 0 \Rightarrow \lambda = -\frac{3}{8}$$

and $\frac{d^2f}{d\lambda^2} = 1.28 > 0$

$\Rightarrow$ f is minimum for $\lambda = -\frac{3}{8} = \lambda^*$ (say)

So, we obtain the new base point
$$Y_{20} = X_3 = X_2 + \lambda^*(X_2 - X_1)$$
$$= (0, 0.8) - \frac{3}{8}(0, 0.8) = (0, 0.5)$$

and $f(X_3) = f(0, 0.5) = -0.25$

Iteration-2.

STEP 1. Let us move in the direction of u_1 with base point $X_3 = (0, 0.5)$

Now, $f^+(X_3 + \Delta x_1 u_1) = f^+[(0, 0.5) + (0.8, 0)] = f^+(0.8, 0.5) = 2.63$ (By (1))
$$\nleq f(X_3)$$

and $f^-(X_3 - \Delta x_1 u_1) = f^-(-0.8, 0.5) = -0.57 < f(X_3) = -0.25$

Therefore, the new temporary base point is given by
$$Y_{21} = (X_3 - \Delta x_1 u_1) = (-0.8, 0.5), f(Y_{21}) = -0.57$$

STEP 2. Now moving in the direction of u_2 from the base point $Y_{21} = (-0.8, 0.5)$ to Y_{22}. Then we have
$$f^+(Y_{21} + \Delta x_2 u_2) = f^+(-0.8, 1.3) = -1.21 < f(Y_{21})$$

So, the new base point is given by
$$Y_{22} = X_4 = Y_{21} + \Delta x_2 u_2 = (-0.8, 1.3)$$
and $f(X_4) = -1.21$ (By (1))

STEP 3. Now, we move along the second pattern search directions $S(p, 2) = X_4 - X_3$ starting from X_4 through optimal step length λ so that
$$f(X_4 + \lambda S_{p2}) = f[(-0.8, 1.3) + \lambda(-0.8, 0.8)]$$
$$= f[-0.8(1 + \lambda), 1.3 + \lambda(0.8)]$$

is minimum.

Now, for the minima of $f(\lambda)$ we must have
$$\frac{df}{d\lambda} = 0 \qquad \Rightarrow \qquad 1.28\lambda - 0.32 = 0$$
$$\Rightarrow \qquad \lambda = 0.25$$

and $\frac{d^2f}{d\lambda^2} = 1.28 > 0$

$\Rightarrow$ f is minimum at $\lambda = 0.25$

Therefore, we get the points $X_5 = Y_{30} = X_4 + \lambda S(p,2) = (-1, 1.5)$
and $f(X_5) = -1.25$ (By (1))

Iteration-3.

STEP 1. First moving in the u_1-direction from $X_5 = (-1, 1.5)$ we have

$$f^+(X_5 + \Delta x_1 u_1) = f^+[(-1,1.5) + 0.8(1,0)]$$

$$= f^+(-0.2,1.5) = 0.03 \nless f(X_5)$$

and $\quad f^-(X_5 - \Delta x_1 u_1) = f^-[(-1,1.5) - 0.8(1,0)]$

$$= f^-(-1.8,1.5) = 0.03 \nless f(X_5)$$

Therefore, $\qquad Y_{31} = X_5$

STEP 2. Now making movement in u_2-direction from $Y_{31} = X_5$. Then

$$f^+(X_5 + \Delta x_2 u_2) = f^+[(-1,1.5) + 0.8(0,1)]$$

$$= f^+(-1,2.3) = -0.61 \nless f(X_5)$$

and $\quad f^-(X_5 - \Delta x_2 u_2) = f^-[(-1,1.5) - 0.8(0,1)]$

$$= f^-(-1,0.7) = -0.61 \nless f(X_5)$$

Thus, we conclude that the movement in any of the axial directions produces no change in the value of the function f at X_5. Therefore, X_5 is an optimal solution.

Hence, minimum of $f = -1.25$ when $x_1 = -1$ and $x_2 = 1.5$

(2) POWELL'S METHOD

This is most successful direct search algorithm, which is an extension of the basic pattern search method. This method is based upon the model of quadratic function and thus have a theoritical basis.

We have the following two reasons for choosing a quadratic model.

(1) It is the simplest type of non-linear function to minimize and so any general technique must work well in a quadratic, if it is to have any success with a general function.

(2) Near the optimum, all non linear functions can be approximated by a quadratic.

Before discussing the Powell's algorithm, let us have the following definitions:

(1) Conjugate Directions: Let $A = [A]$ be an $n \times n$ symmetric matrix. A set of n vectors $\{s_i\}$ is said to be conjugate if

$$s_i^T A s_j = 0 \quad \forall i \neq j, \quad i = 1,2,\ldots,n; j = 1,2,\ldots,n$$

☛ REMARK

• The orthogonal directions are the special case of conjugate directions.

(2) Quadratically Convergent method: If a minimization method, using exact arithmetic can find the minimum point in n steps while minimizing a quadratic function in n-variables, the method is known as quadratically convergent method.

IMPORTANT RESULTS

(1) Given a quadratic function of n variables and two hyperplanes 1 and 2 of dimension $k < n$, let the constrained stationary points of the quadratic function in the hyperplane be X_1 and X_2 respectively, then the line joining X_1 and X_2 is conjugate to any line parallel to the hyperplanes.

(2) If a quadratic function given by

$$f(X) = \frac{1}{2} X^T A X + B^T X + C$$

is minimized sequentially, once along each directions of a set of n mutually conjugate directions, the minimum of the function f will be found at or before the nth step irrespective of the starting point.

POWELL'S ALGORITHM

Let f be a function of n variables which is to be minimized and $u_1, u_2, ..., u_n$ be the axial directions.

In this method, we start from a fixed point, carried out the search sequentially in the direction u_1, $u_2, ..., u_n$ in the first cycle, along $S(p,1), u_2, u_3, ..., u_n, S(p,1)$ in the second cycle and $S(p,2), u_3, u_4,...,u_{n-1}, u_n, S(p,1), S(p,2)$ in the third cycle and so on. Proceed until the minimum point is reached. Here, $S(p,i)'s$ are the pattern search direction which are defined by

$$S(p,i) = X(i) - X(i - n)$$

and n is the number of decision variables.

☞ REMARKS

- The Powell's method is converge in almost two cycles of iterations, *i.e.*, it has quadratic convergence.
- This is also known as Powell's conjugate direction method.
- Since, the no. of cycles n is valid only for quadratic function, it will take generally greater than n cycles for non-quadratic functions.

 SOLVED EXAMPLES

EXAMPLE 1. **Using Powell's method, minimize**
$$f(X) = 2x_1^2 + x_2^2 + 2x_1x_2 + x_1 - x_2$$
starting with $X_1 = (0, 0)$.

SOLUTION. We have
$$f(X) = f(x_1,x_2) = 2x_1^2 + x_2^2 + 2x_1x_2 + x_1 - x_2 \qquad ...(1)$$
and base point (initial or starting point), $X_1 = (0, 0)$

Cycle-1. Here, the search will be made in the direction u_2, u_1, u_2.

Let λ be the step size in the direction u_2 from X_1 to reach X_2 such that

$$f(X_2) = f(X_1 + \lambda_1 u_2) = f\left(\begin{pmatrix} 0 \\ 0 \end{pmatrix} + \lambda_1 \begin{pmatrix} 0 \\ 1 \end{pmatrix} \right) = f \begin{pmatrix} 0 \\ \lambda_1 \end{pmatrix}$$

$$= \lambda_1^2 - \lambda_1 \text{ is minimum}$$

Now, for the minima of f, we must have

$$\frac{df}{d\lambda_1} = 0 \Rightarrow 2\lambda_1 - 1 = 0 \Rightarrow \lambda_1 = \frac{1}{2}$$

and
$$\frac{d^2 f}{d\lambda_1^2} = 2 > 0$$

$$\Rightarrow \quad f \text{ is minimum at } \lambda_1 = \frac{1}{2}$$

Therefore,
$$X_2 = X_1 + \lambda_1 u_2 = (0,0) + \frac{1}{2}(0,1) = \left(0, \frac{1}{2}\right)$$

$$f(X_2) = f\left(0, \frac{1}{2}\right) = -0.25 < f(X_1)$$

Now, moving from X_2 in the direction of $u_1 = (1, 0)$ through step length λ_2 arriving at the point $X_3 = X_2 + \lambda_2 u_1$

We have to find λ_2 so that $f(X_3) = f\left(\lambda_2, \frac{1}{2}\right) = 2\lambda_2^2 + 2\lambda_2 - 0.25$ is minimum.

Now, using the principle of maxima and minima for the minima of f, we must have
$$\frac{df}{d\lambda_2} = 0 \Rightarrow 4\lambda_2 + 2 = 0 \Rightarrow \lambda_2 = -\frac{1}{2}$$

and $\qquad \dfrac{d^2f}{d\lambda_2^2} = 4 > 0$

$\Rightarrow \quad f$ is minimum at $\lambda_2 = -\dfrac{1}{2}$

Therefore, $X_3 = X_2 + \lambda_2 u_1 = \left(0, \dfrac{1}{2}\right) - \dfrac{1}{2}(1,0) = \left(-\dfrac{1}{2}, \dfrac{1}{2}\right)$

and $f(X_3) = -0.75 < f(X_2) = -0.25$

Now, move from X_3 to X_4 taking step length of size λ_3 in the direction of $u_2 = (0, 1)$

$$X_4 = X_3 + \lambda_3(0,1) = \left(-\dfrac{1}{2}, \dfrac{1}{2}\right) + (0, \lambda_3) = \left(-\dfrac{1}{2}, \dfrac{1}{2} + \lambda_3\right)$$

Now, we have to find the value of λ_3 such that the function

$$f(X_4) = f\left(-\dfrac{1}{2}, \dfrac{1}{2} + \lambda_3\right) = \lambda_3^2 - \lambda_3 - 0.75 \text{ is minimum.}$$

For minima of f, we must have

$$\dfrac{df}{d\lambda_3} = 0 \Rightarrow 2\lambda_3 - 1 = 0 \Rightarrow \lambda_3 = \dfrac{1}{2}$$

and $\qquad \dfrac{d^2f}{d\lambda_3^2} = 2 > 0$

$\Rightarrow \quad f$ is minimum at $\lambda_3 = \dfrac{1}{2}$

$\therefore \quad X_4 = \left(-\dfrac{1}{2}, 1\right)$ and $f(X_4) = -1 < f(X_3) = -0.75$

Cycle-2. Let us take $i = 4$

Here, search will be done in the direction $S(p,1)$, u_2, $S(p,2)$

Now, pattern search direction

$$S(p,1) = X_i - X_{i-n} = X_4 - X_2 = \left(-\dfrac{1}{2}, 1\right) - \left(0, \dfrac{1}{2}\right) = \left(-\dfrac{1}{2}, \dfrac{1}{2}\right)$$

From X_4, we move to X_5 by taking step length λ_4 in the direction of $S(p,1) = \left(-\dfrac{1}{2}, \dfrac{1}{2}\right)$

$$X_5 = X_4 + \lambda_4 S(p,1) = \left(-\dfrac{1}{2}, 1\right) + \lambda_4\left(-\dfrac{1}{2}, \dfrac{1}{2}\right)$$

$$= \left(-\dfrac{1}{2}(1 + \lambda_4), 1 + \dfrac{1}{2}\lambda_4\right)$$

Now, we have to find the value of λ_4 so that the function
$$f(X_5) = 0.25\lambda_4^2 - 0.5\lambda_4 - 1 \text{ is minimum.}$$
Now, for the minima of f, we must have

$$\dfrac{df}{d\lambda_4} = 0 \Rightarrow 0.5\lambda_4 - 0.5 = 0 \Rightarrow \lambda_4 = 1$$

and $\qquad \dfrac{d^2f}{d\lambda_4^2} = 0.5 > 0$

$\Rightarrow \quad f$ is minimum at $\lambda_4 = 1$

So, $X_5 = X_4 + \lambda_4 S(p,1) = \left(-\dfrac{1}{2}, 1\right) + \left(-\dfrac{1}{2}\right)\left(\dfrac{1}{2}, \dfrac{1}{2}\right) = \left(-1, \dfrac{3}{2}\right)$

and $f(X_5) = -1.25 < f(X_4)$

Now, from X_5, we move to X_4 by taking step length λ_5 in the direction of $u_2 = (0, 1)$.

$$\therefore \quad X_6 = X_5 + \lambda_5 u_2 = \left(-1, \frac{3}{2}\right) + \lambda_5(0,1) = \left(-1, \frac{3}{2} + \lambda_5\right)$$

We have to find the value of λ_5 for which the function f is minimum.

For the minima of f, we must have

$$\frac{df}{d\lambda_5} = 0 \implies 2\lambda_5 = 0 \implies \lambda_5 = 0$$

and

$$\frac{d^2 f}{d\lambda_5^2} = 2 > 0$$

$$\therefore \qquad X_6 = \left(-1, \frac{3}{2}\right) = X_5$$

Finally, $\lambda_5 = 0$ shows that f can not be minimized in the direction of u_2 and there is no other direction to move.

Hence, Min $f = -1.25$ when $x_1 = -1, x_2 = \dfrac{3}{2}$

10.4 SOME MORE UNCONSTRAINED DIRECT SEARCH METHODS

10.4.1 RANDOM SEARCH METHOD

In these methods, we use the random numbers to find the minimum point. There are following two types of random search method:

   (i) Random Jumping method

   (ii) Random Walk method

(I) RANDOM JUMPING METHOD

In random jumping method, we find the bounds l_i and u_i for each decision variables x_i such that

$$l_i \le x_i \le u_i \quad \forall \, i$$

(l_i is called the lower bound and u_i is upper bound)

Also, we generate a set of n numbers $(r_1, r_2, \ldots, r_n)$ which are uniformly distributed between 0 and 1.

Then to find a point X, we use the following formula

$$X = \begin{bmatrix} x_1 \\ x_2 \\ \vdots \\ x_n \end{bmatrix} = \begin{cases} l_1 + r_1(u_1 - l_1) \\ l_2 + r_2(u_2 - l_2) \\ \vdots \\ l_n + r_n(u_n - l_n) \end{cases}$$

and the value of the function at the point X, i.e., $f(X)$

Repeat the above process to generate a large number of random variable X_i and then evaluate $f(X_i)$ for each i.

Finally, take the smallest value of $f(X_i)$.

(II) RANDOM WALK METHOD

This method is based on generating a sequence of improved approximation to the minimum each derived from the previous approximation.

WORKING PROCEDURE

In this method we proceed as follows:

STEP 1. If X_i is the approximation obtained in the $(i-1)^{\text{th}}$ stage (minimum). Then

$$X_{i+1} = X_i + \lambda u_i \qquad \qquad \ldots(1)$$

where λ is the step size and u_i is the unit vector.

STEP 2.   Find $f_1 = f(X_1)$

STEP 3.   Using $\boldsymbol{u} = \dfrac{1}{(r_1^2 + r_2^2 + \ldots + r_n^2)^{1/2}} \begin{Bmatrix} r_1 \\ r_2 \\ \vdots \\ r_n \end{Bmatrix}$ generate a set of n random numbers $r_1, r_2, \ldots, r_n$

in the interval $[-1, 1]$.

Here, $R = (r_1^2 + r_2^2 + \ldots + r_n^2)^{1/2}$

Discard the random numbers if $R > 1$ and accept if $R \leq 1$.

STEP 4.   Find the new vector and $X = X_1 + \lambda u$ and $f = f(X)$.

STEP 5.   Find the value of f and f_1. Observe that
- ➡ If $f < f_1$, set the new values as $X_1 = Y, f_1 = f$ and set the next value of i.
- ➡ If $f \geq f_1$ go to the next step.

STEP 6.   If $i \leq N$, set $i = i + 1$ and go to step 3, otherwise go to next step.

STEP 7.   Deduce the step length λ as $\dfrac{\lambda}{2}$. If it is less than or equal to ε go to next step, otherwise go to step 3.

STEP 8.   Terminate the procedure by taking
$$X_{\text{opt}} = X_1 \text{ and } f_{\text{opt}} = f_1$$

☛ **REMARKS**
- Random search methods can work even if the objective function is discontinuous and non-differentiable at some of the points.
- These methods can be used to find the global minimum when the objective function has several relative minimum.

10.4.2 GRID SEARCH METHOD

Let l_i and u_i be the lower and upper bounds on the ith decision variables respectively. Then divide (l_i, u_i) into $k_i - 1$ equal parts so that $x_1^{(1)}, x_1^{(2)}, \ldots, x_1^{(k)}$ denote the grid point along x_i. Thus there are $k_1 \cdot k_2 \cdot \ldots \cdot k_n$ grid points in the design space.

> This method involves setting up a suitable grid in the design space, evaluating the objective function at all the grid points and finding the grid point corresponding to the lowest function value.

☛ **REMARKS**
- Clearly grid with $k_i = 3$ and 4 are shown in a 2-dimensional space.
- Grid method can be used to find a good starting point for one of the more efficient methods.

10.5 SOME DESCENT (OR INDIRECT) METHODS

In direct search method, we require only objective function values to find the solution. But we observe that, even the best direct methods can require an excessive number of function evaluations to locate the solution. This combined with the quite natural desire to seek stationary points, motivates us to consider methods that are based on gradient. The methods will be iterative, since the elements of the gradient will in general be non-linear functions of decision variables.

SOME IMPORTANT DEFINITIONS

(1) **Gradient of a function:** Let f be a function. Then gradient of f denoted by ∇f is an n-components vector given by

$$\nabla f = \begin{bmatrix} \partial f / \partial x_1 \\ \partial f / \partial x_2 \\ \vdots \\ \partial f / \partial x_n \end{bmatrix}$$

(2) Direction of steepest ascent: We known that, if we move along the gradient direction from any point to n-dimensional space, the function value increases at the fastest rate. Thus, the gradient direction is known as the direction of steepest ascent.

(3) Direction of steepest descent: Since the gradient vector represents the direction of steepest ascent, the negative of the gradient vector denotes the direction of steepest descent.

☞ REMARKS
- A function decreases most rapidly in the negative direction of the gradient.
- Indirect search methods use the derivatives along with determining the value of the function at search point. These methods, therefore are also called 'gradient methods'.

10.6 STEEPEST DESCENT OR CAUCHY METHOD

In steepest descent method we start with an initial point X_1 (base point) and iteratively move along the steepest descent direction until the optimum point is found.

Here, we use the following working procedure.

WORKING PROCEDURE

Let X_1 be the initial point and $f(x_1, x_2)$ be the function to be minimize, then proceed as follows:

STEP 1. Calculate $f(X_1)$ and search direction $S_1 = -\nabla f(X_1)$.

STEP 2. Find the optimum step length λ_1 in this direction to arrive at the point $X_2 = X_1 + S_1\lambda_1$ so that
$$f(X_2) < f(X_1)$$

Proceeding in this way until one of the following conditions is satisfied and then terminate the process.

  (i) $\left|\dfrac{\partial f}{\partial x_i}\right| < \varepsilon, \varepsilon > 0$ is a small quantity.

  (ii) Change in decision vectors in the consecutive iteration is very small, *i.e.,* $|X_{i+1} - X_i| < \varepsilon$.

  (iii) The relative change in the value of f at two consecutive steps is very small, *i.e.,*
$$\left|\frac{f(X_{i+1}) - f(X_i)}{f(X_i)}\right| < \varepsilon$$

☞ REMARKS
- This method is the best unconstrained minimization method.

> The direction of steepest ascent generally varies from point to point and if we make infinitely small moves along the direction of the steepest ascent, the path will be a curved line.

SOLVED EXAMPLES

EXAMPLE 1. *Using steepest descent method, minimize the function*
$$f(x_1, x_2) = 2x_1^2 + x_2^2 + 2x_1 x_2 + x_1 - x_2$$
starting from the point $X_1 = (0, 0)$ $\quad$ *($\varepsilon = 0.01$)*

SOLUTION. We have
$$f(x_1, x_2) = 2x_1^2 + x_2^2 + 2x_1 x_2 + x_1 - x_2 \qquad \dots(1)$$
Differentiating (1) partially w.r.t. x_1 and x_2 we get
$$\frac{\partial f}{\partial x_1} = 4x_1 + 2x_2 + 1$$
$$\frac{\partial f}{\partial x_2} = 2x_1 + 2x_2 - 1$$

Therefore, $\nabla f(x) = \left(\dfrac{\partial f}{\partial x_1}, \dfrac{\partial f}{\partial x_2} \right) = (4x_1 + 2x_2 + 1, 2x_1 + 2x_2 - 1)$ ...(2)

Since it is given that $X_1 = (x_1, x_2) = (0, 0)$. Then from (2)

$$-\nabla f(X_1) = (-1, 1)$$

Iteration-1. We have to find the value of λ which minimize the function

$$f[X_1 - \lambda_1 \nabla f(X_1)] = f[(0,0) + \lambda_1(-1,1)] = f(-\lambda_1, \lambda_1)$$

$$= 2\lambda_1^2 + \lambda_1^2 - 2\lambda_1^2 - \lambda_1 - \lambda_1 = \lambda_1^2 - 2\lambda_1$$

Using the principle of maxima and minima for the minimum of f, we must have

$$\frac{\partial f}{\partial \lambda_1} = 0 \quad \Rightarrow \quad 2\lambda_1 - 2 = 0 \quad \Rightarrow \quad \lambda_1 = 1$$

and $\quad \dfrac{\partial^2 f}{\partial \lambda_1^2} = 2 > 0$

$\Rightarrow \quad f$ is minimum at $\lambda_1 = 1$

So, the new point X_2 is given by

$$X_2 = X_1 - \lambda_1 \nabla f(X_1) = (-1, 1)$$

Then from (2) $\quad \nabla f(X_2) = (-1, -1)$

$\Rightarrow \quad\quad\quad\quad |\nabla f(X_2)| \nleq \varepsilon$

$\Rightarrow \quad X_2$ is not an optimal point.

Iteration-2. We have to find λ_2 which minimize the function

$$f(X_3) = f[X_2 - \lambda_2 \nabla f(X_2)]$$

$$= f[(-1,-1) - \lambda_2(-1,-1)] = f(-1+\lambda_2, 1+\lambda_2)$$

$$= 5\lambda_2^2 - 2\lambda_2 - 1 \quad\quad\quad\quad \text{(Using (1))}$$

i.e., we have to find λ_2 which minimizes $f(X_3)$

For the minimum of f, we must have

$$\frac{\partial f}{\partial \lambda_2} = 0 \quad \Rightarrow \quad 10\lambda_2 - 2 = 0 \quad \Rightarrow \quad \lambda_2 = \frac{1}{5}$$

and $\quad \dfrac{\partial^2 f}{\partial \lambda_2^2} = 10 > 0$

$\Rightarrow \quad f$ is minimum at $\lambda_2 = \dfrac{1}{5}$

$\therefore \quad X_3 = X_2 - \lambda_2 \nabla f(X_2) = \left(-1 + \dfrac{1}{5}, 1 + \dfrac{1}{5} \right) = (-0.8, 1.2)$

$\Rightarrow \quad \nabla f(X_3) = (0.2, -0.2)$ \quad\quad\quad\quad (By (2))

$\Rightarrow \quad |\nabla f(X_3)| \nleq 0.01$

$\Rightarrow \quad X_3$ is not an optimal point.

Iteration-3. We have to find λ_3 which minimize the function

$$f(X_4) = f[X_3 - \lambda_3 \nabla f(X_3)] = f[(-0.8, 1.2) - \lambda_3(0.2, -0.2)]$$

$$= f(-0.8 - 0.2\lambda_3, 1.2 + 0.2\lambda_3)$$

$$= 0.04\lambda_3^2 - 0.08\lambda_3 - 1.2 \quad\quad\quad\quad \text{(By (1))}$$

Now, for the minimum of f, we must have

$$\frac{\partial f}{\partial \lambda_3} = 0 \quad \Rightarrow \quad 0.08\lambda_3 - 0.08 = 0 \quad \Rightarrow \quad \lambda_3 = 1$$

and $\quad \dfrac{\partial^2 f}{\partial \lambda_3^2} = 0.08 > 0$

$\Rightarrow \quad f$ is minimum at $\lambda_3 = 1$

Therefore, $X_4 = X_3 - \lambda_3 \nabla f(X_3) = (-1, 1.4)$

and $\qquad f(X_4) = -1.24$ (By (1))

$\Rightarrow \qquad \nabla f(X_4) = (-0.2, -0.2)$ (By (2))

$\Rightarrow \qquad |\nabla f(X_4)| \nleqslant 0.01$

$\Rightarrow \qquad X_4$ is not an optimal point.

Iteration-4. We have to find λ_4 which minimize the function
$$f(X_5) = f[X_4 - \lambda_4 \nabla f(X_4)]$$

Here, $\qquad X_5 = X_4 - \lambda_4 \nabla f(X_4)$
$$= (-1, 1.4) - 0.2(-0.2, -0.2) = (-0.96, 1.44)$$

$\Rightarrow \qquad f(X_5) = 0.2\lambda_4^2 - 0.08\lambda_4 - 1.24$ (By (1))

Now, for the minimum of f, we must have

$$\dfrac{\partial f}{\partial \lambda_4} = 0 \quad \Rightarrow \quad 0.4\lambda_4 - 0.08 = 0 \quad \Rightarrow \quad \lambda_4 = 0.2$$

and $\quad \dfrac{\partial^2 f}{\partial \lambda_4^2} = 0.4 > 0$

$\Rightarrow \quad f$ is minimum at $\lambda_4 = 0.2$

Since, $f(X_5) = 0.2\lambda_4^2 - 0.08\lambda_4 - 1.24$

$\Rightarrow \qquad \nabla f(X_5) = (0.04, -0.04)$

$\Rightarrow \qquad |\nabla f(X_5)| \nleqslant 0.01$

$\Rightarrow \qquad X_5$ is not an optimal point.

Iteration-5. Proceed same as above, we have to minimize

$$f(X_6) = f[X_5 - \lambda_5 \nabla f(X_5)] = 0.0016\lambda_5^2 - 0.0032\lambda_5 - 1.248$$

Now, for the minimum of f, we must have

$$\dfrac{\partial f}{\partial \lambda_5} = 0 \quad \Rightarrow \quad 0.0032\lambda_5 - 0.0032 = 0 \quad \Rightarrow \quad \lambda_5 = 1$$

and $\quad \dfrac{\partial^2 f}{\partial \lambda_5^2} = 0.0032 > 0$

$\Rightarrow \quad f$ is minimum at $\lambda_5 = 1$

Therefore, $X_6 = X_5 - \lambda_5 \nabla f(X_5) = (-1, 1.48)$

$\Rightarrow \qquad \nabla f(X_6) = (-0.04, 0.04)$

$\Rightarrow \qquad |\nabla f(X_6)| \nleqslant 0.01$

$\Rightarrow \qquad X_6$ is not an optimal point.

Iteration-6. We have to minimize

$$f(X_7) = f[X_6 - \lambda_6 \nabla f(X_6)] = 0.0016\lambda_6^2 - 1.2496$$

For the minimum of f, we must have

$$\dfrac{\partial f}{\partial \lambda_6} = 0 \quad \Rightarrow \quad 0.0032\lambda_6 = 0 \quad \Rightarrow \quad \lambda_6 = 0$$

and $\quad \dfrac{\partial^2 f}{\partial \lambda_6^2} = 0.0032 > 0$

$\Rightarrow \quad f$ is minimum at $\lambda_6 = 0$

$\therefore \quad X_7 = f[X_6 - \lambda_6 \nabla f(X_6)] = (-1, 1.48)$

Also, $\lambda_6 = 0$ implies that, further improvement in f is not possible.

Hence, $\min f = f(X_6) = f(X_7) = -1.2496 \approx -1.25$ when $x_1 = -1$ and $x_2 = 1.48$

10.7 FLETCHER-REEVES METHOD

We know that any minimization method that make use of the conjugate directions is quadratically convergent, which ensure that the method will minimize a quadratic function in n steps or less. Since, any general function can be approximated by a quadratic near the optimum point, any quadratically convergent method is expected to find the optimum point in a finite number of iteration. The construction of conjugate directions and development of the Fletcher-Reeves method are given below.

WORKING PROCEDURE

The iterative procedure of Fletcher-Reeves method consist the following steps:

STEP 1. Take any arbitrary small point X_1.

STEP 2. Obtain the first search direction using $S_1 = -\nabla f(X_1) = -\nabla f_1$

STEP 3. Find the next point X_2 such that $X_2 = X_1 + \lambda_1^* S_1$ where λ_1^* is the optimum step length in S_1-direction. Set $i = 2$ and go to the next step.

STEP 4. Obtain $\nabla f(X_i)$ and set

$$S_i = -\nabla f_i + \frac{|\nabla f_i|^2}{|\nabla f_{i-1}|^2} S_{i-1}$$

STEP 5. Find the optimum step length λ_i^* in S_i-direction and find the new point

$$X_{i+1} = X_i + \lambda_i^* S_i$$

STEP 6. Test the optimality of the point X_{i+1}.

If X_{i+1} is optimum, stop the process otherwise set $i = i + 1$ and go to step 4.

☛ REMARKS

- The Fletcher-Reeves method is superior to the steepest descent method and pattern search method.
- The Fletcher-Reeves method was originally proposed by Hestenes and Stiefel as a method of solving the system of linear equation derived from the stationary conditions of a quadratic.

SOLVED EXAMPLES

__EXAMPLE 1.__ *Using Fletcher-Reeves method, minimize the function*

$$f(x_1, x_2) = 2x_1^2 + x_2^2 + 2x_1 x_2 + x_1 - x_2$$

starting from the point $X_1 = (x_1, x_2) = \begin{bmatrix} 0 \\ 0 \end{bmatrix}$.

__SOUTION.__ We have

$$f(x_1, x_2) = 2x_1^2 + x_2^2 + 2x_1 x_2 + x_1 - x_2 \qquad \qquad ...(1)$$

$$X_1 = \begin{bmatrix} 0 \\ 0 \end{bmatrix}$$

Iteration-1. Now $\dfrac{\partial f}{\partial x_1} = 1 + 4x_1 + 2x_2$

and $\dfrac{\partial f}{\partial x_2} = -1 + 2x_1 + 2x_2$

$\therefore \quad \nabla f = (1 + 4x_1 + 2x_2, -1 + 2x_1 + 2x_2)$

Now, $\quad \nabla f(X_1) = \nabla f(0,0) = \begin{bmatrix} 1 \\ -1 \end{bmatrix}$

$\Rightarrow \qquad\qquad S_1 = -\nabla f_1 = \begin{bmatrix} -1 \\ 1 \end{bmatrix}$

Now, we have to find optimal step length λ_1^* along S_1

For this we have to minimize $f(X_1 + \lambda_1 S_1)$ w.r.t. λ_1

Here, we have $\quad f(X_1 + \lambda_1 S_1) = f(-\lambda_1, \lambda_1) = \lambda_1^2 - 2\lambda_1$ $\hfill$ (By (1))

Now for the minimum of f, we must have

$$\frac{df}{d\lambda_1} = 0 \quad \Rightarrow \quad 2\lambda_1 - 2 = 0 \quad \Rightarrow \quad \lambda_1^* = 1$$

and $\quad \dfrac{d^2 f}{d\lambda_1^2} = 2 > 0$

$\Rightarrow \quad f$ is minimum at $\lambda_1^* = 1$

Therefore, $\quad X_2 = X_1 + \lambda_1^* S_1 = \begin{bmatrix} 0 \\ 0 \end{bmatrix} + 1\begin{bmatrix} -1 \\ 1 \end{bmatrix} = \begin{bmatrix} -1 \\ 1 \end{bmatrix}$

Iteration-2.

$\because \qquad\qquad \nabla f_2 = \nabla f(X_2) = \begin{bmatrix} -1 \\ 1 \end{bmatrix}$

Now, $\qquad\qquad S_2 = -\nabla f_2 + \dfrac{|\nabla f_2|^2}{|\nabla f_1|^2} S_1$

Here, $|\nabla f_1|^2 = 2$ and $|\nabla f_2|^2 = 2$

$\therefore \qquad\qquad S_2 = -\begin{bmatrix} -1 \\ 1 \end{bmatrix} + \left(\dfrac{2}{2}\right)\begin{bmatrix} -1 \\ 1 \end{bmatrix} = \begin{bmatrix} 0 \\ 2 \end{bmatrix}$

Now, we have to minimize $f(X_2 + \lambda_2 S_2)$

Here, we have $\qquad f = f(-1, 1 + 2\lambda_2)$

$$= -1 - (1 + 2\lambda_2) + 2 - 2(1 + 2\lambda_2) + (1 + 2\lambda_2)^2$$
$$= 4\lambda_2^2 - 2\lambda_2 - 1$$

For the minimum of f, we must have

$$\frac{df}{d\lambda_2} = 0 \quad \Rightarrow \quad 8\lambda_2 - 2 = 0 \quad \Rightarrow \quad \lambda_2^* = \frac{1}{4}$$

and $\quad \dfrac{d^2 f}{d\lambda_2^2} = 8 > 0$

$\Rightarrow \quad f$ is minimum at $\lambda_2^* = \dfrac{1}{4}$

Therefore, $\quad X_3 = X_2 + \lambda_2^* S_2 = \begin{bmatrix} -1 \\ 1 \end{bmatrix} + \dfrac{1}{4}\begin{bmatrix} 0 \\ 2 \end{bmatrix} = \begin{bmatrix} -1 \\ 1.5 \end{bmatrix}$

Iteration-3.

$$\nabla f_3 = \nabla f(X_3) = \begin{bmatrix} 0 \\ 0 \end{bmatrix}$$

$$|\nabla f_2|^2 = 2 \text{ and } |\nabla f_3|^2 = 0$$

Therefore, $\quad S_3 = -\nabla f_3 + \dfrac{(|\nabla f_3|^2)}{|\nabla f_2|^2} \cdot S_2 \quad = -\begin{bmatrix} 0 \\ 0 \end{bmatrix} + \begin{bmatrix} 0 \\ 2 \end{bmatrix}\begin{bmatrix} 0 \\ 0 \end{bmatrix} = \begin{bmatrix} 0 \\ 0 \end{bmatrix}$

$\Rightarrow$ There is no search direction to reduce f further.

Hence, X_3 is the optimal point and minimum of $f(x)$ can be obtained from (1) by putting $x_1 = -1$ and $x_2 = 1.5$.

EXERCISE 10.1

1. Using univariate method minimize the function

$$f(X) = f(x_1, x_2) = x_1^2 + x_2^2 - 2x_1 - 4x_2$$

starting with $X_1 = \begin{bmatrix} 0 \\ 0 \end{bmatrix}$.

2. Using Hooke and Jeeve's method, minimize the function

$$f(X) = 3x_1^2 + x_2^2 - 2x_1 x_2 - 4x_1 - 3x_2$$

taking the starting base point as $X_1 = (0, 0)$ and $\Delta x_1 = \Delta x_2 = 1$.

3. Using Powel's method minimize the function

$$f(X) = f(x_1, x_2) = 4x_1^2 + 3x_2^2 - 5x_1 x_2 - 8x_1$$

taking $X_1 = (0, 0)$ as starting point.

4. Using Steepest-descent method, minimize the function

$$f(x_1, x_2) = 2x_1^2 + x_2^2$$

starting from the point $X_1 = (1, 2)$.

ANSWERS

1. $x_1 = 1, x_2 = 2$

2. $x_1 = 1.75, x_2 = 3.25, f(X) = -8.375$

3. $x_1 = 48/23, x_2 = 40/23$

4. $f(X) = 1.0336$

11 Constrained Optimization Techniques

11.1 INTRODUCTION

In previous chapter, we have studied some search unconstrained optimization technique. This section deals with the techniques that are applicable to the solution of the constrained optimization problems. The general form of constrained non-linear programming problem is given as below

$$\text{Min. } f(X)$$

subject to the constraints

$$\left. \begin{array}{ll} g_i(X) \leq 0 \; ; & i = 1, 2, ..., n \\ h_k(X) = 0; & k = 1, 2, ..., p \end{array} \right] \qquad \qquad ...(1)$$

In this section we shall discuss some methods to solve the constrained non-linear programming problems given by (1).

11.2 CHARACTERISTICS OF A CONSTRAINED PROBLEM

Following are the main characteristics of constrained non-linear programming problem.

(i) The constraints may have no effect on the optimum point.

(ii) The optimum solution occurs on a constraints boundary.

(iii) The negative of the gradient must be expressible as a positive linear combinations of the gradient of the active constraints.

(iv) If the objective function has two or more unconstrained local minima, then the constrained problem may have multiple minima.

(v) If the objective function has a single unconstrained minimum, then the constraints may introduce multiple local minima.

11.3 CONSTRAINED MINIMIZATION METHODS

There are many methods available for the solution of a constrained non-linear programming problems which are classified into following two categories :

(i) Direct methods

(ii) Indirect methods

11.4 DIRECT METHODS

In solving non-linear programming problem, if the function to be minimized is not differentiable but requires computational work, then direct methods are easy to use.

Let us discuss some direct methods one by one.

11.4.1 RANDOM SEARCH METHOD

In this method, we can use methods of unconstrained minimization with minor changes, to solve constrained optimization problem.

In this method we have the following procedure.

WORKING PROCEDURE

Here we use the following steps:

STEP 1. Set a trial design vector using one random number for each design variable.

STEP 2. Verify, whether the constraints are satisfied by the trial design vector (of step-1)
(if not, continue generating new trial vectors)

STEP 3. If all the constraints are satisfied retain the current trial vector as the best design if it gives a reduced objective function value compared to the previous best available design. Otherwise go to the step-1 for new trial design vector.

At the end of generating a specified maximum number of trials design vectors is taken as the solution of the constrained optimization problem.

☞ **REMARK**

- The random search methods are not efficient compared to the other methods.

11.4.2 COMPLEX METHOD

In this method we assume that an initial feasible point X_1, which satisfies all the constraints is available. This method is an extension upon the simplex method of unconstrained minimization to solve constrained optimization problem.

It deals with the constrained optimization problem of the type given below:

$$\text{Minimize } Z = f(X) \qquad \qquad ...(1)$$

subject to the constraints

$$g_j(X) \le 0, \ j = 1, 2, ..., m \qquad \qquad ...(2)$$

$$X = (x_1, x_2, ..., x_n)^T$$

and x_i (l) and $x_i(u)$ are the lower and upper bounds of x_i respectively so that

$$x_i(l) \le x_i \le x_i \ (u) \ \ \forall \ i = 1, 2, ..., n \qquad \qquad ...(3)$$

The condition given in (3) are called "side constraints".

GENERAL PROCEDURE

Given the set of points, the objective function is evaluated at each point and the point corresponding to the highest value is rejected. A new point is generated by reflecting the rejected point a certain distance through the centroid of the remaining point. At the new point, the performance function and the constraints are evaluated with the following alternatives :

(i) The new point is feasible and its function value is not the highest of the set of points. Then select the point that does not correspond to the highest and continue with a reflection.

(ii) The new value is feasible and its function value is the highest of the current set of points. Rather than reflecting back again, retract the point by half the distance to the previously calculated centroid.

(iii) The new point is infeasible. Retract the point by half the distance to the previously calculated centroid.

☞ **REMARK**

- The search is terminated when the pattern of points has shrunk so that the points are sufficiently close together and when the differences between the function values at the points becomes small enough.

➡ For a minimization problem in n variables, if we consider k points where $k \leq n + 1$, then the figure formed on joining them is called '**complex**'.

➡ If we consider the minimization problems in two variables, then in 2-dimension, it will have $4(n=2)$ vertices. In any dimension n, we will have $2n$ as an even number. So, when $n = 2$, k should be 4 and hence take $k = 2n = 4$. These four points will be the vertices of the complex.

WORKING PROCEDURE

To solve the problems of two variables, we use the following steps :

STEP 1. Let one point X_1 be given, obtain remaining $(2n–4) = 3$ points X_2, X_3 and X_4 are at a time by using random number r_{ij}, $0 < r_{ij} < 1$.

STEP 2. Calculate $x_{ij} = x_i(l) + r_{i,j} [(x_i(u) - x_i(l)]$

when $x_{ij} = i^{th}$ component of the function X_j

$$i = 1, 2 \text{ and } j = 2, 3, 4$$

The obtained points X_2, X_3 and X_4 satisfies the constraints (3) but may not satisfy all the constraints given by (2).

If, X_4 is not satisfying all the constraints in (2), then obtained a new point $X_4^{(1)}$ by moving X_4 halfway towards the centroid

$$X_0 = \frac{1}{3}(X_1 + X_2 + X_3)$$

of the remaining points X_1, X_2 and X_3

i.e., $$X_4^{(1)} = \frac{X_0 + X_j}{2}$$

Now check whether $X_4^{(1)}$ satisfies all the constraints in (2) or not. If not obtain, then another point $X_4^{(2)}$ by moving $X_4^{(1)}$ halfway towards the centroid X_0.

Proceed in the same manner until a feasible point X_4 satisfying (2), is obtained.

⇒ Finally four points X_1, X_2, X_3, X_4 all satisfying the constraints given by (2). These points are the vertices of the starting complex.

STEP 3. Calculate value of the function at X_1, X_2, X_3 and X_4 i.e., $f(X_1), f(X_2), f(X_3)$ and $f(X_4)$.

Find the largest and smallest value among them. Find a new point by using (process of reflection)

$$X_r = (1+\alpha)X_0 - \alpha X_h$$

where $\alpha \leq r$ and X_0 is the centroid of all vertices except X_h

i.e., $$X_0 = \frac{(X_1 + X_2 + X_3)}{3}$$

STEP 4. Now we check the feasibility of X_r.

(i) If X_r is feasible and $f(X_r) < f(X_h)$, then X_h is replaced by X_r and carry on step-3.

(ii) If $f(X_r) \leq f(X_h)$, then a new trial point X_r is found by taking a new value of

$$\alpha = \left(\frac{\text{previous value of } \alpha}{2} \right)$$

and tested further until the condition $f(X_r) < f(X_h)$ is satisfied.

Proceeding in this way we will make the value of α smaller and smaller.

(iii) If the value of X_r does not satisfy $f(X_r) < f(X_h)$ in any way, neglect the whole reflection process and start a new reflection process by taking X_h which gives the second largest value of the function.

The procedure terminate when the distance between any two vertices among X_1, X_2, X_3 and X_4 becomes smaller than the prescribed value of ε.

11.4.3 SOME FACTS ABOUT COMPLEX METHOD

(1) If the feasible region is non convex, there is no guarantee that the centroid of all feasible points is also feasible. If the centroid is not feasible then we can not apply the above procedure to find the new point.

(2) Complex method can not be used to solve problem having equality constraints.

(3) This method becomes inefficient rapidly as the number of variables increases.

SOLVED EXAMPLES

EXAMPLE 1. *Using complex method, minimize the function*
$$f(X) = f(x_1, x_2) = (x_1-1)^2 + (x_2-1.5)^2 - 0.25$$
subject to the constraints
$$x_1 + x_2 \le 4, \ 0 \le x_1 \le 2 \ and \ 1 \le x_2 \le 3$$
Starting with $X_1 = \begin{pmatrix} 0.7 \\ 1.1 \end{pmatrix}$

SOLUTION. We have
$$f(X) = f(x_1, x_2) = (x_1-1)^2 + (x_2-1.5)^2 - 0.25 \qquad ...(1)$$
$$g(X) = x_1 + x_2 - 4 \le 0$$
and side constraints are given by
$$0 \le x_1 \le 2 \ and \ 1 \le x_2 \le 3$$
Let X_1, X_2, X_3 and X_4 be four points such that $X_1 = \begin{pmatrix} 0.7 \\ 1.1 \end{pmatrix}$

and choose the random numbers
$$r_{1,2} = 0.4, r_{1,3} = 0.6, \ r_{1,4} = 0.8$$
$$r_{2,2} = 0.5, r_{2,3} = 0.7, \ r_{2,4} = 0.9$$
Now, we have to find the value of X_2, X_3 and X_4 by using the following formula
$$x_{i,j} = x_i(l) + r_{i,j}[x_i(u) - x_i(l)] \ \ i = 1, 2, \ j = 1, 2, 3, 4$$
where $x_{i,j}$ is the ith component of vector X_j.
Clearly $x_1(l) = 0$ and $x_1(u) = 2$
and $x_2(l) = 1$ and $x_2(u) = 3$
Therefore, we have
$$x_{1,2} = x_1(l) + r_{1,2}[x_1(u) - x_1(l)] = 0.8$$
$$x_{1,3} = x_1(l) + r_{1,3}[x_1(u) - x_1(l)] = 1.2$$
$$x_{1,4} = x_1(l) + r_{1,4}[x_1(u) - x_1(l)] = 1.6$$
$$x_{2,2} = x_2(l) + r_{2,2}[x_2(u) - x_2(l)] = 2.0$$
$$x_{2,3} = x_2(l) + r_{2,3}[x_2(u) - x_2(l)] = 2.4$$
$$x_{2,4} = x_2(l) + r_{2,4}[x_2(u) - x_2(l)] = 2.8$$
Therefore, the vertices of the first complex are given by
$$X_1 = \begin{pmatrix} 0.7 \\ 1.1 \end{pmatrix}, X_2 = \begin{pmatrix} 0.8 \\ 2 \end{pmatrix}, X_3 = \begin{pmatrix} 1.2 \\ 2.4 \end{pmatrix} \ and \ X_4 = \begin{pmatrix} 1.6 \\ 2.8 \end{pmatrix}$$

So $\qquad g(X_1) = x_1 + x_2 = 0.7 + 1.1 = 1.8 \le 4$

Similarly $\qquad g(X_2) = 2.8 \le 4$

$\qquad\qquad\qquad g(X_3) = 3.6 \le 4$

and $\qquad\qquad g(X_4) = 4.8 > 4$

$\Rightarrow$ $g(X)$ is not satisfies at X_4

So, we have to replace it by some point in the feasible region as follows.

The centroid of the vertices be given by

$$X_0 = \frac{X_1 + X_2 + X_3}{3} = \begin{pmatrix} 0.9 \\ 1.83 \end{pmatrix}$$

New value of X_4 i.e., $\qquad {}^{(1)} = \dfrac{X_0 \quad X_4}{} = \begin{pmatrix} 1.25 \\ 2.315 \end{pmatrix}$

and $\qquad\qquad g\left(X_4^{(1)}\right) = 3.565 \le 4$

$\Rightarrow X_4^{(1)}$ lies in the feasible region and hence initial complex has the vertices given by

$$X_1 = \begin{pmatrix} 0.7 \\ 1.1 \end{pmatrix}, X_2 = \begin{pmatrix} 0.8 \\ 2 \end{pmatrix}, X_3 = \begin{pmatrix} 1.2 \\ 2.4 \end{pmatrix} \text{ and } X_4 = X_4^{(1)} = \begin{pmatrix} 1.25 \\ 2.315 \end{pmatrix}$$

$\Rightarrow \qquad f(X_1) = 0, \; f(X_2) = 0.04, \; f(X_3) = 0.60, \; f\left(X_4^{(1)}\right) = 0.4767,$ [Using (1)]

Here, $f(X_3) = 0.60$ gives the maximum value and $f(X_1) = 0$ is the minimum value.

Let us take $X_3 = X_h$ with $f(X_h) = 0.6$ and $X_1 = X(l)$ and $f(X(l)) = 0$

Then, new centroid X_0 is given by

$$X_0 = \frac{X_1 + X_2 + X_4^{(1)}}{3} = \begin{pmatrix} 0.917 \\ 1.805 \end{pmatrix}$$

$\Rightarrow \qquad f(X_0) = 0.15$ $\qquad\qquad\qquad$ [Using (1)]

Now, $f(X_0) < f(X_h) \Rightarrow f(X)$ is a decreasing function from $X_h \; (= X_3)$ towards X_0.

To find X_r, we use the following formula

$$X_r = (1+\alpha)X_0 - \alpha X_h$$

Here $\alpha = 1$ so $X_r = 2X_0 - X_h \begin{pmatrix} 0.634 \\ 1.21 \end{pmatrix}$ and $f(X_r) = -0.034944$

Since X_r is feasible and $f(X_r) < f(X_h)$, we proceed further for the different values of α as given below:

(i) Taking $\alpha = 0.1$; then, we have

$$X_r^{(1)} = 1.1X_0 - 0.1X_h = (1.1)\begin{pmatrix} 0.917 \\ 1.805 \end{pmatrix} - 0.1\begin{pmatrix} 1.2 \\ 2.4 \end{pmatrix} = \begin{pmatrix} 0.8887 \\ 1.7455 \end{pmatrix}$$

and $f\left(X_r^{(1)}\right) = -0.1773$

(ii) Taking $\alpha = 0.2$; then, we have

$$X_r^{(2)} = 1.2X_0 - 0.2X_h = (1.2)\begin{pmatrix} 0.917 \\ 1.805 \end{pmatrix} - 0.2\begin{pmatrix} 1.2 \\ 2.4 \end{pmatrix} = \begin{pmatrix} 0.8604 \\ 1.686 \end{pmatrix}$$

and $f\left(X_r^{(2)}\right) = -0.1959$

(iii) Taking $\alpha = 0.3$; then, we have

$$X_r^{(3)} = 1.3X_0 - 0.3X_h = (1.3)\begin{pmatrix} 0.917 \\ 1.805 \end{pmatrix} - 0.3\begin{pmatrix} 1.2 \\ 2.4 \end{pmatrix} = \begin{pmatrix} 0.8321 \\ 1.6265 \end{pmatrix}$$

and $f\left(X_r^{(3)}\right) = -0.2058$

(iv) Taking $\alpha = 0.4$; then, we have
$$X_r^{(4)} = 1.4 X_0 - 0.4 X_h = (1.4)\begin{pmatrix} 0.917 \\ 1.805 \end{pmatrix} - 0.4\begin{pmatrix} 1.2 \\ 2.4 \end{pmatrix} = \begin{pmatrix} 0.8038 \\ 1.567 \end{pmatrix}$$
and $f\left(X_r^{(4)}\right) = -0.2070$

(v) Taking $\alpha = 0.5$; then, we have
$$X_r^{(5)} = 1.5 X_0 - 0.5 X_h = (1.5)\begin{pmatrix} 0.917 \\ 1.805 \end{pmatrix} - 0.5\begin{pmatrix} 1.2 \\ 2.4 \end{pmatrix} = \begin{pmatrix} 0.7755 \\ 1.5075 \end{pmatrix}$$
and $f\left(X_r^{(5)}\right) = -0.1995$

We observe that, the values of f continue to decrease upto $X_r^{(4)}$ replace $X_h = X_3$ with the maximum value of $X_r^{(4)}$ to obtain the new complex with vertices given by

$$X_1 = \begin{pmatrix} 0.7 \\ 1.1 \end{pmatrix} \text{ and } f(X_1) = 0$$

$$X_2 = \begin{pmatrix} 0.8 \\ 1.1 \end{pmatrix} \text{ and } f(X_2) = 0.04$$

$$X_3 = X_r^{(4)} = \begin{pmatrix} 0.8038 \\ 1.567 \end{pmatrix} \text{ and } f(X_3) = f\left(X_r^{(4)}\right) = -0.207016$$

and $\quad X_4 = \begin{pmatrix} 1.25 \\ 2.315 \end{pmatrix}$ and $f(X_4) = f\left(X_4^{(1)}\right) = 0.476725$

Therefore,
$$f(X_4) = f\left(X_4^{(1)}\right) = 0.476725 \text{ gives the maximum values}$$

and $f(X_3) = f\left(X_r^{(4)}\right) = -0.207016$ gives the minimum value.

Thus $\quad X_h = X_4 = \begin{pmatrix} 1.25 \\ 2.315 \end{pmatrix}$ and $f(X_4) = 0.476725$

and $\quad X(l) = X_3 = X_r^{(4)} = \begin{pmatrix} 0.8038 \\ 1.567 \end{pmatrix}$ and $f(X_3) = f(X(l)) = -0.207016$

The centroid $X_0 = \dfrac{(X_1 + X_2 + X_3)}{3} = \begin{pmatrix} 0.7679 \\ 1.557 \end{pmatrix}$ with $f(X_0) = -0.19303$

$\Rightarrow \quad |f(X(l)) - f(X_0)| = 0.014$

If the desired accuracy is $\varepsilon = 0.01$, then the above solution is accepted and hence min. $f(X) = -0.2070$ at X_3 i.e., when $X_1 = 0.8038$ and $X_2 = 1.5557$.

11.4.4 METHOD OF FEASIBLE DIRECTIONS: ZOUTENDIJK'S METHOD

In the methods of feasible directions, we select a starting point satisfying all the given constraints and move to a better point by using the following iterative scheme
$$X_{i+1} = X_i + \lambda S_i$$
Here, $\qquad X_i =$ starting point (base point) for the ith iteration
$\qquad S_i =$ direction of movement
$\qquad \lambda =$ step length (distance of movement)
$X_{i+1} =$ Final point obtained at the end of ith iteration

To find the search direction S_i, we have the following point keep in mind

Prop. (i) a small move in that direction violates no constraints

Prop. (ii) the value of the objective function can be reduced in that direction.

The new point X_{i+1} is taken as the starting point for the next iteration and we will repeat the whole procedure until a point is obtained such that no direction satisfying both (i) and (ii). Such a point denotes the constrained local minimum of the given problem.

☛ REMARKS
- A direction satisfying the prop (i) is called **feasible**.
- A direction satisfying both the above properties ((i) and (ii)) is called a **usable feasible direction**.

11.4.5 ZOUTENDIJK'S METHOD

In this method, the usable feasible direction is taken as the negative of the gradient direction if the initial point of the iteration lies in the interior of the feasible region.

But, if the initial point lies on the boundary of the feasible region, some constraints will be active.

WORKING PROCEDURE

Let us consider the problem given by

$$\text{Minimize } f(X) \qquad \qquad \text{...(1)}$$

subject to the constraints

$$g_j(X) \le 0, \ j = 1, 2, ..., m \qquad \qquad \text{...(2)}$$
$$X = (x_1, x_2, ..., x_m)^T$$

To solve the above problem, we use the following steps:

STEP 1. Choose the starting point X_i ($i = 1$, for starting point) which satisfies all the constraints in (2).

STEP 2. Move to a better point X_{i+1} using the following formula

$$X_{i+1} = X_i + \lambda_i S_i$$

(Better point is the point where value of the function in (1) is less than it has at X_i)

where

$X_i =$ starting point (base point) for i^{th} iteration

$S_i =$ direction to move

$\lambda_i =$ step length (in the direction S_i)

$X_{i+1} =$ Point obtained at the end of ith iteration

STEP 3. Find the search direction such that

(i) even a small movement in that direction violates none of the constraints in (2).

(ii) the value of the function $f(X)$ in (1) decreases.

STEP 4. Test for the convergence. If $\left| \dfrac{f(X_i) - f(X_{i+1})}{f(X_i)} \right| \le \varepsilon_1$ and $|| X_i - X_{i+1} || \le \varepsilon_2$ terminate the iteration by taking $X_{\text{opt}} = X_{i+1}$.

☛ REMARKS
- Care should be taken to choose λ_i such that the new point X_{i+1} is a feasible region.
- For feasible direction:

$$\left[\frac{d}{d\lambda} [g_j(X_i + \lambda_i S_i)] \right]_{\lambda_i = 0} = S_i^T \nabla g_j(X_i) \le 0$$

- For usable direction:

$$\left[\frac{d}{d\lambda} [f(X_i + \lambda_i S_i)] \right]_{\lambda_i = 0} = S_i^T \nabla f(X_i) < 0$$

and

$$\left[\frac{d}{d\lambda} [g_j(X_i + \lambda_i S_i)] \right]_{\lambda_i = 0} = S_i^T \nabla g_j(X_i) \le 0$$

SOLVED EXAMPLES

EXAMPLE 1. *Using Zoutendijk's method minimize the function*

$$f(X) = x_1^2 + x_2^2 - 2x_1 - 3x_2 + 3$$

subject to the constraints

$$g_1(X) = x_1 + x_2 \leq 4, \text{ taking the starting point } X_1 = \begin{pmatrix} 0 \\ 0 \end{pmatrix}$$

SOLUTION.   We have

$$f(X) = f(x_1, x_2) = x_1^2 + x_2^2 - 2x_1 - 3x_2 + 3 \qquad \qquad ...(1)$$

$$X_1 = \begin{pmatrix} 0 \\ 0 \end{pmatrix}$$

Let $\qquad g_1(X) = x_1 + x_2 - 4 \qquad \qquad ...(2)$

Then $f(X_1) = 3$ and $g(X_1) = -4 < 0$ [Using (1) and (2)]

Since $g_1(X) < 0$ so search direction S_1 is given by

$$S_1 = -\nabla f(X_1) = -\begin{bmatrix} \partial f / \partial x_1 \\ \partial f / \partial x_2 \end{bmatrix}_{X_1 = (0,0)} = \begin{pmatrix} 2 \\ 3 \end{pmatrix} \quad \left[\because \frac{\partial f}{\partial x_1} = 2x_1 - 2 \text{ and } \frac{\partial f}{\partial x_2} = 2x_2 - 3 \right]$$

Therefore, $\qquad S_1 = \begin{pmatrix} 2/3 \\ 1 \end{pmatrix}$

Now, to find a new point X_1, we take a step of length λ_1 in the direction of $-\nabla f(X_1)$ to arrive at

$$X_2 = X_1 + \lambda_1 S_1 = \left(\frac{2}{3}\lambda_1, \lambda_1 \right)$$

$$f(X_2) = f\left(\left(\frac{2}{3} \right)\lambda_1, \lambda_1 \right) = \left(\frac{13}{9} \right)\lambda_1^2 - \frac{13}{3}\lambda_1 + 3$$

To minimize f, we must have

$$\frac{df}{d\lambda_1} = 0 \Rightarrow \frac{26}{9}\lambda_1 - \frac{13}{3} = 0 \Rightarrow \lambda_1 = 1.5$$

and $\qquad \dfrac{d^2 f}{d\lambda_1^2} = \dfrac{26}{9} > 0$

$\Rightarrow f$ is minimum when $\lambda_1 = 1.5$

Therefore, $\qquad X_2 = (0, 0) + 1.5 \ (2/3, 1) = (1, 1.5)$

and $\qquad g(X_2) = -1.5 < 0$

Therefore, the new search direction S_2 is given by

$$S_2 = -\nabla f(X_2)$$

$$= \begin{bmatrix} \partial f / \partial x_1 \\ \partial f / \partial x_2 \end{bmatrix}_{X_1 = (1,1.5)} = \begin{pmatrix} 0 \\ 0 \end{pmatrix}$$

$\Rightarrow$ There is no search direction available to obtain minimum f.

Hence, minimum of $f = -0.25$ at $X = X_2 = (1, 1.5)$ *i.e.,* $x_1 = 1, x_2 = 1.5$

EXAMPLE 2. *Using Zoutendijk's method, minimize the function*

$$f(X) = f(x_1, x_2) = x_1^2 + x_2^2 - 4x_1 - 4x_2 + 8$$

subject to the constraints

$$g_1(x_1, x_2) = x_1 + 2x_2 - 4 \leq 0, \text{ with the starting point } X_1 = \begin{pmatrix} 0 \\ 0 \end{pmatrix}.$$

Take $\varepsilon_1 = 0.001, \ \varepsilon_2 = 0.001 \ \text{and} \ \varepsilon_3 = 0.01$

SOLUTION. We have

$$f(X) = f(x_1, x_2) = x_1^2 + x_2^2 - 4x_1 - 4x_2 + 8 \qquad \ldots(1)$$

and $g_1(x_1, x_2) = x_1 + 2x_2 - 4$ $\qquad \ldots(2)$

Iteration-1 :

Since $g_1(X_i) < 0$, therefore search direction given by

$$S_1 = -\nabla f(X_1)$$

$\Rightarrow$

$$S_1 = -\begin{bmatrix} \partial f / \partial x_1 \\ \partial f / \partial x_2 \end{bmatrix}_{x_1} = \begin{pmatrix} 4 \\ 4 \end{pmatrix} \quad (\because \text{ from (1)} \; \frac{\partial f}{\partial x_1} = 2x_1 - 4 \; \text{ and } \; \frac{\partial f}{\partial x_2} = 2x_2 - 4 \;)$$

$\Rightarrow$ normalized vector $S_1 = \begin{bmatrix} 1 \\ 1 \end{bmatrix}$

Now to find the new point X_2, we have to find a suitable step length along S_1. For this we have to minimize

$$f(X_1 + \lambda S_1) \; \text{ w.r.t. } \lambda$$

We have $f(X_1 + \lambda S_1) = f(0 + \lambda, 0 + \lambda) = f(\lambda, \lambda) = 2\lambda^2 - 8\lambda + 8$ (by (1))

Now, for the minima of f, we must have

$$\frac{df}{d\lambda} = 0 \Rightarrow 4\lambda - 8 = 0 \Rightarrow \lambda = 2$$

and

$$\frac{d^2 f}{d\lambda^2} = 4 > 0$$

$\Rightarrow f$ is minimum at $\lambda = 2$

So, we have a new point given by $X_2 = \begin{bmatrix} 2 \\ 2 \end{bmatrix}$ and $g_1(X_2) = 2$

Since, $g_1' = g_1$ (at $\lambda = 0$) = -4 and $g_1'' = g_1$ (at $\lambda = 2$) = 2 so, new step length given by

$$\bar{\lambda} = -\frac{g_1'}{g_1'' - g_1'} \lambda = \frac{4}{3}$$

$\Rightarrow g_1$ at $\lambda = \bar{\lambda} = 0$ and hence $X_2 = \begin{bmatrix} 4/3 \\ 4/3 \end{bmatrix} = \begin{bmatrix} 1.333 \\ 1.333 \end{bmatrix}$

Also from (1), $f(X_2) = \dfrac{8}{9}$

Now, we check the convergence.

Here, $\left| \dfrac{f(X_1) - f(X_2)}{f(X_1)} \right| = \left| \dfrac{8 - 8/9}{8} \right| = \dfrac{8}{9} > \varepsilon_2$

and $||X_1 - X_2|| = \left[\left(0 - \dfrac{4}{3}\right)^2 + \left(0 - \dfrac{4}{3}\right)^2 \right]^{1/2} = 1.887 > \varepsilon_2$

$\Rightarrow$ Convergence criterion is not satisfied.

Iteration-2

Since $g_1 = 0$ at $X = X_2$, so we have to find a usable feasible direction.

Here direction finding problem can be stated as

$$\text{Minimize } f = -\alpha$$

subject to $t_1 + 2t_2 + \alpha + y_1 = 3$

$$-\frac{4}{3}t_1 - \frac{4}{3}t_2 + \alpha + y_2 = -\frac{8}{3}$$
$$t_1 + y_3 = 2$$
$$t_2 + y_4 = 2$$
$$t_1, t_2, \alpha \geq 0$$

which is a linear programming problem and having the solution

$$t_1^* = 2, \ t_2^* = \frac{3}{10}, \ \alpha^* = \frac{4}{10}, \ y_4^* = \frac{17}{10}, \ y_1^* = y_2^* = y_3^* = 0$$

$$-f_{min} = -\alpha^* = -\frac{4}{10}$$

$\Rightarrow$ usable feasible direction is given by

$$S = \begin{bmatrix} t_1^* - 1 \\ t_2^* - 1 \end{bmatrix} = \begin{bmatrix} 1.0 \\ -0.7 \end{bmatrix}$$

Now, we have to move along the direction $S_2 = \begin{bmatrix} 1.0 \\ -0.7 \end{bmatrix}$

from
$$X_2 = \begin{bmatrix} 1.333 \\ 1.333 \end{bmatrix}$$

$\therefore$
$$f(X_2 + \lambda S_2) = f(1.333 + \lambda, \ 1.333 - 0.7\lambda)$$
$$= 1.49\lambda^2 - 0.4\lambda + 0.889$$

Using the principle of maxima and minima, for the minimum of f, we must have

$$\frac{df}{d\lambda} = 0 \Rightarrow 2(1.49)\lambda - 0.4 = 0$$

$\Rightarrow$
$$\lambda = 0.134$$

and
$$\frac{d^2 f}{d\lambda^2} = 2.98 > 0$$

$\Rightarrow f$ is minimum at $\lambda = 0.134$

Hence, new point is given by

$$X_3 = X_2 + \lambda S_2 = \begin{bmatrix} 1.333 \\ 1.333 \end{bmatrix} + 0.134 \begin{bmatrix} 1.0 \\ -0.7 \end{bmatrix} = \begin{bmatrix} 1.467 \\ 1.239 \end{bmatrix}$$

Clearly $g_1(X_3) = -0.005$

$\Rightarrow X_3$ lies in the interior of the feasible domain.

Proceeding same as above, after some iteration

we get $X^* = \begin{bmatrix} 1.6 \\ 1.2 \end{bmatrix}$ and $f_{min} = 0.8$.

11.4.6 ROSEN GRADIENT PROJECTION METHOD

Rosen gradient projection method is an efficient method of solving constrained non-linear programming problem. It does not require the solution of an auxiliary linear optimization problem to find the usable feasible direction. This method can be used for general non-linear programming problems where even the constraints can be non-linear along with the non-linear objective function.

However, it is more effectively used to solve the problem where all constraints are linear.

In this method, we uses the projection of the negative of the objective function gradient onto the constraints that are currently active.

Consider a problem given by

$$\left.\begin{array}{l} \text{Minimize } f(X) \\[6pt] g_j(X) = \sum_{i=1}^{n} a_{ij} x_i - b_j \le 0,\ j = 1, 2, ..., m \end{array}\right] \qquad ...(1)$$

subject to the constraints

where $X = (x_1, x_2, ..., x_n)$

Before finding the solution of (1), let us define the following terms

(i) Feasible point : The set of all X such that $g_j(X), j = 1, 2, ..., m$ is called the set of feasible solution of the above equation (1). It is denoted by S_F

i.e., $$S_F = \left[X : g_j(X) \le 0,\ j = 1, 2, ..., m\right]$$

Then a point $X \in S_F$ is called a feasible point.

(ii) Active Constraints : A constraints $\{g_j(X) \le 0\}$ is called active at a feasible point X if $g_i(X) = 0$.

(iii) Projection matrix : Let $j_1, j_2, ..., j_p$ be the indices of active constraints at a feasible point X, then the matrix of order $n \times p$ of the gradient of these active constraints.

$$N_p = [\nabla g_{j,1}, \nabla g_{j,2}, ... \nabla g_{j,p}]$$

where

$$\nabla g_j(X) = \begin{bmatrix} a_{1j} \\ a_{2j} \\ \vdots \\ a_{nj} \end{bmatrix},\ j = j_1, j_2, ..., j_p$$

and the projection matrix P is given by

$$P = I - N\,[N^T\,N]^{-1}\,N^T$$

where I is the identity matrix.

Now we shall discuss the procedure, by which we solve the given constrained problem by Rosen's gradient projection method.

WORKING PROCEDURE

To solve a non-linear programming problem using Rosen's gradient projection method, we use the following steps :

STEP 1. Take an initial point X_1. Check the feasibility of X_1 i.e., X_1 has to be feasible if $g_j(X_1) \le 0$, $j = 1, 2, ..., m$. We have following two cases:

Case-1 : If $g_j(X_1) < 0$ for $j = 1, 2, ..., m$ then find the normalized search direction using

$$S_i = \frac{-\nabla f(X_i)}{||\nabla f(X_i)||}$$

Case-2 : If $g_j(X_1) = 0$ for $j = j_1, j_2, ..., j_p$, then find the normalized search direction using

$$S_i = \frac{-P_i \nabla f(X_i)}{|P_i \nabla f(X_i)||}$$

where $P_1 = I - N_1[N_1^T N_1]^{-1} N_1^T$

and $N_p = \left[\nabla g_{j1}(X_i) \nabla g_{j2}(X_i) .... \nabla g_{jp}(X_i)\right]$

STEP 2. Here, we also have following two cases :

Case-1 : If $S_j = 0$, find the vector λ at X_i

such that $$\lambda = -\left[N_P^T N_P\right]^{-1} N_P^T \nabla f(X_i)$$

Here, if each component of λ is non-negative then X_i is the optimum value and then stop the iterations.

If some of the components of λ are negative then identify the component λ_q with the most negative value and form a new matrix N_p as

$$[\nabla g_{j1}(X_i)\nabla g_{j2}(X_i)...\nabla g_{j(q-1)}(X_i)\nabla g_{j(q+1)}(X_i)]$$

and go to step-1.

Case-2 : If $S_j \neq 0$, then evaluate the maximum step length λ_m where

$$\lambda_m = \min\{\lambda_k\}, \ \lambda_k > 0 \text{ and } k \text{ is any integer between 1 to } m$$
$$\text{other than } j_1, j_2, ..., j_p$$

Then calculate $\dfrac{df}{d\lambda}$ at $\lambda = \lambda_m$

If at $\lambda = \lambda_m$, $\dfrac{df}{d\lambda} = 0$ or negative, take the step length $\lambda_i = \lambda_m$.

If at $\lambda = \lambda_m$, $\dfrac{df}{d\lambda}$ is positive, then find the minimum step length $\lambda = \lambda_i$ by putting $\dfrac{df}{d\lambda} = 0$

Then optimum point becomes $X_{i+1} = X_i + \lambda_i S_i$

Further,

(i) If $\lambda = \lambda_m$ or $\lambda_m \leq \lambda_i^*$ some new constraints become active at X_{i+1}. Then generate the new matrix N_e which include the gradient of all the active constraints at the point X_{i+1}. Fix the new iteration number as $i = i+1$ and go to case-2 of step-1.

(ii) If $\lambda_i = \lambda_i^*$ and $\lambda_i^* < \lambda_m$, no new constraints will be active at X_{i+1}. Then the matrix N_p remains unchanged. Fix the new iteration number as $i = i +1$ and go to case-1 of step-1.

SOLVED EXAMPLES

EXAMPLE 1. *Use the Rosen's gradient projection method to solve the following problem*

$$\text{Minimize } f(X) = x_1^2 + x_2^2 - 2x_1 - 4x_2$$

subject to the constraints

$$g_1(x_1, x_2) = x_1 + 4x_2 - 5 \leq 0$$
$$g_2(x_1, x_2) = 2x_1 + 3x_2 - 6 \leq 0$$
$$g_3(x_1, x_2) = -x_1 \leq 0$$
$$g_4(x_1, x_2) = -x_2 \leq 0$$

starting from the point $X = \begin{bmatrix} 1 \\ 1 \end{bmatrix}$

SOLUTION. We have $\qquad f(X) = x_1^2 + x_2^2 - 2x_1 - 4x_2 \qquad$ ...(1)

$$\left. \begin{array}{l} g_1(x_1, x_2) = x_1 + 4x_2 - 5 \\ g_2(x_1, x_2) = 2x_1 + 3x_2 - 6 \\ g_3(x_1, x_2) = -x_1 \\ \text{and } g_4(x_1, x_2) = -x_2 \end{array} \right\} \qquad \text{...(2)}$$

From (2)

$$g_1(1, 1) = 1+4-5 = 0 \text{ for } j = 1$$

$\Rightarrow$ only first constraint is active, $p=1, j = 1$

$\therefore$ normalized search direction is given by

$$S_1 = \frac{-P_i \nabla f(X_i)}{||P_i\nabla f(X_i)||}$$

where

$$P_i = I - N_P[N_P^T N_P]^{-1} N_P^T$$

and

$$N_P = [\nabla g_{j1}(X_i) \nabla g_{j2}(X_i) .... \nabla g_{jp}(X_i)]$$

$$\Rightarrow \qquad N_1 = [\nabla g_1(X_1)] = \begin{bmatrix} 1 \\ 4 \end{bmatrix}$$

$$P_1 = I - N_1[N_1^T N_1]^{-1} . N_1^T$$

$$= \begin{bmatrix} 1 & 0 \\ 0 & 1 \end{bmatrix} - \begin{bmatrix} 1 \\ 4 \end{bmatrix} \left[\begin{pmatrix} 1 \\ 4 \end{pmatrix} [1 \ 4] \right]^{-1} [1 \ 4]$$

$$= \begin{bmatrix} 1 & 0 \\ 0 & 1 \end{bmatrix} - \frac{1}{17} \begin{bmatrix} 1 \\ 4 \end{bmatrix} [1 \ 4] = \begin{bmatrix} 1 & 0 \\ 0 & 1 \end{bmatrix} - \frac{1}{17} \begin{bmatrix} 1 & 4 \\ 4 & 16 \end{bmatrix}$$

$$= \frac{1}{17} \begin{bmatrix} 16 & -4 \\ -4 & 1 \end{bmatrix}$$

and therefore, $\nabla f(X_1) = \begin{bmatrix} 2x_1 - 2 \\ 2x_2 - 4 \end{bmatrix}_{(1,1)} = \begin{bmatrix} 0 \\ -2 \end{bmatrix}$

Thus, $\qquad S_1 = \dfrac{-P_1 \nabla f(X_1)}{||P_1 \nabla f(X_1)||}$

$$= \frac{-\dfrac{1}{17} \begin{bmatrix} 16 & -4 \\ -4 & 1 \end{bmatrix} \begin{bmatrix} 0 \\ -2 \end{bmatrix}}{\left\| -\dfrac{1}{17} \begin{bmatrix} 16 & -4 \\ -4 & 1 \end{bmatrix} \begin{bmatrix} 0 \\ -2 \end{bmatrix} \right\|} = \begin{bmatrix} -0.9701 \\ 0.2425 \end{bmatrix}$$

$$\Rightarrow \qquad S_1 \neq 0$$

Therefore, we find $\lambda_m = \min \{\lambda_k\}$, $\lambda_k > 0$ and k is any integer among 2, 3 and 4.

Set $\qquad X = \begin{bmatrix} x_1 \\ x_2 \end{bmatrix} = X_1 + \lambda S_1 = \begin{bmatrix} 1 - 0.9701\lambda \\ 1 + 0.2425\lambda \end{bmatrix}$

Now,

for $j = 2$, $\qquad g_2(X) = (2 - 1.9402\lambda) + (3 + 0.7275) - 6 = 0$ at $\lambda = \lambda_2 = -0.8245$

For $j = 3$, $\qquad g_3(X) = -(1 - 0.9701\lambda) = 0$ at $\lambda = \lambda_3 = 1.03$

For $j = 4$, $\qquad g_4(X) = -(1 + 0.2425\lambda) = 0$ at $\lambda = \lambda_4 = -4.124$

Since λ_3 is the minimum positive value, thus we take

$$\lambda_m = \lambda_3 = 1.03$$

Further,

$$f(X) = f(\lambda) = (1 - 0.9701\lambda)^2 + (1 + 0.2425\lambda)^2 - 2(1 - 0.9701\lambda)$$
$$- 4(1 + 0.2425\lambda) - 0.9998\lambda^2 - 0.4850\lambda - 4 \quad \text{[By (1)]}$$

Now, $\qquad \dfrac{df}{d\lambda} = -1.9996\lambda - 0.4850$

$\therefore$ At $\lambda = \lambda_m$, $\qquad \dfrac{df}{d\lambda} = 1.9996(1.03) - 0.4850 = 1.5746 > 0$

Therefore, we find minimum step length λ_1^* by putting

$$\frac{df}{d\lambda} = 0$$

which gives $\lambda_1 = \lambda_1^* = \dfrac{0.4850}{1.9996} = 0.2425$

Then minimum point will be given by

$$X_2 = X_1 + \lambda_1 S_1$$

$$= \begin{bmatrix} 1 \\ 1 \end{bmatrix} + 0.2425 \begin{bmatrix} -0.9701 \\ 0.2425 \end{bmatrix} = \begin{bmatrix} 0.764701 \\ 1.0588 \end{bmatrix}$$

Here $\lambda_1 = \lambda_1^*$ and $\lambda_1^* < \lambda_m \Rightarrow$ No new constraints will be active at X_2 and hence the matrix N_p remains unchanged.

Further, fix the iteration number as $i = 2$.

Now, since $g_2(X_2) = 0$, we set $p = 1, j = 1$, then normalized search direction S_2 is given by

$$S_2 = \frac{-P_2 \nabla f(X_2)}{\|P_2 \nabla f(X_2)\|}$$

Here, we have

$$N_2 = \begin{bmatrix} 1 \\ 4 \end{bmatrix}, P_2 = \frac{1}{17} \begin{bmatrix} 16 & -4 \\ -4 & 1 \end{bmatrix}$$

$$\nabla f(X_2) = \begin{bmatrix} 2x_1 - 2 \\ 2x_2 - 4 \end{bmatrix}_{X_2} = \begin{bmatrix} -0.4706 \\ -1.8824 \end{bmatrix}$$

Therefore

$$P_2 \nabla f(X_2) = \frac{1}{17} \begin{bmatrix} 16 & -4 \\ -4 & 1 \end{bmatrix} \begin{bmatrix} -0.4706 \\ -1.8824 \end{bmatrix} = \begin{bmatrix} 0 \\ 0 \end{bmatrix}$$

$$\Rightarrow \qquad S_2 = 0$$

Thus we compute the step length λ at X_2 such that

$$\lambda = -[N_P^T N_P]^{-1} N_P^T \nabla f(X_i) = -\frac{1}{17}[1 \; 4] \begin{bmatrix} -0.4706 \\ -1.8824 \end{bmatrix}$$

$$= 0.4707 > 0$$

So, the non-negative value of λ indicates that we have obtained the optimum value and so stop further iteration.

Hence,

$$X_2 = \begin{bmatrix} 0.7647 \\ 1.0588 \end{bmatrix} \text{ and } f_{min} = -4.059$$

11.5 INDIRECT METHODS

In using the technique of indirect-methods for solving non-linear programming problem, the function to be minimized must be differentiable.

In this section, we will discuss the following indirect methods.

11.5.1 TRANSFORMATION TECHNIQUES

Consider the constraints $g_i(X)$ of a non-linear programming problem, which have some simple forms in independent variables, then it may be possible to make use of the transformation of these variables in such a way that constraints are satisfied automatically.

Hence, it is possible to convert a constrained optimization problem into an unconstrained one by transforming the variables.

Here, we are giving some typical transformations of independent variables as given below

(i) Let l_i and u_i are the lower and upper bound of x_i such that $l_i \le x_i \le u_i$

Then these can be satisfied by transforming the variables x_i as given below

$$x_i = l_i + (u_i - l_i) \sin^2 y_i \qquad \qquad ...(1)$$

where y_i is the new variable, which may take any value.

(ii) If $x_i \in (0,1)$, then use any one of the following transformation

$$
\left.\begin{array}{ll}
\text{(a)} & x_i = \sin^2 y_i \\[6pt]
\text{(b)} & x_i = \cos^2 y_i \\[6pt]
\text{(c)} & x_i = \dfrac{e^{y_i}}{e^{y_i} + e^{-y_i}} \\[10pt]
\text{(d)} & x_i = \dfrac{y_i^2}{1 + y_i^2}
\end{array}\right\} \qquad \ldots(2)
$$

(iii) If decision (design) variable is restricted to assume only positive values then use any one of the following transformations

$$
\left.\begin{array}{ll}
\text{(a)} & x_i = |y_i| \\[4pt]
\text{(b)} & x_i = y_i^2 \\[4pt]
\text{(c)} & x_i = e^{y_i}
\end{array}\right\} \qquad \ldots(3)
$$

(iv) If $x_i \in (-1,1)$, then we can use any one of the following transformations :

$$
\left.\begin{array}{ll}
\text{(a)} & x_i = \sin y_i \\[4pt]
\text{(b)} & x_i = \cos y_i \\[6pt]
\text{(c)} & x_i = \dfrac{2y_i}{1 + y_i^2}
\end{array}\right\} \qquad \ldots(4)
$$

☛ **REMARKS**

- To use the transformed transformation, the constrained function $g_i(X)$ should be simple.
- If it is not possible to eliminate all the constraints by changing the variables, then its better not to use the transformation method.

SOLVED EXAMPLES

EXAMPLE 1. *A courier service does not accept rectangular packets of more than 42 cm length. If the maximum [length + 2(width + height)] is 72 cms. Compute the maximum volume of the rectangular packet.*

SOLUTION. Let us suppose that x_1, x_2 and x_3 be the length, width and height of a rectangular packets respectively. According to given question, we formulate the problem as given below

$$\text{Max. } f(X) = x_1 x_2 x_3 \qquad \ldots(1)$$

subject to the constraints

$$x_1 + 2x_2 + 2x_3 \le 72 \qquad \ldots(2)$$

$$x_1 \le 42 \qquad \ldots(3)$$

and

$$x_1, x_2, x_3 \ge 0 \qquad \ldots(4)$$

Let us use the following transformation

$$y_1 = x_1, \; y_2 = x_2 \text{ and } y_3 = x_1 + 2x_2 + 2x_3$$

$$\Rightarrow \qquad x_3 = \frac{1}{2}(y_3 - y_1 - 2y_2) \qquad \ldots(5)$$

Then constraints (1), (2) and (3) can be written as

$$
\left.\begin{array}{l}
0 \le y_1 < 42 \\[4pt]
0 \le y_2 \le 36 \\[4pt]
0 \le y_3 \le 72
\end{array}\right\} \qquad \ldots(6)
$$

Clearly the upper bound for y_i's in (6) can be easily obtained say for y_2 taking $x_1 = x_2 = 0$ in (2), we get

$$2x_2 \le 72 \qquad \text{so} \quad x_2 = y_2 \le 36.$$

Now, constraints in (6) are automatically satisfied, if we define z_1, z_2 and z_3 as given below

$$y_i = l_i + (u_i - l_i) \sin^2 z_i \qquad \text{...(7)}$$

Then from (6) and (7), we get

$$y_1 = l_1 + (u_1 - l_1) \sin^2 z_1 = 42 \sin^2 z_1 \qquad \text{...(8)}$$

$$y_2 = l_2 + (u_2 - l_2) \sin^2 z_2 = 36 \sin^2 z_2 \qquad \text{...(9)}$$

$$y_3 = l_3 + (u_3 - l_3) \sin^2 z_3 = 72 \sin^2 z_3 \qquad \text{...(10)}$$

Then original problem reduces to

$$\text{Max. } f = \frac{1}{2} y_1 y_2 (y_3 - y_1 - 2y_2)$$

$$= \frac{1}{2}(42 \sin^2 z_1)(36 \sin^2 z_2).(72 \sin^2 z_3 - 42 \sin^2 z_1 - 72 \sin^2 z_2)$$

subject to the constraints

$$0 \le \sin^2 z_i \le 1, \quad i = 1, 2, 3 \qquad \text{...(11)}$$

Now, using the principle of maxima and minima, for the minimum of f, we must have

$$\frac{\partial f}{\partial z_1} = 0 \Rightarrow \sin z_1 \cos z_1 \sin^2 z_2 \left(\sin^2 z_3 - \frac{7}{6} \sin^2 z_1 - \sin^2 z_2 \right) = 0 \qquad \text{...(12)}$$

$$\frac{\partial f}{\partial z_2} = 0 \Rightarrow \sin^2 z_1 \sin z_2 \cos z_2 \left(\sin^2 z_3 - \frac{7}{12} \sin^2 z_1 - 2 \sin^2 z_2 \right) = 0 \qquad \text{...(13)}$$

and

$$\frac{\partial f}{\partial z_3} = 0 \Rightarrow \sin^2 z_1 \sin^2 z_2 \sin z_3 \cos z_3 = 0 \qquad \text{...(14)}$$

Therefore, we have

$$\sin^2 z_1 = 0 \Rightarrow z_1 = 0 \Rightarrow y_1 = 0 \Rightarrow f = 0 \qquad \text{(Not acceptable)}$$

$$\sin^2 z_2 = 0 \Rightarrow z_2 = 0 \Rightarrow y_2 = 0 \Rightarrow f = 0 \qquad \text{(Not acceptable)}$$

$$\sin z_3 = 0 \Rightarrow z_3 = 0 \Rightarrow y_3 = 0 \Rightarrow f = 0 \qquad \text{(Not acceptable)}$$

$$\cos z_3 = 0 \Rightarrow z_3 = \frac{\pi}{2}$$

Then from (12)

$$\sin^2 z_3 - \frac{7}{6} \sin^2 z_1 - \sin^2 z_2 = 0$$

$$\Rightarrow \qquad 7 \sin^2 z_1 + 6 \sin^2 z_2 = 6 \qquad \text{...(15)}$$

and from (13)

$$\sin^2 z_3 - \frac{7}{12} \sin^2 z_1 - 2 \sin^2 z_2 = 0 \Rightarrow 7 \sin^2 z_1 + 24 \sin^2 z_2 = 12 \quad \text{...(16)}$$

On solving (15) and (16), we get

$$\sin^2 z_2 = \frac{1}{3}$$

and $\qquad 7 \sin^2 z_1 + 6 \left(\frac{1}{3} \right) = 6 \Rightarrow \sin^2 z_1 = \frac{4}{7}$

Hence,
$$y_1 = 42\left(\frac{4}{7}\right) = 24$$

$$y_2 = 36\left(\frac{1}{3}\right) = 12$$

$$y_3 = 72$$

which implies that $x_1 = 24$, $x_2 = 12$, $x_3 = 12$
and
$$\text{max.} f = 3456 \text{ cm}^3.$$

11.5.2 PENALTY FUNCTION METHODS

Penalty function methods transforms the basic optimization problem into alternative formulation. In these methods, we transform the given problem into a sequence of problems each with no constraints.

To understand it, consider the following problem

$$\text{Minimum } f(X)$$

subject to the constraints

$$g_j(X) \leq 0, \ j = 1, 2, ..., m$$

and

$$h_i(X) = 0, \ i = 1, 2, ..., p$$ ...(1)

We know that penalty function method transform (1) into a sequence of problems each with no constraints. Here, penalty function method create the effect of constraints by bringing in the modification in the objective function (as done in Big-M method of artificial variable technique).

Therefore, to solve (1), let us introduce an auxiliary unconstrained function given by

$$F(X, r_k) = f(X) + P(X, r_k)$$ ...(2)

where $P(X, r_k)$ is a function of constraints $g_j(X)$ and $h_i(X)$ and r_k is a positive parameter such that

$$\lim_{r_k \to 0} \min F(X, r_k) = \min f(X)$$

Now, we will formulate the penalty function.
There are following two types of penalty function method.

(i) **Interior Penalty function method :** In interior penalty function method, the initial point as well as each of the subsequent points generated, lie inside the acceptable region of the design space.

In this method, the form of the penalty function $P(X, r_k)$, which is a function of constraints functions $g_j(X)$ and $h_i(X)$ and a positive parameter r_k is given by

$$P(X, r_k) = -r_k \sum_{j=1}^{m} \frac{1}{g_j(X)} + \frac{1}{\sqrt{r_k}} \sum_{i=1}^{p} h_i^2(X)$$

Here, the minimum of the auxiliary function $F(X, r_k)$ as defined by (2) approaches the minimum of the objective function $f(X)$ from point inside the feasible region as $r_k \to 0$.

☛ REMARKS
- Interior penalty function method is also known as barrier method.
- Once the unconstrained minimization of $P(X, r_k)$ is started from any feasible point X_i the subsequent points generated will always lie within the feasible domain.
- The value of the function P will always be greater than f, since $g_j(X)$ is negative for all feasible point X.

(ii) **Exterior Penalty function method :** In the exterior penalty function method, the function P is generally taken as follows

$$P(X, r_k) = \frac{1}{r_k} \sum_{j=1}^{m} \left[\max(0, g_j(X))\right]^2 + \frac{1}{r_k} \sum_{i=1}^{p} h_i^2(X)$$

and minimize the auxiliary function $F(X, r_k)$ for a sequence of decreasing values of r_k i.e., the minimum of $F(X, r_k)$ approaches to minimum of $f(X)$ from the point of infeasible region as $r_k \to 0$.

The value max. $\{0, g_j(X)\}$ can be defined as follows

$$\langle g_j(X) \rangle = \max.\{0, g_j(X)\} = \begin{cases} g_j(X), & \text{if } g_j(X) > 0 \text{ (constraint is violent)} \\ 0, & \text{if } g_j(X) \le 0 \text{ (constraint is satisfied)} \end{cases}$$

☞ REMARKS
- Usually the function $P(X, r_k)$ possesses a minimum as a function of X in the infeasible region.
- From above, we observe that there will be a penalty for violating the constraints and the amount of penalty will increase at a faster rate than the amount of violation of a constant that is why it is called exterior penalty function method.

 SOLVED EXAMPLES

EXAMPLE 1. *Using interior penalty method, solve*
$$\text{Minimize } f(X) = 2x$$
subject to the constraints
$$x \ge 3$$

SOLUTION. We have to find
$$\text{Min } f(X) = 2x$$
subject to $g_i(x) = 3 - x \le 0$

Clearly this is the problem of one variable.

Using the formula of unconstrained interior penalty function method, the auxiliary unconstrained problem can be written as

$$\text{Min } F(X, r_k) = 2x - r_k \left(\frac{1}{3-x} \right)$$

such that $\lim\limits_{r_k \to 0} \min F(X, r_k) = \min f(X)$

Using the principle of maxima and minima, for the minimization of F, we must have

$$\frac{\partial F}{\partial x} = 0 \Rightarrow 2 - \frac{r_k}{(3-x)^2} = 0 \Rightarrow x = 3 + \sqrt{\frac{r_1}{2}}$$

and

$$\frac{\partial^2 F}{\partial x^2} = \frac{-2r_2}{(3-x)^3} > 0 \text{ at } x = 3 + \sqrt{\frac{r_1}{2}}$$

$$\Rightarrow F \text{ is minimum at } x = 3 + \sqrt{\frac{r_1}{2}}$$

Now as $r_k \to 0$ at $x = 3$

$$\lim\limits_{r_k \to 0} \min F(X, r_k) = \min f(X) = 6 \text{ when } x = 3$$

EXAMPLE 2. *Using the exterior penalty function method, minimize the function f(X) given by*
$$f(X) = x_1^2 + 2x_2^2$$
subject to the constraints $2x_1 + 5x_2 \le 10$

SOLUTION. We have $\qquad f(X) = x_1^2 + 2x_2^2 \qquad \qquad$ ...(1)

and $\qquad \qquad g_i(x) = 2x_1 - 5x_2 - 10 \le 0 \qquad \qquad$ ...(2)

Using the formula of exterior penalty function method the auxiliary function $F(X, r_k)$ can be written as

$$\text{Min. } F(X, r_k) = x_1^2 + 2x_2^2 + \frac{1}{r_k} [\max(0, 2x_1 + 5x_2 - 10)]^2$$

Using the principle of maxima and minima, for the minimization of $F\left(r = \dfrac{1}{r_k}\right)$, we must have

$$\frac{\partial F}{\partial x_1} = 0 \Rightarrow 2x_1 + \frac{4}{r_k}(2x_1 + 5x_2 - 10) = 0$$

$$\Rightarrow \qquad (2 + 8r)x_1 + 20rx_2 - 40r = 0 \qquad\qquad\qquad ...(3)$$

and

$$\frac{\partial F}{\partial x_2} = 0 \Rightarrow 4x_2 + 10r(2x_1 + 5x_2 - 10) = 0$$

$$\Rightarrow \qquad 20rx_1 + (4 + 50r)x_2 - 100r = 0 \qquad\qquad\qquad ...(4)$$

On solving (3) and (4), we get

$$\frac{x_1}{160r} = \frac{x_2}{200r} = \frac{1}{8 + 132r}$$

$$\Rightarrow \quad x_1 = \frac{40r}{2 + 33r} = \frac{40}{2r_k + 33} \quad \text{and} \quad x_2 = \frac{50r}{2 + 33r} = \frac{50}{2r_k + 33}$$

As $r_k \to 0$, we get $x_1 = \dfrac{40}{33}$ and $x_2 = \dfrac{50}{33}$

and Minimum of $f(X) = \left(\dfrac{40}{33}\right)^2 + 2\left(\dfrac{50}{33}\right)^2 = \dfrac{200}{33}$

EXAMPLE 3. *Using penalty function method minimize the function*
$$f(X) = f(x_1, x_2) = (x_1 + 2)^3 + 3x_2 + 1$$
subject to the constraints
$$x_1 \geq 2, \ x_2 \geq 0$$

SOLUTION. We can write the given problem as
$$f(X) = (x_1 + 2)^2 + 3x_2 + 1$$
$$\text{subject to } 2 - x_1 \leq 0, \ -x_2 \leq 0$$
Therefore, the auxiliary unconstrained problem becomes

$$\text{Minimize } F(X, r_k) = (x_1 + 2)^3 + 3x_2 + 1 - r_k\left(\frac{1}{2 - x_1} - \frac{1}{x_2}\right)$$

Now, for the minimization of F, we must have

$$\frac{\partial F}{\partial x_1} = 0 \Rightarrow 3(x_1 + 2)^2 - \frac{r_k}{(2 - x_1)^2} = 0 \quad \Rightarrow \quad x_1 = \left(4 + \sqrt{\frac{r_k}{3}}\right)^{1/2}$$

and

$$\frac{\partial F}{\partial x_2} = 0 \Rightarrow 3 - \sqrt{\frac{r_k}{x_2^2}} = 0 \quad \Rightarrow \quad x_2 = \sqrt{\frac{r_k}{3}}$$

Therefore, $x_1 = \left(4 + \sqrt{\dfrac{r_k}{3}}\right)^{1/2}$ and $x_2 = \sqrt{\dfrac{r_k}{3}}$ are the possible feasible values of x_1 and x_2.

Now as $r_k \to 0$, we get $x_1 = 4$, $x_2 = 0$ and min. $f(X) = 65$ and $x_1 = 4$, $x_2 = 0$

EXAMPLE 4. *Using penalty function method,*
$$\text{Minimize } f(x_1, x_2) = \frac{1}{3}(x_1 + 1)^3 + x_2$$
subject to the constraints
$$g_1(x_1, x_2) = 1 - x_1 \leq 0 \qquad \text{and} \qquad g_2(x_1, x_2) = -x_2 \leq 0$$

SOLUTION. Proceeding same as above, we get the auxiliary unconstrained function $F(X, r_k)$ as given below
$$F(X, r_k) = \frac{1}{3}(x_1 + 1)^3 + x_2 + r_k[\max(0, 1 - x_1)]^2 + r_k[\max(0, -x_2)]^2$$

For the minimum of F, we must have

$$\frac{\partial F}{\partial x_1} = 0 \Rightarrow (x_1+1)^2 - 2r_k[\max(0,1-x_1)] = 0$$

and $\quad \frac{\partial F}{\partial x_2} = 0 \Rightarrow 1 - 2r[\max(0,-x_2)] = 0$

The above equations can be written as

$$\min\left[(x_1+1)^2, (x_1+1)^2 - 2r_k(1-x_1)\right] = 0 \qquad \text{...(1)}$$

and $\quad \min[1, 1+2r_k x_2] = 0 \qquad\qquad\qquad\qquad\qquad \text{...(2)}$

In equation (1), if $(x_1+1)^2 = 0 \Rightarrow x_1 = -1$, which violates the first constraints and if

$$(1+1)^2 - 2r_k(1-x_1) = 0 \Rightarrow x_1 = -1 - r_k + \sqrt{r_k^2 + 4r_k}$$

while in equation (2), the only possibility is that

$$1 + 2r_k x_2 = 0 \Rightarrow x_2 = -\frac{1}{2r_k}$$

Therefore, the solution of the unconstrained minimization problem is given by

$$x_1^*(r_k) = -1 - r_k + r_k\left(1 + \frac{4}{r_k}\right)^{1/2} \qquad \text{...(3)}$$

and $\quad x_2^*(r_k) = -\frac{1}{2r_k} \qquad\qquad\qquad\qquad\qquad \text{...(4)}$

Hence, the optimum points are

$$x_1^* = \lim_{r_k \to \infty} x_1^*(r_k) = 1 \text{ and } x_2^* = \lim_{r_k \to \infty} x_2^*(r_k) = 0$$

and $\quad$ Min. $f(X) = \frac{8}{3} \qquad\qquad\qquad\qquad\qquad$ (see remark)

☛ **REMARK**

- If the function $F(X, r_k)$ is minimized for an increasing sequence of values of r_k, the unconstrained minima x_k^* converges to the optimum solution of the constrained problem $r_k \to \infty$.

EXERCISE 11.1

1. Minimize $f(X) = x_1^2 + x_2^2 - 6x_1 - 8x_2 + 10$
   subject to the constraints
   $$4x_1^2 + x_2^2 \le 16, \ 3x_1 + 5x_2 \le 15$$
   Using
   (i) interior penalty function method.
   (ii) exterior penalty function method.

2. Using complex method
   Minimize $f(X) = (x_1-1)^2 + (x_2-2)^2$
   subject to the constraints
   $$x_1 + x_2 \le 4, x_1 - x_2 \le 2, x_1 \ge 0, x_2 \ge 0 \quad \text{with}$$
   $$X_1 = \begin{bmatrix} 0.5 \\ 1.5 \end{bmatrix}$$

3. Find the dimensions of a rectangular prism

type box that has the largest volume when the sum of its length, width and height is limited to a maximum value of 60 cm and its length is restricted to a maximum value of 36 cms.

4. Consider the problem
   Minimize $f = x_1^2 + x_2^2 - 6x_1 - 8x_2 + 15$
   subject to the constraints
   $$4x_1^2 + x_2^2 \ge 16, 3x_1 + 5x_2 \le 15$$
   Normalize the constraints and find a suitable value of r_1 for use in the interior penalty function method at the starting point
   $$X_1 = \begin{bmatrix} 0 \\ 0 \end{bmatrix}.$$

ANSWERS

1. $x_1 = 3, x_2 = 4$

2. $x_1 = 1, x_2 = 2$

3. $x_1 = 20, x_2 = 20, x_3 = 20$, max. $f = 8000$ cm^3

4. $\frac{1}{4}x_1^2 + \frac{1}{16}x_2^2 - 1 \le 0, \frac{x_1}{5} + \frac{x_2}{3} - 1 \le 0, r_1 = 1.5$

▯▯▯▯

12 Dynamic Programming

Dynamic programming is a mathematical technique which is useful for solving a multistage decision problem *i.e.* for making a sequence of inter-related decision. It provides a dynamic procedure for determining the combination of decision which maximize over all effectiveness. Dynamic programming technique decompose the original problem in n-variables into n-subproblems each in one variable. We want to take such a decision at every stage so that the total effectiveness (the objective function) defined over all stage is optimal. Thus dynamic programming is a technique of recursive optimization. In this technique one has to obtain the solution in an orderly manner by starting from one stage to the next and is completed after the final stage is reached. The technique of dynamic programming was developed by **Richard Bellman** in 1950.

12.2 BELLMAN'S PRINCIPLE OF OPTIMALITY

[MEERUT–1998, 2003, 05, 06, 07, 08, 10, 14, 15; ROHILKHAND–1999; AGRA–2000]

It states that

"An optimal policy (*i.e.* set of decision) has the property that whatever be the initial state and initial decision, the remaining decisions must constitute an optimal policy for the state resulting from the first decision."

☞ REMARK

- A problem which does not satisfy the principle of optimality cannot be solved by dynamic programming.

12.3 MULTISTAGE DECISION PROBLEM

A problem in which the decision have to be made at successive stages is called a multistage decision problem.

Though the multistage problem occur themselves sufficient frequently, it is often possible to introduce the multistage nature in the problem so that the dynamic programming may be used. It can be classified on the basis of the following properties:

(i) The outcome of a decision may be deterministic or probabilistic. In case of deterministic, given the stage of the process, the outcomes of the decision at any stage is uniquely determined and known. In probabilistic case, there is a set of possible outcomes given by a known probability distribution.

(ii) The possible decision at any stage from which we are to choose one are called 'states'. These may be finite or infinite (states are the possible situations in which the system may be at any stage).

(iii) The toal number of stages in the process may be finite or infinite and may be known and unknown.

12.4 CHARACTERSTIC OF A DYNAMIC PROGRAMMING PROBLEM [MEERUT–1998, 99]

The characteristic of a dynamic programming problem may be outlined as follows:

(1) The problem can be divided into stages with a policy decision required at each stage.
(2) Each stage has a number of states associated with it.
(3) The effect of the policy decision at each stage is to transform the current state into a state associated with the next stage.
(4) Given the current stage, an optimal policy for remaining stages is independent of the policy adopted in the previous stage.
(5) The solution procedure begins by finding the optimal policy for each state of the last stage.
(6) A functional equation is available which identify the optimal policy for each state with n-stages remaining, given the optimal policy for each state with $(n – 1)$ stages left.
(7) Using the functional equation, the solution procedure moves backward stage by stage each time finding the policy when starting at the initial stage.

12.5 SOLUTION OF A MULTI-STAGE PROBLEM OF DYNAMIC PROGRAMMING WITH FINITE NUMBER OF STAGES [MEERUT–2001, 07]

The solution of problems by dynamic programming is usually done in three stages.

(i) Mathematical formulation of the problem.
(ii) The development of functional equations for the problem.
(iii) To solve functional equations for determining the optimal policy.

12.5.1 TO DEVELOP A FUNCTIONAL RELATIONSHIP: THE RECURSIVE EQUATIONS APPROACH

At this stage we have to develop a recurrence relation connecting the optimal decision function for n-stage problem with the optimal decision function for the $(n–1)$-stage sub-problem, $n = 1, 2, ..., n$.

12.5.2 TO SOLVE THE FUNCTIONAL EQUATION

Firstly we write the optimal decision function for one stage subproblem and then solve it.

Then we solve the optimal decision functions for 2-stage, 3-stage,... $(n – 1)$-stage and n-stage problem.

WORKING PROCEDURE

STEP 1. Identify the problem decision variables and specify objective function to be optimized.

STEP 2. Divide the given problem into a number of sub-problems (stages). Identify the state variables at each stage and write down the transformation function as a function of the state variables and decision variables at the next stage.

STEP 3. Write a general recursive formula for computing the optimal policy. Also decide whether to follow the forward or the backward method (given at the end of this procedure) to solve the problem.

STEP 4. Determine the overall optimal policy or decision and its value at each stage.

12.5.3 FORWARD AND BACKWARD COMPUTATIONS

If the recursive equations involved in a dynamic programming problem, are solved in the order,

$$f_1 \to f_2 \to .... \to f_n$$

then the computation involved is called forward computational procedure.

But if the recursive equations may be solved in the order $f_n \to f_{n-1} \to .... \to f_1$

Then the computation involved is called backward computational procedure.

The solution of a recursive equation involve two types of computations according as the system is discrete or continuous. If the system is discrete, a tabulator computational scheme is followed at any stage. In each table, the no. of rows and columns are equal to the no. of corresponding feasible states values and the no. of possible decision respectively. In case of continuous system the optimal decisions at each stage are obtained by using the usual classical techniques.

12.6 TYPES OF PROBLEMS

(I) SINGLE ADDITIVE CONSTRAINTS, MULTIPLICATIVELY SEPERABLE RETURN

Consider the following problem

$$\text{Max}.Z = \prod_{j=1}^{n} f_j(y_j)$$

subject to the constraints

$$\sum_{j=1}^{n} a_j y_j = b$$

and $\qquad y_j \geq 0, a_j \geq 0$

Now introduce state variables i.e. $s_n = \sum a_j y_j = b$

$s_{j-1} = s_j - a_j y_j, j = 2,3,...n$

Let $F_j(s_j) = \max_{y_1,y_2,...y_j 1} \prod^{j} f_j(y_j)$

Then the general recursion formula becomes

$F_j(s_j) = \max_{y_j}\left[f_i(y_i)F_{j-1}(s_j - 1)\right], \; j = n,n-1,...2$

$F_1(s_1) = f_1(y_1)$

(II) SINGLE ADDITIVE CONSTRAINTS ADDITIVELY SEPERABLE RETURN

Let us consider a problem in which objective function Z is an additively seperable function of n variables y_i and $f_i(y_i)$ is a function of y_i. Then we have to find $y_j, 1 \leq j \leq n$ which Minimize

$Z = \sum_{j=1}^{n} f_j(y_j)$

subject to the constraints

$$\sum_{j=1}^{n} a_j y_j \geq b$$

where a_j and b are real numbers and $a_j \geq 0, y_j \geq 0, b > 0$. In this problem each decision y_j is associated with a return function $f_j(y_j)$.

Now introduce state variables $s_0, s_1...s_n$ such that

$$s_n = a_1 y_1 + a_2 y_2 + ... + a_n y_n \geq b$$
$$s_{n-1} = a_1 y_1 + a_2 y_2 + ...a_{n-1}y_{n-1} = s_n - a_n y_n$$
$$s_{n-2} = a_1 y_1 + a_2 y_2 + ... + a_{n-2}y_{n-2} = s_{n-1} - a_{n-1}y_{n-1}$$
$$...$$
$$s_1 = a_1 y_1 = s_2 - a_2 y_2$$

Also, $\qquad s_{j-1} = T_j(s_j, y_j), 1 \leq j \leq n$ is the stage transformation function and $F_n(s_n)$ denote the minimum value of Z for any feasible value of s_n such that

$$F_n(s_n) = \min_{y_1, y_2, \dots, y_n} \left[f_1(y_1) + f_2(y_2) + \dots + f_n(y_n) \right], s_n \geq b$$

Now, first choose a particular value of y_n and minimize Z over the remaining $n-1$ variables, then

$$F_n(s_n) = \min_{y_1, y_2, \dots, y_{n-1}} \left[\sum_{j=1}^{n-1} f_j(y_j) = f_n(y_n) + F_{n-1}(s_{n-1}) \right]$$

Here, values of $y_1, y_2, \dots, y_{n-1}$ for which $\sum_{j=1}^{n-1} f_j(y_j)$ is minimum keeping y_n fixed, they depend upon s_{n-1} which is a function of s_n and y_n. So, minimum over all y_n for any feasible s_n would now become

$$F_n(s_n) = \min_{y_n} \left[f_n(y_n) + F_{n-1}(s_{n-1}) \right]$$

If the value of $F_{n-1}(s_{n-1})$ is known for all y_n, the function to be minimized would involve only a single variable y_n and can be solve easily.

Similarly the recursion formula is

$$F_j(s_j) = \min_{y_j} \left[f_j(y_j) + F_{j-1}(s_{j-1}) \right]; 1 \leq j \leq n, \quad F_1(s_1) = f_i(y_i)$$

So, starting with $F_1(s_1)$ and recursively optimizing to get $F_2(s_2), F_3(s_3)\dots$, we obtain $F_n(s_n)$ for each s_n. Here, each time optimization occur over a single variable.

(III) SINGLE MULTIPLICATIVE CONSTRAINTS, ADDITIVELY SEPERABLE RETURN

Consider the problem

$$\min. Z = f_1(y_1) + f_2(y_2) + \dots + f_n(y_n)$$

subject to the constraints

$$y_1 . y_2 \dots y_n \leq p, \qquad p \geq 0, \quad y_i \geq 0 \quad \forall i$$

Now state variables are defined as

$$s_n = y_n y_{n-1} \dots y_2 y_1 \geq p$$

$$s_{n-1} = s_n / y_n = y_{n-1} y_{n-2} \dots y_2 y_1$$

$$\dots\dots$$

$$\dots\dots$$

$$s_2 = s_3 / y_3 = y_2 y_1$$

$$s_1 = s_2 / y_2 = y_1$$

Let $F(s_n)$ be the minimum value of the objective function for s_n. Then recursion formula is given by

$$f_j(s_j) = \min_{y_j} \left[f_j(y_j) + F_{j-1}(s_{j-1}) \right], 2 \leq j \leq n$$

(IV) SYSTEM INVOLVING MORE THAN ONE CONSTRAINTS

Dynamic programming methods can be applied to problems involving more than one constraints. In single constraint problem, there has to be single state variable for each stage while in multi constrained problem, there has to be one state variable per constraint per stage.

SOLVED EXAMPLES

EXAMPLE 1. *Use dynamic programming to find the maximum value of $y_1, y_2, y_3, y_4, \dots y_n$, when $y_1 + y_2 + y_3 + y_4 + \dots + y_n = c$ and $y_j \geq 0$ for $j = 1, 2, 3, \dots, n$, where c is a positive number.* [MEERUT-1993, 98, 2002, 03, 10; IAS-1994; AGRA-2000; ROHILKHAND-1995]

SOLUTION.   To solve this problem by dynamic programming, we proceed the two steps: (i) To obtain functional equations (recursive relation) (ii) solution of the functional equation (recursive relation)

Step 1. To find the Recursive relation: Let y_j be the jth parts of positive number c where $j = 1, 2, 3,...,n$. Here each j corresponding to part y_j may be regarded as a stage. Now since y_j may assume any non-negative value which satisfying the constraints.

$$y_1 + y_2 + y_3 + ... + y_n = c$$

Therefore, alternatives at each stage are infinite. This means that y_j may be considered to be continuous. Thus, it is a problem of continuous system.

Hence the optimal decisions are obtained by using the differentiation at each stage.

Let $f_n(c)$ denote the maximum attainable product when the positive quantity c is divided into n parts. Since the quantity c is fixed so $f_n(c)$ depends upon n. The recursive relations of this problem stage by stage are as follows.

For stage 1, *i.e.*, $n = 1$, If c is divided into one part. Then we have $y_1 = c$

so,   we have $f_1(c) = c$ ...(1)

For stage 2, *i.e.*, $n = 2$ if c is divided into two parts then

Let $y_1 = z$ $\qquad \therefore y_2 = c - z$

therefore $f_2(c) = \max\{y_1, y_2\} = \max_{0 \le z \le c} \{z(c - z)\}$

$\therefore \qquad f_2(c) = \max_{0 \le z \le c} \{z f_1(c - z)\}$ ...(2) [using (1)]

For stage 3, *i.e.*, $n = 3$, If c is divided into three parts

Let $y_1 = z$ then we take $y_2 + y_3 = c - z$

$\therefore \qquad f_3(c) = \max\{y_1, y_2, y_3\} = \max_{0 \le z \le c} \{z \cdot f_2(c - z)\}$ ...(3)

Hence, in general, for n stage problem, the recursive relation is given by

$$f_n(c) = \max_{0 \le z \le c} \{z f_{n-1}(c - z)\}$$ ...(4)

Step 2. Solution of the functional equations:

For stage 1, we have $f_1(c) = c$

For stage 2, we have $f_2(c) = \max_{0 \le z \le c} \{z f_1(c - z)\}$

$$= \max_{z \le c} \{z.(c - z)\} \qquad\qquad [\because f_1(c - z) = (c - z)]$$

Since, the function $z(c - z)$ has the maximum value at $z = \dfrac{c}{2}$ satisfying the restriction

$0 \le z \le c$. therefore

$$f_2(c) = \left\{\frac{c}{2} \cdot \frac{c}{2}\right\} = \left\{\frac{c}{2}\right\}^2 \qquad\qquad [\text{since } c - z = \frac{c}{2}]$$

Hence, the optimal policy of two parts is $(\dfrac{c}{2}, \dfrac{c}{2})$.

For stage 3, we have

$$f_3(c) = \max_{0 \le z \le c} \{z f_2(c - z)\} = \max_{0 \le z \le c} \left\{z\left(\frac{c - z}{2}\right)^2\right\}$$

Since the maximum value of the function $z\left(\dfrac{c - z}{2}\right)^2$ is attained for $z = \dfrac{c}{3}$ satisfying the restriction $0 \le z \le c$. Hence,

$$f_3(c) = \frac{c}{3}(c - \frac{c}{3})^2 = (\frac{c}{3})^3$$

So, the optimal policy of three parts is $(\frac{c}{3}, \frac{c}{3}, \frac{c}{3})$.

Therefore, in general, we can assume for n stage problem

$$f_n(c) = \left(\frac{c}{n}\right)^n$$

Hence, the optimal policy of n parts is $\left(\frac{c}{n}, \frac{c}{n}, \frac{c}{n}, ..., n \text{ time}\right)$.

Using the principle of mathematical induction, we have

$$f_{n+1}(c) = \max_{0 \le z \le c} \{z f_n(c - z)\} = \max_{0 \le z \le c} \left\{ z \left(\frac{c-z}{n}\right)^n \right\}$$

Since the function $z\left(\frac{c-z}{n}\right)^n$ has its maximum value at $z = \frac{c}{n+1}$.

So, $$f_{n+1}(c) = \left(\frac{c}{n+1}\right)^{n+1}$$

Hence, by the principle at mathematical induction, the required policy for n stage

problem is $\left(\frac{c}{n}, \frac{c}{n}, ..., \frac{c}{n}\right)$, and $f_n(c) = \left(\frac{c}{n}\right)^n$.

EXAMPLE 2. *Obtain the minimum value of* $x_1 + x_2 + x_3 + ... + x_n$, *when* $x_1 \cdot x_2 \cdot x_3 \cdot ... \cdot x_n = d$, $x_1, x_2, x_3, ..., x_n \ge 0$.

[MEERUT 1981, 82, 93, 94, 2007, 08, AGRA 2007; ROHILKHAND–1995, 2005]

SOLUTION. To solve this problem by dynamic programming, we proceed the following two steps:
(i) To find the functional equations (Recursive relation) (ii) Solution of functional equations (Recursive relations)

Step 1. To find the Functional equations: Let $f_n(d)$ be the minimum attainable sum $z = x_1 + x_2 + x_3 + ... + x_n$, when d is factorized in n factors, i.e., $x_1, x_2, x_3, ..., x_n$.

For stage 1, i.e., $n = 1$ we have $x_1 = d$, where d is factorized into one part only

$\therefore \qquad f_1(d) = \min z = \min\{x_1\} = d$ ...(1)

For stage 2, i.e., $n = 2$, If d is factorized into two factors.

Let $x_1 = y$ and $x_2 = \dfrac{d}{y}$

$\therefore \qquad f_2(d) = \min\{x_1 + x_2\} = \min_{0 \le y \le d} \{y \cdot \dfrac{d}{y}\}$

$\qquad\qquad = \min_{0 \le y \le d} \{y + f_1(\dfrac{d}{y})\}$ ...(2) $\left[\because f_1(\dfrac{d}{y}) = \dfrac{d}{y}\right]$

For stage 3, i.e., $n = 3$, If d is factorized into three parts.

Let $x_1 = y$ and $x_2 x_3 = \dfrac{d}{y}$

$\therefore \qquad f_3(d) = \min\{x_1 + x_2 + x_3\}$

$\qquad\qquad = \min_{0 \le y \le d} \left\{y + f_2(\dfrac{d}{y})\right\}$ ...(3)

Hence, in general for n stage problem the recursive relation is given by

$$f_n(d) = \min_{0 \le y \le d} \left\{y + f_{n-1}(\dfrac{d}{y})\right\}$$ ...(4)

Step 2 Solution of functional equations (recursive relation):

For stage 1, we have $f_1(d) = d$

For stage 2, we have $f_2(d) = \min_{0 \le y \le d} \left\{ y + f_1\left(\frac{d}{y}\right) \right\} = \min_{0 \le y \le d} \left\{ y + \frac{d}{y} \right\}$

Since, the function $\left(y + \frac{d}{y} \right)$ has the minimum value at $y = d^{1/2}$ satisfying the restriction

$0 \le z \le d$. So that $\frac{d}{dy}(y + \frac{d}{y}) = 0 \Rightarrow 1 - \frac{d}{y^2} = 0 \Rightarrow y = d^{1/2}$

and $\frac{d^2}{dy^2}\left(y + \frac{d}{y} \right)$ has a positive value at $y = d^{1/2}$

Therefore, $f_2(d) = d^{1/2} + \frac{d}{d^{1/2}} = d^{1/2} + d^{1/2} = 2d^{1/2}$ ...(5)

Hence, the optimal policy of two parts is $(d^{1/2}, d^{1/2})$.

For stage 3, we have $f_3(d) = \min_{0 \le y \le d} \left\{ y + f_2\left(\frac{d}{y}\right) \right\} = \min_{0 \le y \le d} \left\{ y + 2\left(\frac{d}{y}\right)^{1/2} \right\}$

Since $y + 2\left(\frac{d}{y}\right)^{1/2}$ has the minimum value at $y = d^{1/3}$ satisfying the restriction $0 \le y \le d$

so that

$\frac{d}{dy}\left\{ y + 2\left(\frac{d}{y}\right)^{1/2} \right\} = 0 \Rightarrow y = d^{1/3}$

and $\frac{d^2}{dy^2}\left\{ y + 2\left(\frac{d}{y}\right)^{1/2} \right\}$ has a positive value at $y = d^{1/3}$.

$\therefore \quad f_3(d) = d^{1/3} + 2\left(\frac{d}{d/3}\right)^{1/2} = d^{1/3} + 2d^{1/3} = 3d^{1/3}$ ...(6)

therefore, the optimal policy for three parts is $(d^{1/3}, d^{1/3}, d^{1/3})$.

Now, we can assume for n stage problem, we have

$f_n(d) = nd^{1/n}$

and the optimal policy for n parts is $(d^{1/n}, d^{1/n}, ..., d^{1/n})$

And, we will prove the result by induction, for this

$f_{n+1}(d) \quad = \min_{0 \le y \le d} \left\{ y + f_n\left(\frac{d}{y}\right) \right\}$

$= \min_{0 \le y \le d} \left\{ y + n\left(\frac{d}{y}\right)^{1/n} \right\}$.

Since, $y + n\left(\frac{d}{y}\right)^{1/n}$ has the minimum value at $y = d^{1/n+1}$

$\therefore \quad f_{n+1}(d) = d^{1/n+1} + n\left(\frac{d}{d^{1/n+1}}\right)^{1/n} = (n+1)d^{1/n+1}$

and the optimal policy for $(n+1)$ stage is $(d^{1/n+1}, d^{1/n+1}, ..., d^{1/n+1})$.

Hence, by mathematical induction, $f_n(d) = nd^{1/n}$ and the required policy for n stage problem is $(d^{1/n}, d^{1/n}, ... d^{1/n})$.

EXAMPLE 3. *Use dynamic programming to show that* $-\sum\limits_{i=1}^{n} p_i \log p_i$ *is maximum subjected*

to $\sum\limits_{i=1}^{n} p_i = 1$ *when* $p_1 = p_2 = p_3 = ... = p_n = 1/n$

[MEERUT-1980, 82(P), 2003 ROHILKHAND-1994, 2006, KANPUR–2000, AGRA-1996, 97, 98, 2013]

SOLUTION. **Step 1 Functional equations (recursive relation):**

In this problem we have

$p_1 + p_2 + p_3 + ... + p_n = 1$ such that $-\sum\limits_{i=1}^{n} p_i \log p_i$ is maximum.

Let $f_n(1)$ be the maximum attainable value of $-\sum\limits_{i=1}^{n} p_i \log p_i$ when $p_1 + p_2 + p_3 + ... + p_n = 1$

Since p_i may assume any non-negative value. So, $f_n(1)$ is a function of discrete variable.

For stage 1, *i.e.*, $n = 1$. If we divide 1 into one part p_i, then $p_1 = 1$

∴ $f_1(1) = \max\{-p_1 \log p_1\} = -1 \log 1$ ...(1)

For stage 2, *i.e.*, If we divide 1 into two parts p_1 and p_2 then

let $p_1 = z$, $p_2 = 1 - z$

∴ $f_2(1)$ $= \max\{-p_1 \log p_1, -p_2 \log p_2\}$

$= \max\limits_{0 \leq z \leq 1} \{-z \log z - (1-z) \log(1-z)\}$

$= \max\limits_{0 \leq z \leq 1} \{-z \log z + f_1(1-z)\}$ ...(2)

$[\because f_1(1-z) = -(1-z)\log(1-z)]$

For stage 3, *i.e.*, If we divide 1 into three parts p_1, p_2 and p_3 then

let $p_1 = z$ $\therefore p_2 + p_3 = 1 - z$

∴ $f_3(1)$ $= \max\{-p_1 \log p_1, -p_2 \log p_2, -p_3 \log p_3\}$

$= \max\limits_{0 \leq z \leq 1} \{-z \log z + f_2(1-z)\}$ ...(3)

Hence, in general, for n stage problem the recursive relation is given by

$f_n(1) = \max\limits_{0 \leq z \leq 1} \{-z \log z + f_{n-1}(1-z)\}$ ...(4)

Step 2 Solution of functional equations (recursive relation):

For stage 1, we have $f_1(1)$ $= -1 \log 1$

For stage 2, we have $f_2(1)$ $= \max\limits_{0 \leq z \leq 1} \{-z \log z + f_1(1-z)\}$

$= \max\limits_{0 \leq z \leq 1} \{-z \log z - (1-z) \log(1-z)\}$

Since the function $\{-z \log z - (1-z) \log(1-z)\}$ has the maximum value at $z = \dfrac{1}{2}$

satisfying the restriction $0 \leq z \leq 1$ so that

$\dfrac{d}{dz}\{-z \log z - (1-z) \log(1-z)\} = 0 \Rightarrow z = \dfrac{1}{2}$

and $\dfrac{d^2}{dz^2}\{-z \log z - (1-z) \log(1-z)\} = -\dfrac{1}{z} - \dfrac{1}{1-z} = -4$ (negative)

therefore $f_2(1) = -\dfrac{1}{2}\log\dfrac{1}{2} - \dfrac{1}{2}\log\dfrac{1}{2} = 2\left(-\dfrac{1}{2}\log\dfrac{1}{2}\right)$ ...(5)

$\left[\because 1 - z = 1 - \dfrac{1}{2} = \dfrac{1}{2}\right]$

and the optimal policy for two parts is $\left(\dfrac{1}{2}, \dfrac{1}{2}\right)$.

For stage 3, $f_3(1)$ $= \max\limits_{0 \leq z \leq 1} \{-z \log z + f_2(1-z)\}$

$$= \max_{0 \le z \le 1} \left\{ -z \log z + 2 \left(-\frac{(1-z)}{2} \log \frac{(1-z)}{2} \right) \right\}$$

Since, the function $\left\{ -z \log z - (1-z) \log \frac{(1-z)}{2} \right\}$ is maximum at $z = \frac{1}{3}$ satisfying the restriction $0 \le z \le 1$ so that

$$\frac{d}{dz} \left\{ -z \log z - (1-z) \log \frac{(1-z)}{2} \right\} = 0 \Rightarrow -\log z + \log \frac{(1-z)}{2} - 1 + 1 = 0$$

$$\Rightarrow \qquad\qquad z = \frac{1}{3}$$

and $\frac{d^2}{dz^2} \left\{ -z \log z - (1-z) \log \frac{(1-z)}{2} \right\}$ is negative at $z = \frac{1}{3}$.

$$\therefore \quad f_3(1) = -\frac{1}{3} \log \frac{1}{3} + 2 \left(-\frac{2/3}{2} \log \frac{2/3}{2} \right) \qquad\qquad \left[\because 1 - z = 1 - \frac{1}{3} = \frac{2}{3} \right]$$

$$= -\frac{1}{3} \log \frac{1}{3} + 2 \left(-\frac{1}{3} \log \frac{1}{3} \right) = 3 \left(-\frac{1}{3} \log \frac{1}{3} \right) \qquad\qquad \text{...(6)}$$

and the optimal policy for three parts is $\left(\frac{1}{3}, \frac{1}{3}, \frac{1}{3} \right)$.

In general, for n stage problem, we have

$$f_n(1) = n \left(-\frac{1}{n} \log \frac{1}{n} \right) \qquad\qquad \text{...(7)}$$

Now, we will prove the result by induction.

For this, we have

$$f_{n+1}(1) = \max_{0 \le z \le 1} \left\{ -z \log z + f_n(1-z) \right\}$$

$$= \max_{0 \le z \le 1} \left\{ -z \log z + n \left(-\frac{1}{n} \log \frac{1}{n} \right) \right\}$$

Since $\left\{ -z \log z + n \left(-\frac{1}{n} \log \frac{1}{n} \right) \right\}$ is maximum at $z = \frac{1}{n+1}$ satisfying the restriction $0 \le z \le 1$ so that

$$\frac{d}{dz} \left\{ -z \log z + n \left(-\frac{1}{n} \log \frac{1}{n} \right) \right\} = 0 \quad \Rightarrow z = \frac{1}{n+1}$$

and $\frac{d^2}{dz^2} \left\{ -z \log z + n \left(-\frac{1}{n} \log \frac{1}{n} \right) \right\}$ is negative for $z = \frac{1}{n+1}$.

Now, $f_n(1-z)$

$$= f_n \left(\frac{n}{1+n} \right) = n \left\{ -\frac{\frac{n}{1+n}}{n} \log \frac{\frac{n}{1+n}}{n} \right\} = n \left[-\frac{1}{1+n} \log \frac{1}{1+n} \right]$$

$$\therefore \quad f_{n+1}(1) = -\frac{1}{1+n} \log \frac{1}{1+n} + n \left(-\frac{1}{1+n} \log \frac{1}{1+n} \right)$$

$$= (1+n) \left(-\frac{1}{1+n} \log \frac{1}{1+n} \right)$$

and the optimal policy is $\left(\frac{1}{n+1}, \frac{1}{n+1}, \frac{1}{n+1}, ..., \frac{1}{n+1} \right)$.

Hence, the maximum value of $f_n(1)$ is $n \left(-\frac{1}{n} \log \frac{1}{n} \right)$ and the optimal policy is $\left(\frac{1}{n}, \frac{1}{n}, \frac{1}{n}, ..., \frac{1}{n} \right)$, i.e., $p_1 = p_2 = ... p_n = \frac{1}{n}$.

EXAMPLE 4. *Find the functional equations for maximizing*

$$z = g_1(x_1) + g_2(x_2) + g_3(x_3) + \ldots + g_n(x_n)$$

subjected to $x_1 + x_2 + \ldots + x_n = c$, *where* $x_1, x_2, x_3, \ldots x_n \geq 0$. [DELHI–1995]

SOLUTION. Let $f_n(c)$ denote the maximum attainable value of

$z = g_1(x_1) + g_2(x_2) + g_3(x_3) + \ldots + g_n(x_n)$ when $x_1 + x_2 + \ldots + x_n = c$,

$x_1, x_2, x_3, \ldots x_n \geq 0$

Since, x_i may assume any non-negative value. So, $f_n(c)$ is a function of discrete variable and it is a continuous system problem.

For stage 1, *i.e.*, $n = 1$, If we divide c into one part *i.e.*, $x_1 = c$

$\therefore \quad f_1(c) = \max\{g_1(x_1)\} = g_1(c)$ ...(1)

For stage 2, *i.e.* $n = 2$. If we divide c into two parts x_1 and x_2, then let $x_2 = z$, then $x_1 = c - z$.

$$\therefore \quad f_2(c) = \max\{g_1(x_1) + g_2(x_2)\}$$

$$= \max_{0 \leq z \leq c} \{g_2(z) + g_1(c - z)\}$$

$$= \max_{0 \leq z \leq c} \{g_2(z) + f_1(c - z)\} \quad \ldots(2)$$

$$[\because f_1(c - z) = c - z]$$

For stage 3, *i.e.*, If we divide c into three parts x_1, x_2 and x_3.

Let $x_3 = z$ then $x_1 + x_2 = c - z$

$$\therefore \quad f_3(c) = \max\{g_1(x_1) + g_2(x_2) + g_3(x_3)\}$$

$$= \max_{0 \leq z \leq c} \{g_3(z) + f_2(c - z)\} \quad \ldots(3)$$

Hence, in general for n stage problem, we have

$$f_n(c) = \max_{0 \leq z \leq c} \{g_n(z) + f_{n-1}(c - z)\} \quad \ldots(4)$$

which are required functional equations.

EXAMPLE 5. *Obtain the maximum value of* $b_1 x_1 + b_2 x_2 + \ldots + b_n x_n$, *when* $x_1 + x_2 + \ldots + x_n = c$,

$x_1, x_2, \ldots, x_n \geq 0$, $i = 1, 2, \ldots, n$ [MEERUT 1990, 97(P), 97, 2001, 09]

SOLUTION. **Step 1 To find the Functional equations (recursive relations):**

Let $f_n(c)$ denote the maximum attainable sum of $b_1 x_1 + b_2 x_2 + \ldots + b_n x_n$ when we divide the positive quantity c into n parts. It is a n stage problem.

For stage 1, *i.e.*, $n = 1$, If we divide c into one part then $x_1 = c$

$\therefore \quad f_1(c) = \max\{b_1 x_1\} = b_1 c$ ...(1)

For stage 2, *i.e.*, $n = 2$ If we divide c into two parts x_1 and x_2 then

$x_2 = z \quad \therefore \quad x_1 = c - z$

$$\therefore \quad f_2(c) = \max\{b_1 x_1 + b_2 x_2\} = \max_{0 \leq z \leq c} \{b_2 z + b_1(c - z)\}$$

$$= \max_{0 \leq z \leq c} \{b_2 z + f_1(c - z)\} \qquad [\because f_1(c - z) = b_1(c - z)] \quad \ldots(2)$$

For stage 3, *i.e.*, $n = 3$, If we divide c into three parts x_1, x_2 and x_3,

Let $x_3 = z$ then $x_1 + x_2 = c - z$

$$\therefore \quad f_3(c) = \max\{b_1 x_1 + b_2 x_2 + b_3 x_3\} = \max_{0 \leq z \leq c} \{b_3 z + f_2(c - z)\} \quad \ldots(3)$$

Hence, in general, for n stage problem, the recursive relation is

$$f_n(c) = \max_{0 \leq z \leq c} \{b_n z + f_{n-1}(c - z)\} \quad \ldots(4)$$

Step 2 Solution of functional equations (recursive relations):

For stage 1, we have $f_1(c) = b_1 c$

For stage 2, we have $f_2(c) = \max_{0 \leq z \leq c} \{b_2 z + f_1(c - z)\}$

$$= \max_{0 \leq z \leq c} \{b_2 z + b_1(c - z)\} \qquad [\because f_1(c - z) = c - z]$$

$$= \max_{0 \leq z \leq c} \{b_1 c + (b_2 - b_1)z\}$$

It is obvious that if $b_2 > b_1$, then $\{b_1 c + (b_2 - b_1)z\}$ is maximum for $z = c$, satisfying the restriction $0 \leq z \leq c$, otherwise, it will be minimum.

$$\therefore \qquad f_2(c) = \{b_1 c + (b_2 - b_1)c\} = b_2 c$$

Therefore, the optimum policy for two parts is $x_1 = 0, x_2 = c$.

For stage 3, we have $f_3(c) = \max_{0 \leq z \leq c} \{b_3 z + f_2(c - z)\}$

$$= \max_{0 \leq z \leq c} \{b_3 z + b_2(c - z)\} = \max_{0 \leq z \leq c} \{b_2 c + (b_3 - b_2)z\}$$

Here, it is obvious that if $b_3 > b_2$, then $\{b_2 c + (b_3 - b_2)z\}$ is maximum for $z = c$ satisfying the restriction $0 \leq z \leq c$, otherwise, it will be minimum.

$$\therefore \qquad f_3(c) = \{b_2 c + (b_3 - b_2)c\} = b_3 c$$

therefore, the optimal policy for three stage problem is $x_1 = 0, x_2 = 0, x_3 = c$

Hence, in general we can say

$$f_n(c) = b_n c$$

and the optimal policy for n stage problem is

$$x_1 = 0, x_2 = 0, x_3 = 0, ..., x_n = c$$

EXAMPLE 6. *Use dynamic programming technique, obtain the minimum value of* $x_1^2 + x_2^2 + x_3^2$, *when* $x_1 + x_2 + x_3 \geq 15$, $x_1, x_2, x_3 \geq 0$.

[MEERUT 1984, 88, 2005, 08, AGRA–2003, 07; IAS–1995]

SOLUTION. **Step 1 To find the Functional equations:**

Since, there are three decision variables x_1, x_2 and x_3, so it is a three stage problem which can be defined as follows:

Consider $\quad s_3 = x_1 + x_2 + x_3 \geq 15$

$$s_2 = x_1 + x_2 = s_3 - x_3$$

and $\quad s_1 = x_1 = s_2 - x_2$

Since, it is a three stage problem. Let $f_3(s_3)$ denote the minimum attainable value of $x_1^2 + x_2^2 + x_3^2$ at the third stage, $s_3 = x_1 + x_2 + x_3$.

then,

$$f_1(s_1) = \min_{0 \leq x_1 \leq s_1} \{x_1^2\} = (s_2 - x_2)^2 \qquad ...(1)$$

$$f_2(s_2) = \min_{0 \leq x_2 \leq s_2} \{x_1^2 + x_2^2\}$$

$$= \min_{0 \leq x_2 \leq s_2} \{x_2^2 + f_1(s_1)\} \qquad ...(2)$$

and $f_3(s_3) = \min_{0 \leq x_3 \leq s_3} \{x_3^2 + f_2(s_2)\} \qquad ...(3)$

Step 2 Solution of the functional equations:

From equation (1), we have

$$f_1(s_1) = (s_2 - x_2)^2$$

From equation (2), we have

$$f_2(s_2) = \min_{0 \leq x_2 \leq s_2} \{x_2^2 + f_1(s_1)\}$$

$$= \min_{0 \leq x_2 \leq s_2} \{x_2^2 + (s_2 - x_2)^2\}$$

Since, $\{x_2^2 + (s_2 - x_2)^2\}$ is minimum for $x_2 = \dfrac{s_2}{2}$ so that

$$\frac{d}{dx_2}\{x_2^2 + (s_2 - x_2)^2\} = 4x_2 - 2s_2 = 0 \Rightarrow x_2 = \frac{s_2}{2}$$

and $\dfrac{d^2}{dx_2^2}\{x_2^2 + (s_2 - x_2)^2\} = 4$ at $x_2 = \dfrac{s_2}{2}$, which is positive.

$\therefore$

$$f_2(s_2) = \left\{\left(\frac{s_2}{2}\right)^2 + \left(s_2 - \frac{s_2}{2}\right)^2\right\}$$

$$= \left\{\frac{s_2^2}{4} + s_2^2 + \frac{s_2^2}{4} - s_2^2\right\}$$

$$= \frac{s_2^2}{2}$$

Now, from equation(3), we have

$$f_3(s_3) = \min_{0 \le x_3 \le s_3} \{x_3^2 + f_2(s_2)\}$$

$$= \min_{0 \le x_3 \le s_3} \left\{x_3^2 + \frac{(s_3 - x_3)^2}{2}\right\} \qquad \left[\because f_2(s_2) = \frac{(s_3 - x_3)^2}{2}\right]$$

Since, $\left\{x_3^2 + \dfrac{(s_3 - x_3)^2}{2}\right\}$ is minimum for $x_3 = \dfrac{s_3}{3}$ so that

$$\frac{d}{dx_3}\left\{x_3^2 + \frac{(s_3 - x_3)^2}{2}\right\} = 0 \Rightarrow 3x_3 - s_3 = 0 \Rightarrow x_3 = \frac{s_3}{2}$$

and $\dfrac{d^2}{dx_3^2} = 3$ at $x_3 = \dfrac{s_3}{3}$, which is positive.

$\therefore$

$$f_3(s_3) = \left\{\left(\frac{s_3}{3}\right)^2 + \frac{1}{2}\left(s_3 - \frac{s_3}{3}\right)^2\right\}$$

$$= \left\{\frac{s_3^2}{9} + \frac{1}{2}\left(s_3^2 + \frac{s_3^2}{9} - \frac{2}{3}s_3^2\right)\right\} = \frac{s_3^2}{3}$$

But, it is given that $s_3 \ge 15$, so the minimum value of s_3 is 15.

$\therefore$

$$x_3 = \frac{s_3}{3} = \frac{15}{3} = 5$$

$$x_2 = \frac{s_2}{2} = \frac{s_3 - x_3}{2} = \frac{15 - 5}{2} = 5$$

and $x_1 = 5$, $f_3(s_3) = \dfrac{s_3^2}{2} = \dfrac{(15)^2}{2} = 75$

Hence, the optimal policy for this problem is

$x_1 = 5$, $x_2 = 5$, $x_3 = 5$ and minimum value of $x_1^2 + x_2^2 + x_3^2$ is 75.

EXAMPLE 7. *Use dynamic programming. Solve*

Max $z = x_1^2 + 2x_2^2 + 4x_3$, when $x_1 + 2x_2 + x_3 \le 8$, $x_1, x_2, x_3, \ge 0$

[MEERUT 1999(O) KANPUR 2007, AGRA 2002]

SOLUTION.   **Step 1 To find the Functional equations (recursive relations):**

Since, there are three decision variables x_1, x_2 and x_3. So, It is a three stage problem.

This problem can be defined as follows:

Consider
$$s_3 = x_1 + 2x_2 + x_3 \le 8$$
$$s_2 = x_1 + 2x_2 = s_3 - x_3$$
and
$$s_1 = x_1 = s_2 - 2x_2$$
...(1)

Since, it is a three stage problem, Let $f_i(s_i)$ denote the maximum attainable value of $z = x_1^2 + 2x_2^2 + 4x_3$ where $s_i = x_1 + x_2 + x_3 + ... + x_i$, $i = 1, 2, 3$.

Then

$$f_1(s_1) = \max_{0 \le x_1 \le s_1} \{x_1^2\} = (s_2 - 2x_2)^2 \qquad ...(2)$$

$$f_2(s_2) = \max_{0 \le 2x_2 \le s_2} \{x_1^2 + 2x_2^2\}$$

$$= \max_{0 \le 2x_2 \le s_2} \{2x_2^2 + (s_2 - 2x_2)^2\}$$

$$= \max_{0 \le x_2 \le \frac{s_2}{2}} \{2x_2^2 + f_1(s_1)\} \qquad ...(3)$$

and $$f_3(s_3) = \max_{0 \le x_3 \le s_3} \{x_1^2 + 2x_2^2 + 4x_3\}$$

$$= \max_{0 \le x_3 \le s_3} \{4x_3 + f_2(s_2)\} \qquad ...(4)$$

Step 2 Solution of functional equations:

From equation (2), we have

$$f_1(s_1) = (s_2 - 2x_2)^2 \qquad ...(5)$$

From equation (3), we have

$$f_2(s_2) = \max_{0 \le x_2 \le \frac{s_2}{2}} \{2x_2^2 + f_1(s_1)\}$$

$$= \max_{0 \le x_2 \le \frac{s_2}{2}} \{2x_2^2 + (s_2 - 2x_2)^2\}$$

$$= \max_{0 \le x_2 \le \frac{s_2}{2}} \{6x_2^2 - 4x_2 s_2 + s_2^2\}$$

Here, differential method is fail, so we check the max or min by the following procedure.

Since, x_2 lies within the range 0 to $\frac{s_2}{2}$. So the maximum value of $f_2(s_2)$ satisfying the restriction will be at $x_2 = 0$, $f_2(s_2) = s_2^2$

and at $x_2 = \frac{s_2}{2}$: $f_2(s_2) = 6\frac{s_2^2}{4} + s_2^2 - 4\frac{s_2^2}{2} = \frac{s_2^2}{2}$

It is obvious that $s_2^2 > \frac{s_2^2}{2}$, therefore $f_2(x_2)$ has maximum value s_2^2 at $x_2 = 0$

Again from (4), we have

$$f_3(s_3) = \max_{0 \le x_3 \le s_3} \{4x_3 + f_2(s_2)\} = \max_{0 \le x_3 \le s_3} \{4x_3 + (s_3 - x_3)^2\}$$

Since, x_2 lies within the range 0 to s_3. So, the maximum value of $f_3(s_3)$ satisfying the restriction will be

at $x_3 = 0$ $f_3(s_3) = s_3^2$
at $x_3 = s_3$ $f_3(s_3) = 4s_3$ ...(6)

Since $s_3 \le 8$, therefore maximum value of s_3 is 8.

From equation (6), at $x_3 = 0$, $f_3(s_3) = s_3^2 = (8)^2 = 64$

and at $x_3 = s_3 = 8$, $f_3(s_3) = 4 \times 8 = 32$.

Clearly, $64 > 32$, therefore, the maximum value of $f_3(s_3)$ is 64 at $x_3 = 0$

Now, $s_2 = s_3 - x_3 = 8 - 0 = 8$ and $x_2 = 0$

and $s_1 = s_2 - 2x_2 = 8 - 2 \times 0 = 8$ and $x_1 = s_2 - 2x_2 = 8 - 2 \times 0 = 8$

Hence, the solution is $x_1 = 8, x_2 = 0, x_3 = 0$ and max $z = 64$

EXAMPLE 8. *Use dynamic programming, solve*

$$\min z = x_1^2 + 2x_2^2 + 4x_3, \text{ when } x_1 + 2x_2 + x_3 \geq 8, x_1, x_2, x_3 \geq 0 \quad \text{[MEERUT–1989, 94(P), 99]}$$

SOLUTION. **Step 1 To find the Functional equation:**

Since there are three decision variable x_1, x_2 and x_3, so it is a three stage problem. This problem can be defined as:

Consider
$$\left.\begin{array}{l} s_3 = x_1 + 2x_2 + x_3 \geq 8 \\ s_2 = x_1 + 2x_2 = s_3 - x_3 \end{array}\right\}$$

and $\qquad s_1 = x_1 = s_2 - 2x_2$ ...(1)

Since, It is a three stage problem. Let $f_i(s_i)$ denote the minimum attainable value of $z = x_1^2 + 2x_2^2 + 4x_3$ at i^{th} stage where $i = 1, 2, 3$.

Then
$$f_1(s_1) = \min_{0 \leq x_1 \leq s_1} \{x_1^2\} = (s_2 - 2x_2)^2 \qquad \text{...(2)}$$

$$f_2(s_2) = \min_{0 \leq x_2 \leq s_2} \{x_1^2 + 2x_2^2\} = \min_{0 \leq x_2 \leq s_2} \{2x_2^2 + f_1(s_1)\} \qquad \text{...(3)}$$

and $\quad f_3(s_3) = \min_{0 \leq x_3 \leq s_3} \{x_1^2 + 2x_2^2 + 4x_3\} = \min_{0 \leq x_3 \leq s_3} \{4x_3 + f_2(s_2)\}$ ...(4)

Step 2 Solution of functional equations:

From equation (1), we have

$$f_1(s_1) = (s_2 - 2x_2)^2$$

From equation (2), we have

$$f_2(s_2) = \min_{0 \leq x_2 \leq s_2} \{2x_2^2 + f_1(s_1)\}$$

$$= \min_{0 \leq x_2 \leq s_2} \{2x_2^2 + (s_2 - 2x_2)^2\}$$

Since, the function $\{2x_2^2 + (s_2 - 2x_2)^2\}$ will be minimum at $x_2 = \dfrac{s_2}{3}$ satisfying the restriction so that

$$\frac{d}{dx_2}\{2x_2^2 + (s_2 - 2x_2)^2\} = 0 \Rightarrow 12x_2 - 4s_2 = 0 \Rightarrow x_2 = \frac{s_2}{3}$$

and $\quad \dfrac{d^2}{dx_2^2}\{2x_2^2 + (s_2 - 2x_2)^2\} = 12$ (Positive) at $x_2 = \dfrac{s_2}{3}$

Therefore, the minimum value of $f_2(s_2)$ is,

$$f_2(s_2) = 2\left(\frac{s_2}{3}\right)^2 + \left(s_2 - 2 \cdot \frac{s_2}{3}\right)^2 = \frac{s_2^2}{3} \text{ at } x_2 = \frac{s_2}{3}.$$

Now, from equation (3), we have

$$f_3(s_3) = \min_{0 \leq x_3 \leq s_3} \{4x_3 + f_1(s_1)\}$$

$$= \min_{0 \leq x_3 \leq s_3} \{4x_3 + \frac{s_2^2}{3}\}$$

$$= \min_{0 \leq x_3 \leq s_3} \left\{4x_3 + \frac{(s_3 - x_3)^2}{3}\right\}$$

Since, the function $\left\{4x_3 + \dfrac{(s_3 - x_3)^2}{3}\right\}$ is minimum at $x_3 = s_3 - 6$. So that

$\dfrac{d}{dx_3}\left\{4x_3 + \dfrac{(s_3 - x_3)^2}{3}\right\} = 0 \Rightarrow x_3 = s_3 - 6$ and $\dfrac{d^2}{dx_3^2} = 0$

therefore, the minimum value of $f_3(s_3)$ is

$$f_3(s_3) = \left\{4(s_3 - 6) + \dfrac{(s_3 - s_3 + 6)^2}{3}\right\} = 4s_3 - 12$$

Since, $x_3 \geq 8$. So the minimum value of s_3 is 8.

Then $f_3(s_3) = 4s_3 - 12 = 4 \times 8 - 12 = 20$

and $x_3 = s_3 - 6 = 8 - 6 = 2$ *i.e.,* $x_3 = 2$

$s_2 = s_3 - x_3 = 8 - 2 = 6$ *i.e.,* $s_2 = 6$

$\therefore \quad x_2 = \dfrac{s_2}{2} = \dfrac{6}{2} = 3$ *i.e.,* $x_2 = 3$

and $x_1 = s_2 - 2x_2 = 6 - 2 \times 2 = 2$

$\therefore \quad x_1 = 2$

Hence, the solution is

$x_1 = 2, x_2 = 3, x_3 = 2$ and Min $z = 20$

12.7 SOLUTION OF L.P.P. BY DYNAMIC PROGRAMMING

[MEERUT–2001, 06]

A general maximization linear programming problem is

Max $z = c_1x_1 + c_2x_2 + c_3x_3 + \ldots + c_nx_n$

subject to the constraints

$$a_{11}x_1 + a_{12}x_2 + a_{13}x_3 + \ldots + a_{1n}x_n \leq b_1$$
$$a_{21}x_1 + a_{22}x_2 + a_{23}x_3 + \ldots + a_{2n}x_n \leq b_2$$
$$a_{31}x_1 + a_{32}x_2 + a_{33}x_3 + \ldots + a_{3n}x_n \leq b_3$$
$$\vdots \qquad \vdots \qquad \quad \vdots \qquad \qquad \vdots$$
$$a_{m1}x_1 + a_{m2}x_2 + a_{m3}x_3 + \ldots + a_{mn}x_n \leq b_m$$
$$x_1, x_2, x_3, \ldots x_n \geq 0$$

It is n stage problem. Since we consider each activity j ($j=1, 2,\ldots,n$) as a individual stage, the level of activities $(x_j \geq 0)$ are the decision variables (alternatives) at stage j. Being x_j continuous, each activity has an infinite number of alternatives with in the feasible region.

We know that the allocation problems are the special case of L.P.P. which requires the allocation of available resources to the activities.

Each constraints of the L.P.P. represents the limitation of different resources and the amount of available resources are $b_1, b_2, b_3, \ldots, b_m$. Since there are m resources in this problem, therefore this problem has m state variables.

Let $(B_{1j}, B_{2j}, B_{3j}, \ldots, B_{mj})$ be the state of the system at j^{th} stage, i.e., $B_{1j}, B_{2j}, B_{3j}, \ldots, B_{mj}$ are the amounts of resources 1, 2, 3, ..., m to be allocated to stage j. Let the optimum value of objective function at j^{th} stage is $f_j(B_{1n}, B_{2n}, B_{3n}, \ldots, B_{mn})$.

Here, we shall use the backward computational procedure. The recurrence equation,

$$f_n(B_{1n}, B_{2n}, B_{3n}, \ldots, B_{mn}) = \max_{0 \leq a_{in}x_n \leq B_{in}} \{c_n x_n\} \quad i = 1, 2, \ldots, m$$

and $$f_j(B_{1j}, B_{2j}, B_{3j}, \ldots, B_{mj}) = \max_{0 \leq a_{ij}x_j \leq B_{ij}} \{c_j x_j + f_{j+1}(B_{1j} - a_{1j}x_1, B_{2j}$$
$$- a_{2j}x_2, B_{3j} - a_{3j}x_3, \ldots, B_{mj} - a_{mj}x_j\}$$

$$i = 1, 2, \ldots, m, \quad j = 1, 2, 3, \ldots (n-1).$$

for $\quad 0 \leq B_{ij} \leq b_i$ for all i and j.

For the better understanding see the following examples.

SOLVED EXAMPLES

EXAMPLE 1.   *Solve the L.P.P.*

$$\text{Max } z = 2x_1 + 5x_2$$

subjected to $2x_1 + x_2 \leq 43, 2x_2 \leq 46, x_1, x_2 \geq 0$

by using dynamic programming technique.   [MEERUT–1998, AGRA–2002]

SOLUTION.   In the given problem, there are two interrelated decision variables so it is a two stage linear programming problem. Due to the two resources b_1, b_2 with available amount 43 and 46, the state vector will have two components.

Let (B_{1j}, B_{2j}) is the state of the system and let us suppose that $f_j(B_{1j}, B_{2j})$ be the optimal value of the objective function for stage 1 and 2. Using backward computational procedure, we have

$$f_2(B_{12}, B_{22}) = \max_{\substack{0 \leq x_2 \leq 43 \\ 0 \leq 2x_2 \leq 46}} \{5x_2\} = 5 \max_{\substack{0 \leq x_2 \leq 43 \\ 0 \leq x_2 \leq 23}} \{x_2\}$$

Since $\max\{x_2\}$ which satisfying $x_2 \leq 43$ and $x_2 \leq 23$ is minimum of $B_{12} = 43$ and $B_{22} / 2 = 23$.

therefore, $\max\{x_2\} = x_2{}^* = \min\{B_{12}, B_{22} / 2\}$

so, $f_2(B_{12}, B_{22}) = 5 \min\left\{B_{12}, \dfrac{B_{22}}{2}\right\}$ ...(1)

We know that the value of B_{12}, B_{22} is given by at the primary stage.

Now $f_1(B_{11}, B_{21}) = \max_{\substack{0 \leq x_1 \leq \frac{B_{11}}{2} \\ 0 \leq x_1 \leq \frac{B_{21}}{2}}} \{2x_1 + f_2(B_{11} - 2x_1, B_{21} - 0)\}$ ...(2)

From equation (2), we have

$$f_2(B_{12}, B_{22}) = 5 \min\left\{B_{12}, \frac{B_{22}}{2}\right\}$$

So, $f_2(B_{11} - x_1, B_{21} - 0) = 5 \min\left\{B_{11} - x_1, \dfrac{B_{21}}{2}\right\}$

On putting this value in equation (2), we get

$$f_1(B_{11}, B_{21}) = \max_{\substack{0 \leq x_1 \leq \frac{B_{11}}{2} \\ 0 \leq x_1 \leq \frac{B_{21}}{2}}} \left\{2x_1 + 5\min\left(B_{11} - 2x_1, \frac{B_{21}}{2}\right)\right\}$$

or $f_1(B_{11}, B_{21}) = \max\limits_{\substack{0 \leq x_1 \leq \frac{B_{11}}{2} \\ 0 \leq x_1 \leq \frac{B_{21}}{2}}} \{2x_1 + 5\min(43 - 2x_1, 23)\}$

$$= \max_{0 \leq x_i \leq 21.5} \{2x_1 + 5\min(43 - 2x_1, 23)\}$$

$$f_1(B_{11}, B_{21}) = \begin{cases} 2x_1 + 5(43 - 2x_1) & 10 \leq x_1 \leq 23 \\ 2x_1 + 5 \times 23 & 0 \leq x_1 \leq 10 \end{cases} = \begin{cases} 215 - 8x_1 & 10 \leq x_1 \leq 23 \\ 2x_1 + 115 & 0 \leq x_1 \leq 10 \end{cases}$$

Now at $x_1 = 0$ the value will be maximum and is given by $= 215 - 80 = 135$

and $x_2^* = \min\left\{B_{12}, \dfrac{B_{22}}{2}\right\}$

Now, $B_{12} = B_{11} - 2x_1$, $B_{22} = B_{21} - 0$, $B_{12} = 43 - 20 = 23$, $B_{22} = 46$

$$x_2^* = \min\{23, 23\} = 23$$

Hence, the optimal solution is $x_1 = 10, x_2 = 23$ and $\max z = 135$

EXAMPLE 2. *Solve the following L.P.P. by using dynamic programming techniques*

$$\text{Max } z = 3x_1 + 5x_2$$

subjected to $\quad x_1 \le 4$

$$x_2 \le 6$$

$$3x_1 + 2x_2 \le 18$$

and $\quad x_1, x_2 \ge 0$ [MEERUT 1998 (P), 2005, AGRA 1998, 99, 2004]

SOLUTION. In this problem there are two interrelated decision variables, therefore it is a two stage problem. Due to the three resources b_1, b_2 and b_3 with available amount 4, 6 and 18, the state vector has three components.

Let (B_{1j}, B_{2j}, B_{3j}) are the state of the system and let us suppose that $f_j(B_{1j}, B_{2j}, B_{3j})$ be the optimal value of the objective function for stage $j = 1$ and 2. Using backward computational procedure, we have

$$f_2(B_{12}, B_{22}, B_{32}) = \max_{\substack{0 \le x_2 \le B_{22} \\ 0 \le 2x_2 \le B_{32}}} \{5x_2\} = 5 \max_{\substack{0 \le x_2 \le B_{22} \\ 0 \le x_2 \le \frac{B_{32}}{2}}} \{x_2\}$$

Since, $\max\{x_2\}$ is minimum of B_{22} and $\dfrac{B_{32}}{2}$ which satisfies $0 \le x_3 \le B_{22}$ and $0 \le x_2 \le \dfrac{B_{32}}{2}$

therefore, $\max\{x_2\} = x_2^* = \min\left\{B_{22}, \dfrac{B_{32}}{2}\right\}$

$$\therefore \quad f_2(B_{12}, B_{22}, B_{32}) = 5\min\left[B_{22}, \frac{B_{32}}{2}\right] \qquad \qquad ...(1)$$

We know that the value of B_{12}, B_{22}, B_{32} is given by at primary stage, Now

$$f_1(B_{11}, B_{21}, B_{31}) = \max_{\substack{0 \le x_1 \le B_{11} \\ 0 \le 3x_1 \le B_{31}}} \{3x_1 + f_2(B_{11} - x_1, B_{21}, B_{31} - 3x_1)\}$$

$$= \max_{\substack{0 \le x_1 \le 4 \\ 0 \le x_1 \le 6}} \{3x_1 + f_2(B_{11} - x_1, B_{21}, B_{31} - 3x_1)\} \qquad ...(2)$$

From equation (1), we know that

$$f_2(B_{12}, B_{22}, B_{32}) = 5\min\left\{B_{21}, \frac{B_{31} - 3x_1}{2}\right\}$$

so, $f_2(B_{11} - x_1, B_{21}, B_{31} - 3x_1) = 5\min\left\{B_{21}, \dfrac{B_{31} - 3x_1}{2}\right\}$

On putting this value in equation (2), we get

$$f_1(B_{11}, B_{21}, B_{31}) = \max_{\substack{0 \le x_1 \le 4 \\ 0 \le x_1 \le 6}} \left\{3x_1 + 5\min\left(B_{21}, \frac{B_{31} - 3x_1}{2}\right)\right\}$$

$$= \max_{\substack{0 \le x_1 \le 4 \\ 0 \le x_1 \le 6}} \left\{3x_1 + 5\min\left(6, \frac{18 - 3x_1}{2}\right)\right\}$$

$$= \max_{0 \le x_1 \le 4} \left\{3x_1 + 5\min\left(6, \frac{18 - 3x_1}{2}\right)\right\} \qquad ...(3)$$

Since, $\min_{0 \le x_1 \le 4}\left(6, \dfrac{18 - 3x_1}{2}\right) = \begin{cases} 6 & if \ \ 0 \le x_1 \le 2 \\ 9 - \dfrac{3x_1}{2} & if \ \ 2 \le x_1 \le 4 \end{cases}$

Therefore, from (3), we have

$$f_1(B_{11}, B_{21}, B_{31}) = \max_{0 \le x_1 \le 4}\begin{cases} 3x_1 + 30 & if \ \ 0 \le x_1 \le 2 \\ 45 - \dfrac{9x_1}{2} & if \ \ 2 \le x_1 \le 4 \end{cases}$$

Since, $3x_1 + 30$ is maximum at $x_1 = 2$ satisfying the restriction $0 \le x_1 \le 2$

$\therefore \qquad f_1(B_{11}, B_{21}, B_{31}) = 3 \times 2 + 30 = 36$ at $x_1^* = 2$

and $\quad x_2^* = \min\left(B_{22}, \dfrac{B_{32}}{2}\right) = \min(6,6) = 6$

Since, $B_{22} = B_{21} - 0 = 6$, $\qquad B_{32} = B_{31} - 3x_1 = 12$

Hence, the optimal solution is $x_1 = 10, x_2 = 23$ and $\max z = 36$

EXAMPLE 3. Solve the L.P.P.

$$\text{Max } z = 3x_1 + x_2$$

s.t. $\qquad 2x_1 + x_2 \le 6, x_1 \le 2 \text{ and } x_2 \le 4, x_1, x_2, x_3 \ge 0$

by dynamic programming technique. [MEERUT 1994, 99 (O)]

SOLUTION. In this problem there are two interrelated decision variables, therefore it is a two stage problem. Due to the three resources $b_1 = 6, b_2 = 2$ and $b_3 = 4$, the state vector has three components. Let (B_{1j}, B_{2j}, B_{3j}) be the state of the system and let us suppose that $f_j(B_{1j}, B_{2j}, B_{3j})$ be the optimal value of the objective function for stage $j = 1$ and 2. Using Backward computational procedure, we have

$$f_2(B_{12}, B_{22}, B_{32}) = \max_{\substack{0 \le x_2 \le B_{12} \\ 0 \le x_2 \le B_{32}}} \{x_2\}$$

since, $\max\{x_2\}$ is minimum of B_{12} and B_{32} which satisfies $0 \le x_2 \le B_{12}$ and $0 \le x_2 \le B_{32}$.

Therefore, $\quad \max\{x_2\} = x_2^* = \min\{B_{12}, B_{32}\}$

$\therefore \qquad f_2(B_{12}, B_{22}, B_{32}) = \min\{B_{12}, B_{32}\}$ ...(1)

Now, $f_1(B_{11}, B_{21}, B_{31}) = \max_{\substack{0 \le 2x_1 \le B_{11} \\ 0 \le x_1 \le B_{21}}} \{3x_1 + f_2(B_{11} - 2x_1, B_{21} - x_1, B_{31} - 0)$

$$= \max_{\substack{0 \le x_1 \le 3 \\ 0 \le x_1 \le 2}} \{3x_1 + f_2(B_{11} - 2x_1, B_{21} - x_1, B_{31})\} \qquad \text{...(2)}$$

From equation (1), we know that

$$f_2(B_{12}, B_{22}, B_{32}) = \min\{B_{12}, B_{32}\}$$

Therefore, $\quad f_2(B_{11} - 2x_1, B_{21} - x_1, B_{31}) = \min\{B_{11} - 2x_1, B_{31}\}$

On putting this value in equation (2).

$$f_1(B_{11}, B_{21}, B_{31}) = \max_{\substack{0 \le x_1 \le 3 \\ 0 \le x_1 \le 2}} \{3x_1 + \min(B_{11} - 2x_1, B_{31})\}$$

$$= \max_{0 \le x_1 \le 2} \{3x_1 + \min(6 - 2x_1, 4)\} \qquad \text{...(3)}$$

$$[\because \text{ if } x_1 \le 2 \text{ then it is obvious that } x_2 \le 3]$$

$$= \max \begin{cases} 3x_1 + 4 & 0 \le x_1 \le 1 \\ 3x_1 + (6 - 2x_1) & 1 \le x_1 \le 2 \end{cases}$$

Since, the value of $3x_1 + 4$ will be maximum at $x_1 = 1$ and the value will be 7, and the value of $6 + x_1$ will be maximum at $x_1 = 2$ and the value will be 8. So, 8 will be the maximum value of objective function.

Therefore, $\qquad x_1^* = 2$ and $\max z = 8$,

and $\qquad x_2^* = \min\{B_{12}, B_{32}\} = \min\{B_{21} - 2x_1, B_{31}\}$

$$= \min\{6 - 4, 4\} = 2$$

Hence, the optimum solution is given by $x_1 = 2, x_2 = 2$ and $\max z = 8$.

EXAMPLE 4. *Solve the problem:*

$$\text{Max } z = x_1 + 9x_2$$

$$s.t. \ 2x_1 + x_2 \leq 25, \ x_2 \leq 11, \ x_1, x_2 \geq 0$$

SOLUTION. There are two decision variables x_1 and x_2. So it is a two stage problem. Due to the two resources b_1 and b_2 with available amount 25 and 11, the state vector has two components.

Let (B_{1j}, B_{2j}) be the state of the system and let us suppose that $f_j(B_{1j}, B_{2j})$ be the optimal value of objective function. Using backward computational procedure, we have, $f_2(B_{12}, B_{22}) = \max_{\substack{0 \leq x_2 \leq B_{12} \\ 0 \leq x_2 \leq B_{22}}} \{9x_1\} = 9 \max_{\substack{0 \leq x_2 \leq 25 \\ 0 \leq x_2 \leq 11}} \{x_1\}$

Since, $\max\{x_2\}$ is minimum of $B_{12} = 25$ and $B_{22} = 11$ which satisfies $0 \leq x_2 \leq 25$ and $0 \leq x_2 \leq 11$, *i.e.*, therefore $\max\{x_2\} = x_2^* = \min\{B_{12}, B_{22}\}$

$$\therefore \qquad f_2(B_{12}, B_{22}) = 9 \min\{B_{12}, B_{22}\} \qquad \qquad \qquad ...(1)$$

Now, $f_1(B_{11}, B_{21}) = \max_{0 \leq x_1 \leq \frac{B_{11}}{2}} \{x_1 + f_2(B_{11} - 2x_1, B_{21} - 0)\}$

$$= \max_{0 \leq x_1 \leq \frac{25}{2}} \{x_1 + 9\min(B_{11} - 2x_1, B_{21})\}$$

$$\therefore \qquad f_1(B_{11}, B_{21}) = \max_{0 \leq x_1 \leq \frac{25}{2}} \{x_1 + 9\min(25 - 2x_1, 11)\} \qquad \qquad ...(2)$$

Since, $\min_{0 \leq x_1 \leq \frac{25}{2}} (25 - 2x_1, 11) = \begin{cases} 11 & 0 \leq x_1 \leq 7 \\ 25 - 2x_1 & 7 \leq x_1 \leq 25/2 \end{cases}$

Therefore, $f_1(B_{11}, B_{21}) = \max \begin{cases} x_1 + 99 & 0 \leq x_1 \leq 7 \\ 225 - 17x_1 & 7 \leq x_1 \leq 25/2 \end{cases}$

The value of $x_1 + 99$ will be maximum at $x_1 = 7$ and the value will be 106.

Therefore, $x_1^* = 7$ and $f_1(B_{11}, B_{21}) = \max z = 106$

and $x_2^* = \min\{B_{12}, B_{22}\} = \min\{11, 11\} = 11$ $[\because B_{12} = B_{11} - 2x_1 = 11, B_{22} = B_{21} - 0 = 11]$

Hence, the optimal solution is $x_1 = 7, x_2 = 11$ and $\max z = 106$

EXAMPLE 5. *Solve the L.P.P.*

$$\text{Min } z = -2x_1 - 18x_2$$

subject to $2x_1 + x_2 \leq 25, x_2 \leq 15$

and $x_1, x_2 \geq 0$

by dynamic programming technique. [MEERUT 1996 (BP)]

SOLUTION. Before solving this problem, we reduces the problem into maximization as follows:

$$\text{Max. } z' = 2x_1 + 18x_2$$

s.to $2x_1 + x_2 \leq 25, x_2 \leq 15$

$$x_1, x_2 \geq 0$$

There are two interrelated decision variables in this problem, therefore, it is a two stage problem. Due to the two resources b_1 and b_2 with available amount 25 and 15, the state vector has two components.

Let (B_{1j}, B_{2j}) are the state of the system and let us suppose that $f_j(B_{1j}, B_{2j})$ be the optimal value of the objective function for stage 1 and 2. Then using backward computational procedure, we have

$$f_2(B_{12}, B_{22}) = \max_{\substack{0 \leq x_2 \leq B_{12} \\ 0 \leq x_2 \leq B_{22}}} \{18x_2\} = 18 \max_{\substack{0 \leq x_2 \leq 25 \\ 0 \leq x_2 \leq 15}} \{x_2\}$$

Since, $\max\{x_2\}$ is minimum of B_{12} and B_{22} which satisfies the restriction $0 \le x_2 \le B_{12}$ and $0 \le x_2 \le B_{22}$.

$\therefore \qquad \max\{x_2\} = x_2^* = \min\{B_{12}, B_{22}\}$

Therefore, $f_2(B_{12}, B_{22}) = \min\{B_{12}, B_{22}\}$ $\qquad\qquad$ ...(1)

Now, $\qquad f_1(B_{11}, B_{21}) = \max_{0 \le 2x_1 \le 25}\{2x_1 + f_2(B_{11} - 2x_1, B_{21} - 0)\}$

$\qquad\qquad\qquad\qquad = \max_{0 \le x_1 \le \frac{25}{2}}\{2x_1 + f_2(B_{11} - 2x_1, B_{21})\}$

On putting this value in equation (1)

$\qquad f_2(B_{12}, B_{22}) = 18\min\{B_{11} - 2x_1, B_{21}\} = 18\min\{25 - 2x_1, 15\}$

Then, $\qquad f_1(B_{11}, B_{21}) = \max_{0 \le x_1 \le \frac{25}{2}}\{2x_1 + 18\min(25 - 2x_1, 15)\}$

$\qquad\qquad\qquad\qquad = \max \begin{cases} 2x_1 + 18(25 - 2x_1) & 5 \le x_1 \le 25/2 \\ 2x_1 + 270 & 0 \le x_1 \le 5 \end{cases}$

$\qquad\qquad\qquad\qquad = \max \begin{cases} 450 - 34x_1 & 5 \le x_1 \le 25/2 \\ 2x_1 + 270 & 0 \le x_1 \le 5 \end{cases}$

Since, the value of $450 - 34x_1$ will be maximum at $x_1 = 25/2$ and the value will be 25 and the value of $2x_1 + 270$ will be maximum at $x_1 = 5$, the value will be 280.

So, the maximum value of objective function is 280.

Therefore, $\qquad x_1^* = 5$ and $\max z' = 280$

and $\qquad\qquad x_2^* = \min\{B_{12}, B_{22}\}, \qquad B_{12} = B_{11} - 2x_1 = 25 - 10 = 15$

$\qquad\qquad\quad x_2^* = \min\{15, 15\} = 15 \qquad\qquad B_{22} = 15$

Hence, the optimal solution is $x_1 = 5$, $x_2 = 15$ and Min $z = -\text{Max } z' = -280$.

12.8 SOLUTION OF INVENTORY PROBLEM BY A DYNAMIC PROGRAMMING TECHNIQUE

If we consider the inventory models in which demand is exactly known but different in each period. We can obtain the solution of such kind of models by dynamic programming technique.

SOLVED EXAMPLES

EXAMPLE 1. *If there are n machines, which can do two jobs. If x of n machines do the first job then they produce goods worth $\phi(x) = 3x$ and if y of n machines do the second job then they produce goods worth if $\psi(y) = 2.5y$. The machines are subjected to depriciation so that after doing the first job only $a(x) = \dfrac{x}{3}$ machines remain available and after doing second job $b(y) = \dfrac{2}{3}y$ machines remain available in the begining of the second year. The process is repeated with the remaining machines. Calculate the maximum return after 3 years and obtain the optimal policy in each year.*

[MEERUT–1991(S)]

SOLUTION. In this problem, we denotes the years as periods. Suppose x_j are the no. of machines devoted to job 1 in j^{th} year. M_j are the total number of machines in the begining of jth year, and $f_n(M)$ is the maximum return when there are n years left with initial number of available machines is M.

Here, we proceed the following three steps.

Step 1: Let in the third year starting with M_3 machine, x_3 and y_3 are the machines assigned to the jobs, then we have

$f_1(M_3)$

$= \max_{x_3, y_3}\{3x_3 + 2.5y_3\}$

s. t. $x_3 + y_3 \le M_3$,

$\quad x_3, y_3 \ge 0$

which is a linear programming problem. If we solve this problem graphically, we have the optimal policy given by

$x_3^* = M_3, y_3^* = 0$

and $f_1(M_3)$

$= 3M_3 + 2.5 \times 0$

$= 3M_3 \quad ...(1)$

Step 2: Let in the begining in the second year with machine M_2, x_2 and y_2 are

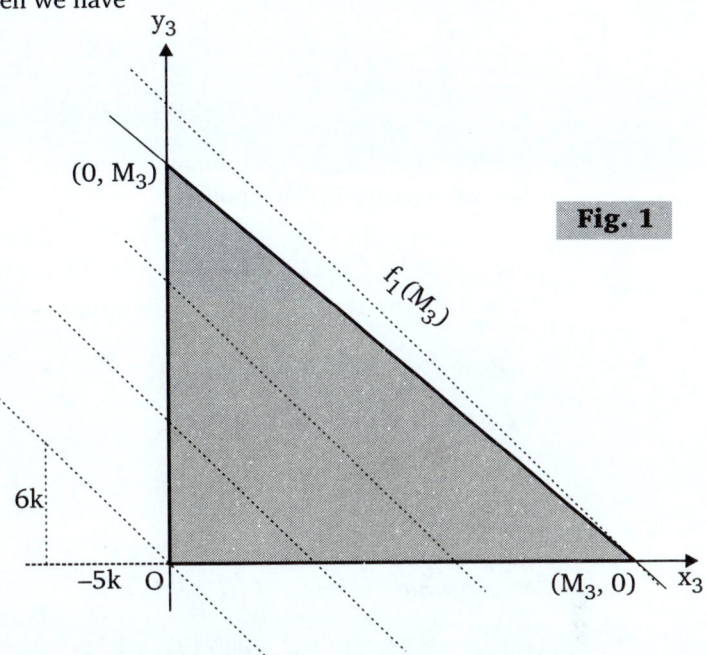

Fig. 1

the machines assigned to the jobs, then $\dfrac{x_2}{3} + \dfrac{2}{3}y_2 = M_3$ machines will remain available in the starting of the next year, then we have

$$f_2(M_2) = \max_{x_2, y_2}\left\{3x_2 + 2.5y_2 + f_1\left(\frac{x_2}{2} + \frac{2}{3}y_2\right)\right\}$$

s.t. $\qquad x_2 + y_2 \le M_2$ with $x_2, y_2 \ge 0$

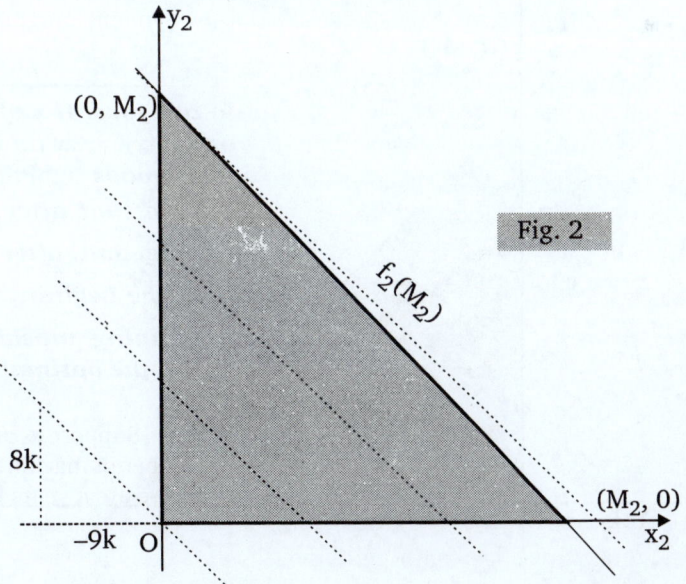

Fig. 2

We can also write

$$f_2(M_2) = \max_{x_2,y_2}\left\{3x_2 + 2.5y_2 + 3\left(\frac{x_2}{3} + \frac{2}{3}y_2\right)\right\}$$

$$= \max_{x_2,y_2}\{4x_2 + 4.5y_2\}$$

which is also a linear programming problem, we can solve this problem graphically, then we have the optimal policy.

$$x_2^* = 0,\ y_2^* = M_2$$

and $$f_2(M_2) = 4.5M_2$$ ...(2)

Step 3: In this step, let in the begining of first year with machine M_1; x_1 and y_1 are the

machines assigned to the jobs, then $\frac{x_1}{3} + \frac{2}{3}y_1 = M_2$ machines will remain available at the starting of the next year then

$$f_3(M_1) = \max_{x_1,y_1}\left\{3x_1 + 2.5y_1 + f_2\left(\frac{x_1}{3} + \frac{2}{3}y_1\right)\right\}$$

s.t. $x_1 + y_1 \leq M_1,\ x_1,y_1 \geq 0$

We can write,

$$f_3(M_1) = \max_{x_1,y_1}\left\{3x_1 + 2.5y_1 + 4.5\left(\frac{x}{3} + \frac{2}{3}y_1\right)\right\}$$

$$= \max_{x_1,y_1}(4.5x_1 + 5.5y_1)$$

s.t. $x_1 + y_1 \leq M_1,\ x_1,y_1 \geq 0$

After solving graphically, we obtain the policy

$$x_1^* = 0,\ y_1^* = M_1 \text{ and } f_3(M_1) = 5.5M_1$$

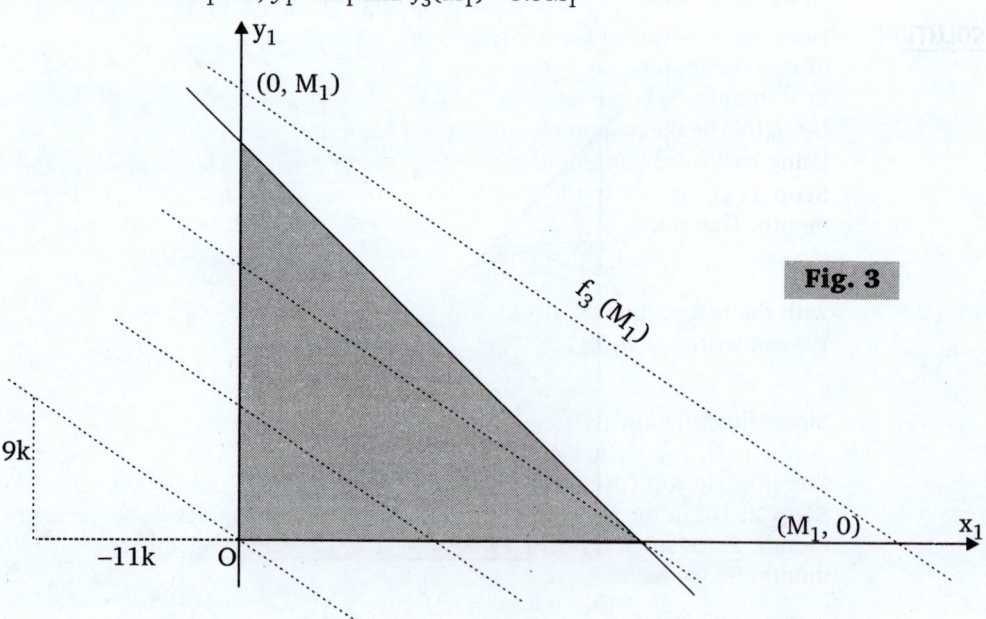

Fig. 3

But initially, we have $\qquad M_1 = n$

$$x_1^* = 0, \ y_1^* = M_1 = n, \qquad M_2 = \frac{x_1^*}{3} + \frac{2y_1^*}{3} = \frac{2n}{3}$$

$$x_2^* = 0, \ y_2^* = M_2 = \frac{2}{3}n, \qquad M_3 = \frac{x_2^*}{3} + \frac{2}{3}y_2^* = \frac{4}{9}n$$

$$x_3^* = M_3 = \frac{4n}{9}, \ y_3^* = 0$$

Hence, the optimal solution for three years is given by

Ist year: $\qquad x_1^* = 0, \ y_1^* = n \qquad M_1 = n$

IInd year: $\qquad x_2^* = 0, \ y_2^* = \frac{2n}{3} \qquad M_2 = \frac{2n}{3}$

IIIrd year: $\qquad x_3^* = \frac{4n}{9}, y_3^* = 0 \qquad M_3 = \frac{4n}{9}$

EXAMPLE 2. *A man engaged in buying and selling identical stems. He operates from a warehouse that can hold 500 items. Each month he can sell any quantity that he choose upto the stock at the begining of the month. Each month, he can buy as such as he wishes for delivery at the end of the month so long as his stock does not exceed 500 items. For the next four months he has the following error free forecasts of cost, sales price:*

Month	i	1	2	3	4
Cost	c_i	27	24	26	28
Sales price	p_i	28	25	25	27

If he currently has a stock of 200 units, what quantities should be sell and buy in the next four month? Obtain the solution using dynamic programming. [MEERUT 1996, 99]

SOLUTION. Here, let x_i is the amount for sell during the month i, y_i is the amount to be ordered during the month i, b_i is the stock level in the begining of month i, p_i is the sale price in i^{th} month, c_i is the purchase price in the ith month and H is the warehouse capacity. Let $f_n(b_n)$ be the maximum return when there are n months with initial stock b_n.

Using backward computational procedure, we will proceed the following steps:

Step 1: Let we are in the starting of fourth month with stock b_4 at the begining of this month. Therefore,

$$f_1(b_4) = \max_{x_4, y_4} \{p_4 x_4 - c_4 y_4\}$$

with the restrictions $\quad 0 \le x_4 \le b_4, \ y_4 \ge 0$ and $b_4 - x_4 + y_4 \le 0$

We can write $\quad f_1(b_4) = \max_{\substack{0 \le x_4 \le b_4 \\ 0 \le y_4 \le x_4 - b_4}} \{27 - 28 y_4\}$

Since, the function $(27 x_4 - 28 y_4)$ will be maximum at $x_4 = b_4, y_4 = 0$

$\therefore \qquad f_1(b_4) = 27 b_4 - 28 \times 0 = 27 b_4$

therefore, in this case, the optimal policy are $x_4^* = b_4, y_4^* = 0$ and $f_1(b_4) = 27 b_4$

Step 2: Let in the starting of third month with initial stock b_3 at the beginning of this month. Since stock level $b_4 = b_3 - x_3 + y_3$ will remain available at the begining of next month. So we have,

$$f_2(b_3) = \max_{x_3, y_3} \{p_3 x_3 - c_3 y_3 + f_1(b_3 - x_3 + y_3)\}$$

with the restrictions $0 \leq x_3 \leq b_3$, $y_3 \geq 0$ and $b_3 - x_3 + y_3 \leq H(= 500)$

We can write
$$f_2(b_3) = \max_{\substack{0 \leq x_3 \leq b_3 \\ 0 \leq y_3 \leq 500 - b_3 + x_3}} \{25x_3 - 26y_3 + 27(b_3 - x_3 + y_3)\}$$

$$= \max_{\substack{0 \leq x_3 \leq b_3 \\ 0 \leq y_3 \leq 500 - b_3 + x_3}} \{27b_3 - 2x_3 + y_3\}$$

$$= \max_{0 \leq x_3 \leq b_3} \{27b_3 + (500 - b_3 + x_3) - 2x_3\}$$

$$= \max_{0 \leq x_3 \leq b_3} \{26b_3 + 500 - x_3\}$$

which is maximum at $x_3 = 0$, therefore, the optimal policy are $x_3^* = 0$, $y_3^* = 500 - b_3 + x_3 = 500 - b_3$ and $f_2(b_3) = 26b_3 + 500$.

Step 3: Let we are in second month, three months are left with stock b_3 at the starting of this month. Since stock level $b_3 = b_2 - x_2 + y_2$ will remain available at the starting of the next month. So, we have
$$f_3(b_2) = \max_{x_2, y_2} \{p_2 x_2 - c_2 y_2 + f_2(b_2 - x_2 + y_2)\}$$

with the restrictions $0 \leq x_2 \leq b_2$, $y_2 \geq 0$ and $b_2 - x_2 + y_2 \leq H(= 500)$

We can write
$$f_3(b_2) = \max_{\substack{0 \leq x_2 \leq b_2 \\ 0 \leq y_2 \leq 500 - b_2 + x_2}} \{25x_2 - 24y_2 + 26(b_2 - x_2 + y_2) + 500\}$$

$$= \max_{\substack{0 \leq x_2 \leq b_2 \\ 0 \leq y_2 \leq 500 - b_2 + x_2}} \{26b_2 - x_2 + 2y_2 + 500\}$$

$$= \max_{0 \leq x_2 \leq b_2} \{24b_2 + x_2 + 1500\}$$

which is maximum at $x_2 = b_2$. Therefore, the optimal policy are $x_2^* = b_2$, $y_2^* = 500 - b_2 + x_2 = 500$ and $f_3(b_2) = 25b_2 + 1500$.

Step 4: Suppose, we are in the first month, four months are left with stock b_1 at the starting of this month. Since $b_2 = b_1 - x_1 + y_1$ will remain available at the starting of the next month. So, we have
$$f_4(b_1) = \max_{x_1, y_1} \{p_1 x_1 - c_1 y_1 + f_3(b_1 - x_1 + y_1)\}$$

with the restriction $0 \leq x_1 \leq b_1$, $y_1 \geq 0$ and $b_1 - x_1 + y_1 \leq H(= 500)$

we can write
$$f_4(b_1) = \max_{\substack{0 \leq x_1 \leq b_1 \\ 0 \leq y_1 \leq 500 - b_1 + x_1}} \{28x_1 - 27y_1 + 25(b_1 - x_1 + y_1) + 1500\}$$

$$= \max\{25b_1 + 3x_1 - 2y_1 + 1500\}$$

which is maximum at $x_1 = b_1, y_1 = 0$. Therefore, the optimal policy are $x_1^* = b_1$, $y_1^* = 0$ and $f_4(b_1) = 28b_1 + 1500$.

Since, at the starting of first month, stock is 200 units.

$\therefore \quad b_1 = 200$, so $\qquad\qquad x_1^* = 200, y_1^* = 0$

$b_2 = b_1 - x_1^* + y_1^* = 0$, $\qquad\qquad x_2^* = b_2 = 0, y_2^* = 500$

$b_3 = b_2 - x_2^* + y_2^* = 500$, $\qquad\qquad x_3^* = 0, y_3^* = 500, b_3 = 0$

$b_4 = b_3 - x_3^* + y_3^* = 500 - 0 + 0 = 500$, $\qquad x_4^* = b_4 = 500, y_4^* = 0$

Hence, the optimal solution is

Month (i)	1	2	3	4
Purchase (y_i^*)	0	500	0	0
Sale (x_i) *	200	0	0	500

and the maximum return $f_4(b_1) = 28b_1 + 1500 = 28 \times 200 + 1500 = ₹ 7100$

12.9 DIFFERENCE BETWEEN DYNAMIC AND LINEAR PROGRAMMING PROBLEM

Following table shows the basic difference between dynamic and linear programming problem

S. No.	Characteristics	Dynamic programming Problem	Linear Programming Problem
1.	Objective function	may be linear or non linear	must be linear
2.	Constraints	may be linear or non linear	must be linear
3.	Technological coefficients	must be positive	may be positive
4.	Solution Procedure	by breaking it into different stages	treated as single problem

EXERCISE 12.1

1. Develop the functional equations to determine $m_j (\geq 0)$ so as to Maximize

$$Z = \sum_{i=1}^{n} m_i \left(\frac{p_i}{m_n} \right)^2$$ subjected to the constraints

$$m_1 + m_2 + m_3 + ... + m_n = M$$

2. Solve the following problem by Dynamic programming technique.

$$\max Z = 12x_1^2 + 27x_2^2 + 147x_3^2$$

subject to $x_1 + x_2 + x_3 = 1$, $x_1, x_2, x_3 \geq 0$.

3. Solve the problem $\max Z = p_1 \cdot p_2 \cdot p_3 \cdots p_n$ subjected to the constraints

$$c_1 p_1 + c_2 p_2 + ... c_n p_n \leq x$$
$$0 \leq p_i \leq 1 \text{ for } i = 1, 2, 3 ... n.$$

4. Using dynamic programming, show that

$$\sum_{i=1}^{n} p_i \log p_i$$ subjected to the constraints

$$\sum_{i=1}^{n} p_i = 1$$ is minimum, when

$$p_1 = p_2 = p_3 = ... = p_n = \frac{1}{n}.$$

[MEERUT–2000, 07; AGRA–1998]

5. Solve the problem $\max Z = x_1^2 + x_2^2 + x_3^2$ subjected to the constraints $x_1 x_2 x_3 \leq 4$, $x_1, x_2, x_3 \geq 0$, by dynamic programming technique.

6. Solve the linear programming problem
$\max Z = -x_1^2 - 2x_2^2 + 3x_2 + x_3$
subject to $x_1 + x_2 + x_3 \leq 1$ with $x_1, x_2, x_3 \geq 0$
by dynamic programming technique.

7. Solve the following linear programming problem by Dynamic programming technique.

$$\max Z = x_1^3 + x_2^3 + x_3^3$$

subject to $x_1 + x_2 + x_3 \leq 6$ with $x_1, x_2, x_3 \geq 0$.

8. Using dynamic programming technique,

solve
$$\min Z = y_1^2 + y_2^2 + y_3^2$$
subject to $y_1 + y_2 + y_3 \geq 10$ with $y_1, y_2, y_3 \geq 0$.

9. Let the function $f_n(a) = \min_{R} \sum_{i=1}^{n} x_i^p$, $p > 0$ where R is defined by

(i) $\sum_{i=1}^{n} x_i \geq a, a > 0$ (ii) $x_i \geq 0 \ \forall i$

(a) Show that $f_n(a)$ satisfies the recurence relation.

$$f_n(a) = \min_{x \leq a} \{x^p + f_{n-1}(a - x)\}$$

(b) If $0 < p < 1$ then show that $f_n(a) = a^p$

(c) If $p > 1$ then show that $f_n(a) = n\left(\frac{a}{n}\right)^{p-1}$

10. Using dynamic programming technique, Maximize $Z = 8x_1 + 7x_2$
subject to $2x_1 + x_2 \leq 8$
$$5x_1 + 2x_2 \leq 15, \ x_1, x_2 \geq 0.$$

11. Solve the following L.P.P. by dynamic programming technique.
$$\max Z = 2x_1 + 3x_2$$
subject to $x_1 - x_2 \leq 1$
$$x_1 + x_2 \leq 3, \ x_1, x_2 \geq 0$$

12. Solve the following by dynamic programming technique.
(i) $\max Z = 3x_1 + 7x_2$
subject to $x_1 + 4x_2 \leq 8$
$$x_2 \leq 2$$
$$x_1, x_2 \geq 0.$$

(ii) $\max Z = 4x_1 + 3x_2$
subject to $x_1 + 2x_2 \leq 4$
$$6x_1 + x_2 \leq 6$$
$$x_1, x_2 \geq 0.$$

13. Solve the L.P.P. by dynamic programming technique

$$\max Z = 8y_1 + 7y_2$$

subject to $2y_1 + y_2 \le 8$, $5y_1 + 2y_2 \le 15$ with

$$y_1, y_2 \ge 0.$$

14. Use dynamic programming to solve

$$\min Z = 3y_1 + 5y_2$$

subject to $-3y_1 + 4y_2 \le 12$, $-2y_1 + y_2 \le 2$, $2y_1 + 3y_2 \ge 12$, $0 \le y_1 \le 4$, $y_2 \ge 2$.

15. Use dynamic programming to solve.

$$\max Z = 2x_1 + 5x_2$$

subject to $3x_1 + x_2 \le 2$, $x_2 \le 3$ and $x_1, x_2 \ge 0$.

16. Solve the following L.P.P. by dynamic programming technique.

(i) $\min Z = y_1^2 + y_2^2 + y_3^2 + \ldots + y_n^2$

subject to $y_1 . y_2 . y_3 \ldots y_n = c$,

$y_1, y_2, y_3, y_4, \ldots y_n \ge 0$.

[MEERUT, 2001 (BP), 02, 04]

(ii) $\max Z = x_1 \cdot x_2 \cdot x_3$

subject to $x_1 + x_2 + x_3 = 5$ with $x_1, x_2, x_3 \ge 0$.

[MEERUT-2002, 06; RAMPUR-2000; IAS-1998]

17. Solve the L.P.P.

$$\max Z = 5x_1 + 7x_2$$

subject to $x_1 + x_2 \le 4$

$$3x_1 + 8x_2 \le 24$$

$$10x_1 + 7x_2 \le 35 \text{ with } x_1, x_2 \ge 0$$

by dynamic programming technique.

[MEERUT 1994 (P)]

18. Solve the following L.P.P. by Dynamic programming technique.

$$\max Z = 4x_1 + 14x_2$$

subject to $2x_1 + 7x_2 \le 21$

$$7x_1 + 2x_2 \le 21 \text{ and } x_1, x_2 \ge 0$$

19. A firm of manufacturers stock up every two months with certain basic material in order to carry out its production schedule. The purchase price p_n and the demand d_n, $n = 1, 2, \ldots, 6$ are given for the next 6 bimonthly periods in the following table:

Period n	1	2	3	4	5	6
Demand d_n	8	5	3	2	7	4
Purchase price p_n	11	18	13	17	20	10

owing to limited storage space the stock must never exceed a certain value 5. The initial stock is 2 and the final stock must be nill. Use dynamic programming to ascertain the quantity to be bought at the begining of each period in such a way that the total cost will be minimum.

20. The total volume available in an air craft for 3 types of item is $13 ft^3$. The unit volume of item A is $2 ft^3$ that of item B is $3 ft^3$, and that of item C is $2 ft^3$. The cost of having a demand that occur when the system is out of stock is ₹600 for item A, ₹1200 for item B and ₹800 for item C. The demand for each item is Poisson distributed with mean being 5, 2 and 2 for item A, B and C respectively. How many of each item should be loaded in order to minimize to expected stock out cost?

ANSWERS

5. $x_1 = 1, x_2 = 1, x_3 = 4$, max. $z = 18$

6. $x_1 = 0, x_2 = \frac{1}{2}, x_3 = \frac{1}{2}$ and $\max Z = \frac{3}{2}$

10. $x_1 = 0$, $x_2 = \frac{15}{2}$ $\max Z = \frac{105}{2}$

11. $x_1 = 0, x_2 = 3$ $\max Z = 9$

12. (i) $x_1 = 8, x_2 = 0$, Max $Z = 24$

(ii) $x_1 = \frac{7}{10}, x_2 = \frac{17}{10}$ $\max Z = \frac{525}{10}$

13. $x_1 = 0, x_2 = \frac{75}{10}, \max Z = \frac{525}{10}$

15. $x_1 = x_2 = 3$, max $Z = 21$

16. (i) $c^{1/n}, c^{1/n}, \ldots, c^{1/n}$, $\min Z = nc^{2/n}$

(ii) $x_1 = \frac{5}{3}, x_2 = \frac{5}{3}, x_3 = \frac{5}{3}$ and max $Z = \frac{125}{27}$

17. $x_1 = \frac{8}{5}, x_2 = \frac{12}{5}$ and $\max Z = \frac{124}{5}$

18. $x_1 = 0, x_2 = 3, \max Z = 42$

19. $x_1 = 7, x_2 = 4, x_3 = 9$, $x_4 = 3, x_5 = 0, x_6 = 4$ and minimum purchase cost is ₹ 357.

13 Maximum Flow and Minimum Potential in Networks

13.1 INTRODUCTION

In recent years, graph theory has found more and more applications in operations research. Most of the applications of graphs in operations research have been in problems involving network flows. Further, a large number of mathematical problems can be presented by networks. The networks are easily seen in communication and transportation system. In this chapter we shall discuss flow and potential problems which are frequently seen in engineering problems.

> Since, the problems of finding an optimal solution can be viewed as a problem of selecting the best sequence of operations from the list of finite no. of alternatives which can be represented by a graph.

13.2 GRAPHS AND THEIR REPRESENTATION

A graph can be thought of as a drawing or diagram consisting of a collection of vertices (dots or points) together with edges (lines) joining certain pair of these vertices.

Definition: *A graph G consists of following two things :*

(i) *A set $V = V(G)$, whose elements are called vertices, points or nodes of G.*

(ii) *A set $E = E(G)$ of unordered pairs of distinct pairs called edges of G.*

Such a graph is denoted by $G(V, E)$.

☛ Remarks

- Vertices are also sometimes called points, nodes or dots.
- If e is an edge with end vertices u and v, then e is said to join u and v.
- The definition of a graph allows the possibility of the edge having identical end vertices, *i.e.*, it is possible to have a vertex u joined to itself by an edge.

13.2.1 SOME EXAMPLES ON GRAPHS

(1) Let $G = (V, E)$, where $V = \{a, b, c, d, e\}$, $E = \{e_1, e_2, ..., e_8\}$ and the ends of the edges are given by

$e_1 \leftrightarrow (a, b)$, $e_2 \leftrightarrow (b, c)$, $e_3 \leftrightarrow (c, c)$, $e_4 \leftrightarrow (c, d)$

$e_5 \leftrightarrow (b, d)$, $e_6 \leftrightarrow (d, e)$, $e_7 \leftrightarrow (b, e)$, $e_8 \leftrightarrow (e, b)$

Then, this graph is given by

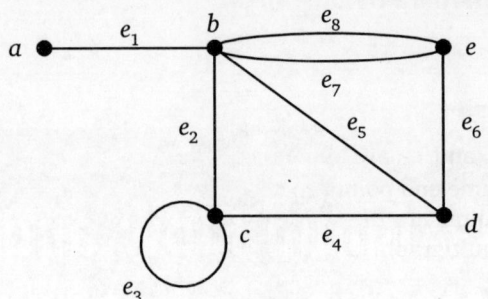

Fig. 1

(2) The following figure is a graph with five vertices and seven edges :

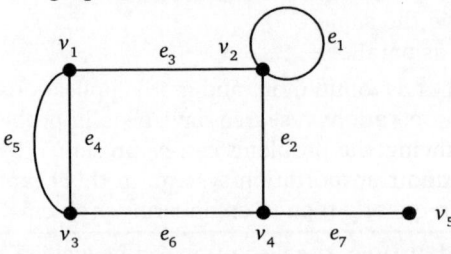

Fig. 2

13.2.2 FORMAL DEFINITION OF A GRAPH

The definition of graph given above works quite well if we have the visual representation of the graph before us to show which edges connect with which vertices. Without the picture, however, we need a concise way to convey this information. In this way, we define the graph formally as follows :

Definition : *A graph G is an ordered triplet* (V, E, g), *where*

V = a non-empty set of vertices

E = a set of edges

g = a function associating with each edge a an ordered pair (x, y) of vertices called the end points of a.

For Example : Consider the following graph

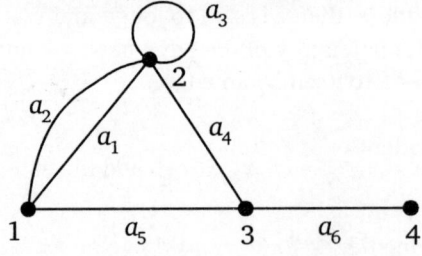

Fig. 3

In this graph the function g associating edges with endpoints perform the following mappings

$$g(a_1) = 1-2 \qquad g(a_2) = 1-2 \qquad g(a_3) = 2-2$$

$$g(a_4) = 2-3 \qquad g(a_5) = 1-3 \quad \text{and} \qquad g(a_6) = 3-4$$

13.3 MULTIGRAPH

Consider the following graph :

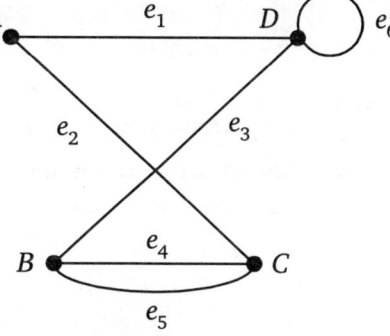

In this graph, the edges e_4 and e_5 are known as multiple edges since they connect the same end points and the edge e_6 is called a loop since its end points are the same vertex. Such type of diagram is known as multigraph.

13.4 GRAPH TERMINOLOGY

(1) Parallel-edges: More than one edge associated with a given pair of vertices, *i.e.*, if two (or more) edges of a graph G have the same end points, then these edges are known as parallel edges.

For example: In Fig. 5, e_4 and e_5 are parallel edges.

Fig. 4

(2) Self-loop: An edge having the same vertex as both its end vertices and edges to be associated with a vertex pair (v_i, v_j).

For example: In Fig. 5, e_1 is self loop.

(3) Isolated Vertex: A vertex of graph G, which is not the end of any edge is known as isolated vertex. (MEERUT(B.C.A.)–2008)

For example : In Fig. 5, v_6 is an isolated vertex.

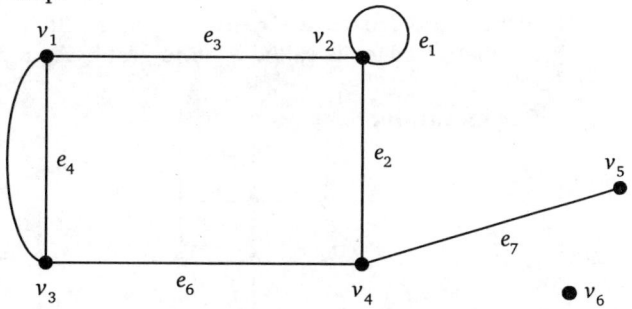

(4) Adjacent Vertex: Two vertices which are joined by an edge are called adjacent or neighbours. The set of all such neighbours of vertex v is called neighbourhood set of v and it is denoted by $N(v)$.

For example: In Fig. 5, v_3 and v_4 are adjacent.

Fig. 5

(5) Incidence and Degree: When a vertex v_i is an end vertex of some edge e_j, v_i and e_j are said to be incident with each other. In Fig. 5, e_2, e_6 and e_7 are incident with vertex v_4. Two non-parallel edges are said to be adjacent if they are incident on a common vertex.

For example: In Fig. 5, e_2 and e_4 are adjacent.

Further, the number of edges incident on a vertex v_i with self-loop counted twice is called the degree $d(v_i)$ of vertex v_i.

For example: In Fig. (5), $d(v_1) = d(v_3) = d(v_4) = 3$, $d(v_2) = 4$, $d(v_5) = 1$

(6) Pendant Vertex: A vertex having degree one is called pendant vertex.

13.4.1 IMPORTANT CONCLUSIONS ABOUT THE DEGREE OF THE GRAPH

1. If the degree of a vertex v is zero, then v is called an isolated vertex of graph G.

2. If the degree of a vertex v is one, then the vertex v is called a pendant vertex or a leaf and the edge incident with the pendant vertex of G is called a pendant edge.

3. Let $v_1, v_2, ..., v_n$ be the vertices of a graph G with $\deg(v_i) \leq \deg(v_{i+1})$ for all i $(1 \leq i \leq n)$. Then the sequence $\deg(v_1), \deg(v_2), ..., \deg(v_n)$ is called a degree sequence of a graph G and two graphs may have a same degree sequence with different graphical representation.

4. A vertex of a graph is called odd or even depending on whether its degree is even or odd.
5. We can define

$$\delta(G) = \min\{\deg(v_1) : v_1 \in V\}$$
$$= \text{minimum of all the degrees of the vertices of a graph } G$$

and
$$\Delta G = \max\{\deg(v_1) : v_1 \in V\}$$
$$= \text{maximum of all the degrees of the vertices of a graph } G$$

13.5 TYPES OF GRAPHS

1. **Simple Graph:** A graph is called simple if it has no loops and no parallel edges.
   For example: The following graph is an example of simple graph.

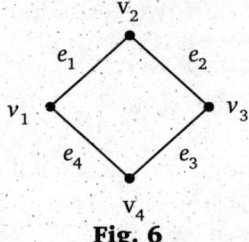

Fig. 6

2. **Finite and Infinite Graph:** A graph with a finite number of vertices as well as a finite number of edges is called a finite graph. A graph which is not finite is known as infinite graph.   (MEERUT B.C.A. 2008)
   For example:

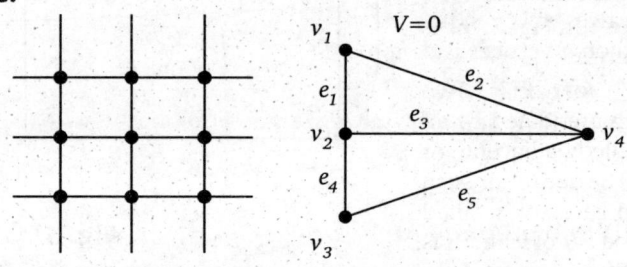

Portion of Infinite Graph   Finite Graph

Fig. 7

3. **Trivial Graph or Null Graph:** A finite graph with one vertex and no edges, i.e., a single point is called the trivial graph.   (MEERUT(B.C.A.)–1999, 2008)
4. **Pseudo Graph:** A graph having loops but no multiple edges is known as pseudo graph.
5. **Labelled Graph:** A graph G is called a labelled graph if its edges are labelled with name or data. We can write labels in place of an ordered pair in the edge set.
   For example: $G = [\{1,2,3,4,5\}, \{e_1, e_2, e_3, e_4, e_5\}]$

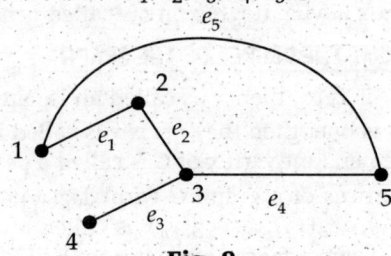

Fig. 8

6. k-Regular Graph: If for some positive integer k, $\deg(v) = k$ for every vertex v of the graph G, then G is called k-regular graph. (MEERUT(B.C.A.)–2005, 2006)

IMPORTANT FACTS

(i) A 3-regular graph is known as cubic graph.

(ii) $\delta(G) = \Delta(G) = k$, *i.e.* all the vertices of G have the same degree k.

(iii) A regular graph of degree zero has no edge.

(iv) In a regular graph of degree 1, every component has exactly one line.

(v) If G is a 2-regular graph, then every component has a cycle.

(vi) Every cubic graph has an even number of vertices.

7. Complete Graph: A complete graph is a simple graph in which each pair of distinct vertices is joined by an edge. Thus, a graph with n vertices is said to be complete if it has as many edges as possible provided there are no loops and no parallel edges.

If the complete graph has vertices $v_1, v_2, ..., v_n$, then the edge set E can be written as

$$E = \{(v_i, v_j) : v_i \neq v_j; \ i, j = 1, 2, ..., n\}$$

8. Bipartite Graph: Let G be a graph. If the vertex set V of G partitioned into two non-empty subsets X and Y (*i.e.*, $X \cup Y = V$ and $X \cap Y = \phi$) in such a way that each edge of G has one end in X and one end in Y, then G is called bipartite. The partition $V = X \cup Y$ is known as bipartition of G.

9. Complete Bipartite Graph: It is a simple bipartite graph G with bipartition $V = X \cup Y$ in which every vertex in X is joined to every vertex of Y. (MEERUT(B.C.A.)–2002, 03, 04, 05, 06, 07)

If X has m vertices and Y has n vertices, such a graph is denoted by $k_{m,n}$. A complete bipartite graph on n vertices is denoted by k_n.

☛ REMARKS

- Since each of the m vertices in the partition set X of $k_{m,n}$ is adjacent to each of the n vertices in the partition set Y, $k_{m,n}$ has mn edges.
- The graph $k_{1,n}$ is called a star graph.

For Example: **(i) Some Complete Bipartite Graphs**

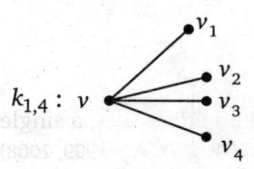

$k_{1,4} : v$

Fig. 9

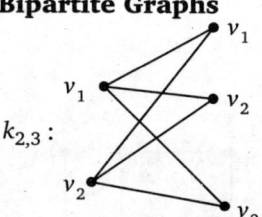

$k_{2,3} :$

Fig. 10

(ii) Some Bipartite Graphs

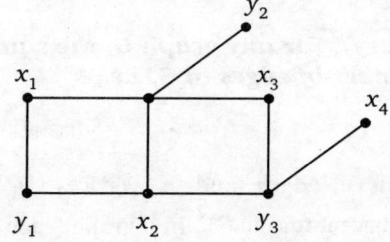

Fig. 11

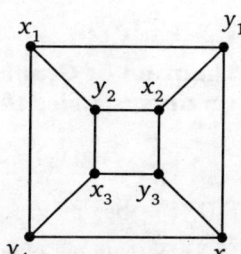

Fig. 12

10. Cycle Graph: For $n \geq 3$, the cycle C_n consists of n vertices $v_1, v_2, ..., v_n$ and edges $\{v_1, v_2\}, \{v_2, v_3\}, ..., \{v_{n-1}, v_n\}$ and $\{v_n, v_1\}$.

For Example: The cycles C_3, C_4, C_5 and C_6 are given as follows :

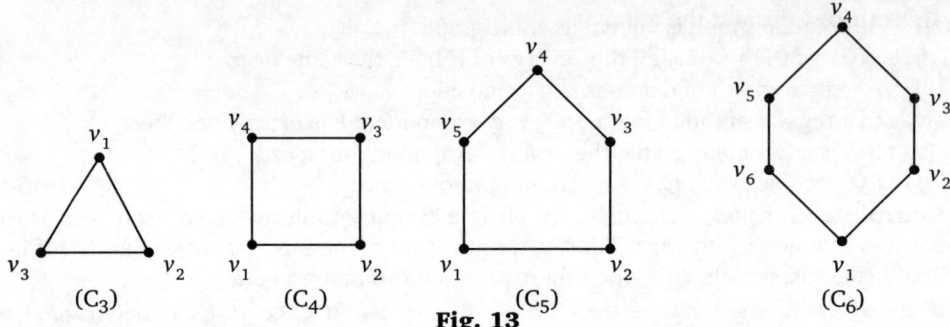

Fig. 13

11. Wheel Graph: We can obtain wheel W_n by adding an extra vertex to the cycle C_n, for $n \geq 3$. We connect this vertex to each of the n vertices in C_n by new edges.

For Example: The wheels W_3, W_4, W_5 and W_6 are given as follows :

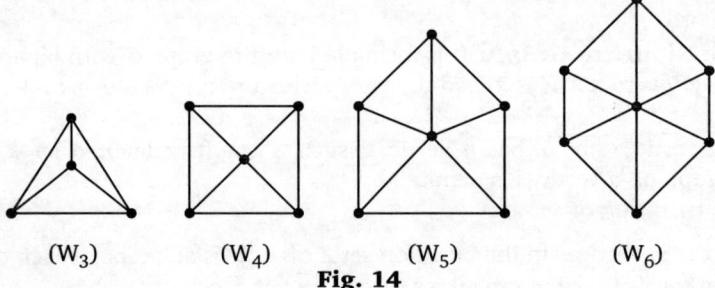

Fig. 14

12. Path Graph: We can obtain a path graph of order n by deleting an edge from a cycle graph C_n. It is denoted by P_n.

For Example:

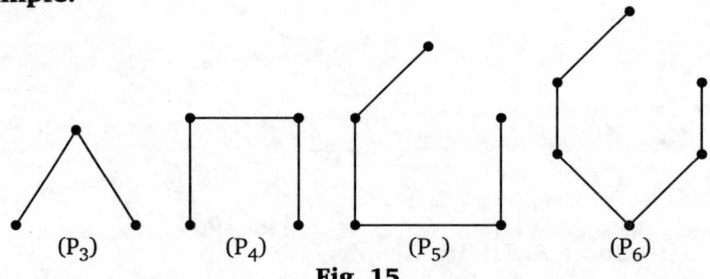

Fig. 15

THEOREM 1   **(First Theorem of Graph Theory).** *For any graph G, the sum of the degree of all vertices is twice the number of edges in G, i.e.,*

$$\sum_{i=1}^{n} \deg(v_i) = 2e$$

                                                      (ROHILKHAND(B.C.A.)–2006)

PROOF.   Let $G = (V, E)$ be a graph with e edges and n vertices $v_1, v_2, ..., v_n$. Then $\sum_{i=1}^{n} \deg(v_i) = \sum_{i=1}^{n}$ (Number of edges adjacent to v_i in G). In a graph, each edge is adjacent to exactly two vertices of G and while taking the sum in right hand side of the above equation, each edge count twice and every edge of G contributes to the degree of a vertex of G. Hence, the right hand side of above equation is equal to $2e$, where e is the number of edges of graph G.

☞ REMARK
- The above theorem is also known as **hand shaking lemma**.

For Example: Consider the following graph

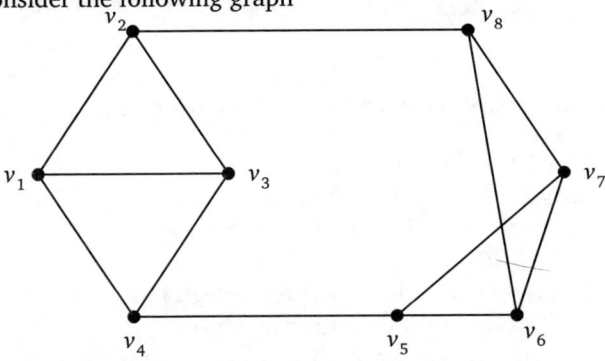

Fig. 16

We observe that

$$d(v_1) = 3, d(v_2) = 3, d(v_3) = 3, d(v_4) = 3, d(v_5) = 3, d(v_6) = 3, d(v_7) = 3, d(v_8) = 3$$

$$\sum_{i=1}^{8} d(v_i) = d(v_1) + d(v_2) + d(v_3) + d(v_4) + d(v_5) + d(v_6) + d(v_7) + d(v_8)$$

$$= 3 + 3 + 3 + 3 + 3 + 3 + 3 + 3 = 24 = 2 \times 12 = 2 \times \text{number of edges}$$

THEOREM 2. *The number of vertices of odd degree in a graph is always even.*

(MEERUT(B.C.A.)–2002, 2004)

PROOF. If we consider the vertices with odd and even degree separately, the quantity in the left side of $\sum_{i=1}^{n} d(v_i) = 2e$ can be expressed as the sum of two sums, each taken over vertices of even and odd degrees, as follows :

$$\sum_{i=1}^{n} d(v_i) = \underset{\text{even}}{\sum} d(v_i) + \underset{\text{odd}}{\sum} d(v_i)$$

Since left hand side of this equation is even and the first expression on the right hand side is even, the second expression must also be even. Hence, $\underset{\text{odd}}{\sum} d(v_i)$ = an even number, in this each $d(v_i)$ is odd, the total number of terms in the sum must be even to make the sum an even number. This proves the theorem.

☞ REMARK
- It is not true in general that a graph must have an odd number of even vertices.

SOLVED EXAMPLES

EXAMPLE 1. *Draw the diagram of the following graph $G(V, E)$ where*

  (i) $V = \{A, B, C, D\}$, $E = \{AB, AC, BC, BD, CD\}$

  (ii) $V = \{A, B, C, D, E\}$, $E = \{AB, AC, BC, DE\}$

SOLUTION. We have the following diagrams :

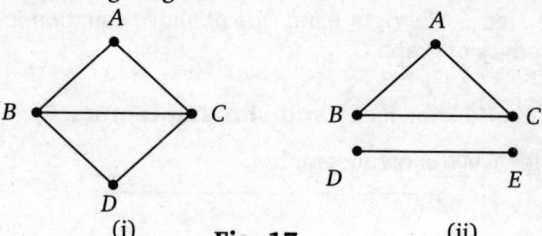

(i)   **Fig. 17**   (ii)

EXAMPLE 2. *Describe the graph G in the adjoining diagram.*

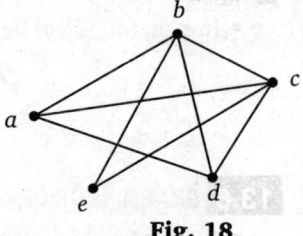

Fig. 18

SOLUTION. We observe that

The set of vertices, $V = \{a, b, c, d, e\}$

and set of edges, $E = \{ab, ac, ad, bc, bd, be, cd, ce\}$

Also, $\deg(a) = 3$, $\deg(b) = 4$,

$\deg(d) = 3$, $\deg(e) = 2$, $\deg(c) = 4$

Clearly, $\deg(a) + \deg(b) + \deg(c) + \deg(d) + \deg(e)$

$= 16 = 2 \times$ number of edges

EXAMPLE 3. *Write down the degree of each vertices of the following graph.*

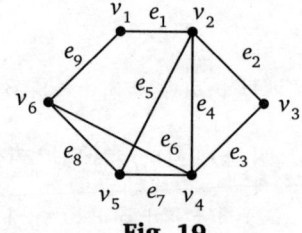

Fig. 19

SOLUTION. We observe that

$\deg(v_1) = 2$, $\deg(v_2) = 4$, $\deg(v_3) = 2$ $\deg(v_4) = 4$, $\deg(v_5) = 3$, $\deg(v_6) = 3$

EXAMPLE 4. **(Utility Problem).** *Nine members of a new club meet each day for lunch at a round table. They decide to sit so that every member has different neighbours at each lunch. How many days can this arrangement last?*

(MEERUT(B.C.A.)–2004)

SOLUTION. The given situation can be represented by a graph. Each member can be represented by a vertex and an edge joining two vertices represent the relationship of sitting next to each other.

In the following figure, two sitting arrangements have been shown.

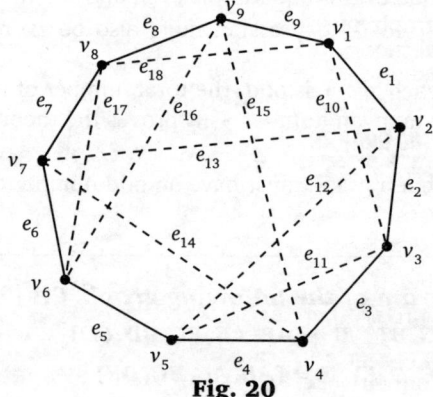

Fig. 20

These are

$$v_1 e_1 v_2 e_2 v_3 e_3 v_4 e_4 v_5 e_5 v_6 e_6 v_7 e_7 v_8 e_8 v_9 e_9 v_1 \text{ (Solid line)}$$

and $v_1 e_{10} v_3 e_{11} v_5 e_{12} v_2 e_{13} v_7 e_{14} v_4 e_{15} v_9 e_{16} v_6 e_{17} v_8 e_{18} v_1$ (Dotted line)

Further, there are only two more arrangements possible by graph theoretic consideration. They are

$$v_1 v_5 v_7 v_3 v_9 v_2 v_8 v_4 v_6 v_1$$

and $\qquad v_1 v_7 v_9 v_5 v_8 v_3 v_6 v_2 v_4 v_1$

☛ **REMARK**

- In general, it can be shown that for n people, the number of such possible arrangement is

  $\dfrac{n-1}{2}$, if n is odd

  and $\qquad \dfrac{n-2}{2}$, if n is even.

13.6 GRAPH ISOMORPHISM

(GARHWAL(B.C.A.)–2004)

Two graphs $G_1 = (V_1, E_1)$ and $G_2 = (V_2, E_2)$ are said to be isomorphic to each other if they have one-to-one correspondence between their edges and their vertices such that their incidence relationship must be preserved.

For example:

(1) (2)

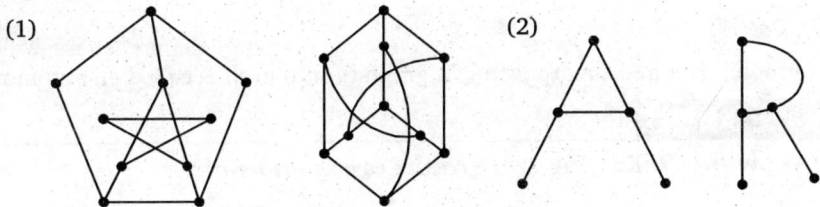

Fig. 21 Isomorphic Graphs

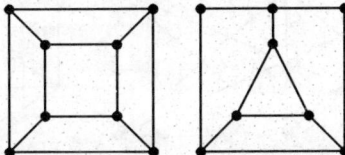

Fig. 22 These graphs are not Isomorphic Graphs

☛ **REMARK**

- For isomorphism, both graph must have
  - The same number of vertices
  - The same number of edges
  - An equal number of vertices with a given degree.

13.6.1 PROPERTIES OF ISOMORPHIC GRAPHS

If G_1 is isomorphic to G_2, i.e., $G_1 \cong G_2$, then there exists a function $\phi : V(G_1) \rightarrow V(G_2)$ with following properties :

(i) $\phi(x) = \phi(y) \Rightarrow x = y, \quad \forall x, y \in V(G_1)$

   i.e., ϕ is one-one.

(ii) For every vertex $y \in V(G_2)$, there exists a vertex $x \in V(G_1)$ such that $\phi(x) = y$

   i.e., ϕ is onto.

(iii) $(x, y) \in E(G_1)$ if and only if $(\phi(x), \phi(y)) \in E(G_2)$

i.e., structure preserving property is satisfied.

Self Complementary Graph: A graph $G = (V, E)$ is said to be self-complementary if it is isomorphic to its complements.

13.7 HOMEOMORPHIC GRAPH

Given any graph G, we can obtain a new graph by dividing an edge of G with additional vertices. Two graphs are said to be homeomorphic if they can be obtained from the same graph or isomorphic graphs by this method.

Definition. *Two graphs G_1 and G_2 are said to be homeomorphic graphs if G_2 can be obtained from G_1 by a sequence of subdivision of the edge of G.*

For example:

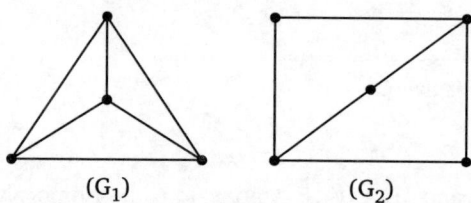

(G$_1$)   (G$_2$)

Fig. 23

Here, G_1 and G_2 are homeomorphic because G_1 can be obtained from G_2 by introducing vertices of degree 2 on edges (v_1, v_3) and (v_2, v_4).

13.8 AUTOMORPHISM

Let G be a graph. Then an isomorphism of a graph G onto itself is called an automorphism of G.

SOLVED EXAMPLES

EXAMPLE 1. *Show that following two graphs are isomorphic.*

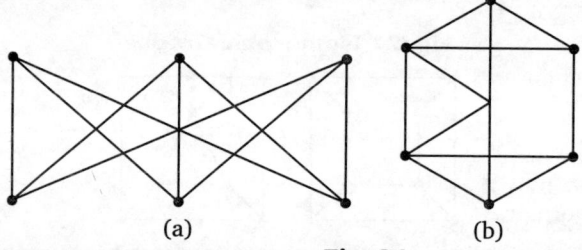

(a)   (b)

Fig. 24

SOLUTION. We observe that
   (i)   Both have same number of vertices, *i.e.*, six.
   (ii) Both have same number of edges, *i.e.*, nine.
   (iii) All vertices of both the graphs are of three degree.
   Hence, both graphs (a) and (b) are isomorphic to each other.

EXAMPLE 2. *Check this pair of graphs is isomorphic or not?*

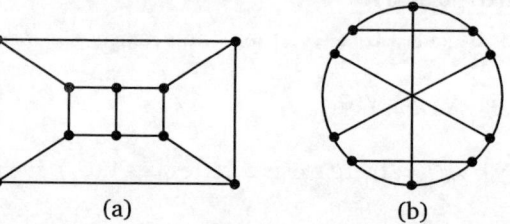

(a)   (b)

Fig. 25

SOLUTION. We know that for isomorphism, following conditions must be satisfied :
  (i) Both graphs have same number of vertices, *i.e.*, 10.
  (ii) Both graphs have same number of edges, *i.e.*, 15
  (iii) All vertices of both the graphs are of same degree, *i.e.*, 3.
  Hence, given graphs (a) and (b) are isomorphic.

EXAMPLE 3. *Show that the following graphs are not isomorphic to each other*

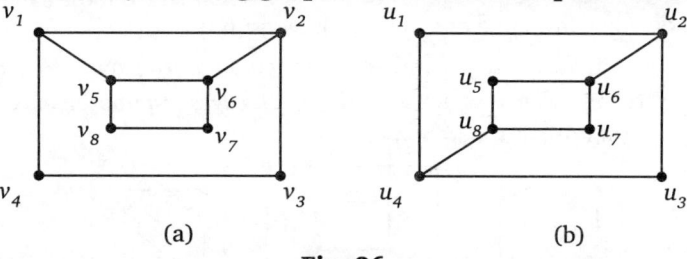

(a)   (b)

Fig. 26

SOLUTION. In the above two graphs, four vertices are of degree three and four vertices are of degree two. Also both graphs have equal number of vertices (*i.e.*, 8) and equal number of edges (*i.e.*, 10). Therefore, we can say that two graphs satisfy all the conditions of isomorphism, but they are not isomorphic to each other because in graph (a), vertices v_5 and v_6 are of degree 3 while in graph (b) vertices u_6 and u_8 are of degree 3. Hence, they can not be isomorphic to each other until v_5 corresponding to u_8.

EXAMPLE 4. *Show that the graphs given below are homeomorphic.*

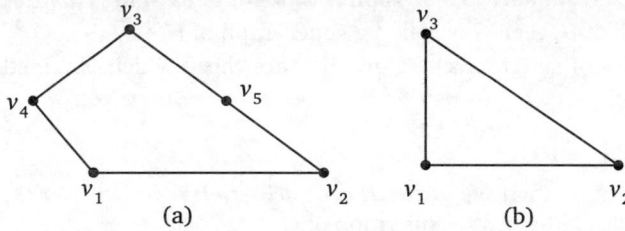

(a)   (b)

Fig. 27

SOLUTION. The graph (a) can be obtained from the graph (b) by adding vertices v_4 and v_5 of degree 2 on the edges (v_1, v_3) and (v_2, v_3) respectively. Hence, above two graphs are homeomorphic.

EXAMPLE 5. *Show that a graph G is self complementary if it has 4n or 4n + 1 vertices, where n is a non-negative integer.*

SOLUTION. Let $G = (V, E)$ be a self complementary graph with m points. It is given that G is self complementary, therefore G is isomorphic to $\bar{G}$.

$$\Rightarrow \qquad |E(G)| = |E(\bar{G})|$$

$$\Rightarrow \qquad |E(G)| + |E(\bar{G})| = \frac{m(m-1)}{2}$$

$$\Rightarrow \qquad 2|E(G)| = \frac{m(m-1)}{2}$$

$$\Rightarrow \qquad |E(\bar{G})| = \frac{m(m-1)}{4}$$

Now, $\dfrac{m(m-1)}{4}$ is an integer and one of m or $(m-1)$ is odd.

Therefore, m or $(m-1)$ is a multiple of G. Hence, n is of the form $4n$ or $4n + 1$.

13.9 SUBGRAPHS
(MEERUT(B.C.A)–1999)

Definition 1. *Let H be a graph with vertex set $V(H)$ and edge set $E(H)$. Also, let G be a graph with vertex set $V(G)$ and edge set $E(G)$. Then H is said to be a subgraph of G if*

$$V(H) \subseteq V(G) \quad and \quad E(H) \subseteq E(G)$$

In other words, a graph H is said to be subgraph of G if all the vertices and all the edges of H are in G and each edge of H has the same end vertices in H as in G.

Definition 2. *A subgraph H of a graph G is any graph obtained from G by a deletion of edges and / or vertices from G. Note that when a vertex v is deleted, all edges terminating at v must also be deleted.*

For example: Consider the following graphs

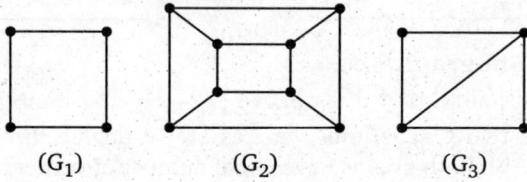

$$(G_1) \qquad\qquad (G_2) \qquad\qquad (G_3)$$

Fig. 28

Clearly, G_1 is a subgraph of both G_2 and G_3, but G_3 is not a subgraph of G_2.

☞ REMARKS
- The concept of subgraph is similar to the concept of subset in set theory. As a subset, it is contained in the set, similarly a subgraph is thought of as being contained in another graph.
- If H is a subgraph of G, then G is called a supergraph of H.
- The simplest types of subgraphs of a graph G are those which obtained by the deletion of a vertex or an edge.

13.9.1 PROPER SUBGRAPH

If H is a subgraph of G, then we write $H \subseteq G$. When $H \subset G$ but $H \neq G$, i.e., $V(H) \neq V(G)$ or $E(H) \neq E(G)$, then H is called a proper subgraph of G.

The subgraph, other than proper is known as improper subgraph.

13.9.2 SOME IMPORTANT CONCLUSIONS

1. Every graph is its own subgraph.
2. A single vertex in G is a subgraph of G.
3. A single edge in G together with its end vertices, is a subgraph of G.
4. A subgraph of a subgraph of G is also a subgraph of G.

13.10 TYPES OF SUBGRAPHS

13.10.1 DISJOINT SUBGRAPHS

(a) Vertex Disjoint Subgraph : Let $G = (V, E)$ be a graph, then two subgraphs H_1 and H_2 of G are called vertex disjoint if H_1 and H_2 have no vertex in common (Also, they do not have any edge in common).

Mathematically, for vertex disjoint subgraphs, we have

$$V(H_1) \cap V(H_2) = \phi \quad and \quad E(H_1) \cap E(H_2) = \phi.$$

For Example:

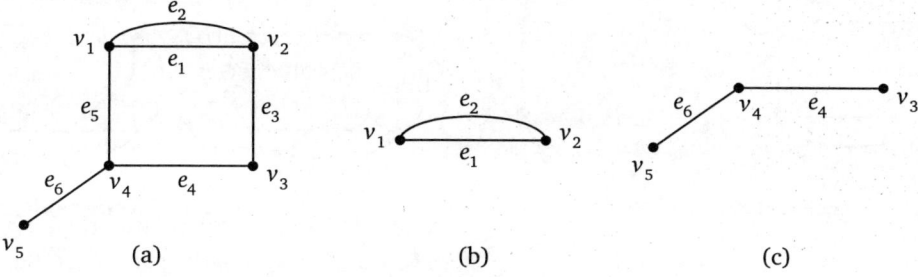

Fig. 29

Clearly (b) and (c) are vertex disjoint subgraphs of graph (a).

(b) Edge Disjoint Subgraph: Let $G = (V, E)$ be a graph. Two subgraphs H_1 and H_2 of G are said to be edge disjoint if they do not have any edge in common (they may have vertices in common).

Mathematically,

If $E(H_1) \cap E(H_2) = \phi$, then H_1 and H_2 are edge disjoint subgraphs.

For Example: Consider the following graphs

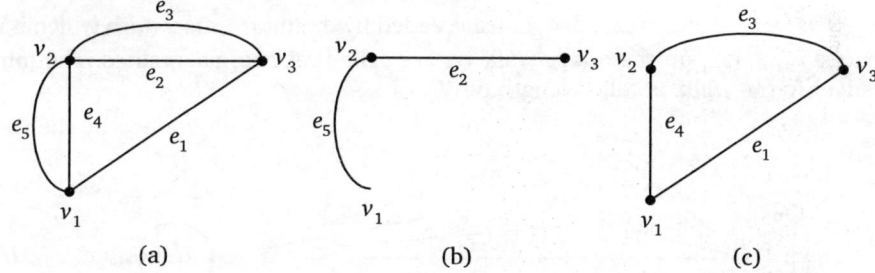

Fig. 30

Clearly (b) and (c) are edge disjoint subgraphs of graph (a).

13.10.2 INDUCED SUBGRAPH

Let $G = (V, E)$ be a graph. Also, let U be a subset of V. Then the subgraph $G(U)$ of G induced by U is defined to be the graph having vertex set U and edge set containing those edges of G that have both ends in U. Similarly, if F is a non-empty subset of the edge set E of G, then the subgraph $G(F)$ of G induced by F is the graph whose vertex set is the set of ends of edges in F and whose edge set is F.

13.10.3 SPANNING SUBGRAPH

A subgraph of G is said to be spanning subgraph if it contains all the vertices of G.

13.10. 4 VERTEX DELETED SUBGRAPH AND EDGE DELETED SUBGRAPHS

Let $G(V, E)$ be a graph. If from G, we delete a subset U of V and all the edges which have a vertex in U as an end point, then $(G - U)$ is called a vertex deleted subgraph and similarly, if a subset F of E is deleted from G, then $(G - F)$ denotes the subgraph of G with vertex set V and edge set $E - F$, then $(G - F)$ is called edge deleted subgraph.

For example :

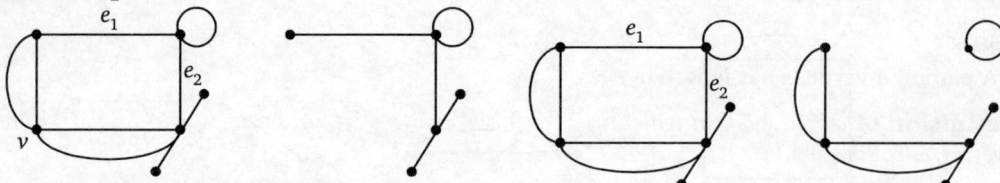

Edge deleted subgraph   **Vertex deleted subgraph**

Fig. 31

☞ REMARK

- By deleting from a graph G all loops and in each collection of parallel edges all edges but one in the collection, we get a simple spanning subgraph of G, called the underlying simple graph of G.

13.11 WALKS, PATHS AND CIRCUITS

Definition. *A walk in a graph G is a finite sequence*

$$W = v_0 \, e_1 \, v_1 \, e_2 \, v_2 \, ... \, v_{k-1} \, e_k \, v_k$$

whose terms are alternately vertices and edges such that for $1 \le i \le k$, *the edge* e_i *has ends* v_{i-1} *and* v_i.   (MEERUT(B.C.A.)–2005, 06, 08)

Each edge e_i is immediately preceded and succeeded by the two vertices with which it is incident.

The vertices $v_1, ..., v_{k-1}$ in the above walk W, are called its internal vertices. The integer k, the number of edges in the walk, is called length of W.

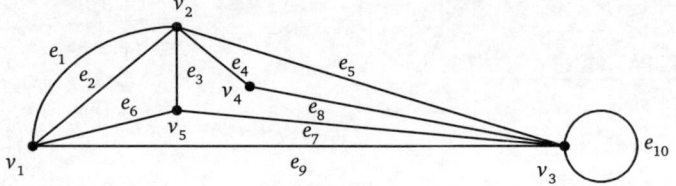

Fig. 32

In the figure, $W_1 = v_1 \, e_1 \, v_2 \, e_5 \, v_3 \, e_{10} \, v_3 \, e_5 \, v_2 \, e_3 \, v_5$ and $W_2 = v_1 \, e_1 \, v_2 \, e_1 \, v_1 \, e_1 \, v_2$ are both walks of length 5 and 3 respectively from v_1 to v_5 and from v_1 to v_2 respectively.

Given two vertices u and v of a graph G, a $u - v$ walk is called closed or open depending on whether $u = v$ or $u \ne v$.

Definition. *If the vertices* $v_0, v_1, ..., v_k$ *of the walk* $W = v_0 \, e_1 \, v_1 \, e_2 \, v_2 \, ... \, e_k \, v_k$ *are distinct then W is called path.*   (MEERUT(B.C.A.)–2002, 08)

$W = v_2 \, e_4 \, v_4 \, e_8 \, v_3 \, e_7 \, v_5 \, e_6 \, v_1$ is a path in a graph G.

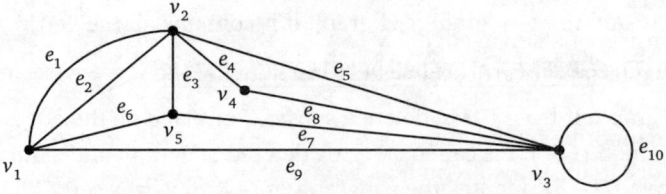

Fig. 33

☞ REMARK
- A path of n vertices has length $n - 1$.

Definition. *A closed walk in which no vertex appears more than once, except the initial and the final vertex, is called circuit.*

☞ REMARK
- A circuit is closed and non-intersecting walk.

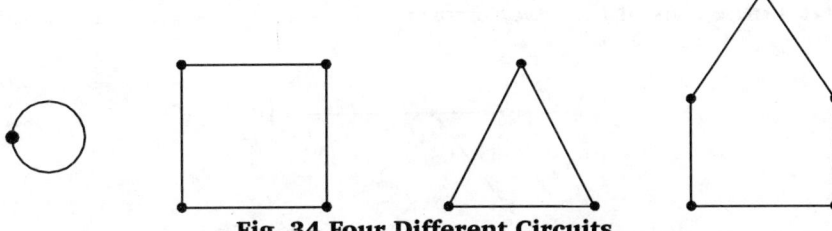

Fig. 34 Four Different Circuits

☞ REMARKS
- A trivial walk is one containing no edge.
- If the edges $e_1, e_2, ..., e_k$ of the walk $W = v_0\, e_1\, v_1\, e_2\, v_2...e_k\, v_k$ are distinct then W is called a trail.
- Circuit is also known as cycle, elementary cycle, circular path and polygon.
- Circuit has at least one edge.
- In circuit, every vertex has degree two.

FLOWCHART OF WALK, PATH AND CIRCUITS

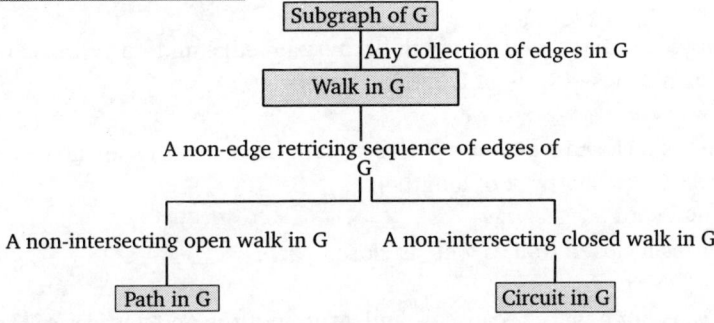

Fig. 35

SOME IMPORTANT FACTS

1. If a path consists of a single edge, then its length is one.
2. In a path, the terminal vertices are of degree 1 while each intermediate vertex is of degree 2.
3. A self loop cannot be included in a path while a self loop can appear in a walk.
4. If the path or circuit is a subgraph of some graph, then the degree of vertices are counted with respect to the edge included in the path or circuit only.
5. A circuit (or cycle) is also called an elementary cycle or a circular path or a polygon.
6. Every self loop is a circuit but the converse is not true.
7. A graph without cycles is called an acyclic graph.
8. A path does not intersect itself.

MORE DEFINITIONS

Definition 1. *The distance between any two vertices u, v in a graph to be denoted by* $d(u,v)$ *is the length of the shortest path between u and v.*

Definition 2. *The diameter of a connected path is the maximum distance between any two vertices in a graph.*

SOLVED EXAMPLES

EXAMPLE 1. *Find all walks of the given graph*

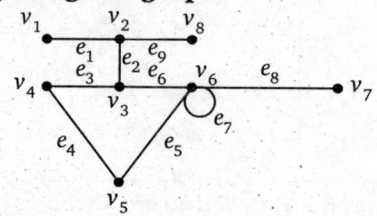

Fig. 36

SOLUTION.   We observe that

(i) w_1 is an open walk from v_1 to v_4 of length 7, *i.e.,* $l(w) = 7$

   $w_1 : v_1e_1v_2e_9v_8e_9v_2v_3e_6v_6e_6v_3e_3v_4$

   w_1 is not a trail because edge e_6 is repeated and not a path.

(ii) w_2 is an open walk from v_5 to v_7 of length 3

   $w_2 : v_5e_5v_6e_7v_6e_8v_7$

   w_2 is also a trail but not a path because the vertex v_6 is repeated.

(iii) w_3 is an open walk from v_1 to v_7 of length 4

   $w_3 : v_1e_1v_2e_2v_3e_6v_6e_8v_7$

   w_3 is a trail as well as a path because no edge and no vertex is repeated.

(iv) w_4 is a closed walk of length 5.

   $w_4 : v_4e_3v_3e_6v_6e_7v_6e_5v_5e_4v_4$

   w_4 is a closed trail but not closed path because v_6 is repeated two times.

(v) w_5 is a closed walk of length 4.

   $w_5 : v_3e_3v_4e_4v_5e_5v_6e_6v_3$

   w_5 is a closed trail as well as closed path.

(vi) $w_6 : v_1e_1v_2e_8v_7$

   w_6 is not a walk because v_2 and v_7 are not the end vertices of the edge e_8 that is cited between them.

   Since w_6 is neither a trail nor a path, therefore we can not define the length.

(vii) $w_7 : v_1e_1v_2e_9v_8e_9$

   w_7 is not a walk as sequence of w_7 is ended at an edge e_9 of G

(viii) $w_8 : v_1e_1v_2v_8e_9v_2$

   w_8 is not a walk as no edge occur between the vertex v_2 and v_8.

EXAMPLE 2. *Find all walks of the given graph*

   (i) Find all paths from the vertex A to the vertex F.

   (ii) All trails from A to F.

   (iii) The distance between A and F.

   (iv) Diameter of the graph.

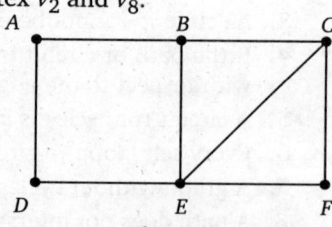

Fig. 37

SOLUTION. (i) Paths from A to F are

$P_1 = \{AB, BE, EF\}$ $\qquad$ $P_2 = \{AB, BC, CF\}$

$P_3 = \{AD, DE, EF\}$ $\qquad$ $P_4 = \{AB, BE, EC, CF\}$

$P_5 = \{AD, DE, EB, BC, CF\}$ $\qquad$ $P_6 = \{AD, DE, EC, CF\}$

$P_7 = \{AB, BC, CE, EF\}$

(ii) All the above paths and two more are

$T_8 = \{AD, DE, EB, BC, CE, EF\}$ $\qquad$ $T_9 = \{AD, DE, EC, CB, BE, EF\}$

(iii) The distance between A and F is 3, $d(A, F) = \{AB, BE, EF\}$.

(iv) $d(A, B) = 1$ $\quad$ $d(B, C) = 1$ $\quad$ $d(C, E) = 1$ $\quad$ $d(A, C) = 2$

$d(B, D) = 2$ $\quad$ $d(C, F) = 1$ $\quad$ $d(A, D) = 1$ $\quad$ $d(B, E) = 1$

$d(D, E) = 1$ $\quad$ $d(A, E) = 2$ $\quad$ $d(B, F) = 2$ $\quad$ $d(D, F) = 2$

$d(A, F) = 3$ $\quad$ $d(C, D) = 2$ $\quad$ $d(E, F) = 1$

diameter is 3.

EXAMPLE 3. *Draw all the circuits of the following graph.*

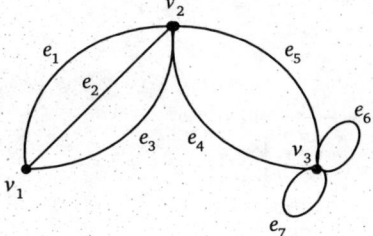

SOLUTION. We have

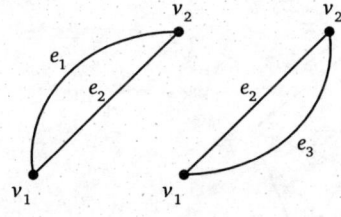 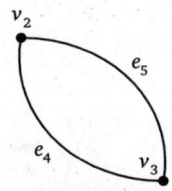  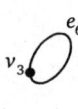

Fig. 38

(i) $v_1 e_1 v_2 e_2 v_1$ $\quad$ (ii) $v_1 e_2 v_2 e_3 v_1$ $\quad$ (iii) $v_2 e_5 v_3 e_4 v_2$ $\quad$ (iv) $v_3 e_7 v_3$ $\quad$ (v) $v_3 e_6 v_3$

EXAMPLE 4. *List all path from v_1 to v_8.*

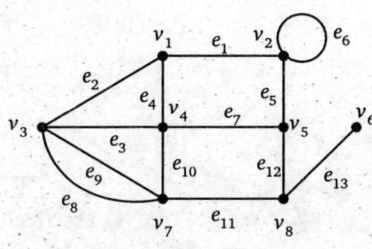

Fig. 39

SOLUTION.

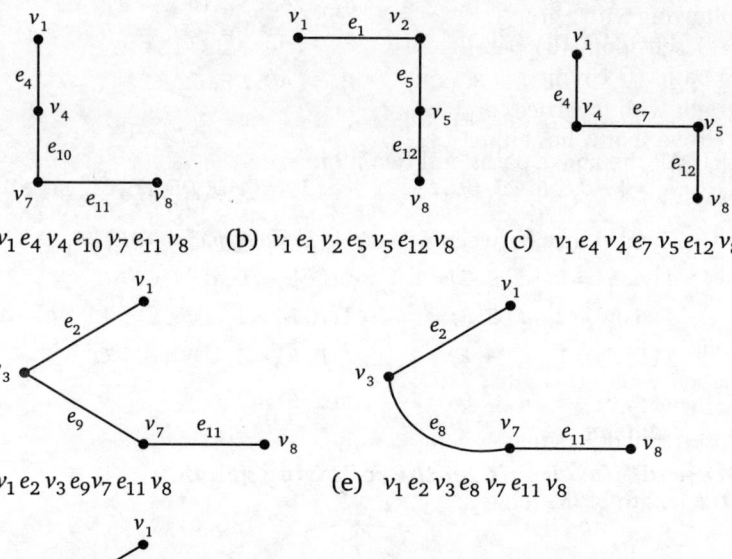

(a) $v_1 e_4 v_4 e_{10} v_7 e_{11} v_8$ (b) $v_1 e_1 v_2 e_5 v_5 e_{12} v_8$ (c) $v_1 e_4 v_4 e_7 v_5 e_{12} v_8$

(d) $v_1 e_2 v_3 e_9 v_7 e_{11} v_8$ (e) $v_1 e_2 v_3 e_8 v_7 e_{11} v_8$

(f) $v_1 e_2 v_3 e_3 v_4 e_7 v_5 e_{12} v_8$

EXAMPLE 5.   *Draw a circuit from the graph which is of length nine.*

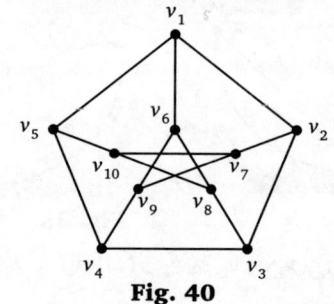

Fig. 40

SOLUTION.

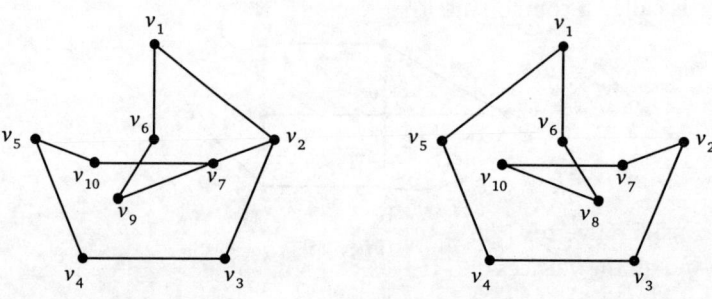

(a) $v_1 v_2 v_3 v_4 v_5 v_{10} v_7 v_9 v_6 v_1$ (b) $v_1 v_5 v_4 v_3 v_2 v_7 v_{10} v_8 v_6 v_1$

These are the only two cycles of length nine of the given graph.

EXERCISE 13.1

1. Define the following with a graph : (a) graph, (b) vertex, (c) self loop, (d) parallel edge, (e) walk, (e) path, (f) circuit.

2. Let G be a graph with n vertices and exactly $n-1$ edges. Prove that G has either a vertex of degree 1 or an isolated vertex.

3. Let G be a graph :

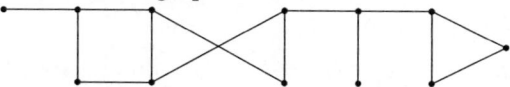

   (a) Find a closed walk of length 6. Is your walk a trail?

   (b) Find an open walk of length 12. Is your walk a path?

   (c) Find a closed trail of length 6. Is your trail a cycle?

   (d) What is the length of the longest cycle in G?

   (e) What is the length of a longest path in G? How many paths in G are there of this length?

4. In each of collection of three graphs shown in Fig. (a), there is exactly one isomorphic pair. Find each pair and justify that your answer is correct. Find the odd one out.

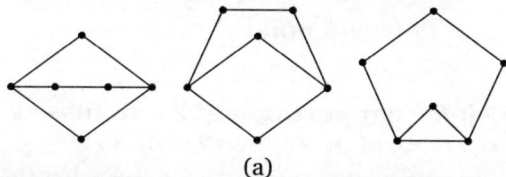

(a)

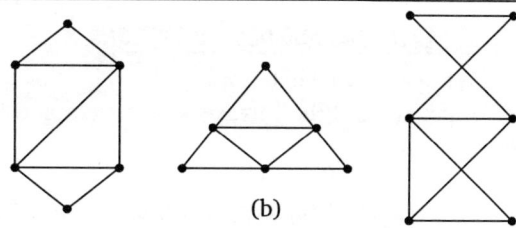

(b)

5. Find which of these are isomorphic pairs ?

(a)

(b)

(c)

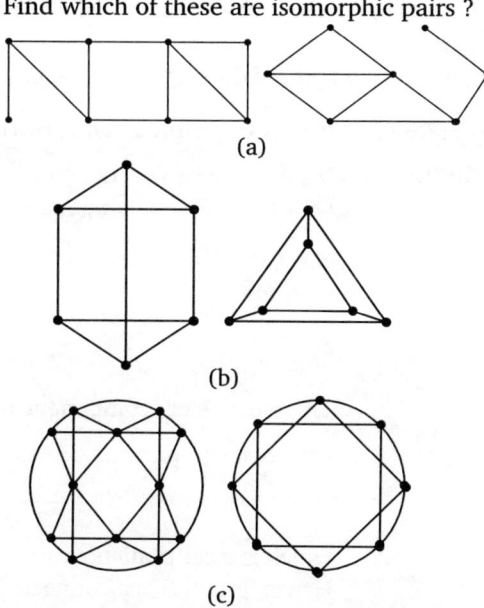

6. Prove that any two simple connected graphs with n vertices all of degree two are isomorphic.

13.12 CONNECTED, DISCONNECTED GRAPHS AND COMPONENTS (MEERUT(B.C.A.)–2002, 2003)

A graph G is said to be connected if there is at least one path between every pair of vertices in G. Otherwise G is disconnected. A null graph of more than one vertex is disconnected.

It is easy to see that a disconnected graph consists of two or more connected graphs. Each of these connected subgraphs is called a component. This graph consists of six components.

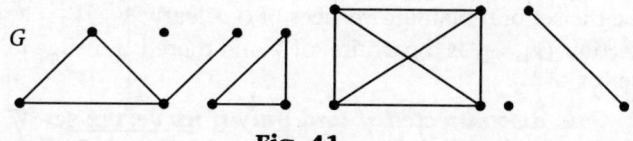

Fig. 41

In another way, consider a vertex v_i in a disconnected graph G, not all vertices of G are joined by paths to v_i. Vertex v_i and all the vertices of G that have paths to v_i together with all the edges incident on them form a component. A component is itself a graph.

☛ **REMARK**
- A single vertex is itself a connected graph.

13.12.1 PARTITIONS AND DECOMPOSITIONS

Let $G = (V, E)$ be a graph.

A partition of the vertex set V of a graph G is a collection $\{V_i\}_{i \le j \le k}$ of non-empty subset of V such that

$$V = V_1 \cup V_2 \cup ... \cup V_k, \quad k \ne 1, \qquad V_i \cap V_j = \phi, \text{ whenever } i \ne j$$

A partition of the edge set E of G is a collection $\{E_i\}_{1 \le i \le k}$ of non-empty subsets of E such that $E = E_1 \cup E_2 \cup ... \cup E_k$, $k \ne 1$ and $E_i \cap E_j = \phi$, whenever $i \ne j$.

☛ **REMARK**
- The partition of the edge set is called decomposition of G.

THEOREM 1. *If G is a graph with n points and* $\delta(G) \ge \dfrac{n-1}{2}$*, then G is connected.*

PROOF.   Let us suppose G is not connected. Then graph G has more than one component.
Let us consider a component $G_1 = (V_1, E_1)$ of graph G.

Let $v_1 \in V_1$, since $\delta(G) \ge \dfrac{n-1}{2}$, there exists at least $\dfrac{n-1}{2}$ points in G_1 which are adjacent to v_1 in G_1.

So, $$|V| \ge \frac{n-1}{2} + 1 \qquad \Rightarrow \qquad |V| \ge \frac{n+1}{2}$$

Therefore, each component of graph G has at least $\dfrac{n+1}{2}$ points and graph G has at least two components. Hence, the number of points in

$$G \ge \left(\frac{n+1}{2}\right), \text{ i.e., } V(G) \ge n+1$$

which is a contradiction.
Hence, graph G is a connected graph.

THEOREM 2. *A graph G is connected if and only if for any partition of V into subsets* V_1 *and* V_2*, there is an edge joining a vertex of* V_1 *to a vertex of* V_2*.*

PROOF.   Let $G = (V, E)$ be a graph which is connected and $V = V_1 \cup V_2$ be a partition of V into two subsets.
Let $u \in V_1$ and $v \in V_2$
Now, since G is connected, there exists a $u - v$ path in G, say

$$u = v_0 v_1 v_2 ... v_n = V$$

Let i be the least positive integer such that $v_i \in V_2$, then $v_{i-1} \in V_1$ and the vertices $v_{i-1} \in V_1$ and $v_i \in V_2$. Conversely, let G is not connected, *i.e.*, G is disconnected, then G contains at least two components. Let V_1 be the set of all vertices of one component and V_2 be the set of remaining vertices of G. Clearly $V_1 \cup V_2 = V$ and $V_1 \cap V_2 = \phi$.
The collection $\{v_1, v_2\}$ is a partition of V and there is no edge joining any vertex of V_1 to any vertex of V_2.

THEOREM 3. *A graph G is disconnected if and only if its vertex set V can be partitioned into two non-empty disjoint subset* V_1 *and* V_2*, such that there exists no edge in G whose one end vertex is in subset* V_1 *and the other in subset* V_2*.*

PROOF.   Suppose that such a partitioning exists. Consider two arbitrary vertices a and b of G such that $a \in V_1$ and $b \in V_2$. No path can exist between vertices a and b, otherwise there would be at least one edge whose one end vertex would be in V_1 and the other in V_2. Hence, if a partition exists, G is not connected.

Conversely, let G be a disconnected graph. Consider a vertex a in G. Let V_1 be the set of all vertices that are joined by paths to a. Since G is disconnected, V_1 doesn't include all vertices of G. The remaining vertices will form a non-empty set V_2. No vertex in V_1 is joined to any in V_2 by an edge. Hence, the partition exists.

THEOREM 4. *If a graph has exactly two vertices of odd degree, there must be a path joining these two vertices.*

PROOF.

Let G be a graph with all even vertices except vertices v_1 and v_2 which are odd. For every component of a disconnected graph no graph can have an odd number of odd vertices. Therefore, in graph G, v_1 and v_2 must belong to the same component and hence must have a path between them.

THEOREM 5. *A simple graph with n vertices and k component can have at most $(n - k)(n - k + 1)/2$ edges.*

PROOF.

Let the number of vertices in each of the k components of graph G be $n_1, n_2, ..., n_k$. Thus, we have $n_1 + n_2 + ... + n_k = n$, where $n_i \geq 1$.

Consider $\sum\limits_{i=1}^{k} (n_i - 1) = (n_1 - 1) + (n_2 - 1) + ... + (n_k - 1)$

$$= (n_1 + n_2 + ... + n_k) - k$$
$$= n - k$$

Squaring both sides, we get

$$\left[\sum\limits_{i=1}^{k} (n_i - 1) \right]^2 = n^2 - 2nk + k^2$$

$$(n_1^2 - 2n_1 + 1) + (n_2^2 - 2n_2 + 1) + ... + (n_k^2 - 2n_k + 1) = n^2 - 2nk + k^2$$

$$\sum\limits_{i=1}^{k} n_i^2 - 2(n_1 + n_2 + ... + n_k) + k = n^2 - 2nk + k^2$$

(terms consisting multiple of non-negative factors)

$$\sum\limits_{i=1}^{k} n_i^2 - 2n + k \leq n^2 - 2nk + k^2$$

$$\sum\limits_{i=1}^{k} n_i^2 \leq n^2 - 2nk + k^2 + 2n - k$$

$$= n^2 - 2n(k - 1) + k(k - 1)$$

$$= n^2 + (k - 2n)(k - 1)$$

Now, the maximum number of edges in the ith component of G is $^nC_2$, i.e., $\dfrac{n_i(n_i - 1)}{2}$. Therefore, the maximum number of edges in G is given by

$$\frac{1}{2} \sum\limits_{i=1}^{k} (n_i - 1) n_i = \frac{1}{2} \left(\sum\limits_{i=1}^{k} n_i^2 \right) - \frac{n}{2}$$

$$\leq \frac{1}{2} [n^2 - (k - 1)(2n - k)] - \frac{n}{2} = \frac{1}{2}(n - k)(n - k + 1)$$

13.13 EULER GRAPH : KÖNIGSBERG BRIDGE PROBLEM

Two islands M and N, termed by the Preyel river in Königsberg were connected to each other and to the banks P and Q with seven bridges, as shown in Fig. (a). The problem was to start at any of the four land areas of the city, M, N, P or Q walk over each of the seven bridges exactly once, and then return to the starting point (without swimming across the river).

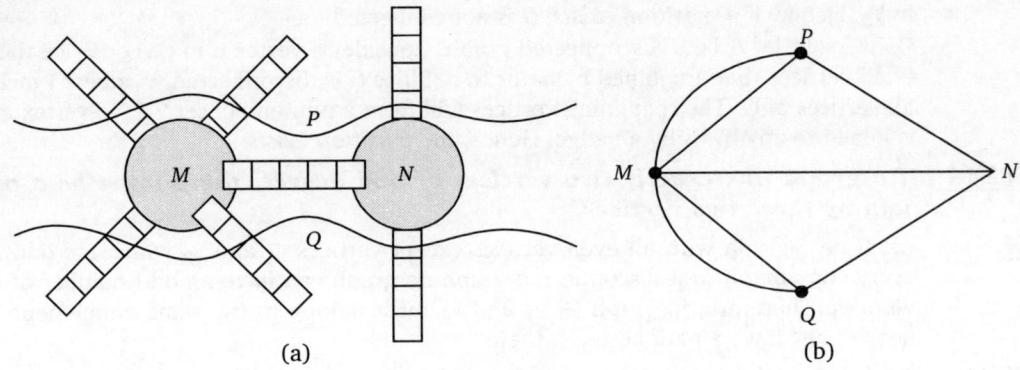

(a)   (b)

Fig. 42

Euler represented this situation by means of a graph as shown in (b). Vertices represent the land area and edges represent the bridges.

The problem is same as the problem of drawing figure without lifting the pen from the paper and without retracing a line.

☞ REMARK
- Utilities problem, seating problems and electrical engineering problems are the other applications of graph theory.

13.13.1 EULERIAN PATH

An Eulerian path is a path which exists between any pair of vertices such that starting from one vertex reaching back to the same after travelling through all the edges once and only once.

In any connected graph, if degree of each vertex is even, then it always possess an Eulerian path.

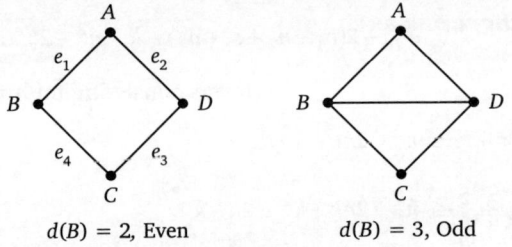

$d(B) = 2$, Even $d(B) = 3$, Odd

Fig. 43

☞ REMARK
- Problem of Euler is not possible because two vertex has odd degree 3 and 5.

13.13.2 EULERIAN CIRCUIT

An Eulerian circuit is a path through a graph in which the initial vertex appears second time as the terminal vertex.

13.13.3 TRAVERSAL GRAPH

A graph G is said to be traversable, if it has a path. A Euler path uses every edge exactly once but vertices may be repeated.

THEOREM I.   *A given connected graph G in an Euler graph iff all vertices of G are of even degree.*

PROOF.   Suppose that G is an Euler graph. It therefore contain an Euler line (which is a closed walk). In tracing this walk, we observe that every time the walk meets a vertex v, it goes through two "new" edges incident on v with one we "entered" v and with other we "exited". This is true not only of all intermediate vertices of the walk but also of the terminal vertex, because we exited and entered the same vertex at the beginning and end of the walk respectively. Thus, if G is an Euler graph, the degree of vertex is even. Conversely, assume that all vertices of G are of even degree. Now, we construct a walk starting at an arbitrary vertex v and going through the edge of G such that no edge is traced more than once. We continue tracing as far as possible. Since vertex is of even degree, we can exit from every vertex we enter. The tracing can not stop at any vertex but v. Since v is also of even degree, we shall eventually reach v when the tracing comes to an end. If this closed walk h we first traced includes all the edges of G, G is an Euler graph. If not, we remove from G all the edges in h and obtain a subgraph h' of G formed by the remaining edges. Since both G and h have all their vertices of even degree the degree of the vertices of h' are also even. Moreover, h' must touch h at least at one vertex a, because a is connected. Starting from a we can again construct a new walk in graph h'. Since all the vertices of h' are of even degree. This walk in h' must terminate at vertex a, but this walk in h' can be combined with h to form a new walk, which start and ends at vertex v and has more edges than h. This process can be repeated until we obtain a closed walk that traverses all the edges of G. Thus, G is an Euler graph.

13.13.4 OPEN EULERIAN OR UNICURSAL LINE

If an open walk in a graph is such that all the edges are traversed once and only once.

A graph containing a universal line is called universal graph. We start from a but not end at a and end at b so it is an unicursal graph.

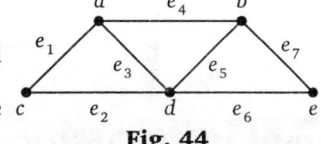

Fig. 44

$$a\,e_1\,c\,e_2\,d\,e_3\,a\,e_4\,b\,e_5\,d\,e_6\,e\,e_7\,b$$

THEOREM 2. *In a connected graph G with exactly $2k$ odd vertices, there exist k edge-disjoint subgraphs such that they together contain all edges of G and that each is a unicursal graph.*

PROOF. Let the odd vertices of the given graph G be named $v_1, v_2, ..., v_k, w_1, w_2, ..., w_k$ in any arbitrary order. Add k edges to G between the vertex points $(v_1, w_1)(v_2, w_2)...(v_k, w_k)$ to form a new graph G'.

Since every vertex of G' is of even degree, G' consists of an Euler line ρ. Now, if we remove from ρ the k edges, we just added, ρ will be split into k walks, each of which is a unicursal line. The first removal will leave a single unicursal line. The second removal will split that into two unicursal lines and each successive removal will split a unicursal line into two unicursal lines, until there are k of them.

13.14 LABELLED GRAPH

A graph G in which each vertex or edge is assumed a unique name or label (no two vertex have same name), then it is called labelled graph. In particular of each edge e of G is assigned a non-negative $l(e)$. The $l(e)$ is called the weight of length of (e).

For Example : (1) Consider the graph

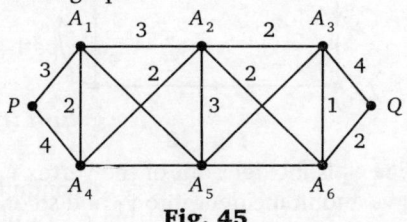

Fig. 45

It is weighted as well as labeled graph.

(2) Consider the graph

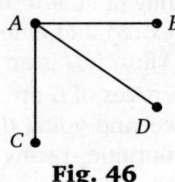

Fig. 46

It is a labeled graph w.r.t. vertices.

SOLVED EXAMPLES

EXAMPLE I. *Construct labeled graph with four vertices.*

SOLUTION. We can construct sixteen labelled graph with four vertices given below :

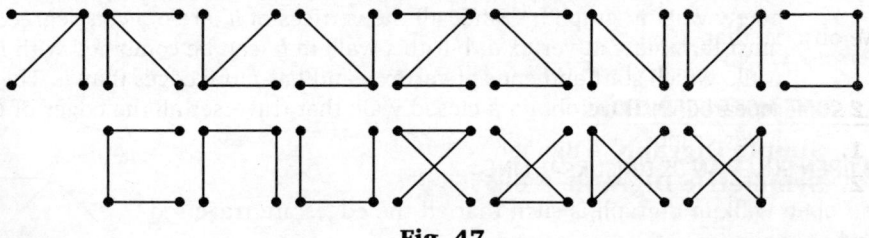

Fig. 47

13.15 DIRECTED GRAPH

(MEERUT(B.C.A.)–2003, 08)

Definition. *A directed graph* (*digraph*) $G = (V, E)$ *consists of a set of vertices* $V = \{v_1, v_2, ...\}$ *and a set of edges* $E = \{e_1, e_2, ...\}$ *such that each edge is identified (or associated) by some ordered pair of vertices* $\{v_i, v_j\}$.

If $e_k = \{v_i, v_j\}$ is an edge then this edge in digraph is represented by a line segment between the vertices v_i and v_j along with an arrow from v_i to v_j, where v_i is called initial vertex and v_j is called the terminal vertex of the edge e_k. Also the edge e_k is incident out of the vertex v_i and is incident into the vertex v_j.

In a directed graph, we also used the following graph:

(i) the vertex v_i is the origin of the edge e_k

(ii) v_i and v_j are adjacent to each other.

(iii) v_j is a successor of v_i.

For Example: Following graph is a directed graph with four vertices and seven edges

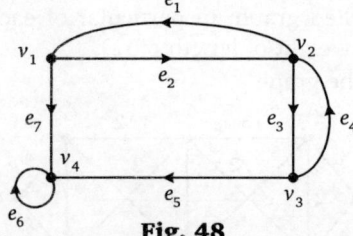

Fig. 48

In this directed graph, the edge e_2 is incident out of the vertex v_1 and is incident into the vertex v_2. The edge e_1 is incident out of v_2 and in incident into v_1 and so on.

☞ REMARK

• A directed graph or digraph is also known as oriented graph.

13.15.1 OUT DEGREE AND INDEGREE OF A VERTEX (GARHWAL B.C.A. 2007)

The number of edges incident out of a vertex v_i is said to be the out degree (also called out valance or outward demidegree) of the vertex v_i and is denoted by $d^+(v_i)$. In a similar way, the number of edges incident into a vertex v_i is said to be the indegree (also called in-valance or inward demidegree) of the vertex v_i and is denoted by $d^-(v_i)$.

For Example: In the above directed graph

$$d^+(v_1) = 2 \qquad\qquad d^-(v_1) = 1$$
$$d^+(v_2) = 3 \qquad\qquad d^-(v_2) = 1$$
$$d^+(v_3) = 1 \qquad\qquad d^-(v_3) = 2$$
$$d^+(v_4) = 1 \qquad\qquad d^-(v_4) = 3$$

We observe that

$$d(v) = d^+(v) + d^-(v) \text{ for every vertex } v \text{ in digraph.}$$

13.15.2 SOME MORE DEFINITIONS

1. **Simple Digraph.** A digraph is said to be simple if it has no self loop or parallel edge.
2. **Symmetric Digraph.** A digraph is said to be symmetric if for every edge (u, v), there is also an edge (v, u) in it.
3. **Complete Symmetric Directed Graph.** A simple directed graph in which there is exactly an edge directed from every vertex to every other vertex is said to be complete symmetric directed graph. The complete symmetric graph contains $n(n - 1)$ edges.

   For Example:

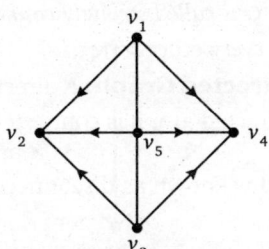

Fig. 49

4. **Asymmetric or Antisymmetric Digraph.** A digraph which has atmost one directed edge between a pair of vertices (self loops are allowed) is called its antisymmetric digraph.
5. **Complete Asymmetric Directed Graph.** If a directed graph has exactly one edge directed between every pair of vertices is called complete asymmetric directed graph. It contains $^nC_2$ number of edges.

   For Example :

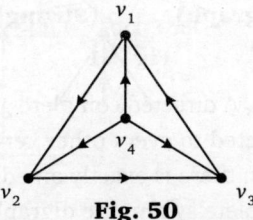

Fig. 50

☞ REMARK

• A complete asymmetric directed graph is also called a tournament or a complete tournament.

6. **Directed Walk.** Let G be a directed graph. A directed walk in G from a vertex v_i to v_j is an alternating sequence of vertices and directed edges, beginning with the vertex v_i and ending with v_j such that each edge is directed from the vertex preceding it to the vertex following it. In a directed walk, no edge will appear more than once whereas a vertex can appear more than once.

7. **Semi Walk.** A walk in a directed graph means either a directed walk or a semiwalk.

8. **Directed Path.** A directed path in a digraph is an open directed walk in which no edge appears more than once. In other words, a directed path P is an alternating sequence of vertices and edges, say $P = (v_0 e_1 v_1 e_2 v_2 e_3 v_3 ... e_n v_n)$ such that each edge e_i begins at the vertex v_{i-1} and ends at the vertex v_i.

☞ REMARKS
- In a directed path, all vertices and edges are distinct.
- The length of the path is defined as the number of edges in path.

9. **Semidirected Path.** In a directed path, a semi directed path is defined as a path in the corresponding undirected graph.

☞ REMARK
- A semidirected path is not a directed path.

10. **Connected Directed Graph.** In directed path, there are two types of paths namely, directed path and semidirected paths. Thus, we define the following:

    (i) Strongly Connected Directed Graph: A directed graph G is said to be strongly connected if for every two vertices u and v in G, there is path from u to v as well as path from v to u.

    Definition. *A diagraph G is called strongly connected if there exists at least one directed path from every vertex to every other vertex.*

    (ii) Weakly Connected Directed Graph: A directed graph G is called weakly connected if its corresponding undirected graph is connected, but G is not strongly connected.

☞ REMARK
- A digraph which is not connected is known as disconnected.

For example:

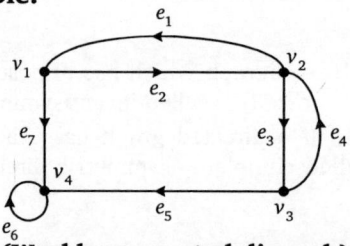 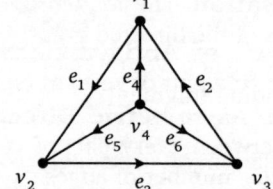

    (Weakly connected digraph)   **(Strongly connected digraph)**

Fig. 51

11. **Directed Complete Graph.** A directed complete graph $G = (V, E)$ of n vertices is a graph in which each vertex is connected to every other vertex by an arrow. It is denoted by k_n. A simple directed graph in which there is exactly an edge directed from every vertex to every other vertex is said to be complete symmetric digraph.

For example : k_3

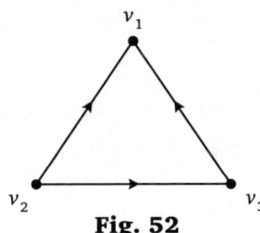

Fig. 52

12. **Directed Labeled Graph.** A graph $G = (V, E)$ is said to be labeled if its edges are labeled with some name or data. We can write these labels in place of an ordered pair in the edge set.

For example:

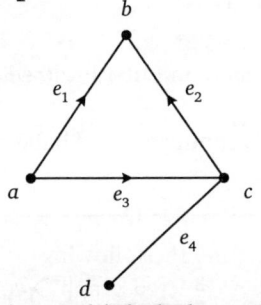

(Directed labeled graph)

$G = [\{a, b, c, d\}, \{e_1, e_2, e_3, e_4\}]$

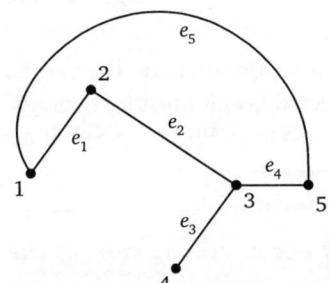

(Undirected labeled graph)

$G = [\{1, 2, 3, 4, 5\}, \{e_1, e_2, e_3, e_4, e_5\}]$

Fig. 53

13. **Balance Directed Graph or Pseudosymmetric Directed.** A directed graph in which for every vertex of it, the in-degree is equal to the out-degree is called a balance directed graph or pseudosymmetric directed graph or isograph.

14. **k-regular Directed Graph.** If the indegree of all the vertices of a balance directed graph is equal to k, then it is known as a k-regular directed graph.

15. **Condensation.** The condensation G_c of a directed graph is a graph whose vertices are the fragments of the directed graph G and there is a directed edge from one vertex to the other vertex in G_c whenever there exists at least one directed edge from one fragment to the other corresponding fragment.

☛ REMARK

- The condensation of a strongly connected directed graph is an isolated vertex and no condensation consists a directed circuit.

16. **Isomorphic Directed Graph.** The directed graphs are said to be isomorphic to each other if there exists an isomorphism from the vertex set of one graph to other, which preserve adjacency and orientation as well as non-adjacency.

17. **Accessibility.** In a directed graph a vertex b is said to be accessible (or reachable) from vertex a if there is a directed path from a to b. Thus, a directed graph G is strongly connected if and only if every vertex in G is accessible from every other vertex.

18. **Euler Directed Graph.** A directed graph D is a closed directed walk (a directed walk that starts and ends at the same vertex) which traverses every edge of D exactly once is called a directed Euler line.

For example:

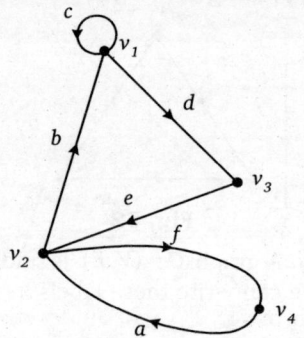

Fig. 54 Euler directed graph

☞ REMARKS
- A directed graph containing a directed Euler line is called an Euler directed graph.
- An Euler directed graph must be strongly connected.
- A directed graph is an Euler digraph if and only if it is connected and balanced.

SOLVED EXAMPLES

EXAMPLE 1. *Find the condensation of the graph*

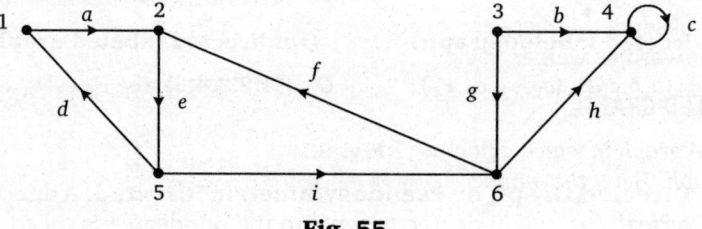

Fig. 55

SOLUTION. The fragments of the given graph are

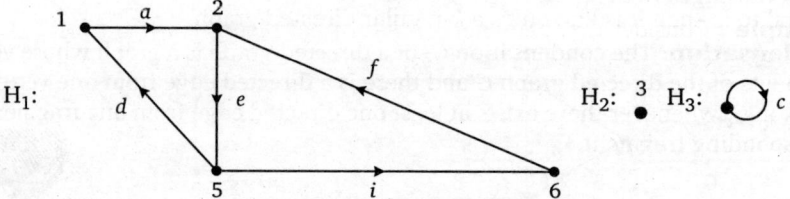

Hence, the required condensation of the graph is given by

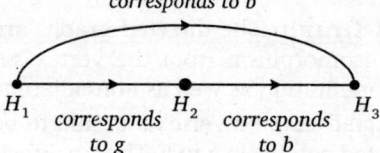

EXAMPLE 2. *Is the following directed graph strongly connected ?*

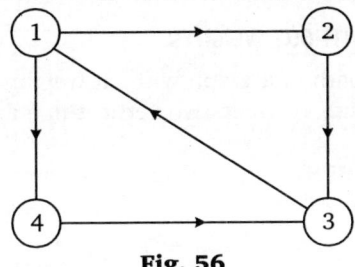

Fig. 56

SOLUTION.   Here, the possible pair of vertices and the forward and backward paths between them are given below :

Pair of vertices	Path	
	Forward	Backward
(1,2)	1-2	2-3-4
(1,3)	1-2-3	3-1
(1,4)	1-4	4-3-1
(2,3)	2-3	3-1-2
(2,4)	2-3-1-4	4-3-1-2
(3,4)	4-3	4-3

We observe that between every pair of distinct vertices of the given graph, there exists a forward as well as backward path. Hence, it is strongly connected.

13.16 WEIGHTED GRAPH

Definition. *A graph in which a non-negative real number $w(e)$ is assigned to its each edge e, is said to be weighted graph. Here, the number $w(e)$ is called the weight or length of the edge e.*

13.16.1 WEIGHT OF A PATH

Let G be a weighted graph. The weight of a path in the weighted graph G is equal to the sum of the weights of the edges in the path.

For example : Consider the following weighted graph

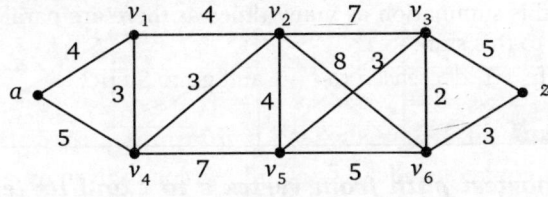

Fig. 57

In the above weighted graph, some path from a to z and their weights are given below:

(i)   Path : $(av_1v_2v_5v_3v_6z)$; Weight : $4 + 4 + 4 + 3 + 2 + 3 = 20$

(ii) Path : $(av_1v_4v_2v_6z)$; Weight : $4 + 3 + 3 + 8 + 3 = 21$

(iii) Path : $(av_4v_5v_6v_3z)$; Weight : $5 + 7 + 5 + 2 + 5 = 24$ and so on.

13.16.2 LABELED GRAPH

A graph G is said to be labeled if its edges and/or vertices are assigned data of one kind or another.

13.17 SHORTEST PATH PROBLEMS

13.17.1 SHORTEST PATH IN A GRAPH WITHOUT WEIGHTS

We know that the length of a path in a graph without weights denote the number of edges in a path and the shortest path is the path between two vertices in u and v that uses the least number of edges.

    1. Breath First Search (BFS) Algorithm. Here, first we process the starting vertex s. Then we process all the neighbours of s. Then we process all the neighbours of neighbours of s and so on.

WORKING PROCEDURE

STEP 1. Label vertex s with 0, set $i = 0$.

STEP 2. Find all unlabeled vertices in G which are adjacent to vertices labeled t. If there are no such vertices, then t is not connected to s. If there are such vertices, label them $i + 1$.

STEP 3. If t is labeled go to Step 4. If not increase i to $i + 1$. Go to Step 2.

STEP 4. The length of the shortest path for s to $i + 1$. Stop.

CONCLUSION. Once the length of the shortest path is found from the previous algorithm. To find the actual shortest path, we use back-tracking algorithm. This algorithm uses the label $\lambda(v)$ which are generated in BFS algorithm.

    2. Back Tracking Algorithm for a Shortest Path

WORKING PROCEDURE

STEP 1. Set $i = \lambda(t)$ and assign $v_i = t$.

STEP 2. Obtain a vertex u adjacent to v_i and with $\lambda(u) = i - 1$, assign $v_{i-1} = u$.

STEP 3. If $i = 1$, stop. If not increase i to $i - 1$ go to Step 2.

    3. The Back Tracking Algorithm for the Number of Shortest Path

WORKING PROCEDURE

STEP 1. Set $i = \lambda(t)$ and $\mu(t) = 1$. All other vertices v for which $\lambda(v) = \lambda(t)$ are assigned $\mu(v) = 0$.

STEP 2. For each vertex v which satisfies $\lambda(v) = i - 1$, find the sum $\Sigma\mu(u)$ over all u's which satisfy the following conditions $\lambda(u) = i$ and v is adjacent to u, if there are parallel edges, $\mu(u)$ is repeated in this summation as many times as there are parallel edges. For each set v, set $\mu(v)$ equal to this sum.

STEP 3. If $i = 1$, stop. If $i \neq 1$, decrease i to $i - 1$ and go to Step 2.

SOLVED EXAMPLES

EXAMPLE 1. *Find the shortest path from vertex s to t and its length from the graph given below:*

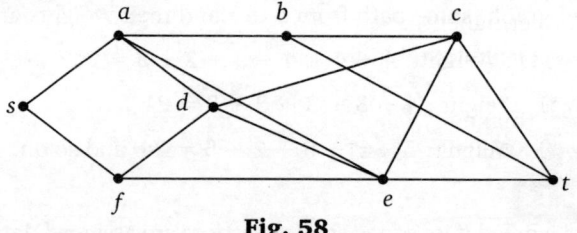

Fig. 58

SOLUTION.   Using Breadth First Search, label $s = 0$.

Then, we labelled a and f by $0 + 1 = 1$

and b,d,e are labelled $1 + 1 = 2$

also c and t are labelled $2 + 1 = 3$.

Now, since t is labelled 3, so length of a shortest path from s to t is 3.

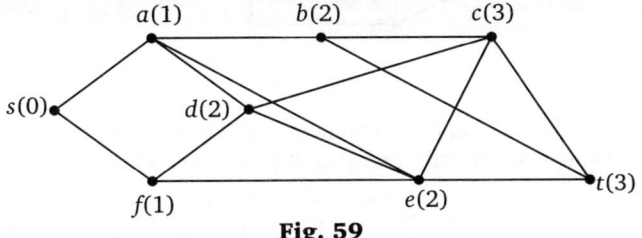

Fig. 59

Now, using the second algorithm, since $\lambda(t) = 3$, we start with $i = 3$ and $v_i = t$.

Choose e (or b) adjacent to $v_3 = t$ with $\lambda(e) = 2$ and assign $v_2 = e$. Further, we choose f adjacent to $v_2 = e$ with $\lambda(f) = 1$ and assign $v_1 = f$. Finally, we take s adjacent to f with $\lambda(s) = 0$ and assign $v_0 = s$ which gives the shortest path $v_0, v_1, v_2, v_3 = s$ f e t from s to t.

13.17.2 SHORTEST PATH IN A WEIGHTED GRAPH

We know that the length of a path in a weighted graph is the sum of the weights of the edges of this path and the shortest path between the two vertices is the minimum length of the path.

There are several different algorithm to find the shortest path between two vertices in a weight graph. Here, we shall discuss an algorithm given by Dutch mathematician Dijkstra.

13.17.3 DIJKSTRA ALGORITHM

This algorithm is used to obtain the shortest path from a specified vertex to another specified vertex. In this algorithm the vertices of the given graph are labelled. First of all the starting vertex A is assigned a permanent label O and the remaining vertices are assigned temporary label ∞. Then, at each iteration another vertex gets a permanent label according to the following procedure.

WORKING PROCEDURE

STEP 1. Every vertices j which is not permanently labelled yet gets a new temporary label whose value is given by

$$\min[(\text{old value of } j), (\text{old value of } i + ij)]$$

where, i is the vertex permanently labelled in the previous iteration and d_{ij} is the length of the edge between i and j vertices. If i and j are not connected by an edge then $d_{ij} = \infty$.

STEP 2. The smallest value among all the temporary labels is obtained which is assigned as the permanent label to the corresponding vertex. In case of a tie any one of the vertices may be selected for permanent labelling.

We repeat these two steps alternately till the destination vertex L gets a permanent label.

In this procedure the second vertex to get a permanent label is the vertex nearest to the vertex A. The next vertex which is permanently labelled will be the next nearest vertex to A. In this way, each permanently labelled vertex will be nearest to A. Thus we will get the shortest distance and path.

SOLVED EXAMPLES

EXAMPLE 1. *Find the shortest path from v_1 to v_7 in the following weighted graph.*

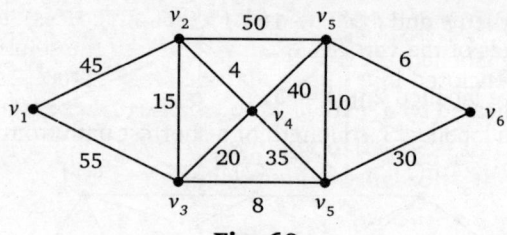

Fig. 60

SOLUTION. We construct table of temporary and permanent labels of vertices. The permanent labels are enclosed in squares and latest permanent label is marked as ⬜°. The labelling procedure is given below:

i	v_1	v_2	v_3	v_4	v_5	v_6	v_7
0	0	45 °	∞	∞	∞	∞	∞
1	0	45	55	∞	∞	∞	∞
2	0	45	55	49 °	95	∞	∞
3	0	45	55 °	49	89	84	∞
4	0	45	55	49	89	63 °	∞
5	0	45	55	49	73 °	63	93
6	0	45	55	49	73	63	93 °

This shows that the shortest distance from v_1 to v_7 is 93. This method gives the shortest distance.

The shortest path can be obtained by going in backward from the terminal vertex such that we go to the predecessor vertex whose label differs by the length of the connecting edge. In case of tie, there are more than one shortest paths. The shortest path can also be obtained by keeping a record of vertices from which each vertex was labelled permanently. Here, the shortest path is $v_1 \rightarrow v_3 \rightarrow v_6 \rightarrow v_7$.

EXAMPLE 2. *Find the shortest path from v_1 to v_{12} in the following weighted graph.*

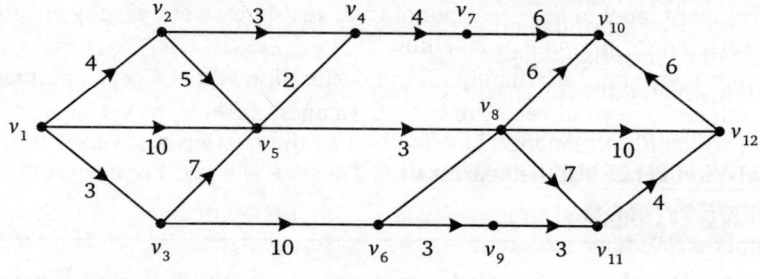

Fig. 61

SOLUTION. We shall use an array of length eleven (number of vertices), to show the temporary and permanent labels of the vertices as we go through the solution. The permanent labels will be shown enclosed in a square and the latest vertex assigned permanent label in the array is indicated by a mark □. The labelling proceeds as following:

i	v_1	v_2	v_3	v_4	v_5	v_6	v_7	v_8	v_9	v_{10}	v_{11}	v_{12}
0	[0]	∞	∞	∞	∞	∞	∞	∞	∞	∞	∞	∞
1	[0]	4	[3]	∞	10	∞	∞	∞	∞	∞	∞	∞
2	[0]	[4]	[3]	∞	10	13	∞	∞	∞	∞	∞	∞
3	[0]	[4]	[3]	[7]	9	13	∞	∞	∞	∞	∞	∞
4	[0]	[4]	[3]	[7]	[9]	13	11	∞	∞	∞	∞	∞
5	[0]	[4]	[3]	[7]	[9]	13	[11]	12	∞	∞	∞	∞
6	[0]	[4]	[3]	[7]	[9]	13	[11]	[12]	∞	17	∞	∞
7	[0]	[4]	[3]	[7]	[9]	[13]	[11]	[12]	∞	17	19	22
8	[0]	[4]	[3]	[7]	[9]	[13]	[11]	[12]	[16]	17	19	22
9	[0]	[4]	[3]	[7]	[9]	[13]	[11]	[12]	[16]	[17]	19	22
10	[0]	[4]	[3]	[7]	[9]	[13]	[11]	[12]	[16]	[17]	[19]	22
11	[0]	[4]	[3]	[7]	[9]	[13]	[11]	[12]	[16]	[17]	[19]	[22]

Thus the shortest distance from v_1 to v_{12} is 22. Note that this method gives only the shortest distance. The shortest path can be easily obtained by going backward from the terminal vertex such that we go to that predecessor (vertex) whose label differs exactly by the length of the connecting edge. A tie indicates more than one shortest path. We can also determine the shortest path by keeping a record of the vertices from which each vertex was labelled permanently. This record can be stored in another array of length n, such that whenever a new permanent label is assigned to vertex j, the vertex from which j directly reached is recorded in the jth position to this array. In the above example, the shortest path is $v_1 \to v_2 \to v_5 \to v_8 \to v_{12}$.

13.17.4 ALTERNATIVE FORM OF DIJKSTRA ALGORITHM

Let $G = (V, E)$ be a weighted graph with w as a weight function from the set E to the set of positive real numbers. We have to determine the shortest path from the vertex a to z.

WORKING PROCEDURE

STEP 1. Consider two subsets P_1 and T_1 of v such that $P_1 = \{a\}$ and $T_1 = V - P_1$. Now a shortest path from $a \in P_1$ to one of the vertices in T_1 is determined as follows :

For each vertex $x \in T_1$, let $l(x)$ be the length of a shortest path among all paths from a to x such that these path do not include any other vertex in T_1. If a and x are joined by the edge, then we take $l(x) = \infty$. Here $l(x)$ is known as index of x with regard to P_1. Let $t_1 \in T_1$ has the smallest index.

STEP 2. Take $P_2 = \{a, t_1\}$, $T_2 = v - P_2$

Now, we find the index $l(x)$ of every vertex $x \in T_2$ w.r.t. P_2 as follows :

$l(x) = $ min {index of x w.r.t. P_1, sum of the lengths of joining a to x through t_1}

Let the vertex $t_2 \in T_2$ has the minimum index $l(t_2)$ with respect to P_2.

STEP 3. Let t_2 be the required vertex. We want to reach a, then process is stopped. If t_2 is not the required vertex, then take $P_3 = \{a, t_1, t_2\}$, $T_3 = v - P_3$.

Now, we find the index $l(x)$ of every vertex $x \in T_3$ w.r.t. P_3 as follows:

$l(x) = $ min {index of x w.r.t. P_2, sum of the lengths of joining a to x through t_1 and t_2}

Let $t_3 \in T_3$ has the minimum index $l(t_3)$ w.r.t. P_3.

STEP 4. Take $P_4 = \{a, t_1, t_2, t_3\}$, $T_4 = v - P_4$ and repeat the process till we get the required length of the shortest path from the vertex a to z.

SOLVED EXAMPLES

EXAMPLE I. *Find the shortest path between a and z for the graph given below where numbers associated with edges are the weights*

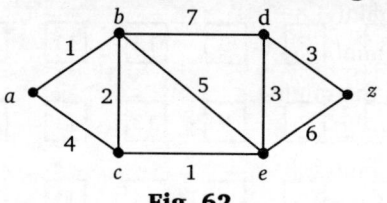

Fig. 62

SOLUTION. Let $G = (V, E)$ be the graph given above such that

$$V = \{a, b, c, d, e, z\}$$

Then we use Dijkstra algorithm as follows:

STEP 1. Let $P_1 = \{a\}$, $T_1 = \{b, c, d, e, z\}$

$$l(b) = 1, l(c) = 4, l(d) = \infty, l(e) = \infty, l(z) = \infty$$

$\Rightarrow b \in T_1$ has the minimum index 1.

STEP 2. Taking $P_2 = \{a, b\}$, $T_2 = \{c, d, e, z\}$

Then, $l(c) = $ min $(4, 1 + 2) = $ min$(4, 3) = 3$, $l(d) = $ min$(\infty, 1 + 7) = 8$

$l(e) = $ min$(\infty, 1 + 5) = 6$, $l(z) = $ min$(\infty, 1 + \infty) = \infty$

$\Rightarrow c \in T_2$ has the minimum index 3.

STEP 3. Taking $P_3 = \{a, b, c\}$, $T_3 = \{d, e, z\}$

Then, $l(d) = $ min$(8, 3 + \infty) = 8$, $l(e) = $ min$(6, 3 + 1) = 4$

$l(z) = $ min$(\infty, 3 + \infty) = \infty$

$\Rightarrow e \in T_3$ has the minimum index 4.

STEP 4. Taking $P_4 = \{a, b, c, e\}$, $T_4 = \{d, z\}$

$$l(d) = \min(8, 4+3) = 7 \ , \ l(z) = \min(\infty, 4+6) = 10$$

$\Rightarrow d \in T_4$ has the minimum index 7.

STEP 5. Taking $P_5 = \{a,b,c,e,d\}$, $T_5 = \{z\}$

$$l(z) = \min(10, 7+3) = 10 \ .$$

Hence, the length of the shortest path from a to z is 10 with shortest path $a \to b \to c \to e \to d \to z$.

EXAMPLE 2. *Find the shortest path from a to z in the following weighted graph*

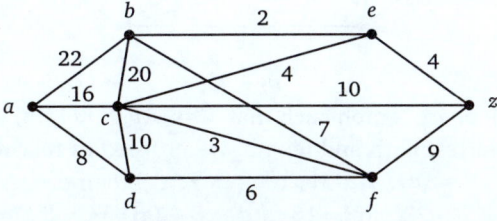

Fig. 63

SOLUTION. To find the shortest path from a to z, we proceed as follows :

Let $G = (V, E)$, where $V = \{a,b,c,d,e,f,z\}$

STEP 1. Let $P_1 = \{a\}$, $T_1 = \{b,c,d,e,f,z\}$

Then, $l(b) = 22$, $l(c) = 16$, $l(d) = 8$

$$l(e) = l(f) = l(z) = \infty$$

$\Rightarrow d \in T_1$ has the minimum index, *i.e.*, 8.

STEP 2. Taking $P_2 = \{a,d\}$, $T_2 = \{b,c,e,f,z\}$

Then, $l(b) = \min(22, 8+\infty) = 22$, $l(c) = \min(16, 8+10) = 16$

$$l(e) = \min(\infty, 8+\infty) = \infty \ , \ l(f) = \min(\infty, 8+6) = 14$$

$$l(z) = \min(\infty, 8+\infty) = \infty$$

$\Rightarrow f \in T_2$ has the minimum index, *i.e.*, 14.

STEP 3. Taking $P_3 = \{a,d,f\}$, $T_3 = \{b,c,e,z\}$

Then, $l(b) = \min(22, 8+6+7) = 21$, $l(c) = \min(16, 8+6+3) = 16$

$$l(e) = \min(\infty, 8+6+\infty) = \infty \ , \ l(z) = \min(\infty, 8+6+9) = 23$$

$\Rightarrow c \in T_3$ has the minimum index, *i.e.*, 16.

STEP 4. Taking $P_4 = \{a,d,f,c\}$, $T_4 = \{b,e,z\}$

Then, $l(b) = \min(21, 17+20) = 21$, $l(e) = \min(\infty, 17+4) = 21$

$$l(z) = \min(23, 17+10) = 23$$

$\Rightarrow b \in T_4$ has the minimum index, *i.e.*, 21 (we can also take that $e \in T_4$ has the minimum index 21).

STEP 5. Taking $P_5 = \{a,d,f,c,b\}$, $T_5 = \{e,z\}$

Then, $l(e) = \min(21, 37+2) = 21$

$$l(z) = \min(23, 37+\infty) = 23$$

$\Rightarrow e \in T_5$ has the minimum index 21.

STEP 6. Let $P_6 = \{a,d,f,c,b,e\}$, $T_6 = \{z\}$

Then, $l(z) = \min(23, 39+4) = 23$

Hence, the length of the shortest path from a to z is 23. Also, the shortest path for this graph is $a \to d \to f \to z$.

EXAMPLE 3. *Find the shortest path between a and z in the following graph*

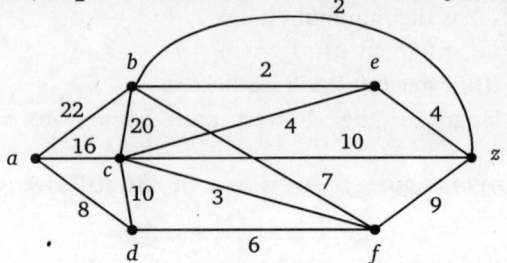

Fig. 64

SOLUTION.   Let $G = (V, E)$ be the graph such that $V = \{a, b, c, d, e, f, z\}$

To find the shortest path and length, we proceed as follows :

STEP 1. Let $P_1 = \{a\}$, $T_1 = \{b, c, d, e, f, z\}$

Then, $l(b) = 22$, $l(c) = 16$, $l(d) = 8$, $l(e) = \infty = l(f) = l(z)$

$\Rightarrow d \in T_1$ has the minimum index 8.

STEP 2. Let $P_2 = \{a, d\}$, $T_2 = \{b, c, e, f, z\}$

Then, $l(b) = \min(22, 8 + \infty) = 22$, $l(c) = \min(16, 8 + 10) = 16$

$l(e) = \min(\infty, 8 + \infty) = \infty$, $l(f) = \min(\infty, 8 + 6) = 14$, $l(z) = \min(\infty, 8 + \infty) = \infty$

$\Rightarrow f \in T_2$ has the minimum index, *i.e.*, 14.

STEP 3. Let $P_3 = \{a, d, f\}$, $T_3 = \{b, c, e, z\}$

Then, $l(b) = \min(22, 14 + 7) = 21$, $l(c) = \min(16, 14 + 3) = 16$

$l(e) = \min(\infty, 14 + \infty) = \infty$, $l(z) = \min(\infty, 14 + 9) = 23$

$\Rightarrow c \in T_3$ has the minimum index, *i.e.*, 16.

STEP 4. Let $P_4 = \{a, d, f, c\}$, $T_4 = \{b, e, z\}$

Then, $l(b) = \min(21, 17 + 20) = 21$, $l(e) = \min(\infty, 17 + 4) = 21$

$l(z) = \min(23, 17 + 10) = 23$

$\Rightarrow e \in T_4$ has the minimum index 21.

STEP 5. Let $P_5 = \{a, d, f, c, e\}$, $T_5 = \{b, z\}$

Then, $l(b) = \min(21, 21 + 2) = 21$, $l(z) = \min(23, 21 + 4) = 23$

$\Rightarrow b \in T_5$ has the minimum index.

STEP 6. Let $P_6 = \{a, d, f, c, e, b\}$, $T_6 = \{z\}$. Then, $l(z) = \min(23, 23 + 2) = 23$

Hence, the length of the shortest path from a to z is 23 and the shortest path for this graph is $a \rightarrow d \rightarrow f \rightarrow z$.

EXERCISE 13.2

1. Use the shortest path algorithm to find the shortest path from A to G in the weighted graph

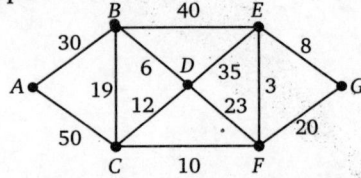

2. Use Dijkstra's algorithm to find the shortest path between the indicated vertices in the given weighted graphs

(i)

(ii)

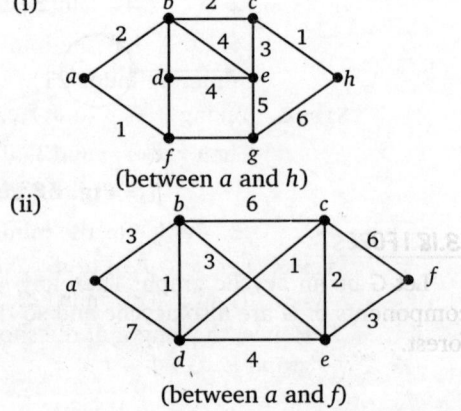

(between a and h)

(between a and f)

1. $A \rightarrow B \rightarrow C \rightarrow F \rightarrow E \rightarrow G$, with length 77.

2. (i) $a \rightarrow b \rightarrow c \rightarrow h$;   (ii) $a \rightarrow b \rightarrow d \rightarrow c \rightarrow e \rightarrow f$

13.18 TREES

Definition 1. *A connected graph is called a tree if it contains no cycles. We can say a tree is a connected acyclic graph.*

Definition 2. *A connected graph having no circuit is called a tree.*

Definition 3. *Any set of branches in the original graph, just sufficient in number to connect all the vertices (nodes) is called a tree.*

Its edges are called branches.

For Example:

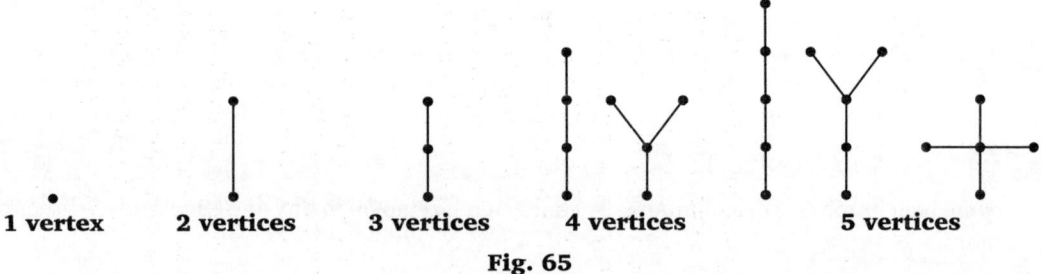

1 vertex   **2 vertices**   **3 vertices**   **4 vertices**   **5 vertices**

Fig. 65

☞ REMARKS

* Since we are considering only finite graph having at least one vertex, therefore a tree is finite and has at least one vertex.
* Also from the definition of the tree it follows that a tree is a simple graph without self loops or parallel edges.
* A tree with only one vertex is called a trivial tree or degenerate tree, otherwise it is a non trivial tree.
* A leaf (or terminal node) in a tree is a vertex of degree 1.
* A vertex of degree more than one is called a branch node.
* A vertex of degree one is called a pendant vertex and a non pendant vertex in a tree is called an internal vertex.

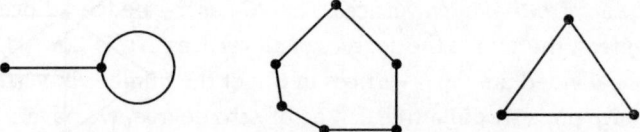

Fig. 66 Few graphs which are not a tree

13.18.1 FOREST

Let G be an acyclic graph. Then any subgraph of G must also contain no cycle. The connected components of G are also acyclic and so they are trees. This is why an acyclic graph is also called a forest.

For Example:

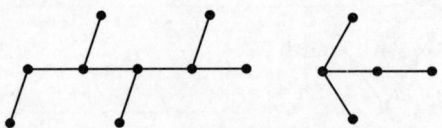

Fig. 67

SOLVED EXAMPLES

EXAMPLE 1. *Which of the following graph are trees?*

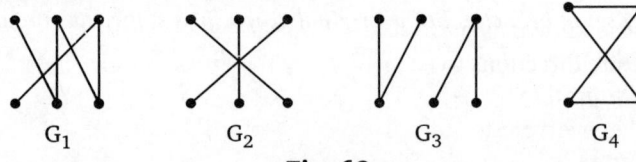

G_1 G_2 G_3 G_4

Fig. 68

SOLUTION. G_1 connected graph with no circuit and hence it is a tree. G_2 is also connected graph with no circuit hence G_2 is also a tree. G_3 is not connected so it is not a tree. G_4 is not a tree because it has a cycle or a circuit.

13.19 MINIMALLY CONNECTED GRAPH

A connected graph G is called minimally connected if removal of any one edge from G disconnects the graph G.

THEOREM 1. *A graph G is a tree iff it is minimally connected.*

PROOF. Suppose that G is a tree. We show G is minimally connected. Since G is a tree, it is connected. If G is not minimally connected, then there must exist an edge e in G such that $G - e$ is connected. Therefore e is in some circuit, which implies that G is not a tree, a contradiction. Thus G is minimally connected.

Conversely, suppose that G is a minimally connected graph. Then G is connected and can not have a circuit, otherwise, we could remove one of the edges in the circuit and still leave the graph connected. Thus a minimally connected graph is a tree.

THEOREM 2. *In any tree (with two or more vertices), there are at least two pendant vertices.*

PROOF. Let G be any tree having n vertices. Then G has $n - 1$ edges. Since each edge contributes two degrees, the sum of the degrees of all vertices in G is $2(n-1)$. Now $2(n-1)$ degrees are to be divided among n vertices in G. Let the number of vertices of degree one in G be x. Since no vertex in a tree can be of zero degree, we have

$$\frac{2(n-1)-x}{n-x} \geq 2$$

$$\Rightarrow \qquad x \geq 2.$$

Thus, we must have at least two vertices of degree one in a tree.

13.20 ARBORESCENCES

A digraph G is to be an arborescences if G contains no circuit (neither directed nor semi-directed) and G has exactly one vertex v of zero indegree.

The vertex v with zero indegree is called root of the arborescence.

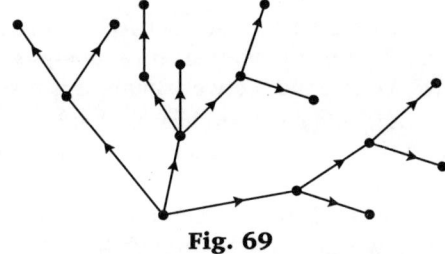

Fig. 69

☛ REMARKS

- Arborescence is also known as out-tree.
- If we assign opposite directions to each edge of an arborescence, we get a tree and this tree is known as intree.
- In an arborescence, every vertex of it is reachable from the root and the root is not accessible from any other vertex.
- A vertex of out degree zero in an arborescence is necessarily a pendant vertex.
- If a connected directed graph has a spanning tree, then it is an arborescence and called a spanning arborescence.

THEOREM 1. *An arborescence is a tree in which every vertex of it other than one (root) has an indegree of exactly one.*

PROOF. Let the number of vertices in an arborescence be n. We know that the graph is a directed tree, then it can have at most $n - 1$ edges. Now, we consider the sum of all in degree of the vertices of the graph D.

$$d^-(v_1) + d^-(v_2) + \ldots + d^-(v_n) \leq n - 1$$

Since there are n vertices in the graph of which one is of indegree zero, the remaining $n - 1$ vertices is of degree at least one. By this inequality, left hand side contains only one zero term (that is for the root vertex). If a vertex has indegree at least 2, then the sum of left hand side of the inequality is greater than $n - 1$ (since there are exactly $n - 1$ non-zero term). This indegree of each vertex of D should be exactly one.

THEOREM 2. *In an arborescence, there is a directed path from the root R to every other vertex.*

PROOF. Consider a path P which starts from the root R and traverses as far as possible in an arborescence. P can end only at a pendant vertex because otherwise we get a vertex whose indegree is two or more. But then it would be a contradiction. Hence, P ends at a pendant vertex. Now, since an arborescence is a connected and every directed path from R ends at a pendant vertex, every vertex lie on some directed path from the root R.

13.21 MAXIMUM FLOW PROBLEMS

Here, we consider the problem of shiping a certain homogeneous commodity available at a given node (called source), is required at another node (called sink). In such type of problem, there may not be a direct link connecting the sources and sinks. It can be routed from the source node to the sink node via a number of possible paths consisting of a sequence of links. Here, the flow network generally consists of some intermediate nodes called transhipment points. Each of these points in turn supply to other points. Therefore, when the shipments pass from destination to destination and from source to source, we have a transhipment problem. Also, several path may share the same link. Here we would like to find the maximum total flow through the network from the source to the sink, hence such problems are called maximum flow problems.

13.22 NETWORK FLOWS

Now a days, we are largely governed by networks, such as network of telephone lines, network of transportation (highways, rails, etc.). Thus the mathematical analysis of such networks has become very important. In this section, we will see that network analysis is essentially a study of graphs.

A network is represented by a weighted connected graph that contains no self loops. The vertices denote the stations (or places) and the edges are links through which the given commodity (such as oil, gas, number of messages, etc.) flows.

The weight associated with each edge represents the capacity of the edge. The capacity of an edge is the maximum amount of material which can flow through the edge.

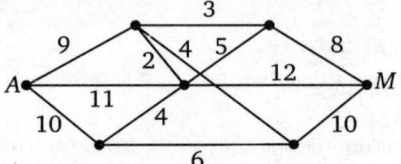

Fig. 70 A network of flow

The following assumptions are made for network problems :

  (i) At any intermediate vertex, the amount of material entering is equal to the amount of material leaving.

  (ii) The flow in any edge can not exceed the capacity of that edge.

  (iii) There is no loss of material during flows.

13.22.1 MAXIMUM FLOW MINIMUM CUT THEOREM

STATEMENT. *The maximum flow possible between two vertices in a network is equal to the minimum of the capacities of all cut sets with respect to two vertices.*

PROOF.   Consider any cut set S with respect to the vertices v_1 and v_2. In the subgraph $G - S$ a graph after removing S from G. There is no path between v_1 and v_2. Hence, every path between v_1 and v_2 must contain at least one edge of S. hence, the total flow rate between v_1 and v_2 can not exceed the capacity of S. Since this holds for all cut sets with respect to v_1 and v_2, the flow rate can not exceed the minimum of their capacities.

EXERCISE 13.3

1. List all cut sets with respect to the vertex v_1 v_2 in the graph.

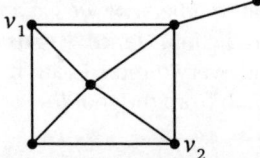

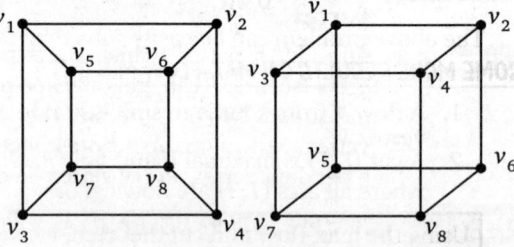

2. Give an example of a simple regular graph of degree three that is separable.

3. Prove that every connected graph with three or more vertices has at least two vertices which are not cut vertices.

4. What is the vertex connectivity of the complete graph of n vertices?

5. Find the edge connectivity and vertex connectivity of the following graphs :

6. (a) Give an example of a graph whose edge connectivity is one.

    (b) Give an example of a graph whose vertex connectivity is one.

    (c) Give an example of a graph whose edge connectivity and vertex connectivity both are equal to one.

7. Obtain the maximum flow in the network.

13.23 FORMULATION OF MAXIMUM FLOW PROBLEM AS A LINEAR PROGRAMMING PROBLEM

A maximum flow problem can be expressed as a linear programming problem. The process of it can be understood using the following example.

Consider the following network

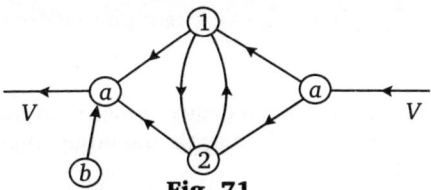

Fig. 71

Let us suppose source node be represented by the symbol a and the sink node by b.

Clearly, nodes 1 and 2 are the intermediate nodes. There are six possible arcs connecting the various nodes denoted by $(a \to 1), (a \to 2), (1 \to 2), (2 \to 1), (1 \to b), (2 \to b)$. Now, suppose the flow in arc $(i \to j)$ is possible only from node i to node j i.e. all the arcs are directed.

Further, let us have the following notations

V = total amount shipped from the source to the sink

x_{ij} = the amount of the flow shipped from the node i to node j

and C_{ij} = the capacity of arc $(i \to j)$ which is the maximum flow possible from i to j

Now, given the capacities C_{ij} on flows on each arc $(i \to j)$ and that the total flow into a node must be equal to the total flow out of that node, we have to find the maximum flow V that can be sent from the source node a to the sink node b.

Then we have the following corresponding linear programming problem.

$$\text{max. } Z = V$$

subject to the constraints

$$x_{a1} + x_{a2} = V$$
$$x_{12} + x_{1b} = x_{a1} + x_{21}$$
$$x_{21} + x_{2b} = x_{a2} + x_{12}$$
$$x_{1b} + x_{2b} = V$$

and

$$0 \le x_{a1} \le C_{a1} \qquad\qquad 0 \le x_{a2} \le C_{a2}$$
$$0 \le x_{12} \le C_{12} \qquad\qquad 0 \le x_{21} \le C_{21}$$
$$0 \le x_{1b} \le C_{1b} \qquad\qquad 0 \le x_{2b} \le C_{2b}$$

The above problem can be easily solved by the method of solution of linear programming problem.

SOME MORE RESULTS ON MAX-FLOW MIN CUT THEOREM

1. A flow V from source to sink is maximal if and only if there is no flow augmenting path with vertex V.
2. A cut (I, T) is maximal if and only if every maximal flow x saturates $(x_{ij} - C_{ij})$ for all arcs (I, T) . where all arcs (T, I) are flowless i.e. $x_{ji} = 0$.

> Using the max-flow min cut theorem, we can find the maximal flow in a network by determining the capacities of all the cuts and selecting the minimum capacity.

13.24 LABELLING ROUTINE ALGORITHM

STEP 1. Let N_1 be the set of all nodes connected in the source by an edge with positive excess capacity. Label each j in N_1 with $[E_j, 1]$, where E_j is the excess capacity e_{1j} of edge $(1, j)$. The 1 in the label indicates that j is connected to the source, node 1.

STEP 2. Let node j in N_1 be the node with smallest node number and let $N_2(j)$ be the set of all unlabeled nodes, other than the source, that are joined to node j and have positive excess capacity.

Suppose that node k is in N_2 (j) and (j, k) is the edge with positive excess capactiy.

Label node k with $[E_k, j]$, where E_k is the minimum of E_j and the excess capacity e_{jk} of edge (j, k).

When all the nodes in N_2 (j) are labeled in this way, repeat this process for the other nodes in N_1.

Let $N_2 = \bigcup_{j \in N_1} N_2(j)$.

Note that after step 1, we have labeled each node j in N_1 with E_j, the amount of material that can flow from the source to j through one edge and with the information that this flow came from node 1.

In step 2, previously unlabeled nodes k that can be reached from the source by a path $\pi : 1, j, k$ are labeled with $[E_k, j]$.

Here E_k is the maximum flow that can pass through π since it is the smaller of the amount that can reach j and the amount that can then pass on to k.

Thus, when step 2, is finished, we have constructed two-step paths to all nodes in N_2.

The label for each of these nodes records the total flow that can reach the node through the path and its immediate predecessor in the path.

We attempt to continue this construction increasing the lengths of the paths until we reach the sink (if possible).

Then the total flow can be increased and we can retrace the path used for this increase.

STEP 3.   Repeat step 2, labelling all previously unlabeled nodes N_3 that can be reached from a node in N_2 by an edge having positive excess capacity.

Continue this process forming sets $N_4, N_5, \ldots$ until after a finite number of steps either

    (i) The sink has not been labeled and no other nodes can be labeled. It can happen that no nodes have been labeled, remember that the source is not labeled, or

    (ii) The sink has been labeled.

STEP 4.   In case (i), the algorithm terminates and the total flow then is a maximum flow.

STEP 5.   In case (ii) the sink, node n, has been labeled with $[E_n, m]$ where E_n is the amount of extra flow that can be made to reach the sink through a path π. We examine π in reverse order. If each $(i, j) \in N$, then we increase the flow in (i, j) by E_n and decrease the excess capacity e_{ij} by the same amount.

Simultaneously, we increase the excess capacity of the (virtual) edge (j, i) by E_n since there is that much more flow in (i, j) to reverse.

If on the other hand, $(i, j) \notin N$, we decrease the flow in (i, j) by E_n and increase its excess capacity by E_n.

We simultaneously decrease the excess capacity in (i, j) by the same amount, since there is less flow in (i, j) to reverse.

We now have a new flow that is E_n units greater than before and we return to step 1.

13.25 MAX-FLOW ALGORITHM

This algorithm will start with a feasible flow on all arcs satisfying capacity restrictions and conservation of flows at all nodes. Here, initially we label source node a and then apply the labelling routine to label another node. When the sink node b is labelled we have a flow augmenting path.

Now we retrace the flow augmenting path with the help of the labels on the nodes and compute the max flow that can be sent in the path.

Repeat the above process by finding another flow augmenting path from a to b using the above labelling routine.

SOLVED EXAMPLES

EXAMPLE 1. *Use the labeling algorithm to find a maximum flow for the network in following figure.*

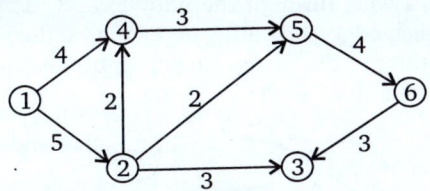

Fig. 72

SOLUTION. Fig. shows the network with initial capacities of all edges in G. Then initial flow in all edges is zero.

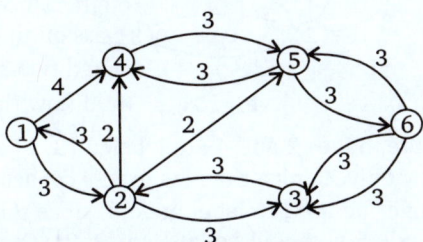

Fig. 73

STEP 1. Starting at the source, we can reach nodes 2 and 4 by edges having excess capacity, so $N_1 = \{2, 4\}$.

We label nodes 2 and 4 with the labels [5, 1] and [4,1], respectively, as shown in Fig.

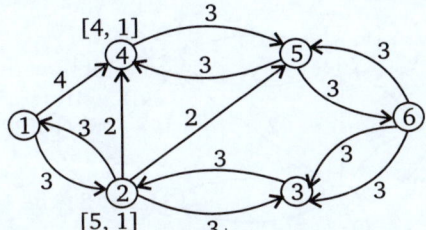

Fig. 74

STEP 2. From node 2 we can reach nodes 5 and 3 using edges with positive excess capacity. Node 5 is labeled with [2, 2] since only two additional units of flow can pass through edge (2, 5).

Node 3 is labeled with [3,2] since only 3 additional units of flow can pass through edge (2, 3). The result of this step is shown in Fig.

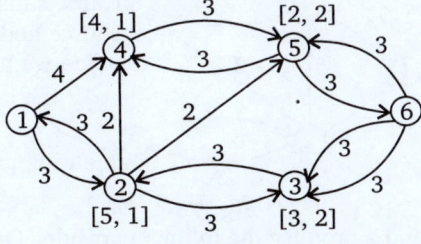

Fig. 75

We cannot travel from node 4 to any unlabeled node by one edge. Thus, $N_2 = \{3,5\}$ and step 2 is complete.

STEP 3. We repeat step 2 using N_2. We can reach the sink from node 3 and 3 units through edge (3,6). Thus, the sink is labeled with [3,3].

STEP 4. We work backward through the path 1, 2, 3, 6 and subtract 3 from the excess capacity of each edge, indicating an increased flow through that edge, and adding an equal amount to the excess capacities of the (virtual) edges. We now return to step 1 with the situation shown in Fig.

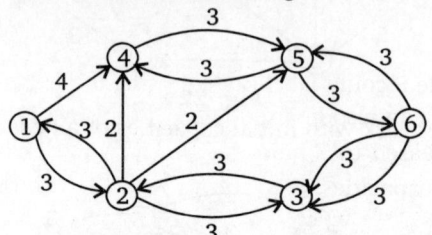

Fig. 76

Proceeding as before, nodes 2 and 4 are labeled [2, 1] and [4, 1] respectively.

Note that E_2 is now only 2 units, the new excess capacity of edge (1, 2).

Node 2 can no longer be used to label node 3, since there is no excess capacity in the edge (2, 3). But node 5 now will be labeled [2, 2]. Once again no unlabled node can be reached from node 4, so we move to step 3.

Here we can reach node 6 from node 5 so node 6 is labeled with [2, 5].

The final result of step 3 is shown in Fig. (4.52), and we have increased the flow by 2 units to a total of 5 units.

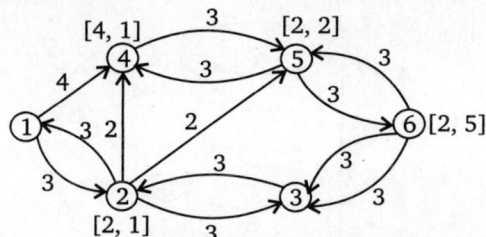

Fig. 77

We move to step 4 again and work back along the path 1, 2, 5, 6, subtracting 2 from the excess capacities of these edges and adding 2 to the capacities of the corresponding (virtual) edges.

We return to step 1 with fig. given below

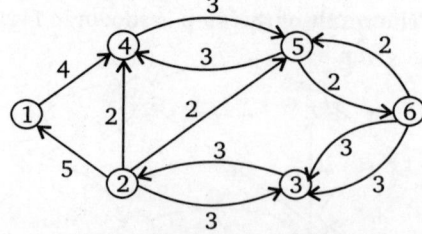

Fig. 78

This time steps 1 and 2 produce the following results. Only node 4 is labeled from node 1, with [4,1]. Node 5 is the only node labeled from node 4, with [3, 4] step 3 begins with Fig. given below.

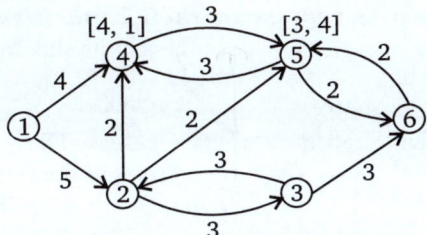

Fig. 79

At this point, node 5 could label node 2 using the excess capacity of edge (5, 2). However, node 5 ean also be used to label the sink. The sink is labeled [2, 5] and the total flow is increased to 7 units. In step 5, we work back along the path 1, 4, 5, 6, adjusting excess capacities. We return to step 1 with the configuration shown in Fig. given below.

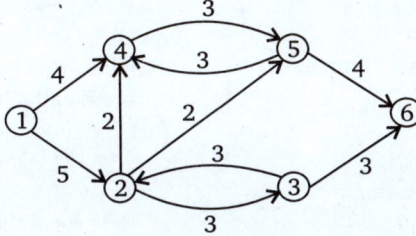

Fig. 80

Verify that after steps 1, 2 and 3, nodes 4, 5 and 2 have been labeled as shown in Fig. and no further labeling is possible. The final labeling of node 2 uses the virtual edge (5, 2).

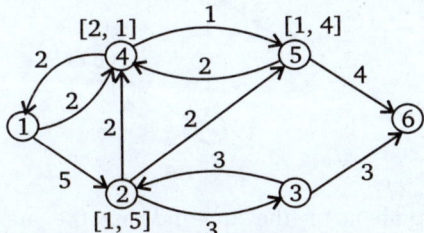

Fig. 81

Thus, the final overall flow has value 7. By subtracting the final excess capacity e_{ij} of each edge (i, j) in N from the capacity C_{ij}, the flow F that produces the maximum value 7 can be see in Fig. given below.

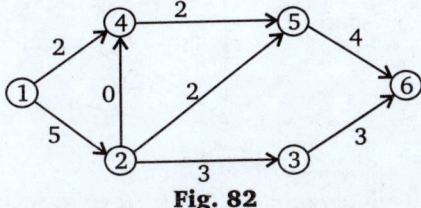

Fig. 82

EXAMPLE 2. *Find the maximal flow V from a to b in the following network, where the numbers on the arcs represents their capacities.*

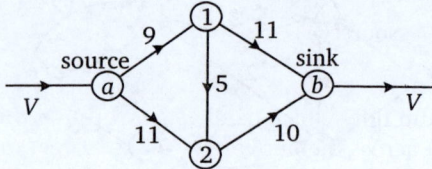

Fig. 83

SOLUTION. We start with zero flows on all arcs. Clearly, the numbers on the arcs (i, j) represented (x_{ij}, C_{ij}). Then we have the following steps.

STEP 1. To find a flow augmenting path from a to b node a is initially labeled by putting *. From a, we can label node 1 since $(a, 1)$ is a forward arc, carrying a flow x_{a1} which is less than its capacity.

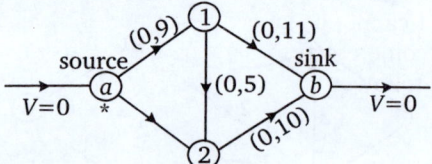

Fig. 84

From node 1, node 2 is labeled through the forward arc $(1, 2)$ and from node 2 the sink b is labeled. So we obtain a flow augmenting path consisting only of forward arcs which are shown in the following graph.

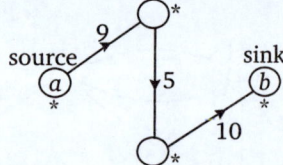

Fig. 85

Here, the numbers on the arcs indicate the maximum flow which is possible in each arc. Therefore, the maximal flow which can be sent through the flow augmenting path is 5 units. This increase V by 5 units and the flow on all forward arcs in the path by 5 units. Therefore, the new flow pattern becomes as shown in the following figure.

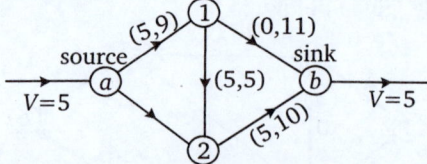

Fig. 86

STEP 2. Repeat the labeling routine and a new flow augmenting path is obtained as below.

Fig. 87

The maximum flow which can be sent through this path is 4 units. This increase the flow (V) across the network to 9 units. Now the new flow pattern is obtained as shown in the following figure.

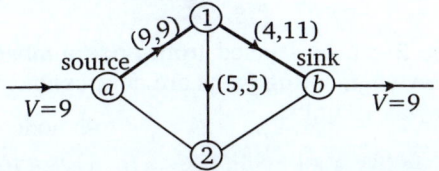

Fig. 88

STEP 3. Here, node 1 cannot be labeled from node a because $(a,1)$ is a forward arc whose flow has become equal to its capacity. But a new flow augmenting path can be obtained as follows:

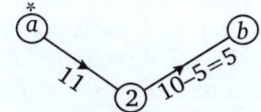

Fig. 89

This increase the total flow from a to b by 5 units as shown in the following figure.

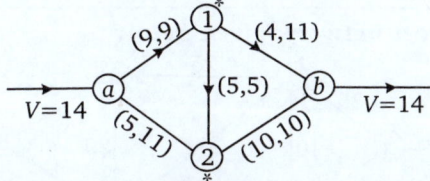

Fig. 90

STEP 4. Now starting from node a node 2 can be labeled but the sink can not be labeled from 2 because the flow in arc $(2, b)$ has reached to its capacity 10. But node 1 cab be labeled from node 2 because $(1, 2)$ is a backward arc carrying a positive flow. From node 1, the sink can be labeled using the forward arc $(1, b)$. So, we obtain a new flow augmenting path consisting of two forward arcs $(a, 2)$ and $(1, b)$ and a backward arc $(1, 2)$ as shown below.

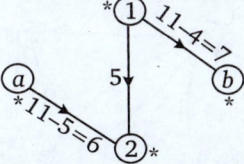

Fig. 91

Now, to increase the flow V through this path we increase the flow on the forward arcs and decrease it in the backward arcs. Therefore, the maximum possible increase in V becomes 5 units and the new assignment of flows is given in the following figure.

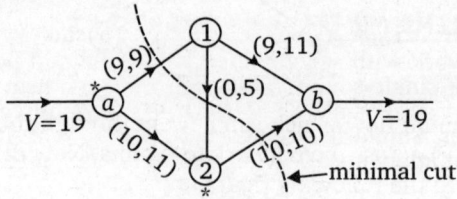

Fig. 92

STEP 5. Since, node 2 can be labeled from node a but the sink can never be labeled. So, now flow augmenting paths are not possible. Hence, the obtained maximum possible flow is 19 units.

13.26 EXTENSION OF MAX-FLOW ALGORITHM

13.26.1 UNDIRECTED ARCS

Consider a network containing arc with no specified direction of flow. If an arc connecting nodes i and j undirected with capacity C, then we can write

$$x_{ij} \leq C, \ x_{ji} \leq C, \ x_{ij} \cdot x_{ji} = 0$$

which shows that a maximum of C units of flow is possible between the nodes i and j in either direction but the flow is allowed in any of the direction. Since, the maximal flow algorithm can be applied to directed network only, therefore to find the maximal flow in an undirected network, it becomes necessary first to convert the undirected network to an equivalent directed network and then to apply labeling technique.

The whole procedure can be understood using the following example.

SOLVED EXAMPLES

EXAMPLE 1. *Consider a street network given below*

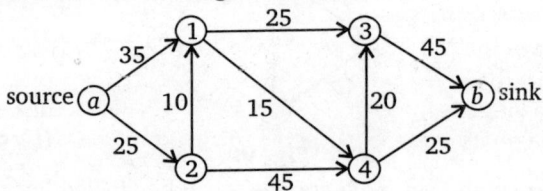

Fig. 93

The numbers on the arcs represents the traffic flow capacities. Place one way signs on streets can already oriented so as to maximize the traffic flow from the point a to b in the network.

SOLUTION. Replace each undirected arc by a pair of oppositely directed arcs with the same capacities as shown in following figure.

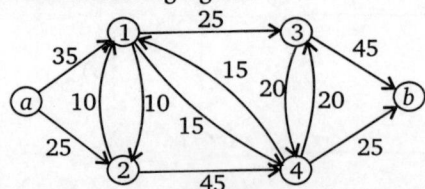

Fig. 94

Therefore, the maximal flow algorithm can be applied to this network to find the largest traffic flow from a to b. When the optimal flow is obtained, cancel the arc flows in the opposite directions, so as to find the direction of flow in each of the undirected arcs. Finally a numerical solution of this problem can be obtained easily.

13.26.2 TRANSHIPMENT PROBLEMS: MULTIPLE SOURCES AND SINKS

Here, we consider a network with several supply and demand points. To solve it by max-flow algorithm, we convert it to a single-source and single sink problem by introducing an imaginary super-source and an imaginary super-sink. Then from the super-source, a directed arc will be created to each one of the real sources. Similarly, from each one of the real sink a directed arc to the super-sink will be created. After that apply max-flow algorithm to maximize the flow from super-source to super-sink.

The whole procedure can be understood using the following example.

SOLVED EXAMPLES

EXAMPLE 1. *Consider the following transhipment problem*

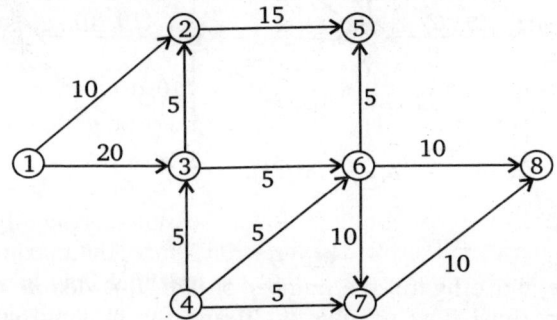

Fig. 95

In the above network, node 1 and 4 are sources with supply $s_1 = 20$, $s_4 = 20$. Nodes 5 and 8 are sinks with demands $d_5 = 15$, $d_8 = 20$. The numbers on the arcs represents the arc capacity. Determine whether the transhipment problem is feasible i.e. whether it will be possible to meet the demands with the available supplies.

SOLUTION. Here, we have the following steps:

STEP 1. Convert the given problem into maximization problem which maximize the flow from a single source to a single sink. An equivalent network is constructed with an imaginary source (a) and an imaginary sink (b) as shown in the following figure.

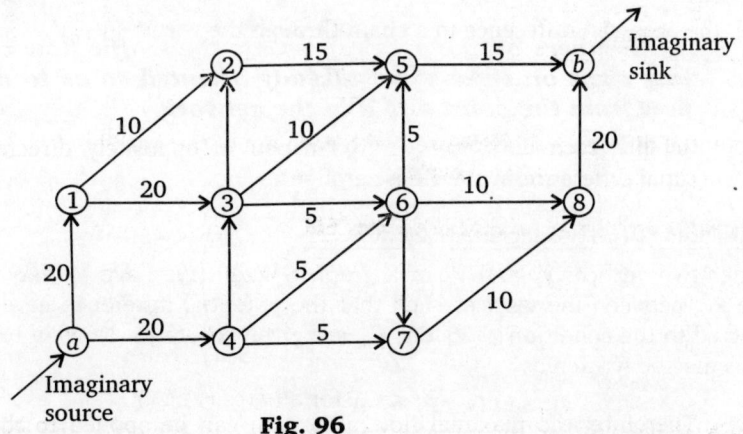

Fig. 96

In the above figure, the imaginary supply arcs $(a, 1)$ and $(a, 4)$ have capacities equal to the supplies in nodes 1 and 4 respectively. Similarly, arcs $(5, b)$ and $(8, b)$ have capacities equal to their respective demands. Although the total supply exceeds the total demands, the trashipment problem may not be feasible because of the capacity restrictions on the immediate nodes.

STEP 2. To maximize the flow (V) from the imaginary source to the imaginary sink in the directed network apply the max-flow algorithm, we get the optimal distribution of the flow as shown below

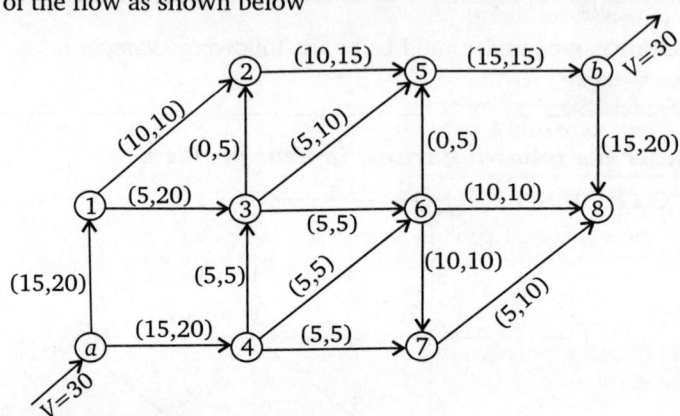

Fig. 97

In the above figure, the number on arc (i, j) denote (x_{ij}, C_{ij}). We observe that the maximal flow possible is 30 and the minimal cut is given by the subset $I = (a, 1, 2, 3, 4, 5)$ and $T = (6, 7, 8, b)$.

Finally, since the maximal flow is less than the total demand of 35 units, it is not possible to satisfy all the demands at the sinks. Hence, the transhipment problem is not feasible.

☛ **REMARK**

• The transhipment problem becomes feasible only when the value of the maximum flow V equals the sum of all the demands at the sinks.

13.27 POTENTIAL DIFFERENCE

Let $G = (V, E)$ be a graph and let a real number p_j be associated with each vertex V_j. Then p_j is called the potential of V_j. If $e_i = (V_j, V_k)$ is an arc then $x_{jk} = p_k - p_j$ is called the potential difference in the arc e_i.

Furthe, the potential difference in a chain through the vertices $V_1, V_2, ..., V_n$ may be defined as

$$x_{12} + x_{23} + ... + x_{n-1,n} = p_n - p_1$$

☛ **REMARKS**

• A potential difference is associated with each arc of the graph.
• The potential difference in a cycle is zero.

13.27.1 MINIMUM POTENTIAL DIFFERENCE PROBLEM

Consider two vertices V_a and V_b of a graph $G(V, E)$. We have to find the minimum potential difference x_{ab} between the vertices such that the potential difference in all arcs (V_j, V_k) of graph G is subjected to the condition $x_{jk} < C_{jk}$, C_{jk} are given constants. Now, in terms of potentials, these restrictions may be written as

$$p_k - p_j < C_{jk} \text{ for all arcs } (V_j, V_k)$$

and then we have to find the minimum value of $p_b - p_a$ subject to the constants.

If we take $p_a = 0$ then above problem can be restated as follows:

"Let p_j be a real valued function associated with the vertex V_j in a graph $G(V, E)$ and let $p_a = 0$ then find

$$\min p_b$$

subject to the the constraints

$$p_k - p_j \leq C_{jk}$$

for all arcs (V_j, V_k) in G"

The above problem is now identical with the minimum path problem when arc lengths are unrestricted in sign. Hence, we have to solve the problem of minimum path from V_a to V_b with C_{jk} as the length of the arc (V_j, V_k).

> Without loss of any generality, we may take $p_a = 0$ because we are interested in potential differences and not in absolute values of potential.

13.27.2 GENERAL PROBLEMS OF MIN-POTENTIAL DIFFERENCES

We can define a more general problem of minimum potential difference in a network, if the constraints are of the type

$$b_{jk} \leq p_k - p_j \leq C_{jk}$$

for all arcs (V_j, V_k)

Here, the method of solution will remain the same because each inequality of the above type can be written as two inequalities

$$p_k - p_j \leq C_{jk}$$
$$p_j - p_k \leq -b_{jk}$$

SOLVED EXAMPLES

EXAMPLE 1. *Find the minimum potential difference x_{14} between V_1 and V_4 of the graph with the following data subject to the conditions that for each arc $x_{jk} \leq C_{jk}$*

V	1	2	3	4		
E	(1, 2)	(1, 3)	(2, 3)	(2, 4)	(3, 4)	(1, 4)
C_{jk}	3	2	−2	1	4	−1

SOLUTION. We can form successive arborescences with the help of minimum path problem as given in the following table.

A	Path	f	Alternate path	f
1	(1, 2, 4)	4	(1, 4)	−1
	(1, 3)	2	(1, 2, 3)	1
2	(1, 4)	−1	(1, 2, 3, 4)	5
	(1, 2, 3)	1		
3	(1, 4)	−1		
	(1, 2, 3)	1		

The third arborescence gives the minimum potential difference $x_{14} = -1$ with optimal path (V_1, V_4).

EXAMPLE 2. *Find the minimum potential difference V_1 and V_4 in the graph $G(V, E)$ where*

V	1	2	3	4		
E	(1, 2)	(1, 3)	(2, 3)	(3, 4)	(4, 2)	(1, 4)

subject to the constraints

$$-2 \leq p_2 - p_1 \leq 3, p_4 - p_3 \leq -2, 1 \leq p_4 - p_1 \leq 6, 6 \leq p_3 - p_2 \leq 10, -2 \leq p_2 - p_4, p_3 - p_1 \leq 7$$

Non-Linear and Dynamic Programming

SOLUTION. The given constraints can be written as

$$p_2 - p_1 \leq 3, \qquad p_1 - p_2 \leq 2, \qquad p_3 - p_2 \leq 10 \qquad p_2 - p_3 \leq -6,$$
$$p_4 - p_3 \leq -2, \qquad p_4 - p_2 \leq 2 \qquad p_1 - p_4 \leq -1, \qquad p_3 - p_1 \leq 7$$

The graph of the problem and arborescence of minimum path are shown in the following diagram.

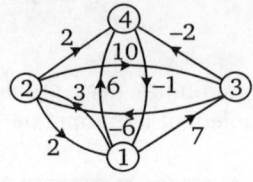
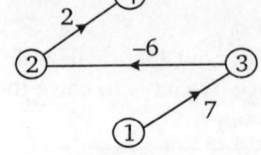

Fig. 98   **Fig. 99**

Hence, the minimum potential $= p_4 - p_1 = 3$.

1. Find the minimum non-negative flow in the network described below are (V_j, V_k) being denoted as (j, k) where V_a is the source and V_b the sink.

Arc	Capacity	Arc	Capacity
$(a, 1)$	8	$(3, 2)$	3
$(a, 2)$	10	$(3, 4)$	4
$(1, 2)$	3	$(4, 3)$	2
$(1, 3)$	4	$(3, b)$	10
$(1, 4)$	2	$(4, b)$	9
$(2, 4)$	8		

2. Convoys of many vehicles have to go from stations a_i: $i = 1, 2, 3, 4$ to b_j: $j = 1, 2, 3$ at night. The maximum number of vehicles leaving a_i or arriving at b_j is different for each station due to limited parking space and is given in the following table. Each a_i is connected to each b_j by road. For secrecy reason no convoy should consists of more than 15 vehicles.

Station	Parking capacity (No. of vehicles)
a_1	40
a_2	30
a_3	25
a_4	55
b_1	50
b_2	30
b_3	45

Find how the vehicles should be sent so that the total number of vehicles moved is maximum. Is the optimal solution unique? If not, find two alternatives.

ANSWERS

1. 12   **2.** 125

Index

❏❏❏❏